Dynamic Zoogeography

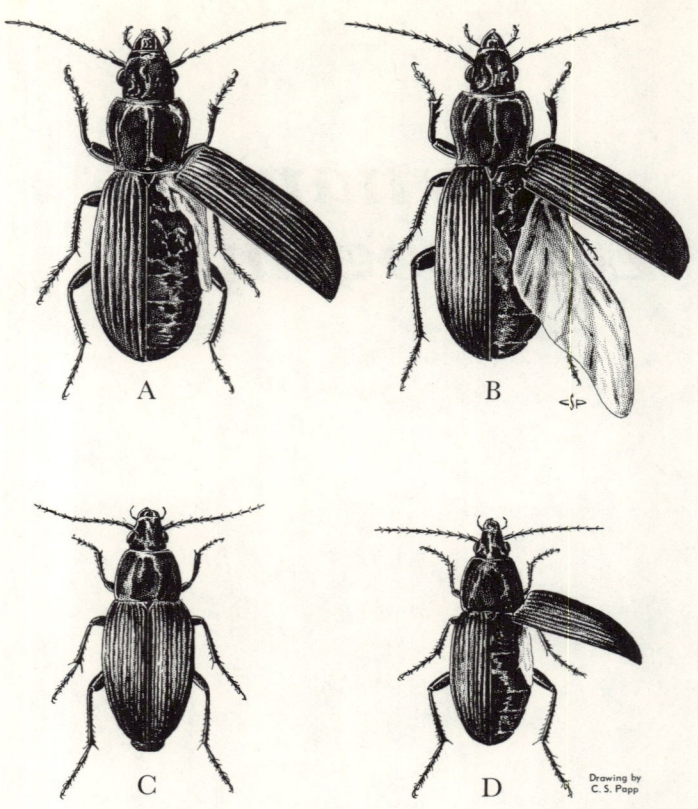

Wing dimorphism as a structural dispersional characteristic.

Above: Brachypterous (A) and macropterous (B) female of the carabid beetle *Pterostichus anthracinus* Ill. from Upland, Sweden.

Below: Macropterous female (C) and brachypterous male (D) of the carabid beetle *Calathus mollis* Marsham from Scania, Sweden. See discussion on pp. 62–64 and Figure 2-15. Specimens collected by C. H. Lindroth.

Dynamic Zoogeography
With Special Reference to Land Animals

MIKLOS D. F. UDVARDY
Sacramento State College

Illustrated by
CHARLES S. PAPP

VAN NOSTRAND REINHOLD COMPANY
New York   Cincinnati   Toronto   London   Melbourne

Van Nostrand Reinhold Company Regional Offices:
New York Cincinnati Chicago Millbrae Dallas

Van Nostrand Reinhold Company Foreign Offices:
London Toronto Melbourne

Copyright © 1969 by Reinhold Book Corporation

Library of Congress Catalog Card Number: 70–81357

All rights reserved. No part of this work covered by the
copyrights hereon may be reproduced or used in any
form or by any means—graphic, electronic, or mechanical,
including photocopying, recording, taping, or information
and retrieval systems—without written permission of the
publisher.

Manufactured in the United States of America.

Published by Van Nostrand Reinhold Company
450 West 33rd Street, New York, N.Y. 10001

Published simultaneously in Canada by
D. Van Nostrand Company (Canada), Ltd.

15 14 13 12 11 10 9 8 7 6 5 4 3 2 1

To
Ralph Soó de Bere
Hungarian geobotanist

who directed my enthusiastic interest in nature toward biogeography at my age of sixteen. His inspiring teaching and the example of his pioneering and synthetizing research led me toward my present goals.

Preface

When one turns the pages of books or papers in ecology, systematics, evolutionary studies, and the like, one finds that almost everyone of them has a chapter, or at least a few paragraphs, about the geography of the subjects. Every practicing field zoologist is a zoogeographer to a certain extent whether he did or did not have training in this discipline. My book is intended to ease his way. Besides, as professional zoogeographers are few and far apart, a book on the 'principles' may inspire more to become zoogeographers. With the present increased interest and publications on the geography of animals, a guideline of principles, methods, and achievements is timely. Having taught zoogeography for many years without adequate text I saw the need to write this book. Its organization follows that developed in the course of my lectures, and the plan of the book is outlined in the first chapter.

The *terminology of zoogeography* which I use is rich in expressions, special usages and phrases that will be explained when they appear for the first time. I do not consider it a sin to make scientific reading material more precise and specific by the use of specific terminology. A few new terms are introduced and other, less known terms will be used because they have attained spread and popularity in some of our sister disciplines or in some of the world's zoogeography schools.

As a field zoogeographer I often find that the word "fauna" is employed in so many different ways that it almost becomes confusing. The zoogeographic meaning of "fauna" is the assemblance of *animal species* occurring in a *particular* area. Its botanical counterpart is "flora" and most botanists agree that this term also denotes a collective abstraction, a regional list of plant species. However, the similarity of usage ends with these classic meanings of the twin terms. When the botanist talks about "a definite quantitative representation of the species available in the flora" (Dansereau 1951, p. 174) he talks about the *vegetation*. This idea, viz., the assemblance of individual organisms in an area (representing simultaneously the species of the regional biota) is not expressed in zoological terminology. We have no counterpart of 'vegetation'. We cannot easily denote the *concrete* entity, viz., the assemblance of organisms (of the animal kingdom) living together in an area. For instance, when a botanist talks about civilisation's impact on a formerly virgin area, he is able to say with Meusel (1943), that "the original vegetation disappeared but the original flora is still to be found there." Now try to substitute 'fauna' for 'flora' and it makes sense—but what do you substitute for 'vegetation'? Most ecologists and zoogeographers who would have need of this concept talk about "the animals," "the animal population" or even worse, "the fauna" and cover two meanings with one word. Elton (1966) also noted this deficiency and attributes it to the fact, rightly, that animals are usually elusive, and do not appear as a solid mass, like the vegetation does. It is true that vegetation appears to the eye as a solid mass, while animals are mobile. Markus (1933) argues that plants *on land* usually form a thick *mat*, animals form, as it were, a net, with tight or wide, sometimes broken, *meshwork*. Where this "meshwork" is thickest, it becomes visible (and sometimes very audible) as it would, in the case of termite- and anthills in Africa combined with the herds of ungulates and their predatory retinue; or rabbit warrens in Australia, bird and seal rookeries in the Arctic, whisting pikas in the Kuku-Nor mountains, noisy caimans in the Orinoco, chorus frogs, or flocks of migratory birds and the like. In contrast, the 'vegetation' that corresponds to the above given definition in an aquatic environment is restricted to the coastal shallows. The producer 'stratum' of the aquatic community consists of a loose, elusive 'meshwork' and there the 'animal world' often appears in masses and its entities have earned the vernacular or scientific names such as "reef," 'benthos,' 'nekton' and 'plankton.' Perhaps it is due to this semantic deficiency that geographically inclined botanists, as we shall see later on, have created two kinds of phytogeographies—floristic and vegetational plant geography. The latter, a very rewarding discipline, is still nonexistent as far as the 'animal world' is considered. I am proposing therefore a word and term, *Faunation, which means the assemblance of animal individuals of all species occurring at a locality.*

A second, but minor matter in terminology is the alternate use of 'to colonise,' 'to settle' and even 'to pioneer.' To pioneer—whether singly or in groups—means to penetrate localities where the pioneer did not live before. Colonizing—as Elton uses this verb (e.g. 1958, p. 26:" . . . rice grass . . . that has colonized . . . mud flats") means the same and implies no social relationship. 'To settle' is the synonym of 'to colonize.' Therefore these three verbs will be used interchangeably, i.e. when a dispersing animal (or animals) establishes a nucleus or a breeding population.

In preparing and writing this book I found that the scientific world is a very understanding, appreciative and helpful community. Dealing with

Table 1 Introduction of the concept of faunation

	Taxonomic entities composing the biota	Assemblance of individuals inhabiting ecological units on geographic localities
Plant	*Flora* (introduced by Linnaeus) group abstraction (list of plant species)	*Vegetation* (Original concept of western languages) mat of plant individuals
Animal	*Fauna* (introduced by Linnaeus) list of animal species	*Faunation* "meshwork" of animal individuals

a many-sided subject I had to enlist the help of an almost endless number of colleagues from near and far—they all contributed gladly with advices, encouragement, reprints, preprints, manuscripts of unpublished works, graphs, photographs, copyrights and whatever else was needed.

However my first thanks are due to my wife, Maud E. Udvardy, who helped in—and endured—my tedious and seemingly unending work of several years and did all the secretarial work involved, including the typing and retyping, several times over, of the whole manuscript.

My old friend and compatriot, Charles S. Papp of Riverside, Calif. illustrated this book with great zeal and artistic ability, a most successful collaboration! Besides him I am grateful to several persons who furnished original, often unpublished illustrative material; each of these is especially acknowledged where used.

I have heavily drawn on the experience of two entomologist-zoogeographer friends: Carl H. Lindroth of Lund, Sweden and the late Robert L. Usinger of Berkeley, California. I am very much in debt to C. G. G. L. van Steenis (Amsterdam) who volunteered to organize for me a rally for clarifying some advances in modern plant geography, and to M. T. Myres (Calgary, Alberta) who let me use his file of zoogeographical literature. Much details of information have been exchanged with, and threshed out by, my colleagues, e.g., B. McK. Bary, C. C. Lindsey and I. McT. Cowan (Vancouver, B. C.), G. A. Bartholomew and T. R. Howell (Los Angeles) and M. R. Brittan and R. L. Livezey (Sacramento). I was helped and inspired by questions, discussions and proofreading by my students, *before all* by J. Bagley, R. B. Bury, J. H. Bédard, F. T. Crase, G. V. E. Crewe and D. Hancock. Early drafts of certain chapters have been read and criticized by G. A. Bartholomew (Los Angeles), W. S. Hoar (Vancouver), T. R. Howell (Los Angeles), M. G. Netting (Pittsburgh), and R. H.

Whittaker (Irvine, Calif.). W. F. Blair (Austin, Texas, Chapter 4), E. M. Hagmeier (Victoria, B. C., Chapter 5), R. S. Hoffmann (Lawrence, Kansas, Chapter 5), S. C. Kendeigh (Champaign, Ill., Chapter 2), C. H. Lindroth (Lund, Sweden, Chapter 3), E. Mayr (Cambridge, Mass., Chapter 6), G. Niethammer (Bonn, Germany, Chapter 2), G. G. Simpson (Cambridge, Mass., part of Chapter 5) R. L. Usinger (Berkeley, Calif., Chapter 2) and K. H. Voous (Amsterdam, Holland, Chapters 4 and 6) critically read part of the complete manuscript and P. Gray (Pittsburgh, Pa.), L. von Haartman (Helsinki, Finland), and C. J. McCoy, Jr. (Pittsburgh, Pa.) undertook the tedious task of reading the whole manuscript. Their criticism improved the text greatly, though the errors that remained are mine.

G. I. Nicks, R. Ring and L. Roche collaborated with me initially on the scientific English and Eugene White combed through the whole text for the sake of clarity of expression. The numerous editorial staff that has been involved at the Publishers was most cooperative and infatigable; I am very grateful to them. I am aware, though, that all this help could not eradicate the discrepancies between my strange—perhaps original—way of expression and the almost unlimited subtilties of the English language.

While I was working on *Dynamic Zoogeography* I enjoyed financial help, leaves of absence and much understanding by the University of British Columbia, Sacramento State College, the National Research Council of Canada (Grants in Wildlife Research). The American Philosophical Society (Penrose Fund), and the Lida Scott Brown Lectureship in 1963/64 at the University of California (Los Angeles), where working facilities were also provided at the same time. The working space offered by the University of California (Santa Barbara) in 1966 is also greatly appreciated.

Sacramento, California
July, 1969

M.D.F. Udvardy

Contents

Preface	vii
1 Introduction	1
Comprehensive Literature of Zoogeography	9
2 The Ecology of Dispersal	10
Barriers	13
Physical Barriers	13
Water Barrier	13
Physical Barriers on Land	17
Ecological Barriers	22
Deserts	23
The Living Environment as Barrier	26
Climatic Zones as Barriers	28
Distance and Time Barriers	28
Durability of Barriers	30
Active Dispersal	32
Passive Dispersal	34
Anemochore Dispersal	34
Hydrochore Dispersal	40
Anemo-hydrochore Dispersal	42
Biochore Dispersal and Phoresy	44
Anthropochore Dispersal; Adventives; Introductions	49
Chance Dispersal	53
Time as a Factor Affecting Dispersal Abilities of Animals	58
Intrinsic Abilities of Dispersal	58
Structural Attributes	59
Physiological Attributes	64
Behavioral Attributes	64
Natural History Attributes	70
Population Attributes	70
Interplay of the Intrinsic Attributes	71
Modes of Dispersal	75
Dispersal and Dispersal Routes of Faunas	85
Conclusions	91
3. The Ecology of Colonization	93

External Factors Limiting Survival	95
Light	101
Temperature	102
Physical and Chemical Properties of the Environment	104
The Principle of Interdependence of Limiting Factors	111
Regional Changes in Ecological Valency	112
The Intrinsic Factors of Existence	116
Reproduction of the Settling Animal	122
Reproductive and Cytogenetic Systems that Facilitate Colonizing of New Areas	124
Other Intrinsic Conditions Influencing Pioneering	129
Food Requirements of the Young as a Condition of Settling	130
Special Habitat for the Young	131
The Combined Effect of Extrinsic Factors Influencing Reproduction and Thereby Colonization	132
Ecology of Settling Groups	135
Ecology of Settling Population	135
Environmental Relationships of the Population and its Amalgamation into the Ecosystem	144
Conclusions	150

4 Areography: The Study of the Distribution Area — 152

Distribution Areas and Their Mapping	153
The Shape of the Area	177
The Structure of the Area	182
The Ecology of the Area	186
The Synecology of the Area	194
The History of the Area	195
Relict Areas	205
Endemicity and Related Historic Concepts	215
Geography of the Area	217
Dynamics of the Area	225
Area and Extinction	230
The Area of Higher Taxa	230
Community Areas	235
Areas of Aquatic Animals	236
Plant Areas	238
Conclusions	239

5 Regional and Analytical Zoogeography — 241

Regional Systems Based on Ecological Distribution	243
Allen's Ecogeographic System	244
Life Zones	247
Ecogeographical Grouping of Plant Distribution	248
Bioecology	253

The Concept of Biotic Provinces	255
Regional Systems Based on Geographic Distribution	258
Wallace's Work on Faunal Regions	262
Floral Regions	266
Subdivisions of the Zoogeographic Regions	268
Statistical Methods in Regional Zoogeography	270
Areographic Analysis	281
Faunal Diversity	293
Conclusions	300

6 Dynamic Zoogeography — 302

Interacting Entities that Influence Zoogeographic Processes	305
Geographic Entities	305
Temporal Entities	307
Organismic Entities	309
Biogeographical Entities	310
Vectors and Modes of Dynamism in Zoogeographical Processes	311
Short-Term Changes of Distribution	311
Changes in Environmental Conditions	315
Climatic Changes and Their Rates	317
Dynamic Changes of the Vegetation	327
Dynamic Changes of Animal Distributions	334
Single Animal Species	334
Dynamism of Faunas	341
Area Analysis	344
Extinction and Zoogeography	349
Singular Extinctions	351
Synthetic Extinctions	353
Dynamism of Communities	357
Man as a Dynamic Agent in Zoogeography	359
Historic Aspects of Dynamic Zoogeography	365
The Application of the Principle of Uniformitarianism in Historical Zoogeography	366
Constancy of Ecological Valency	367
Dynamics of Island Zoogeography	370
Historical and Analytical Dynamics of Islands	379
Insularity of Ecological Archipelagoes on Mainlands	380
Dynamic Zoogeography of the Continents	383
Paleozoogeography	392
Conclusions	400
References	403
Name Index	430
Scientific Name Index	435
Subject Index	439

ILLUSTRATIONS

Figure		Page
	Wing dimorphism as a structural dispersional characteristic *Frontispiece*	ii
1-1	The system of biological science	3
1-2	Zoogeography and its helping disciplines	5
2-1	Relativity of dispersal barriers	12
2-2	Dispersal through mountain passes	19
2-3	Topographical barrier: cliff	20
2-4	Topographical barriers: valleys, mountain ridges	21
2-5	Topographical barriers: lava fields	22
2-6	Distance barriers	29
2-7	Colonization by anemochore dispersal	39
2-8	Hydrochore dispersal by rivers	41
2-9	Hydrochore dispersal by raft	42
2-10	Anemo-hydrochore dispersal	43
2-11	Phoresy	45
2-12	Europeanization of the North American carabid beetle fauna	50
2-13	Anthropochore dispersal	51
2-14	Hydrochore dispersal	55
2-15	Wing dimorphism and distribution	62
2-16	Wing dimorphism, distribution, and pass	63
2-17	Three types of philopatry in passerine birds	65
2-18	Origin of nonphilopatric bird populations in Northern Europe	66
2-19	Philopatry	68
2-20	Expansion by pioneering	76
2-21	Theory of slow penetration	78
2-22	Slow penetration	79
2-23	Dispersal by irruption	81
2-24	Migration and dispersal	83
2-25	Dispersal by wandering	85
2-26	Faunal corridor	88
2-27	Filter route	89
2-28	Sweepstakes route	91
3-1	The concepts of tolerances and requirements I.	97
3-2	The concepts of tolerances and requirements II	98
3-3	The concepts of tolerances and requirements III	99
3-4	The three types of ecological valency and their effect on distribution	100
3-5	Environmental factors limiting distribution	103
3-6	Environmental factors limiting distribution; the roe deer in Northeastern Europe.	105
3-7	Environmental factors limiting distribution: Bat distributions in Eastern Europe	105
3-8	Physico-chemical properties of the substratum influencing microclimate	106

3-9	Soil limiting distribution	107
3-10	Distribution interrelated with a particular environment	108
3-11	Limiting factors and distributional limits I	109
3-12	Limiting factors and distributional limits II	110
3-13	Interdependence of limiting factors	111
3-14	Relations of altitudinal and depth distribution with latitude I	114
3-15	Relations of altitudinal and depth distribution with latitude II	114
3-16	Regional variation of environmental requirements	115
3-17	Habitat selection mechanisms	119
3-18	Geographic parthenogenesis I	126
3-19	Geographic parthenogenesis II	127
3-20	Temperature and reproductive success	133
3-21	Cold limiting distribution through chick survival	134
3-22	Population dynamics and expansion	137
4-1	Seasonal occupation of distribution areas I	154–155
4-2	Seasonal occupation of distribution areas II	156
4-3	Seasonal occupation of distribution areas III	157
4-4	Seasonal occupation of distribution areas IV	158
4-5	Sterile expatriation area	159
4-6	Sterile expatriation area as area of irruption	160
4-7	Extralimital occurrence	161
4-8	The relative exploration of beetles in Fennoscandia	162
4-9	Incompletely known distribution	163
4-10	Mapping of distribution by emphasizing the circumference of the area	164
4-11	Mapping the circumference of the area	165
4-12	Accurate distribution maps: *Columbia palumbus*	166
4-13	Accurate distribution maps: *Stellaria palustris*	167
4-14	Mapping frequency of distribution	168
4-15	Use of symbols in mapping distribution: *Rophalocere* butterflies	169
4-16	Use of symbols in mapping distribution: *Triclads*.	170
4-17	Use of grid systems in distribution mapping: *Atlas of British Flora*	171
4-18	Use of grid systems in distribution mapping: Gray squirrel in Britain	172
4-19	Use of grid systems in distribution mapping: *Malacothrix*	173
4-20	Use of grid systems in distribution mapping. *Nightingale*	173
4-21	Accurate distribution mapping: *Aphelocoma coerulescens* I	174
4-22	Accurate distribution mapping: *Aphelocoma coerulescens* II	175
4-23	Area limits; hypothetical area	176
4-24	Area limits: *Austroicetes* model	177
4-25	Types of distribution area	178
4-26	Disjunct distribution area	179
4-27	Apparent discontinuity of the area	180
4-28	Fallaceous discontinuity of the area	182
4-29	Structure of the area: some schemes of dispersion	184
4-30	Structure of the area: some schemes of local distribution	184

xvi *Illustrations*

4-31	The effects of ecological and historical factors on the area	185
4-32	Models of eight theoretical distribution area types	187
4-33	Cosmopolitan distribution	191
4-34	Vicarism	192
4-35	Ecological vicars	193
4-36	Correlation of species and community areas in the Eurasian taiga	196
4-37	Correlation of species and community areas in Australia	197
4-38	Distribution of a genetic mutation	200
4-39	Some discontinuous distribution patterns in the Holarctic region	202
4-40	Discontinuous distribution due to spreading by long-distance pioneers	204
4-41	Relict concepts	206
4-42	Survivor concept	207
4-43	Reductional relict	208
4-44	Refugional relicts	210
4-45	Tertiary relict	212
4-46	Pleistocene (glacial) relict	213
4-47	Boreo-alpine relict	214
4-48	Bore-montane distribution, boreo-montane relict	215
4-49	Arctic-alpine distribution	216
4-50	Pluvial relict	216
4-51	Xerotherm relicts I	218
4-52	Xerotherm relicts II	219
4-53	Secondary relict distribution	220
4-54	Illustrations of some zoogeographical terms	221
4-55	Panboreal distribution	222
4-56	Amphipacific distribution	223
4-57	Some types of distribution areas	224
4-58	Dynamism of area	227
4-59	Extinction: *Ciconia ciconia*	228
4-60	Extiction: *Thersamonia dispar*	229
4-61	Generic area and center of diversity I	232
4-62	Generic area and center of diversity II	233
4-63	Generic area and "centers of endemism"	235
4-64	Mapping of fish distribution	237
5-1	Clumping of area limits	242
5-2	Allen's ecogeographic system: the North American Temperate Region	245
5-3	Merriam's life zones	248
5-4	Biomes of North America	249
5-5	World classification of plant formations in correlation with climatic factors	251
5-6	The biotic provinces of part of North America	256
5-7	The biotic provinces of Southern Africa	257
5-8	Sclater's zoogeographic regions	260
5-9	Faunal realms and regions; floral kingdoms	262
5-10	Map of the floral regions of the world	269
5-11	Biogeographical provinces of Europe	271

5-12	Mammal provinces of North America	272
5-13	Schilder's representation of faunal resemblances	276
5-14	Percent similarity values of theriofaunal provinces of North America	280
5-15	Examples of Hultén's equiformal progressive areas	284
5-16	Dispersal centers of the arboreal and eremial faunas of the Holarctic	286
5-17	Distribution area of five North American passerine bird species	289
5-18	Ecofaunal groups in California	291
5-19	The composition of the Californian passerine avifauna	292
5-20	Faunal diversity pertaining to mammals	296–297
5-21	Climatic changes, altitudinal zonation and faunal diversity	299
6-1	Duration of time periods	303
6-2	Map of the world with the 600-foot submarine, and 600 foot elevational, contours	306
6-3	Evolution, succession, and dispersal	314
6-4	Post-wisconsin rise of the Laurentian shield	317
6-5	Changing patterns of generalized Tertiary environments	319
6-6	Late Tertiary changes of an environment	320
6-7	Pleistocene climatic extremes in North America	322–323
6-8	Northern Colombia during the Pleistocene	324
6-9	Millennial dynamism in Northern Europe	325
6-10	Vegetational changes in Africa	326–328
6-11	Extreme phases of the Australian environment during the past million years	330–331
6-12	Climatic changes during the past millennium	332
6-13	Evidences of changing environments in southern South America	333
6-14	Parkland in Tierra del Fuego	334
6-15	Profile of the Great Smoky Mountains	335
6-16	Distributional changes of plants and animals	336–337
6-17	Advance of the northern limit of *Melanargia galathea*	338
6-18	Advance in Central Europe of *Senecio vernalis*	339
6-19	Expansion of *Eupithecia sinuosaria*	340
6-20	Former discontinuities revealed by taxonomic study	350
6-21	Man's impact through time in Hawaii	363
6-22	Change of ecological valency through time	369
6-23	Evolution of Himalayan nival species	370
6-24	Correlation of island size and faunal size	373
6-25	Equilibrium model of island faunas	375
6-26	Catching angle of islands	376
6-27	Speciation in an alpine ecological archipelago	381
6-28	Insularity in an ecological archipelago	382
6-29	Coordination of vertebrate and invertebrate distributions	385
6-30	Millennial dynamics of habitats and areas	388–390
6-31	Geographic speciation in *Funambulus* squirrels	391
6-32	Distributional and evolutionary history of the genus *Pachycephala*	392
6-33	Geography and phylogeny of Diamesinae	393

xviii *Illustrations*

6-34	The postulation of land bridges on the basis of recent distribution	395
6-35	Map of the world with postulated land bridges	396
6-36	Distributional history of tapirs	398
6-37	Distribution of certain insects known from the Baltic amber	399

PLATES

I	The zoogeographical regions of the world according to Wallace —	270
II	The fauna types of the Palaearctic region	270
III	Superimposed distribution areas of mammal species from cave deposits	285
IV	Ecofaunal analysis of the birds of California	285

TABLES

1.	Introduction of the concept of faunation	ix
2.	Wing dimorphism of carabid beetles of the Azores	64
3.	Role of polymorphic dispersional systems	73
4.	Some polymorphic dispersional systems	74
5.	Vicarious species and subspecies pairs in the European avifauna	87
6.	Factors and conditions of ecology of colonization	96
7.	Population attributes	136
8.	Success of introduction of mammals and birds to Europe and New Zealand	139
9.	Certain characteristics of ecosystems	147
10.	North American vertebrates exemplifying some distribution area types	189
11.	The geological time scale	304
12.	Temporal units of environmental changes as they affect animal distribution	309
13.	Analysis of the intrinsic dynamic potential of the avifauna in the Carpathian Basin	342
14.	Geographic and taxonomic indications of speciation in the western North American herpetofauna	348

1 / Introduction

Nature has been defined as a "principle of motion and change," and it is the subject of our inquiry. We must therefore see that we understand the meaning of "motion" for if it were unknown, the meaning of "nature" too would be unknown.—Aristotle: Physics, III,1.

Zoogeography is defined as the scientific study of the distribution of animals on the earth, and with this discipline two types of scientist are primarily concerned—the geographer and the zoologist.

The subject matter of geography is the description and explanation of the features which comprise the earth's surface, and of their causal phenomena. Living beings—members of the classic plant and animal kingdoms—occupy nearly all the surface of the globe. They are widely distributed in the sea; they float and swim on the surface or in the water, and they cover the bottom as mats of animal and plant life. There is scarcely a land area that cannot harbor some sort of life. Even such inhospitable areas as deserts, icefields, and windswept mountaintops may show at least enduring diaspores or primitive forms carried there by the elements. Besides the more or less continuous cover of land vegetation and the network of animal life within it, the lower reaches of the air also carry aerial forms of living organisms. The geographer considers the sum total of these environments as forming the *biosphere*. Not only does the physical presence of living beings affect or influence the appearance of the landscape, but also through their activity the physiognomy of the earth is strongly modified. In a constructive way, plants and animals can actively contribute to rock formation—for example, in the form of coral and algal rocks, coal, and guano deposits. Alternatively, changes in the plant and animal composition of a landscape are often followed by important modifications in its physical features, such as those of the microclimate, soil, and drainage. The geographer is, then, fully justified when he directs his interest particularly toward the distribution of plants and animals, and he has come to call this chapter of his discipline "biogeography."

The interest of the biologist, on the other hand, is focused upon the plants and the animals themselves. To trace the process whereby *biogeography* and, strictly speaking, *zoogeography* came to be incorporated into the biological sciences, we must go back to the beginning of the nineteenth century, when the biological sciences started to differentiate from the natural history of the past. This differentiation actually occurred earlier with zoology than with botany, but logical division of the fields of knowledge was accomplished first in botany. De Candolle's system of botany (1813) presented three groups of knowledge. In the first group, he placed the recognition, description, and systematics of plants. In the second, he summarized the study and knowledge of the plant as a living organism; here he considered the structure (organography), physiology, pathology, and geography of plants. In the third major group, he treated the relationships of plants to man.

The first important theorizer of zoological knowledge, the German scholar Ernst Haeckel,* was deeply influenced by Darwin's theory of evolution. Haeckel not only introduced the concept of ecology in his book on general zoology (1866), but in the same work he also gave a new term, definition, and content to zoogeography, which had advanced since de Candolle's time mainly in the descriptive field. Haeckel claimed that the spatial distribution of organisms is properly the object of a special discipline, which he termed "chorology."

Chorology includes not only the geographic and topographic description of the habitats and their limits of distribution but also the vertical distribution—i. e., the depth limits—for aquatic animals and the altitudinal zonation of land forms. Haeckel argued that while these factors had indeed been studied by biogeographers of the past, their true causal explanation was possible only through an understanding of Darwin's theory of evolution. Haeckel thereby sought to establish the necessity of the zoogeographical approach in dealing with biological phenomena.

Alfred Russel Wallace (1876) is often considered the founder of zoogeography. In his book *The Geographical Distribution of Animals* (1876) he established a clear distinction between the points of view of the geographer and the zoologist. For the one, he coined the term

*Haeckel was well known as a philosopher as well as a zoologist.

"zoological geography"; for the other, "geographical zoology." In the area of *geographical zoology,* he advocated a revision of the distribution of the animal world by systematic groups and, at the same time, made due allowance for historical causal relations and for the interactions of the different biota. Wallace's system of zoogeographical regions was based principally on the distribution of a few groups of animals, but his results satisfied geographers because the regional limits coincided with the limits of continents or of major topographic regions within them— that is, with geographic boundaries.

After Wallace, many global or regional zoogeographies were published up to the middle of the twentieth century, but no basically new concept arose concerning the place of zoogeography among the biological disciplines. Zoogeography was pursued by taxonomists, evolutionists, paleobiogeographers, ecologists, and others, each placing emphasis on his own particular point of view. Yet in the theoretical system of biology, as elaborated by the Swiss zoologist Tschulok in 1910, chorology is one of its main divisions, equivalent with classification, morphology, physiology, ecology, chronology (evolutionary history of the biota), and genetics.

To find the place of biogeography, or zoogeography, in the system of biological sciences, we must take into account the fact that those geographic phenomena and attributes which Haeckel united under chorology

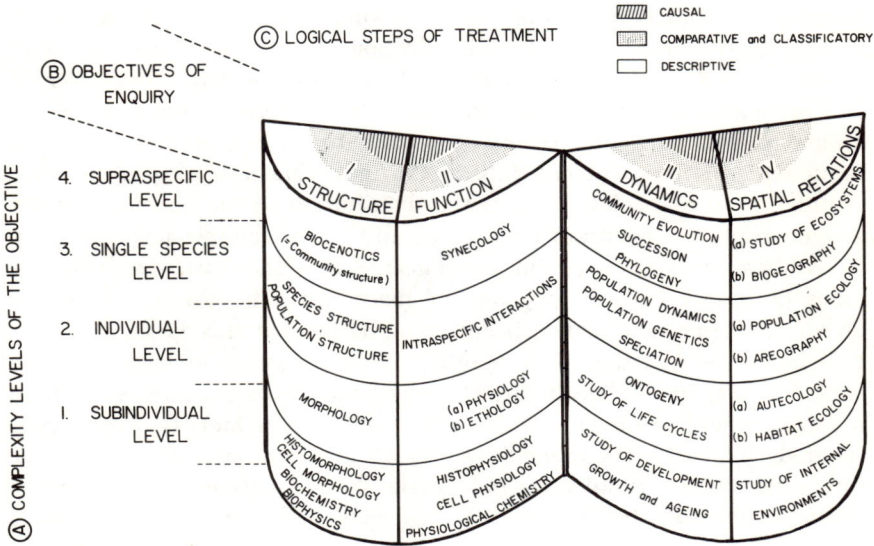

Figure 1-1   The system of biological sciences.

are, as biological characteristics of the species, just as important as the number of their chromosomes, the way they metabolize energy or the way they are related to another group. Figure 1–1 presents a system of the biological disciplines as they pertain to the organism, its parts, and its aggregations. The spatial relations of these objects of study are treated here as biological attributes, and it can be easily seen that different aspects of zoogeography intertwine on a biological rather than a spatial basis.

Several biological characteristics of animals are tangible, such as the facts of their structure, of their internal functioning (physiological functions), and of their functioning in relation to the environment (ecology and ethology); but even among these phenomena, there are some that easily escape our attention as concrete objects of study. Such parameters as number of offspring, longevity, selection of particular habitat and food, and site of reproduction will serve as examples. Many biologists continue to regard the study of these phenomena as belonging to an unclassified area of knowledge known as natural history. On the contrary, such parameters are intrinsic in the animal and are highly specific, since they are inherited following the laws of genetics and are subject to evolution. In 1932 the Russian entomologist Bartenev suggested a collective term—*biological habitus*—for these parameters and attributes of the species. In his interpretation, the biological habitus complements the morphological habitus (appearance, structure, and shape) and modes of functioning. This book will endeavor, in subsequent chapters, to demonstrate that the present and past distribution of animals is based on characteristics of their biological habitus. In this perspective, zoogeography is a branch of the study of the structural and functional reactions of the animal. These particular reactions pertain to the *space* that forms the immediate environment not only of the individual but also of the total population of a species. It is on the distribution of the species that the zoogeographer concentrates his research,* even though he may occasionally study the gross spatial relationships of higher taxa. To define the areas of higher taxa is only one of many ways to group distribution areas; this approach is not strongly emphasized in this book.

It has been said that zoogeography is only a point of view. Yet this statement can be applied to any branch of the biological sciences. Causal research is seldom possible without consideration of facts that seem only remotely related to the immediate subject. There is not a single scientific discipline which can advance independently and without the support of advances in other fields. We can thus view biological knowledge as a

---

*Linnaeus said, "*Omnis vera cognitio cognitione specierum innitatur.*"

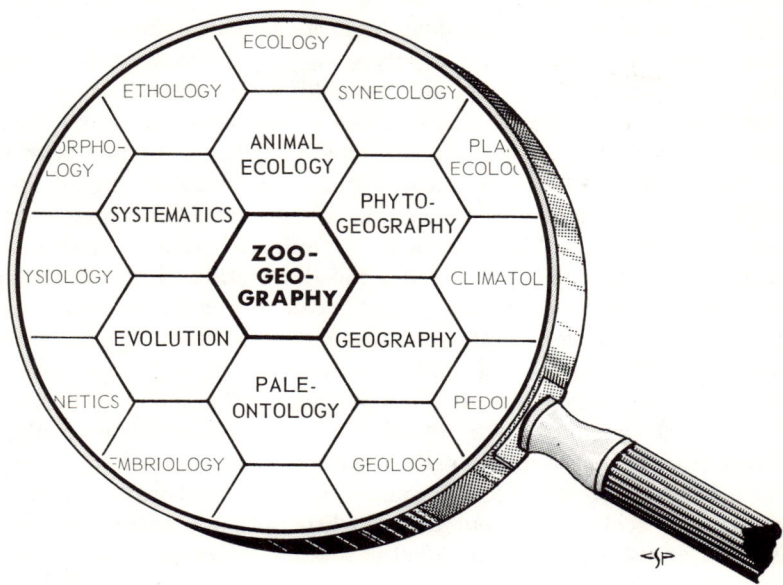

Figure 1-2 Zoogeography and its helping disciplines.

meshing of different disciplines. When the specialist focuses a magnifying glass on a particular mesh (Figure 1–2) this mesh is enlarged and clarified. Within the viewed area, however, are other meshes which, as it were, put the central one in its proper perspective. When zoogeography is viewed this way, the magnifying glass shows a kaleidoscopic picture composed of geography, animal ecology, plant geography, paleontology, evolution, and systematics—the disciplines which immediately surround, and intermesh with, zoogeography. Further away we find demecology (a recent term for population ecology), synecology (the ecology of biotic communities), plant ecology, plant evolution, geology, ethology, physiology, and morphology; these disciplines are also related to zoogeography. Scientific German uses an excellent descriptive term to cover those other disciplines that are helpful in understanding the subject in focus—*Hilfswissenschaften*, "auxiliary disciplines." Using data from these sciences does not mean that the central subject (in our case, zoogeography) is a combination of them. History, for example, is greatly aided by anthropology, linguistics, and paleography. Yet their combination alone does not comprise the discipline of history.

Assuming that our diagram and the above discussion puts zoogeography in its proper place in the net of related disciplines, we may now turn to the question of how it deals with its subject matter.

The first stage of an inquiry in any science is descriptive. In zoogeography, the gathering of facts in the field is usually followed by an enumeration of the animals occurring in the area covered by the scientist in his field work and by the complementing of his own data with data from the literature. Linnaeus termed the assemblage of animal species in a particular area a "fauna"; the study of different local faunas is called "faunistics." Monographic treatment of animal species, genera, families, and other taxa usually includes the description of the distribution range. Zoogeographic chapters in handbooks of geography usually enumerate the faunas of geographic entities—mountain chains, plains, islands, continents, and the like—or of political units. Because of the specialization of most field zoologists, we also find it natural that many useful works deal only with portions of the total fauna. In this way, for example, the avifauna, the herpetofauna, or the orthopteran fauna of an area becomes a basic source of zoogeographical information. Ideally, descriptive zoogeography seeks to present a clear distributional picture of *all* animals in *all* areas, although this ideal is seldom achieved.

The second stage of inquiry is to classify the multitude of data. In zoogeography, this means the grouping of the distributional data according to as many different points of view as a particular investigator may find necessary. In some studies, the distribution ranges of a particular phylogenetically related group are brought together and then compared; otherwise, the fauna of a geographically, ecologically, or historically uniform area is analyzed.

The third and last stage of inquiry, the causal analysis, tries to explain the reason for the present distribution as it has been presented in the literature. It is mainly in this stage that the results of the *Hilfswissenschaften* become most valuable, because here the explanation of the present distribution can often be found in the geographic conditions in which the animal occurs, in its ecology, and in its history and evolution with respect both to the particular area and to the other organisms in it. The principles of zoogeography emerge from this causal analysis. For that reason, the major portion of this book is devoted to the discussion of the principal factors which cause and affect animal distribution.

The study of the status quo of animal distributions and of the grouping of these distributions according to different ordering principles is a study of momentary phenomena. Faunistic and regional zoogeography is basically static zoogeography; present distribution is the result of processes which move the animal in space and as a consequence of which the distributional picture changes with the passage of time. Thus, causal zoogeography is identical with dynamic zoogeography. Causations of distributions are dynamic processes. As long as our inquiry probes into the reasons for the arrival and settling of a species in a certain area, we are within the field of dynamic zoogeography. As soon as we ask ourselves why and how an animal is able to live in a particular area, we leave dynamic zoogeography; as a matter of fact, we leave zoogeography entirely, and our question receives an answer along *ecological* lines.* The answers pertaining to dynamic zoogeography are often ecological, but here ecological cause is followed by movement, individual or population, of the subject. When the ecological cause is followed by a positive or negative response of the organism or population and when a spatial shift is *not* involved, the whole phenomenon is within the sphere of ecology. When distributional phenomena change over a longer time span, and when the functional relations of the organism and the environment result in adaptive changes, we come to the meeting ground of dynamic zoogeography and evolutionary studies. The latter, however, are usually assigned to the taxonomist, for evolutionary changes are most easily discerned by noting morphological changes. As Lindroth (1962) expresses the relationship of these two disciplines, zoogeography always depends on taxonomy to know *what* to study; but taxonomy also depends on zoogeography, since geographic speciation is the accepted norm of the formation of its basic working unit.

Every animal species originated from a few ancestors in a limited area; if a particular species is now found to be widespread, it must of necessity have reached parts of its present range at an earlier period. The first aspect of dynamic zoogeography pertains to *dispersal*. If we know the details of dispersal processes, we can explain much about the presence or absence of animals. Dispersal may result as a by-product of other important phenomena belonging to the biological habitus of the animal, or it can result from distinct, adaptive characteristics of the species that directly assist dissemination into wider areas. In dispersal study, both the intrinsic abilities of an animal to spread and the opportunities the environment offers for spreading are equally important. Though every animal species has the powers of dispersal, dispersing individuals must find a suitable area in which to settle and reproduce through many generations. The principles of settling will form the second item of our future discussion. When we study the process of settling, or colonization, we must scrutinize the ecological factors that make existence possible in a given area as well as the adaptations and limitations of a species—struc-

---

*Lack of an animal species in a certain area may be caused by ecological (it cannot exist there) or zoogeographical (it has not arrived there) reasons.

tural, physiological, behavioral, population dynamical, and so on—which enable it to start a new population and to survive (successfully) in the newly colonized area. The present study is mainly ecological. Factors of dispersal as well as factors of existence in an area influence the size, extent, and dynamism of the distributional range of the animal. Many conclusions can be drawn from the study of the distributional area itself; a thorough discussion of this subject, dealing with the morphology of the area, its ecology, and the different distributional area types—including both single species and the compound of higher taxonomic units —will be found in Chapter 4.

Most of the ecological zoogeographic principles treated by Ekman (1922) Hesse (1924), and Hesse, Allee, and Schmidt (1937, 1951) have been reclassified in my discussions of dispersal, settling, and the nature of the area. In most earlier zoogeographical studies, however, the principles of dispersal, settling, and the nature of the distribution area did not receive adequate treatment. The majority of these studies dealt with the classification of distributional types, and the principles involved were manifold. Ecological, evolutionary, historical, and statistical methods were employed in regional zoogeography, and these are reviewed elsewhere in this book. Regional grouping can be based on distribution areas without involving, a priori, any of the above-mentioned points of view.

The closely related science of plant geography demonstrates that concentrating on grouping and comparison of the areas themselves will produce a satisfactory "synthetic" picture which will contribute greatly to an understanding of the reasons for a particular distribution.

Historical zoogeography has been based on an integration of paleontological, systematic, and evolutionary studies; all of these disciplines are concerned with the evolution of the present zoogeographical picture. The biological habitus of the species—including its "address," or its geographic distribution—is the outcome of evolutionary processes, adaptations, and compromises. Like many other aspects of the habitus of the animal, this is subject to change in time with changing environment and changing genetic constitution. It is in this sense, rather than within the framework of classic, historical zoogeography, that the final chapter of this book will examine the dynamic parameters of distribution, emphasizing the momentariness of the present environmental and distributional picture.

It is particularly important that, in our discussion of the different principles, due emphasis be placed on the origins and the originators of the subject. For this reason, many references to the history of zoogeography and to general biogeography are included in the material which follows.

COMPREHENSIVE LITERATURE OF ZOOGEOGRAPHY

Rensch (1931, 1937, 1935–37, 1950) has published lists of the zoogeographical literature from 1908 to 1950. Publications between 1950 and 1964 are reviewed by Niethammer (1958b, 1966). The following list includes the most important methodological and monographic works and textbooks published in the twentieth century; complete references are listed at the end of the present text.

English
  Bartholomew et al. 1911
  Newbigin 1936
  Hesse et al. 1937, 1951
  de Beaufort 1951
  Ekman 1953
  Simpson 1953
  Dansereau 1957
  Darlington 1957
  Croizat 1958
  George 1962
  Simpson 1965
  Croizat 1966
  MacArthur and Wilson 1967

French
  Prenant 1933
  Jeannel 1942
  Furon 1958

German
  Dahl 1921
  Hesse 1924
  Pax 1930
  Marcus 1933
  Ekman 1935
  Jacobi 1939
  Ekman 1940
  Rensch 1950
  Thienemann 1950
  Schilder 1956
  Niethammer 1958a
  de Lattin 1967

Norwegian
  Økland 1955

Russian
  Bobrinskij 1927
  Bobrinskij 1951
  Geptner 1936
  Isakov and Formosov 1963

Swedish
  Ekman 1922
  Pearson 1955

2 / The Ecology of Dispersal

> *Wollen wir die Verbreitung der Tiere auf der Erde verstehen, so müssen wir vor allen Dingen die Ausbreitungsmittel und Ausbreitungshindernisse kennen . . . (F. Dahl 1921 I,51)*

Dispersal is the process whereby an organism is able to spread from its place of origin to another locality. This faculty of plants and animals is of profound importance in maintaining healthy populations, and the means and extent of spreading are as basically characteristic as are the abilities to reproduce or to bind enough energy for self-maintenance. In sessile organisms—most plants and many bottom-dwelling aquatic animals—dispersal of the offspring is a necessity if their life span overlaps that of the parents. In the majority of animals, mobility of both parents and progeny would allow common utilization of the same locality but not of exactly the same spot in the biosphere. Thus we often find that on a twig or leaf infested by aphids the mother and her offspring are together, forming neat rows or clusters. Colonial insects, such as ants, bees, and termites, live socially, but periodic swarming allows dispersal of any surplus which the environment or the social capacity of the group can no longer support. In higher animals, "behavioral institutions"—home range, territorialism, intraspecific aggressiveness, and the like—may force the offspring to disperse elsewhere.

This *dispersal* movement, then, is the vehicle of the *spacing* process whereby even *dispersion* is brought about. The pressure of too dense a local population is relieved by the loss of individuals dispersing away from that area, who, in turn, may move on to fill gaps where the population had become too thin. This is the role of dispersal, or spacing, movements in population ecology, and we are at present concerned with those dispersing individuals which bring about *expansion* of the population into new areas.

As the process of plant and animal evolution is understood at present, spatial shift of range of populations is, as a rule, the basic necessity for the formation of new species. Where species arise by means other than the isolation-mutation-selection mechanism (e. g., in apomyctic forms, by chromosome mutation, by hybridization), the new form must still spread in order to found a new colony, this again being achieved by the processes of dispersal. Evolution by special creation without dispersal was of course, a cornerstone of the pre-Darwinian view of the origin of diversified floras and faunas. Such theories as "hologenesis"—the simultaneous appearance of a species throughout its present distribution area —were still entertained as late as the 1910s. In the stricter, zoogeographical sense, dispersal is the method of reaching an area not previously inhabited by the individual, the species, or the group. It is therefore the basis of the spreading or colonization process.

Dispersal is undertaken by the individual. Whatever geographic displacement is achieved will be the result of multiple interactions between the individual and the environment. In these interactions, more often than not, the environment offers considerable resistance, and the dispersing individual makes such compromises and adaptations as exemplify the role of selection in the evolutionary process. Therefore, the *environmental factors* hindering or facilitating dispersal will be discussed first. After these extrinsic elements of dispersal ecology have been examined, the *intrinsic dispersal capacities* of the species will be treated. In conclusion, a synthetic view of the *dispersal of faunas* will be presented.

Extrinsic, or environmental, factors of dispersal impose a different set of conditions upon each dispersing animal species and upon each age and sex group within a species—especially because any group may be endowed with a different set of functional and structural capacities which can affect its dispersal across a given piece of land. These intrinsic factors involve all those functions and structures which permit survival and reproduction (we mentioned spacing above). The structural and behavioral means to move about among the varying and often adverse features of the environment is an adaptation which is needed in the everyday life of the individual in finding food, mate, and so on. However, the processes of dispersal and the subsequent processes of establishing a new colony of the species are often so unique and specialized that we must think of them as basic biological characteristics of each species— characteristics which have evolved under strong selection pressure and

12   The Ecology of Dispersal

which adapt each species to respond, throughout time and generations, to the variation of the extent and location of suitable, habitable areas. Elton (1927, p. 146) stated that most dispersal is not directly intended to spread the species. It is a thesis of this present book that without evolved means of dispersal most animal populations would have succumbed, over a period of time, to the vicissitudes of the environment.

Focusing first on the extrinsic factors of dispersal, we are aware of course that the emphasis on the "hindering" or "furthering" role of environmental entities is entirely subjective, and is dependent upon the kind of organism and the general or particular situation. It is evident, for example, that a body of water can for a long time effectively block the dispersal of such a strictly terrestrial mammal as a pocket gopher

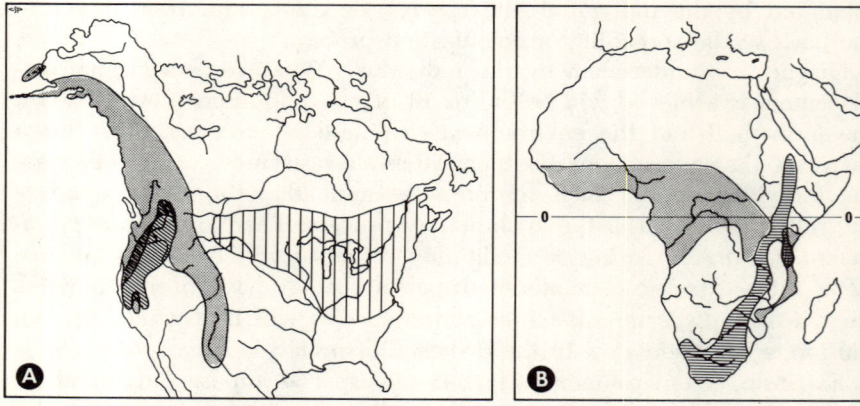

Figure 2-1 Relativity of dispersal barriers. The dispersal highway of one form is the dispersal barrier of another form.

A. Rosy finches (genus *Leucosticte*, Fringillidae, Aves) spread to North America from Asia along the high mountain ranges of the west where three species now occur (dotted area). The same mountain ranges block and separate the western and eastern subspecies of the Nashville warbler *Vermivora ruficapilla* (Wilson) [Parulidae, Aves] from one another. This is a bird of lowland forest edges, and would not settle in the high mountains (hatched area).

B. In East Africa, the malachite sunbird *Nectarinia famosa* (L.) [Nectariniidae, dotted area] is a bird of highland savannas, but in the Cape Province it reaches sea level. (Map after Chapin 1932.) The same mountain chains that served the sunbird as dispersal highways block and divide the distribution area of the black-and-white flycatcher, *Bias musicus* (Vieill.) [Muscicapidae, Aves, hatched area], an inhabitant of the lowland evergreen forest. (Mapped by Benson and Irwin 1965)

while at the same time, can be a dispersal highway for the otter (Figure 2-1).

BARRIERS

Physical Barriers A dispersing animal uses either active locomotion or a carrier to change domicile. Certain environmental features may be entirely unsuitable for either, or they may jeopardize the life or reproductive capacity of the spreading individual. Thus, in discussing the factors which inhibit dispersal, it is well to distinguish between physical obstructions and ecological barriers. In this way controversies—mainly of a semantic nature—can be avoided, such as whether an ecologically unsuitable area should be classified as a geographical (topographical) or an ecological barrier. The ecological or environmental barriers themselves have a geographical character, since the physical environment is an organic part of any type of habitat. Similarly, a given topographical formation may offer an unfavorable environment for any species not endowed with the means of crossing it. Thus, topography also acts as an ecological barrier.

The overlapping nature of these concepts caused a great deal of confusion during the crystallizing process of our ideas on isolating factors active in speciation. Barriers that effectively hinder the contact of subpopulations of a species are prerequisites of successful speciation by mutation and selection. However, the term *barrier* implies isolating mechanisms other than mere physical or ecological unsuitability of a land area.

Water Barriers Bodies of water act as barriers of dispersal in two capacities. First, they prevent the spreading of animals which are not endowed with organs for swimming or with buoyancy. Second, they hinder the spreading of even those animals which can swim. No matter how well a land animal can swim, water is not its natural element and is a strange medium in which it can seldom rest, let alone feed, thermoregulate, display, and avoid predators. Thus, even the best swimmers are unlikely to survive a prolonged stay in water.

The role of a body of water in dispersal is directly related to its size and permanence. Every naturalist or sportsman who observes the flotsam on brooks and small rivers is aware of the small land invertebrates—centipedes, spiders, ants, and other insects—that involuntarily get into and float upon the water. Even at the moment of observation, many of them are dead or dying, and many others will fall prey to fish before reaching dry land. Yet seldom are these minor watercourses totally effective in preventing land animals from crossing to populate suitable habitats on the opposite shore. On the other hand, lakes of considerable length and rivers that are wide and fast or girded by flood-exposed habitats may so separate ecologically uniform areas that they cannot be regularly crossed by most members of the land fauna. In such a geographical situation, the biologist usually notices that toward the source or upstream of a river or toward the lakehead region where their roles as a barrier would naturally decrease, some other physiognomical or ecological formation takes over the role of the water barrier and thus hinders the concerned species from circumventing the body of water. Thus, freshwater bodies may act as barriers for only a relatively short distance. Such barriers, though seldom studied in their entirety, are known to occur at the present time and have most likely occurred in the past also, but their permanence as a dispersal barrier is negligible when compared with that of the sea.

An excellent experimental study of the crossing of a water barrier by land snails was undertaken by Wolda (1963). The barrier in this case was a mere 1.5m-wide ditch in Holland; the experimental objects consisted of 2,420 marked individuals of the snail *Cepaea nemoralis* (L.) [Cepaeidae, Pulmonata]. Of these 2,420 individuals released on their usual habitat, 132 were later found on the other side of the ditch, or, in some cases, recuperating on some emergent plant above water level. However, 152 individuals were retrieved, drowned, from the bottom of the ditch. Thus, more than half of the potential crossers perished in this slight water barrier.

Many of the very small land animals, such as insects, are not effectively barred by rivers, for they float easily or are carried across by the wind. Rivers are, however, effective barriers where active or passive means of crossing is prevented by lack of proper structural, physiological, or behavioral attributes. For subterranean forms, not only the river proper but even the flooded bottomland extends the width of such barriers, as Grinnell (1926) surmised when writing about habitats of pocket gophers [Geomyidae, Mamm.]. Taxonomic study of the Plains pocket gopher, *Geomys bursarius* (Shaw), revealed that the Mississippi River at St. Louis is an absolute barrier for their populations; the opposite banks are populated by different subspecies, and no intergradation has been noticed (MacLaughlin 1958).

Howden (1963), after studying flightless scarabaeid beetles, concluded that for them, "a few miles of flatland may represent an insurmountable barrier." Several beetles of the genus *Mycotrupes* are isolated by the Savannah River, and the eastern limit of the genus itself is formed by the Appalachicola River. This river also bars the distribution of the genus

Gronocarus toward the east, while the Mississippi bottomland forms its western limit, although suitable habitat apparently exists on both sides of these rivers. Some of the best examples come from small terrestrial mammals, even though they are fair swimmers. Zimmermann (1935) investigated the incidence of a mutant form in the vole *Microtus arvalis* (Pallas), in northern Germany (see Figure 4–38 and discussion on page 199), and found that its frequency north of the Elbe estuary was 90 percent, while on the south side it dropped to about 50 percent. A similar drop is apparent at the estuary of the Geeste. Further inland, the difference in frequency is much less, corresponding to the diminished width of the river barriers. (However, Zimmermann did not comment on differences in habitats at the two localities.) Dice (1939, 1949), in studies of the deer mouse *Peromyscus maniculatus* (Wagner) [Cricetidae, Mamm.], focused directly on the same phenomenon. He studied these mice in the basin of the Columbia River and in the Snake River Valley. Two of his collecting stations were about 5 km apart, on opposite sides of the Snake River, and in identical habitats. Marked statistical differences were found in size and pelage color, although here, as in the much wider Columbia, basic adaptations to the similar climate were similar but not identical on both sides of the river. Dice's studies point toward parallelism on the microevolutionary scale and also toward the fact that despite occasional rafting, both rivers were strong enough barriers to allow evolutionary fixation of parallel yet mutually independent adaptations. The California zoologist Joseph Grinnell (1914) was perhaps the first to emphasize the role of the Colorado River and of rivers in general as prime factors in the differentiation of species. His pioneer work was followed by Kelson's study, among others. Kelson (1951) found striking faunal differences along the Colorado River north of the Grand Canyon in eastern Utah. In this area, he discovered that no *genus* of rodents was restricted to but one side of the river. Twelve *species* were common to both sides; two additional species were present west of the river, and eleven east of the river. Only four *subspecies* were common to both shores, another ten being restricted to the east side and twenty more to the west.

Sea water kills obligate freshwater organisms or land animals which are not protected by water-impermeable integuments. Active crossing by such animals is nearly impossible. It might be thought that most amphibians would be effectively barred by the sea, since they are in this category. However, a review of the herpetological literature of the past few decades reveals this assumption to be misleading. Myers (1953) discusses at length the ability of amphibians to cross sea barriers. He finds circumstantial, historical evidence in the distribution patterns of many frog families which indicates accidental crossing of the sea, but finds none for caecilians. Inger (1954) discusses an exceptional frog, *Rana cancrivora* (Gravenhorst) [Ranidae], which occurs in brackish water and has been seen leaping into the sea and climbing ashore unharmed. Even its tadpoles are highly salt-tolerant. It is thus widely distributed along the shores of the Indonesian Archipelago and the Malay Peninsula. Neill (1958) brings together new field data from North America and, in addition, surveys the literature and finds that at least 52 species or subspecies of amphibians—frogs, toads, and salamanders—either dwell habitually in or at times invade brackish or salt-water habitats. He also points out that the mangrove habitats of tropical seashores are poorly known and may still hold some surprises.

Naked slugs (land mollusks) are in the same vulnerable category, with respect to the sea barrier, as are the amphibians. Among the shelled land snails, the dry-adapted forms tolerate salt water better than do the hygrophilic forms from the humid forest floors, which cannot cross sea-water barriers (Solem 1959).

In general, for land animals with protective exoskeletons, the sea plays a role similar to that of the freshwater barrier. However, because the geographical distribution, size, and permanence of sea-water bodies are greater than those of rivers and lakes, the seas and oceans are the more formidable barriers. Furthermore, the thermal properties of seas and oceans add to their effectiveness as barriers, even to those land animals capable of actively attempting to cross. The oceans freeze only near the arctic and antarctic regions, where the general environment is least favorable for prolonged dispersal movements. But in temperate and subtropical regions, the sea water is relatively cold even in the warmest seasons, when the largest numbers of animals would be present as potential crossers. A good case in point is the differential populating by reptiles of islands off the east coast of Canada. Snakes are good swimmers, and offshore islands are generally inhabited by them; however, both Newfoundland and Anticosti Island are devoid of snakes, although the adjacent Labrador coast has at least one species. However, Bleakney (1958) thinks that the cold waters of the straits would block dispersal by numbing any snakes attempting to swim across.

On the other hand, the insect fauna of Oceania is found to be an attenuation of the Oriental Fauna (p. 264), gradually diminishing in numbers from west to east. Gressitt (1956) shows that species which persist in the faunal lists of mid-Oceanian islands are sedentary—living in protected microenvironments—and small, and are thus eminently suitable for passively crossing water.

Taken together, the regional faunistical study of land animals shows that islands always have smaller and more unbalanced faunas than do mainland areas of comparable size, and that oceans separating continents or large islands tend to isolate faunas on opposite shores. Thus there is overwhelming statistical evidence behind the statement that oceans are very effective physical barriers for land animals.

Physical Barriers on Land Precipitous cliffs, chains of mountain peaks, and deep valleys offer formidable obstacles for walking, crawling, running, or even climbing land animals, especially if such barriers are devoid of vegetation in which footholds can be found. Quicksand, moving sand dunes, asphalt pits, volcanic cones, and fresh or cooled rough lava surfaces* similarly obstruct passage. These formations may reach large proportions and serve as barriers of dispersal for many organisms.

Because of their geographical characteristics, it is difficult to select examples of the action of such topographic formations. If they are of limited extent, their barring effect would not manifest itself in faunal differences on their opposite sides or in range limits coinciding with them, since the animals in question could as easily circumvent them as they can limited water barriers. On the other hand, a mountain chain usually presents a barrier of considerable magnitude, possessing characteristics of vegetational zonation, altitudinal climate, and lack of soil structure which themselves serve as ecological barriers, and play a major role as *faunal dividers*. Again, an ecological formation may serve as physical barrier because of its physiognomy which bars locomotion, and thereby crossing, by certain animals. Insects, indigenous to open country, and which are weak flyers, may be blocked in their dispersal by dense, tall forests which they are unable to cross. Wellington (1964) points out that the moths of the western tent caterpillar *Malacosoma pluviale* (Dyar) [Lasiocampidae, Lepid.] are stopped in their ovipository flights by tall trees, on which they alight and lay eggs (see p. 120), while in brushy or woodland areas they are able to fly much longer distances. Mackerras (1962) makes a similar observation concerning tabanid flies, and regards the Australian rain forest as a dispersal barrier for these organisms.

Niethammer (1953a), in discussing the bird fauna of Bolivia, emphasizes that the Cordilleras chain is a barrier of prime importance. However, he states that to the east of this barrier is a zone of *yungas* (rain and cloud forests) covering the eastern slopes, and that to the west of the high mountaintops and ranges there are dry prairie and scrub habi-

---

*The "aa" lava of the volcanic geologist, the term taken from the original Hawaiian name of such rough lava fields.

tats of the *altiplano* (plateau). Some 37 bird families are thus separated, but basically the barrier is ecological. The Scandinavian Alps form a well-known barrier to the animals of Norway and Sweden. Since the Scandinavian Peninsula emerged from its Pleistocene ice cover only some eight or nine thousand years ago, the dispersal phenomena there have a very short history, and distributional limits are still very dynamic. Many habitat zones are ecologically very similar on both sides of the mountain chain, and thus distribution restricted either to the Atlantic or to the Baltic slope can often be attributed to the barring role of the high mountains. By the nature of the habitat—a difficult barrier even for man—there is a scarcity of positive evidence in the form of observed data on individual animals which have tried and failed to cross the wall of snow-covered alps. What evidence there is comes mainly from distribution mapping and analysis, which shows that colonization occurring across the mountains is restricted to the vicinity of a pass. Passes, bridging the vegetation zones of both sides of the mountain chain, act also as bridges for dispersing animals, and actual observations have been made of crossings through a pass by migrating or wandering animals. Beebe (1951 and earlier) and Beebe and Fleming (1951) observed dozens of butterfly and moth species, as well as many Orthoptera, Odonata, Dermaptera, Coleoptera, and other insects, crossing the Andes at the Portachuelo Pass at an elevation of 1,100 m in Venezuela. Ekman (1922) lists seven bird and mammal species entirely blocked by the chain of the Scandinavian Alps; on the other hand, 17 avian and one mammalian species, by establishing bridgeheads on the opposite side, spread more or less permanently across several passes of low altitude and favorable vegetation zone. Interesting is Lindroth's analysis (1949) of the same phenomenon with respect to ground beetles (Carabidae) in which twenty-two species crossed the Scandinavian Alps, all through the low, vegetated passes (Figure 2–2). Eleven species utilized passes in the coniferous forest zone; seven carabid beetles crossed through passes covered only with subalpine birch scrub; and four species used both kinds of passes. Judging from this distribution, no colonization occurred across passes in the alpine tundra zone or across the alps themselves. A morphological grouping of the cross-mountain colonists reveals that out of the 22 species, the majority—17 forms—are good flyers, and that only three species are flightless, with atrophied wings. The remaining two species are dimorphic (see p. 64). The conclusion is that the majority of the barrier crossers fly over the obstacles, but even these spread with the greatest success along a favorable habitat bridge (Figure 2–16). These studies also corroborate the conclusion (see above) that topographic, physical barriers affect most animals by their ecological unfitness to support the dispersing pioneers.

Physical Barriers 19

Figure 2-2 Dispersal through mountain passes.
Wooded passes through the Scandinavian Alps. (After Lindroth 1949.) Arrows with dots indicate passes in the coniferous forest; those with open circles, passes in the subalpine birch wood zone. The hatched areas (oak forest and beech forest zones) do not have mountains with passes. The single thick arrow indicates the subalpine pass leading to Trondheim in Norway. (Cf. Figure 2-16)

An additional argument for the ecological nature of most physical barriers would be that while high-altitude zones of mountains bar lowland animals, the opposite is also true—i.e., that for genuine high-mountain forms, the intervening lowland serves as barrier through its different habitats. This is borne out by Findley and Anderson (1956) in their

20 The Ecology of Dispersal

Figure 2-3 Topographical barrier: cliff.
View of Mount Roraima in southeastern Venezuela, showing barrier between habitats of two races of the Andean sparrow. *Zonotrichia capensis macconnelli* Sharpe [Fringillidae, Aves] occupies the plateau; *Z. c. roraimae* (Chapman), the talus. Their ranges are separated by the intervening cliff, over 420 m high. Chapman 1940; the zoogeography of this area is discussed in Mayr and Phelps 1967. (*Photograph courtesy of W. H. Phelps, Jr. and The American Museum of Natural History*)

study of montane forest mammals of the Rocky Mountains in Colorado, Utah, and Wyoming, in which they discuss the slow crossing of the ecological barrier by the more euryecious forms (see Chapter 3). A major topographic barrier in southeastern Venezuela near the border of Guyana is described by Chapman (1940); Mount Roraima forms a plateau system with steep, sheer cliffs (Figure 2–3) which rise to a height of 2,450 m, while at the base of the cliffs, the elevation is approximately 2,000 m. Studying the rufous-collared sparrow, *Zonotrichia capensis* [Fringillidae, Aves], Chapman found that the subspecies *mcconelli* (Sharpe) lives on the plateau separated from *roraimae* (Chapman) by 400 m of vertical cliff.

On volcanic oceanic islands, there exists a unique geographic and biologic situation which serves as a natural demonstration—in diminished time and geographic scale—of the true topographic barriers as well as of the accompanying speciation phenomena. Geologically speaking, such islands are young and short-lived; their fauna is restricted in numbers and is subject to rapid radiating evolution. Figures 2–4 and 2–5 show some of the striking topographical features of the Hawaiian Islands which bar earthbound land animals from quickly colonizing neighboring but

Figure 2-4 Topographical barriers: valleys, mountain ridges.
Erosion peculiar to volcanic rock in a semitropical climate resulted in steep slopes and rugged ridges that bar land snail distribution from one valley to another in the Hawaiian islands. The Napali Coast of Kauai. (*Courtesy of U.S. Geological Survey*)

geographical research is Welch's work (1938) on the land snails of the genus *Achatinella,* which is also discussed and illustrated by Mayr (1942). More than a hundred races occur in the Waianae Mountains on the island of Oahu alone, where the population of each valley or rain-exposed ridge is separated from the others by an unusually effective topographic barrier (Figure 2–4). Here, as in many other places, "even minor geographical obstacles may act as major dispersal barriers to land snails, and local morphologically distinct populations, sometimes occupying a few square yards or a single tree, can develop" (Solem, 1959, p. 23). McCoy (*in litt.* 1967) mentions the snail genera *Sonorella, Ashmunella, Micrarionta,* and *Humboldtiana* in southwestern North America, which developed species swarms in continental mountain ranges. However, in continental situations, wind dispersal and changes in local topography caused by floods, earthquakes, and the like often mix geographically distinct populations in such a way that the effective isolating action of a microgeographic—i.e., topographic—barrier depends on chance.

22 *The Ecology of Dispersal*

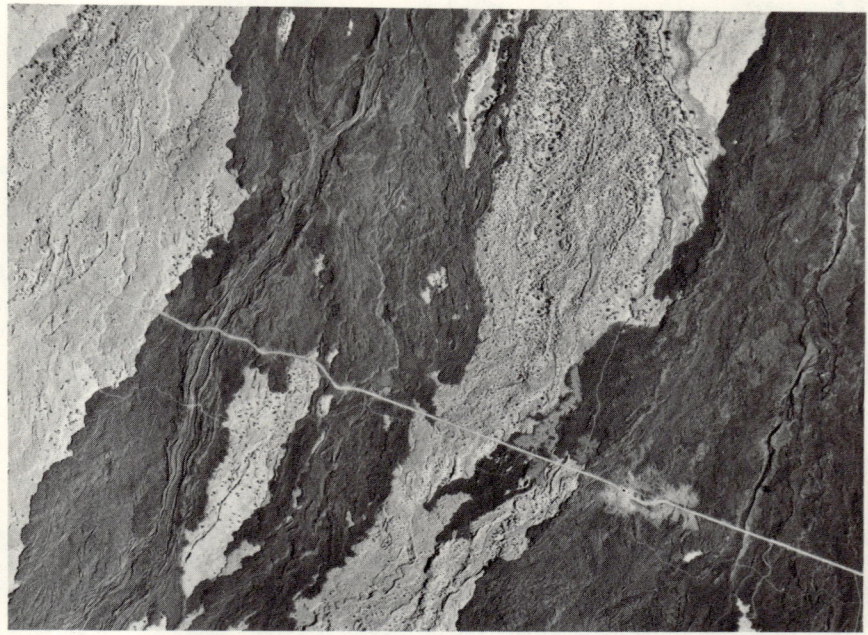

Figure 2-5 Topographical barriers: lava fields.
Vertical air view of recent *aa* lava flows on the slope of Mauna Loa on the island of Hawaii. The gray color is caused by a complete cover of lichens. Note the island of rain forest (*kipuka* in Hawaiian) cut off and surrounded by the lava flows. Many rare members of the endemic Hawaiian biota survive on only one or a few of such habitat isolates. (*Courtesy of U.S. Department of Agriculture*)

Ecological Barriers

Generally, land animals did not evolve special, hardy forms for dispersal, like the diaspores of plants. Animal dispersal is achieved by larvas, juveniles, or adults which exhibit the characteristics of their kind. Even the most eurytopic animals (p. 186) have a limited distribution range, and are thus obstructed by some substantial barrier, usually environmental, where limiting factors prohibit the spreading of those species. The kind and magnitude of such limiting factors—they seldom act alone—are almost as manifold as are the forms of biota; their enumeration—let alone, discussion—is beyond the scope of this book. Instead, there follows a discussion of those environmental factors, and especially those major habitat types, which often serve as substantial barriers for a number of animal species or even for whole faunas over long periods of time.

Deserts. If we classify the different environmental factors under four categories, as Andrewartha and Birch (1954) have done—namely, weather, food, other organisms, and a place to live—then most inhospitable habitats (i.e., those that are lacking in all of these categories) can be described collectively as *desert*. Desert as a dispersal barrier can be best characterized by the broad definition of the plant ecologist, for whom *desert* is a habitat type where, as a consequence of adverse and extreme ecological conditions, less than half of the total area is covered by vegetation, consisting mainly of single plants or plant clusters.

The following types of desert are generally recognized:

Coastal Deserts. Tidal or seasonal water, ice action, or salinity on sea coasts results in barren land. These coastal strips, of varying widths, greatly hinder the spreading of animals which have by some means crossed the water barrier and become stranded on an inhospitable coastal desert. A coastal desert may be only a narrow strip along the coast, or it may adjoin a greater expanse of an inland desert (Meigs 1966).

Cold Desert. Because perennial or seasonal ice cover, arctic and antarctic ice and snow sheets, and mountain glaciers do not support permanent life, they effectively prohibit the crossing of almost all kinds of animals. The cold desert regions of the earth thus serve as primary dispersal barriers. It is difficult to evoke direct, observational evidence without considering the difficulties man himself has experienced in crossing such barriers before the age of mechanized transportation; the fate of Peary, Ross, Franklin, Nansen, Scott, and others comes readily to mind. There is an abundance of negative circumstantial evidence pointing to the effectiveness of long-lasting ice fields (Figures 6-29B and 6-30) in separating animal populations until they show diverging evolutionary trends (Zeuner 1945, Rand 1948, Deevey 1949, Wright and Frey 1965, and others).

Mobile Deserts. Compact vegetation cover—enough to support the animal members of such communities—cannot develop on areas where subsoil or rock structure, frost or other mechanical effects, or other irregularly moving obstacles alter the surface of the ground from time to time. Talus slopes; rock and bank slides; floating logs, seaweed, reed, or other bulky material; and loose, shifting sand—all may make virtual deserts of the areas they occupy. For the spreading animals, they merely add to the obstacles which coastal deserts, steep mountain slopes, and canyons already provide in these generally forbidding areas.

Stone and Rock Deserts. Unweathered rock fields, rocky peaks, plateaus, and islands support little vegetation and lower animal life. Unless these areas are in extreme climatic belts, they are, ecologically speaking, short-lived, for vegetational succession will eventually penetrate them. However, rocky or stony areas of such a barren nature do not provide suitable "stepping stones" in the true meaning of this term, and thus they hinder the spread of animals.

Arid Deserts. Together with the cold deserts, the deleterious temperature of which is often enhanced by an arid atmosphere, arid deserts are at the present time the most extended desert areas of the world. At the pessimal season or time of day, they support, even for crossing, only organisms which have special adaptations to tolerate desiccation. Miller (1963) observed migratory birds trying to cross the Mohave Desert in California in late summer, at temperatures ranging from 30°C to over 40°C. Minimal weights of 102 specimens of small birds of 11 species were between 62 and 85 percent of their normal weight. (Experiments have demonstrated that loss of weight to a level of 70 to 80 percent of normal is lethal for birds [Salt 1952]). Many of the birds which Miller observed were dead or dying after having spent days in attempting to cross the desert, and he vividly describes how some hummingbirds alighted in the afternoon heat and died of exhaustion in his hands.

Cultural Deserts. Last but not least, cultural deserts present substantial ecological barriers possessing many characteristics of other types of deserts and incorporating still other limiting factors. Most of them comprise large areas which have been under cultivation for centuries or even for thousands of years; in extreme cases, they are covered by the artificial, stony, cement, or at least barren habitat of human beings. There is no adequate descriptive term in the English language for these areas, since the climatically governed English way of cultivating the land was traditionally a mixture of small farming and pastoralism which did not exterminate all woodland and which, more often than not, resulted in a green and varied countryside. Mediterranean and Central European farmers, on the other hand, cultivated grain fields of huge extent; in Eastern Europe, climatic and pastoral conditions created treeless prairies and pastureland centuries ago. Central European languages possess in their technical vocabularies the terms: "cultural prairie," an area of grainfields, hayfields, or pastures as artificial grasslands; and "cultural forest," where alien trees are often systematically planted and where these plantations support only a selection of natural or adventitious weeds. Extended tropical plantations, irrigated rice, and other fields produce similarly alien and artificial biotic communities inappropriate to the dispersing animal life. Human ecologists point out that large portions of Africa, Asia, and even the other continents within tropical savanna or subtemperate-temperate dry belts have felt the profoundly altering effects of human agencies.

For the dispersal of natural vegetation and faunation, those acres are most detrimental in which annual crops are raised and the fields tilled, plowed, burned, or otherwise drastically altered by rotational methods of cultivation. From the standpoint of dispersal ecology, these areas are comparable to deserts. Exposure to physical elements, weeding and harvesting methods, draining or irrigation, and treatment by chemical fertilizers and insecticides almost invariably create limiting factors of the utmost severity. The urbanized areas themselves are among the most prohibitive of man-made habitats; few indeed are the organisms that can cross such areas, and even fewer are voluntarily tolerated by men. In surveying the dispersal phenomena of the past few millenia, or even of the immediate past, we must give these barriers the fullest possible consideration.

A study of the distribution of the wapiti of North America (Figure 4–53) will make clear how this important game animal has become extinct on most of its former range because of the deforestation and forest alteration that has taken place there. The presently reintroduced herds—on the Ozark Plateau west of the Mississippi, in Michigan, and in various parts of central California—are isolated from one another by expanses of "cultural steppe" which, for obvious reasons, no wandering individual or small herd is able to cross. The completely foreign and artificial desert habitat of New York City effectively barred a newly introduced European bird, the starling, *Sturnus vulgaris* L., [Sturnidae], between 1892 and 1895, from breaking out of Central Park—the large woodland in the heart of Manhattan—where it had established itself (Cooke, 1928). Los Angeles, California, has many extensive areas of large gardens, weedy, uncultivated lots, and other similar, semiwild habitats where the southern pocket gopher *Thomomys umbrinus* (Richardson) [Geomyidae, Mamm.] occurs. In the Hollywood Hills, there is a small pure albino population living on an area of about three or four city blocks north of Los Feliz Boulevard; these albino gophers are blocked in their dispersal by the busy, wide, paved streets, so that no mixing with the neighboring, normally colored populations has been observed (C. A. MacLaughlin *in litt.* 1966).

In summation, it may be said that water bodies, rugged, rocky mountain peaks, canyons, and deserts of different kinds hinder the spreading of most land animals. Some animals are especially adapted to the living conditions on rocky mountains or in different types of desert; in these cases, however, the nondesert land habitats usually form barriers that are equally effective. The areas under intensive human cultivation or habitation are also absolute barriers for most spreading species.

It is difficult, because of the small amount of work which has been done, to find examples of the dispersal-hindering effects of single ecological factors. Yet it is evident that those land animals possessing a narrow range of tolerance to temperature or humidity cannot readily cross areas outside their tolerance limits. Thus the desert reptile, with high temperature requirements for its active movement, would become sluggish or even torpid in a prolonged sojourn in damp and shady forests; in the midst of an arid desert, the salamander would inevitably perish.

The Living Environment as Barrier In addition to facing the hazards of the physical environment, the dispersing animal during its barrier crossing is exposed to the living environment in a way which is extraordinary. Because it is, quite literally, out of its element, the chances of being attacked by predators or parasites are increased, while opportunities for obtaining food and shelter are decreased. A classic example is found in Errington's study (1943) on muskrat *Ondatra zibethica* L. [Cricetidae, Mamm.] populations, in which pioneering young animals that left their protective marsh habitat and ventured into meadows, fields, and other ecologically strange habitats became regular subjects of devastating mink predation.

The majority of plant formations—or biomes, when the animal component of the communities is also considered—seem effectively to limit the dispersal of many species of different, alien communities, although a universal limiting mechanism is not known. The evidence is circumstantial in that if such a barrier, i.e., a strange community—is artificially removed and a new, connecting habitat is established by man, the mingling of formerly discontinuous elements follows. This indicates that their spread, but not their settling on the other side of the barrier, has been hindered.

The valley of the Jenissei River, in Siberia, intersects woodland country in a south-north direction and, because it is heavily forested, forms an ecological barrier for woodland birds. This barrier limits the distribution of 11 species toward the west and of another 11 species toward the east (Johansen 1955). Intensive deforestation by logging and the establishing of human settlements eventually broke this forest barrier; in about forty years, eight species were able to extend their ranges across the former barrier. In the central part of North America, an opposite habitat situation existed before settlement by the white man. The open, treeless prairie once formed an ecological barrier for forest- and tree-dwelling animals. The establishing of woodlots, gardens, and parks created a woodland corridor—in fact, several such corridors—across which such birds as the Baltimore oriole *Icterus galbula* (L.) [Icteridae] and the rose-breasted grosbeak *Pheucticus ludovicianus* (L.) [Fringillidae], are now spreading—some westward from the gallery forests of the Mississippi

Valley, others eastward from the foothills of the Rocky Mountains (Sibley 1957, Sibley and Short 1959, 1964). Because birds can easily fly over such ecological formations as these, we may assume that strange habitats form a "psychological" or, to use a better term, an *avoidance barrier*.* Animals often avoid pioneering across those habitats which do not correspond to their innate, behavioral releasing mechanism of habitat selection. In the above examples, when the right habitat was created by men's activity, animals pioneered and penetrated it.

The rift valley systems in eastern Africa are significant barriers to bird dispersal (Benson, Irwin, and White 1962). The bottoms of these valleys—sometimes only a few kilometers wide—have a hot, dry climate and a corresponding arid vegetation, which the avifauna of the surrounding plateau woodlands and forest avoids. The following data are self-explanatory (Benson *et al., l.c.*).

Rift Valley:	Number of Species confined to		Number of subspecific differences on the two sides:
	South side:	North side:	
Limpopo	15	31	15
Zambesi	8	35	30
	West side:	East side:	
Luangwa	22	8	9
Nyasa/Shire	19	4	13

We have observed that in flying vertebrates, behavioral mechanisms play a very important role in the motivation of pioneering achievements. We shall see below that such factors are also important in the case of invertebrates, such as insects, but in a different context. Much conjectural evidence is derived both from the study of present distributions and from faunal history, which leads to an assumption that ecological barriers on land are extremely important for earthbound land animals. Ground beetles [Carabidae, Coleoptera] provide a good example, and the barrier is that between New Guinea and Australia. At present there is a sea barrier between the two, but during the Pleistocene they were connected by land. Darlington (1965) summarizes the evidence in support of his belief that this connecting corridor must have been relatively dry, open country. He argues that of the forest carabids of Australian affinities which are common to Australia and New Guinea, none is

---

*The term "ethological barrier" would be more appropriate, if it had not already been used to denote behavioral avoidance of heterospecific mating attempts (cf. Mayr 1963).

flightless; only one flightless New Guinean form reached Australia, but this species occurs also in the Malay Archipelago, and thus probably rafted across the water barrier after the land bridge had been broken.

Climatic Zones as Barriers Climatic zones of the biosphere are ecological barriers of great complexity, as are the multitude of ecological factors which are basically determined by climate: the soil, the dominant plant formations, the microclimate which the latter create, the members of the biotic communities they harbor—all adding to the effectiveness of this major barrier. Earthbound, slow, and ecologically restricted (stenoecious) animals are halted entirely. Thus, corresponding climatic belts of the Northern and Southern Hemispheres excel in harboring whole faunas of ecological vicars, i.e., evolutionarily unrelated forms that fill the same niche and that show convergent structural characteristics.

In most cases where the effect of climatic barriers is discussed, it remains to be clarified by observation or experiment whether this climatic effect inhibits the pioneering individuals only during their dispersal movement or whether it also adversely affects their attempts at colonization. This doubt applies, for instance, to the many excellent bioclimatic studies of insects that Messenger (1959) summarizes. A close knowledge of the life history cycles, and climate ecology of individual insect species would aid in the discovery of the missing evidence.

Distance and Time Barriers

A distance barrier is purely geographic in nature. If an area that provides suitable habitat for the spread of the species is separated from the existing range limit by a distance, across *habitable* grounds, that dispersing individuals of the species habitually cannot overcome during their lifetime, then distance alone forms a major barrier preventing the occupation of this area. Since many subtle environmental factors may bar a species from crossing such an expanse, it is difficult to establish unequivocal examples of his phenomenon. We may reason in the following manner. Environmental changes may alter the terrain beyond the limits at which the species can survive. If the rate of these changes, in time and in geographic progression, is slower than the rate of the dispersal of the species under scrutiny, every area that becomes available will be occupied almost immediately. If the species is slower in its movement than the progression of environmental changes, it will lag behind, and the distance itself will act as a barrier. The larger the newly available area and the slower the rate of dispersal of the animal, the longer the distance barrier will last. Thus, distance is intimately connected with the time factor of dispersal. The collared dove *Streptopelia decaocto* Friv. [Columbidae, Aves] transgressed its relatively stable northern limit in east-Central Europe in the early 1930's, and by the 1960's it had colonized most of Europe, including the British Isles and southern Scandinavia (see Figure 4–40). In this instance we are able to conclude from the evidence that during the 1940's, distance alone prevented this species from populating Ireland. Two North American mammals of southern origin—the opossum *Didelphia marsupialis* L. [Didelphidae], and the armadillo *Dasypus novemcinctus* L. [Dasypodidae]—have vigorously expanded their distribution ranges in recent times. The opossum has been introduced to the Pacific coastal states, and has established thriving and spreading populations there (Figure 4–58). The dispersal of the armadillo was also helped by man when he took it to Florida, where it became well established (Figure 2–6). It is likely that both animals—especially the armadillo—would have eventually reached the areas of introduction by natural spreading. Therefore, we may again cite the distance barrier to explain why they were not occupying these areas until their introduction by man.

In the early stages of dynamic thinking, zoogeographers easily reached

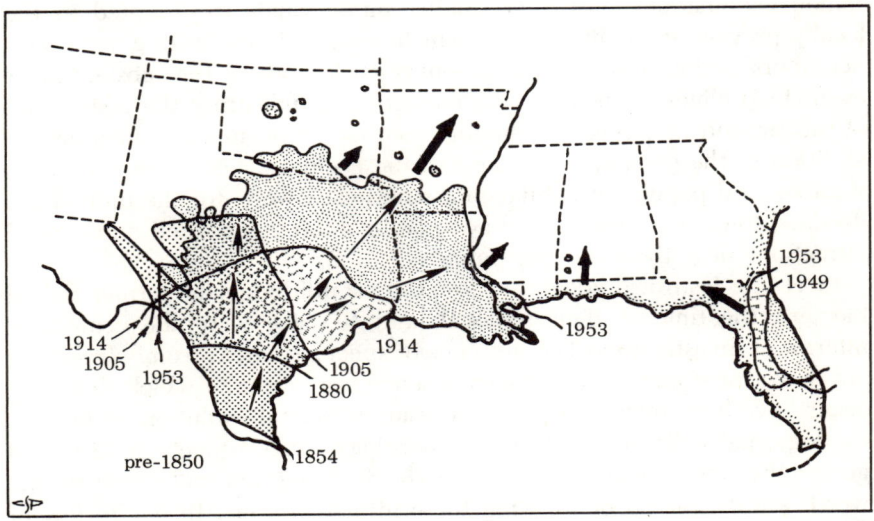

Figure 2-6 Distance barriers.

The nine-banded armadillo *Dasypus novemcinctus* L. [Dasypodidae, Mammalia] has expanded its range northeastward into the U.S.A. during the past one hundred years. In 1949, it had been introduced to eastern Florida, whence it also spread rapidly. It was thus the distance barrier, relative to the spreading potential and speed of the animal, that prevented colonization of Florida by natural spread. (Map from Buchanan and Talmage 1954, modified)

the verdict that this or that species "did not yet have the time" to advance in the short interval during which the glaciers had melted from the subpolar areas. With more detailed knowledge of the sharp fluctuations in climate during the past ten thousand years, one has to be more cautious. Distribution maps (Udvardy 1963b) reveal, for example, that the wheatear *Oenanthe oenanthe* L. [Turdinae, Aves] shows a distribution easily ascribable to a shortage of time. This Holarctic bird nests in places in Alaska and on the coasts of Greenland, Labrador, Baffin Island, and the eastern Canadian Arctic Archipelago. Taxonomic study indicates that these two disjunct areas were populated from separate sources. The Alaskan birds belong to the widespread Eurasian subspecies; the eastern North American birds are endemic, but migrate to the Old World, indicating their European ancestry. The central Canadian Arctic hiatus might be due to a shortage of time; however, some other phenomenon may have barred the wheatear from colonizing that area.

Even within the continuous distribution area of a species, the effect of a distance barrier can be traced—but this phenomenon belongs to the realm of speciation geography, not to general zoogeography. Sewall Wright's concept of "isolation by distance" (1943) indicates that a continuous population consists of smaller units which are adapted to the locally prevailing conditions, and which—though freely mixing with their neighbors within a short distance—nevertheless allow for substantial or even steep clines in their genetic constitution, indicating the restrictions of mixing and of dispersal. This distance factor presumably includes, in addition to the geographical element, the ecological effects of the large, suitable, and populated habitat in a paradoxical way, for the individuals destined to disperse would find adequate habitat, mates, and other social attractions near their starting point.

In classic historical zoogeography, the establishment of former faunal barriers, the time of their removal, and the like were the centers of interest of investigators. The presence of species in areas from which they are now absent and their previous absence from areas in which they now occur have been major areas of discussion in faunal evolutionary history. Geology, paleoclimatology, and paleoecology—three aspects of historical geography and "natural history," in the first but unusual sense of the word—are the means of studying these former barriers. In our review of historical zoogeography, we shall discuss these disciplines further.

Durability of Barriers

Numerous examples of the distribution areas of animals show that these areas are often split by evident barriers or are so widely disrupted that members of each part of a distribution area are not likely to be able to interchange or communicate with one another. It may be deduced that the animals either spread by some means across the barrier or that the barrier was nonexistent during some time in the past history of the species. In either case, our reasoning process brings us to the realization that distribution is a dynamic event and that conditions of dispersal change with time.

All dispersal barriers are of such a nature that their life is limited according to the geological forces that alter the earth's surface.

Oceans are the most stable and reliable water barriers. Opinions of geologists and their followers among biogeographers are still divided regarding the earlier permanence of oceans and continents. Land bridges are admitted to have connected—or, in some cases, still connect—Asia with North America, North America with South America, Eurasia with Africa and with Australia once or several times during this long period in which much of the present land fauna evolved and spread. The existence of a land bridge between North America and Europe across the North Atlantic is still debated (see also p. 360); Antarctica's connections with South America and New Zealand are within the limits of possibility. Across the Pacific, the Indian Ocean, and the major area of the Atlantic Ocean, where suggested land bridges have been thoroughly disproved, chains of islands might have had a wider spread in the past than they now have across the central and south Pacific and the Atlantic. In some cases these might have acted as "stepping stones" between continents for spreading animals, and they certainly did so in interisland dispersal.

Parts of the continental shelf are at times transgressed by the adjacent ocean. Then they emerge again as a result of subsidence and elevation or of the decrease of ocean water following the onset of a glacial period. The alternation of water barriers and land connections or island chains results in large-scale speciation by isolation and extinction, provided sufficiently rich fauna is near at hand. At any rate, present disjunctions of distribution ranges, as well as mergers of formerly distinct populations and faunas, find their explanation in the short life of sea barriers.

Once the discussion of ocean barriers carries us into the consideration of large-scale geohistoric events, it will be easy to understand that the life of freshwater barriers is still shorter. Lakes which fill tectonic troughs or are blocked by such ephemeral formations as moraines, dunes, or alluvial deposits hardly check dispersal for periods longer than a few tens of thousands of years. In their elimination, orographical dynamics as well as biotic filling play a role. Even the most ancient of the present large lakes had fluctuating water levels during their history, and thus were not insuperable barriers for any long period of time. In addition to being subjected to these same dynamic changes, rivers work upon their own destruction by erosion, course alteration, meander cutting, base leveling, watershed capturing, and delta-and-alluvium building.

The geological history of such physiognomic formations of the land as mountain chains, plateaus, and peneplains similarly indicate that barriers of a geographic nature are raised and leveled again. Still more important is the effect of these features and of water bodies on the local climate, for climate is determined not only by the major air circulation over oceans and continents but also by the modifying effect of land and its vegetation.

Every naturalist—if not every thinking man—is aware of year-to-year climatic fluctuations; during a lifetime, even larger wave sequences can be noticed. Until the most recent decades, only the major climatic fluctuations measured in geological epochs or periods were common knowledge to the biologist and zoogeographer. However, intensified biological and anthropological research has revealed those trends and fluctuations which have secular rhythms—for example, major climatic waves the fluctuations of which are measurable in mere centuries or even in decades. Where the number of biota is small but faunistic research is intensive, these trends are known to be followed by faunal fluctuations; thus the ephemeral role of the climatic barrier also becomes evident.

ACTIVE DISPERSAL

Dispersal is caused either by active movement—the locomotion of the dispersing organism—or by means of a carrier. The former is called *active dispersal;* the latter, *passive dispersal.* Both may be involved in the regular process whereby most or all individuals of a species disperse from the parental stock or domicile. They may also be used irregularly and by chance, thereby often leading to barrier crossing or to dispersal into newly available areas beyond the distributional area of the species or population. Hence we may talk about regular or accidental dispersal.

Active dispersal in land plants results from mechanical devices triggered by the drying of the seed pod or by external stimuli; but it is usually the diaspore—fruit, seed, bulbil, adventitious bud, and the like—that is actively dispersed, and only to a distance that brings it more or less out of the competitive reach of the parent plant. Diaspores and parts of plants which are possible organs of vegetative reproduction are carried passively by gravity, by the elements, or by animals; these are the chief methods of dispersal in land plants. Because evolutionary adaptations for the means of being carried and for withstanding long journeys are manifold, long-range dispersal is characteristic of plants. Yet the diaspores of plants have to rely on the carrier; when this means of transport ceases, they are at the mercy of the habitat in which they happen to be. In such instances, many animals can continue their traveling and can actively choose the terminus of their journey. Thus they are better adapted to dispersal than plants are, since they have both active and passive means of locomotion. Plant and animal distribution patterns (Chapter 4) show that these follow common principles and are governed by the same factors. In this sense, the needs of dispersal are the same although the means may be different.

Active dispersal in land animals is as varied as are their organs and modes of locomotion. Slowest are the diggers, creepers, climbers; runners and hoppers are much faster; and of course the flyers are the fastest. In contrast to that of plants, dispersal is only one function of locomotion in animals; often the multiplicity of functions may lead to confusion for the analyzing observer. It is easiest to detect and demonstrate specific dispersal adaptations of locomotor organs and behavior where these adaptations are not identical with those used in vegetative and reproductive activities. The sexual casts of termites and ants emerge with wings but shed them immediately after the nuptial—and simultaneously *dispersing*—flights. Grebes [Podicipediformes] are water birds, principally divers. They feed, court, escape, and hide by swimming and diving; they nest, roost, and socialize while afloat. These birds fly almost solely to disperse from one lake to another and, in temperate regions, for annual migration. They travel a considerable distance by swimming during migration, but they inevitably fly over land obstacles. There is no general survey of locomotor adaptations of dispersal from which more examples could be taken.

The interactions of anatomical and behavioral aspects of the organism and the resistance of the environment in the form of biotic and other barriers complicate the assessement of the means of dispersal and of dispersal achievements. *Speed or ease of locomotion is in no direct relation to the speed or ability of dispersal.* In a classic paper on insect geography, Handlirsch (1913, p. 467) says:

Es gibt sowohl unter den schwerbeweglichen als unter den leichbeweglichen Formen solche mit grosser und solche mit kleiner Verbreitung. Oft sogar sind erstere viel weiter verbreitet—namentlich wenn sie starke passive Verbreitungsmöglichkeit besitzen—als letztere (z. B. gute Flieger), denn diese sind unter Umständen leichter imstande, ihre Areale zu behaupten und dadurch der passiven Verbreitung zu wiederstehen.

Both large and small distribution areas are to be observed both among those forms that move with difficulty and among those that move with ease. Indeed, the former are more often widely distributed—in those cases in which they possess a strong means of passive dispersal—than the latter. Good flyers, for example, are able in many circumstances to maintain their range and thereby to resist passive dispersal.

Croizat introduces the same argument at the beginning of his *Panbiogeography* (1958 I, iv):

> ... while the means of dispersal of a hawk and a land-snail stand ... like 10,000 to 1 according to the immediate visual perception of the two animals performing in space, still the actual distribution of hawks over the face of the modern earth is assuredly not wider than that of land-snails, in which respect the two flatly stand one to the other like 1 to 1.

On page 65 of this book, there is a discussion of a recently observed behavioral mechanism—philopatry—which enables the animal to maintain or to return to its previous domicile despite changes of forceful displacement.

PASSIVE DISPERSAL

Active locomotion is studied and analyzed by the functional anatomist, but there is no general survey of the locomotor adaptations of dispersal. The means of passive dispersal have been studied even less, and they will therefore be considered at length in the pages which follow.

Anemochore Dispersal

In *anemochore* dispersal, wind and air currents are the vehicles of the dispersing individual. The process of aerial dispersal can be discussed in its three components: the take-off, the voyage, and the landing.

Take-off is no problem for those animals which are capable of active flight. Even those which become temporarily airborne by gliding—such as certain squirrels, lemurs, and marsupials (among the mammals) and certain volplaning snakes, lizards, and frogs—are carried farther when a wind is blowing. Caterpillars and mites that live in the canopies of trees take off on silk thread. Small spiders take off from the ground by spinning a thread into the wind until it is sufficiently long to support and lift the weight of the animal. Jumping insects also achieve the start of a wind-induced glide or float. Mealybugs [Coccioidea, Hemipt.] float in the air by their long, silklike dorsal filaments. For the rest of the terrestrial animals which do not regularly fly, there must be a wind strong enough to shake or blow them loose. Arboreal forms and those that live on exposed ground, on rocky areas, or on the surface of vegetation have more opportunty to become airborne than ones living in or on the ground, in crevices of tree trunks or rocks, in the forest interior, or on the bodies of larger animals. Whirl-winds, gale winds, tornadoes, and hurricanes are able to lift even these forms, and of course, in an increasing proportion, the larger animals are also exposed to the force of these strong winds. Darlington (1938*b*) presents a discussion supported by some convincing calculations on this phenomenon. Human judgment is often subjective, and we tend to underestimate the lifting effect of wind on smaller animals. But, since volume (and implicitly, also weight) correlates with the cube of any one dimension, and since surface area correlates only with its square, the lifting power of wind increases tremendously as length (and other dimensions as well) decreases.

Travel close above the surface of vegetation is accomplished easily, and is usually of short duration. When rising currents carry animals into the higher reaches of the atmosphere, strong winds (jet streams) that often blow in an entirely different direction from that of the surface wind, present special problems for the dispersing individuals. Cold and desiccation are factors that undoubtedly affect the animals; and since the smaller the animal the easier it is carried up and away, the relatively large surface area magnifies these problems. However, the importance of this kind of passive dispersal has been demonstrated several times in the zoogeographical literature on the spread of insects to oceanic islands and across sea channels; this is now considered the most important method of dispersal, both for actively flying insects and for those which are distributed passively. We may reason that anemochore dispersal is just as important a means of mixing continental insect faunas of different geographic and climatic regions as dispersal across water barriers. This aspect of faunal mixing is not discussed in contemporary literature.

Landing poses no special problems for airborne animals, but the success of dispersal will of course depend on living conditions at the landing place.

Our first group of examples deals with flying insects. Their power of dispersal is out of all proportion to the limited distance their own energy resources would allow them to cover by unassisted flight; they regularly and habitually—as well as accidentally, but with statistical regularity—drift with the movement of the air masses in which they arrive. Hardy and Milne (1937) describe a survey of insect drift over the North Sea, in which nets—some hung high in the rigging of a vessel; others, from kites—captured 730 insects, most of them alive. The majority were Diptera and aphids, but Hymenoptera, Lepidoptera (only one specimen of the habitual migrant noctuid family), Neuroptera, Hemiptera, and several spiders were also taken. This survey was conducted mainly between

180 and 240 km from the nearest coast of the British Isles or of the European continent. We know, however, from Palmén's study (see below) that insects that fly or drift over water at such low altitudes are often depressed by the winds to the surface of the water and that if there is no land nearby, they drown. Therefore, wind drift at a greater height is a more important vehicle of dispersal. As man himself has only recently become airborne, the aerial habits of insects are just beginning to be known. Several instances of preliminary work are known. Most of these studies have been made by entomologists in North America who were interested in the wind dispersal of introduced insect pests from the port of infestation (Glick 1939, Wolfenbarger 1946). Gressitt (1961, 1963, 1964) applied special catching devices to commercial airlines crossing the Pacific. Over 25 km^3 of air were screened, and 1,075 insects were captured. Besides Diptera, Hymenoptera and Homoptera, which comprised the major share of the catch, 11 other insect orders, together with spiders and mites, were present at distances ranging from 100 to 3,000 km from land. It is not known how many of these insects were dead when captured; assuming that most of them were dead, we must take into account that such drift has been occurring every day over the total surface of the globe ever since the first appearance of land arthropods. Thus the possibilities and varieties of aerial dispersal seem immense, and one wonders why smaller arthropods, especially flying insects, are not more evenly distributed over the land areas of the earth.

The answer is more speculative than the demonstration of anemochore dispersal. The airborne pioneer has to alight on the right spot, in the right habitat; and even if it is a fertilized female carrying viable eggs, these have to survive the hardships of a strange habitat and establish a sizable starting population. Certainly Gressitts' experiments lend credence to the wind-dispersal theory of the origin of the faunas of oceanic islands. Zimmerman (1948) calculated that between 233 and 254 ancestral species could have given rise to the 3,722 endemic insects of the Hawaiian island group. Each species could have started from a single immigrant. How few, then, were the successful landings of airborne pioneers during the approximately ten-million-year history of the island group! The Antilles of the West Indies are considered part of a sunken continental connection between North and South America, but their insect fauna is largely derived from across the ocean and even shows African affinities. Darlington (1938*a* and *b*) bases his theory of faunal similarities in the Antilles and in northern Africa upon the storm tracks that "connect" these two areas across the Atlantic in an east-west direction. Tropical storms easily carry the carabid beetles that take to the air under their own power. Within a continent, very regular and, at times, very strong winds blow up and across the mountain chains. This updraft carries with it dust, leaves, small branches, and all the live material it can lift. Like the white collecting sheet of the entomologist, mountain snowfields are collecting places for wind-borne waifs. Especially on days with moderate winds and much sunshine, when the updraft is particularly strong, the snowfields are blanketed with dead and numb insects. These snowfields are the regular feeding places of alpine birds (Verbeek 1965). Thus, the phenomenon of wind-borne dispersal influences alpine ecosystems and helps carry insects across mountain chains. Niethammer (*in litt.* 1967) observed the same phenomenon among alpine birds of the Apennines in Italy and the Sierra Nevada in Spain. In the latter, the rock bunting *Emberiza cia* L. [Emberizidae] feeds predominantly on snowfield litter. Mani (1962) writes that insect drift is the most important source of food of the nival arthropod fauna of the high Himalayas. When the updraft lifts the lowland insects into the heights, these become chilled—deep-frozen, as it were—and are deposited onto the snowfields in such "preserved" condition.

Recent studies on the biology of aphids give a good account of their dispersal behavior (Kennedy and Stroyan 1959). The winged individuals have a special means of flight—gliding with an air current—but at the sight of the right habitat (green vegetation) they terminate their ride upon the wind and glide downward. In this way they have been known to cross several hundred km of adverse habitat. Observations were taken on the adventive pest of alfalfa fields—the spotted alfalfa aphid *Therioaphis maculata* (Buckton)—by Dickson (1959) in the southern California desert. A single, potted, alfalfa plant was used as bait, about 26 km from the nearest irrigated field, and averaged two aphids per week. This number does not sound impressive. However, glue traps suspended between the ground and a height of 16 m provided a quantitative sample. Dickson estimates that 3.5 million aphids passed through a one-mile (1,600 m) front during 26 days in April, 1956. A single specimen of *T. riehmi* Börner was caught at a minimum of 78 miles (about 135 km) from the nearest alfalfa field where it occurs. This aphid was first found in New England in 1948; thus, in about ten years it had crossed the whole North American continent by going from one alfalfa field to another. Dickson's traps caught only the low-flying aphids. A great number of these insects disperse with the high winds. The multitude of spruce aphids *Dilacnus piceae* Panz. [Aphididae, Homopt.] which an expedition found on the Spitsbergen Islands in 1924 must have drifted in the air high over the Arctic Ocean, presumably originating on the Kola Peninsula, some 1,300 km away (Økland 1955).

Damselflies—those delicate zygopterous Odonata—are weak flyers, but the wind is an active, helping dispersal agent in their life history. Observations on marked populations in Florida (Mitchell 1962) showed that in calm weather there was less than 2 percent loss of the population of imagoes but that during a two-day wind storm there was a tenfold increase in their dispersal to a neighboring pond.*

Birds, though among the most skillful of flying animals, may also drift with exceptionally strong winds over water and across land areas which they would avoid under normal circumstances. Such drifting flocks often pioneer new areas; during the past century there are several well-observed cases of a species establishing itself across a major water barrier. In January, 1937, a storm blew across the North Atlantic from western Europe in the direction of Greenland and carried a number of migrating fieldfares *Turdus pilaris* L. [Turdinae, Aves]. On January 20, they were observed on the volcanic island of Jan Mayen, hundreds of miles northeast of Iceland, and in northeastern Greenland; a week later, they were seen at several points in western Greenland (Salomonsen 1951). Thereafter, these birds started to breed in southern Greenland, where they have thrived and where their population is on the increase (Salomonsen *in litt.* 1964) (Figure 2-7).

There is a group of animals—the spiders [Araneae, Arthropoda]—that do not fly actively but have a direct adaptation, both structural—gossamer—and behavioral, which enables them to utilize regularly aerial drift for dispersion within their range as well as dispersal beyond its limits. By means of a behavioral adaptation, the young spider spins a length of gossamer into the wind until it is long enough to support the tiny animal, and then it takes off and sails away. In temperate regions, such as the Hungarian plains, the mass movement of young spiders is very conspicuous in the early fall. The gossamer tangles with weeds, bushes, and grass, and one may find many aphids, small flies, and other insects trapped therein and dispersing accidentally along with the spiders. Bristowe (1939) discusses this phenomenon at length. Most spiders,

---

*One is likely to consider storms as rare events and their effects as chance-determined phenomena. However, storms, though occurring irregularly, have a regular season of occurrence in every climatic belt. Though the first cyclonic storm of August may occur any time from the first to the 20th, it triggers the fall departure of swifts and nighthawks from northwestern North America. Ornithologists of the Atlantic seaboard of North America have found that traditional migratory routes of small birds do not follow the seacoast but lie somewhat inland. The explanation is that if they fly along the coast, a storm of even medium intensity would carry them over the ocean; as it is, they are wind-drifted toward the coast during the storm and return to their foothill routes after the storm. The effects of storms on bird migration and on subsequent adaptations illustrate the regular nature of storms; we may infer that storms also regularly disperse insects and other small animals.

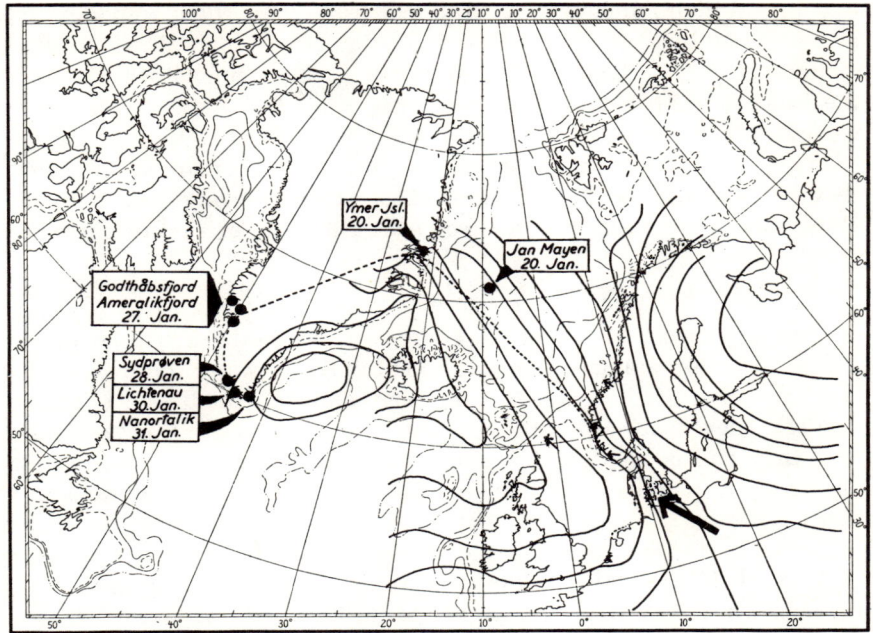

Figure 2-7 Colonization by anemochore dispersal (drift dispersal).
Wind drift of fieldfares *Turdus pilaris* L. [Muscicapidae, Aves] from Europe to Greenland in January, 1937. (From Salomonsen 1951.) Black circles: observed birds or flocks. Broken line: probable course of drift. Unbroken lines: isobars of January 19, 1937. Arrows: wind direction. By this flight, the fieldfare became established in southwestern Greenland, where it is now on the increase. (*Courtesy of F. Salomonsen*)

with such important exceptions as the mygalomorphs (Solem 1959), engage in aerial dispersal; the family Linyphiidae does so extensively.

Leech (1965) draws attention to the possibility that gossamer secretion is subject to selection, based upon the need of the spider to disperse. On remote arctic Ellesmere Island, three species of small spiders have not been observed to produce true gossamer but only a very weak thread which could not support even a young individual. Leech thinks that these species might be Pleistocene relicts and that the other Ellesmere species —those with regular gossamer production—are late postglacial immigrants. Zimmerman (1948) recorded many examples of aerial, passive dispersal from the data of earlier observers and experimenters concerning spiders, insects, and small land snails. Inger (1954) suggested the possibility of aerial dispersal of small amphibians, especially in species that live in high mountains. A small leaf frog *Oreophryne annulata* Stejneger [Microhylidae] is restricted to the cloud forests in the high mountains of New Guinea, the Moluccas, Celebes, and other islands of the Indo-Malayan area. As no continuous mountain range is known by geologists to have existed in the past, it is likely that they spread by "waifing." The adults of these animals are no more than 25 mm long; they lay their eggs in the wet moss of tree trunks, from which pieces of moss, bark, small twigs, and the frog eggs are easily carried by the wind.

Hydrochore Dispersal

Hydrochore dispersal can be an additional aid to the active swimmers and to the other animals that occasionally take to water, because of the carrying effect of the currents. For nonswimmers, small size is an advantage; they float easily and, as long as their respiratory organs function, they will float, alive, for some distance. Their endurance depends on the water permeability of their body surfaces,* on the physiology of their respiration, and on the temperature and the salinity of the water (Palmén 1944). Specimens of the centipede *Pachymerium ferrugineum* C. L. Koch [Geophilidae, Chilopoda] have lived for as long as 25 days immersed in freshwater of from 19°/ to 27°C; they have lived 178 days in colder water from 6° to 12°C (Økland 1955).

Rafting provides dispersal opportunities even for those land animals which would easily succumb to the hazards of hydrochore travel. Rafts originate along the ocean shores and on larger rivers. Calving glaciers set icebergs afloat; if these icebergs originate near the end of side moraines, they will often carry rocks, pieces of logs, debris, and even living, rooted plant material. Snow slides are still more likely carriers of living things picked up on their way to the sea or a river. It takes a long time for an ice float to disintegrate, at which time its passengers may continue floating until they are washed up on another coast or island (see McCabe and Cowan 1945). Ice floats of the arctic seas are regular carriers of polar animals, and it is well known that the arctic fox *Alopex lagopus* (L.) [Canidae], polar bear *Thalarctos maritimus* (Phipps) [Ursidae], caribou *Rangifer tarandus* (L.) [Cervidae], and muskox *Ovibos moschatus* (Zimm.) [Bovidae] are dispersed ice-borne in the arctic archipelagoes.

Ice fields and bridges are regularly used by these same arctic animals in the active crossing of frozen waters; even in subarctic and temperate latitudes, this manner of barrier crossing is well known. Of course only

---

*The body surfaces of many insects are covered with a water-repellent coat of lipids (Beament 1962).

winter-active, endotherm vertebrates are able to use this kind of transportation. In this way the reindeer has been known to travel from Novaya Zemlya to Spitsbergen across the frozen Arctic Ocean (Antevs 1947). The islands in the Gulf of St. Lawrence are populated or at least are occasionally visited across the winter ice by winter-active mammals; the skunk *Mephitis mephitis* (Schreber) [Mustelidae, Mamm.], and the porcupine *Erethizon dorsatum* (L.) [Erethizontidae] are missing from these islands, presumably because both are in hibernation during the winter (Denman 1965). Ice rafts are suspected to have carried the flightless ground beetle *Carabus chamissonis* Fisch [Carabidae, Coleopt.] to Newfoundland across these same waters, because the species also occurs in southern Labrador and on Belle Isle (Lindroth 1963). River rafting and

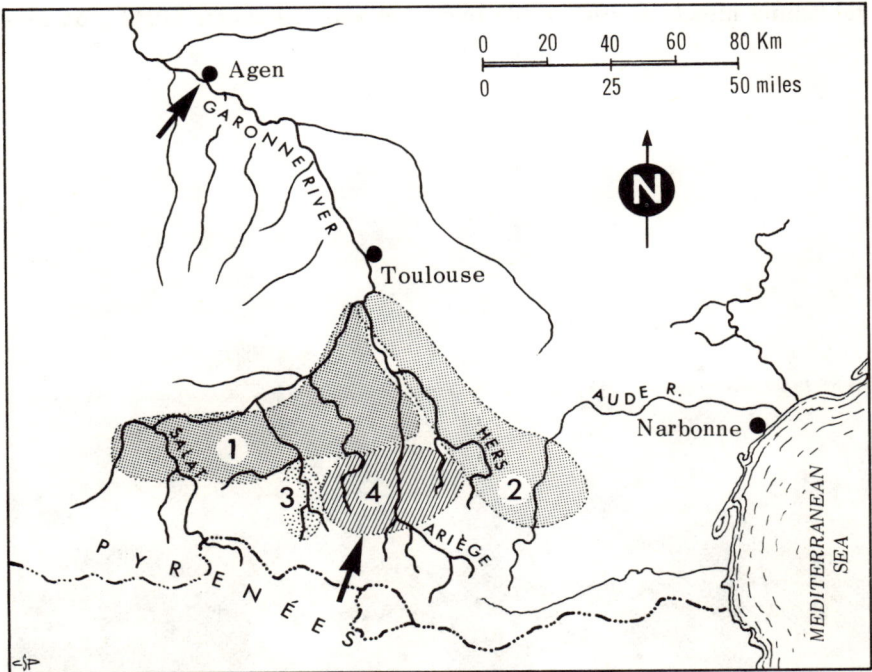

Figure 2-8 Hydrochore dispersal by rivers.
Distribution in the French Pyrenees of the four subspecies of the land isopod *Phymatoniscus tuberculatus* (Racov) [Oniscoidea, Isopoda]. 1. *arbassanus* Vandel; 2. *tolusanus* Vandel; 3. *gironensis* Vandel; 4. *tuberculatus* Racov. *P. t. tuberculatus* (marked with an arrow) has been collected on the plains at Agen (arrow), evidently having floated down the Salat across the range of another subspecies, and then along the Garonne river. After Vandel 1960.

42   The Ecology of Dispersal

floating are the most important means of passive dispersal for small animals of hygric and mesic habitats. Holdhaus (1929) writes that in the Danube valley it is mainly the land invertebrates of the mountains who utilize stream transport. In France, many such animals from the mountains spread by means of river flotsam to the lower valleys (Figure 2–8) and thence along the humid coast. The littoral zone, with tidal water interacting as a carrier, is the most used route of the halophilous and littoral species, such as the very widespread oniscoid isopods (Vandel 1960). Rafting has such a large literature that no additional discussion is included in this book (see Figure 2–9).

Anemohydrochore Dispersal

A particular combination of anemo-hydrochore dispersal (wind- and water-borne) is not generally known; yet it seems very significant in establishing insects across sizable bodies of water. Palmén (1944), on sev-

Figure 2-9   Hydrochore dispersal by raft.
Floating Nipa palm about to land on Daru Island, more than three miles off the mouth of the river Oriomo in South New Guinea. Such rafts carry a whole faunation, and may deposit the animals on islands and remote shores. From Rand and Brass 1940. (*Courtesy of Archbold Expeditions of the American Museum of Natural History*)

eral occasions during the course of a single summer, found huge numbers of live insects, especially beetles and bugs, in flotsam on the southwest-

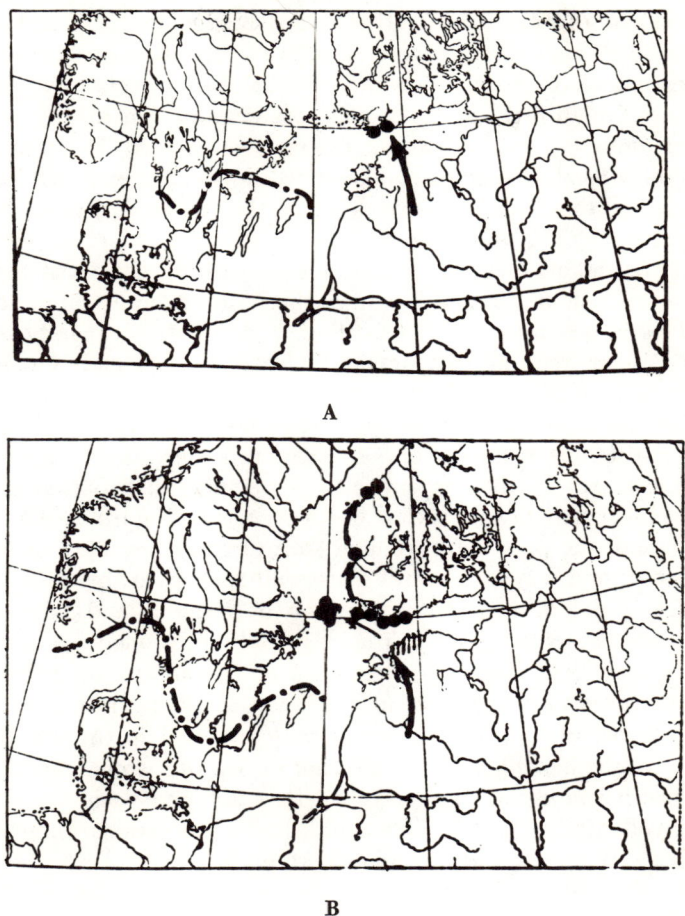

Figure 2-10 Anemo-hydrochore dispersal.
The Fenno-Scandinavian distribution of two beetle species as explained by anemo-hydrochore dispersal across the Baltic to southeastern Sweden and to Finland. (Palmén 1944.) Arrival of the beetles has actually been observed at the Hanko Peninsula, the western of the two dots on map A. Dotted line shows the northern limit of these species in Scandinavia.

A. *Anthracus conspulus* Doft. [Carabidae, Coleopt.] is a typical Baltic immigrant which in Finland is limited to one bridgehead area.

B. *Paederus fuscipes* Curt. [Staphylinidae, Coleopt.] has considerably expanded toward the north from southern bridgehead areas in Finland. (*Courtesy of E. Palmén and the Finnish Zoological-Botanical Society Vanamo*)

ern Baltic coast of Finland. Thorough study has revealed that these insects originated across the Baltic, in Estonia, and that they were probably stimulated to initiate mass flights by the prevailing meteorological conditions. These flights, at low altitude, are carried by the wind and then depressed into the water, which eventually casts them ashore according to the course of the coastal currents. The floating insects literally blanket the beach. On June 17 and 18, 1939, for example, in a 0.25 m^2 sample of beach debris, 1,590 individual insects were found, belonging to 542 species, of which 12 species had over 50 individuals each. Only about 3 percent of all these insects were dead; of the remainder, some had dried and were wriggling about, some had copulated, and some were flying away. Experiments have shown that most insects can exist as long as five days in fresh and brackish water (the Baltic is brackish!) and at least three days in ocean water without their reproductive capacity being impaired. Palmén reasons that anemo-hydrochore dispersal is much more effective than either wind or water dispersal, for the chance factor is greatly diminished. In wind dispersal, chance affects the successful meeting of drifting insects of the opposite sex; in water dispersal, the animals are at the mercy of the sea currents. A peculiar distribution pattern of many Finnish and Scandinavian beetle species (Figure 2–10) can now be explained: the prevailing summer storm track represents an important immigration highway of Central European species to this area. Palmén's work also pointed out the significance of large inland lakes—such as those in central Finland—in aiding dispersal and pioneering through the anemo-hydrochore dispersal system. *Mutatis mutandis*, Palmén's impressive work also proves that many, if not all, flying insects are subject to mass dispersal *by air* and that, having traveled in the air, they may be funneled to beachheads of colonization by the water carrier.

The passive mode of dispersal may be combined with active locomotion as well. Kangas (1953) has published observations about *Carabus cancellatus* Ill. (Carabidae), which is one of the rapidly expanding southern elements in the Finnish beetle fauna. The beetles were crossing bodies of water from 200 to 250 m wide, actively swimming, keeping their direction, and arriving at islands where previous surveys had failed to find this species. In such cases, one has to count with four components of dispersal: flight, aerial drift, swimming, and drifting with strong water currents.

Biochore Dispersal and Phoresy

The passive transporting of one animal by another is termed "biochore dispersal." This method, as we have already noted, is common and important among plants. Plants have served as the requisites and the habitats of members of the higher animal world ever since their first appearances, and such mutual relationship has had ample time to evolve. But animals are apparently much less in contact with one another than plants are with animals; even predator-prey relationships are, ipso facto, more recent. Symbiotic and parasitic relationships are the only kinds that necessarily lead to the evolution of *biochore* methods of dispersal.

It is evident that *parasites* are carried by actively moving hosts. In this way, the host helps in dispersing the parasites, and becomes a *vehicle* (*carrier*). For example, the sheep tick is spread by the sheep (p. 119). Hoogstraal and his associates (1964) investigated over 11,000 migratory birds in Cairo that passed through Egypt on their migration. They found 1,442 ticks on 881 infested birds. In a study of migratory birds in Finland, Nuorteva and Hoogstraal (1963) found ticks on only nine specimens out of 674 swallows they caught, but these ticks belonged to a species hitherto unknown in that country. Evidently, swallows transported *Hyalomma m. marginatum* Koch [Ixodidae, Ixodoidea] to Finland from eastern Eurasia. This tick, like many other species, is a vehicle of virus diseases such as different kinds of encephalitis. In this example, then, as in many others, biochore dispersal is telescoped.

Australian feral rabbits were successfully controlled by the introduction of the virus disease myxomatosis. Here also is the most spectacular example of the rapid spread of a parasite, the *Myxoma* virus, by means of culicid mosquitoes as vectors. Rabbits were experimentally inoculated and then released in December, 1950. By the following summer, the disease had spread to an area 1000 to 1100 miles (1600 by about 1,760 km) in south-north and east-west diameters! Fenner and Ratcliffe (1965) describe the circumstances of the spread. This was due to the wind-aided dispersal of the mosquitoes, some of which are able to spread 40 miles per individual. For instance Woody Island, off the coast of Australia, received myxomatosis after a wind had blown toward it, on January 12, 1952, when the nearest mainland infestation area was 200 miles (about 320 km) away. Biochore dispersal—or phoresy—of the virus was coupled with wind-aided (anemochore) active flight dispersal of the mosquito vectors. Movements of the primary host (the rabbit) were insignificant, considering the time and distance involved.

The more intimate and lasting the relationship, the more the reliance of the parasite on this kind of biochore transportation. Such symbiotic relationships as commensalism may lead, on the one hand, to parasitism or, on the other, to the obligate utilization of the commensal as carrier. This relationship is called *phoresy*. In phoresy, the emphasis is on the absolute reliance of the passenger upon the carrier animal for

46 *The Ecology of Dispersal*

transport. Not all phoretic passengers are parasites of the carrier organism. For example, the *Myxoma* virus is not known to harm the mosquito carrier. All parasites do not use phoretic means of transport; for instance, parasitic wasps fly under their own power. The blister beetle *Meloe* [Meloidae, Coleopt.] lays its eggs on the ground, and the hatchlings crawl on the top of the vegetation by a positively phototactic behavior pattern. Those which happen to reach flowers that later are visited by insects attach themselves to the hairy body of the latter by means of their three specialized tarsal claws. If the carrier happens to be a bee, the bee carries the blister beetle to the bee's nest where the beetle feeds on both the larval bee and its food. The imago leaves the nest and deposits its eggs on the ground. Though this phenomenon is described as the best example of phoresy (Schaller 1960), it is a combination of parasitism, predation, and—as perhaps a by-product—phoretic dispersal. In the case of the scarabaeid beetle of the genus *Canthon,* the phoretic relation is a useful by-product of commensalism in which the carrier pro-

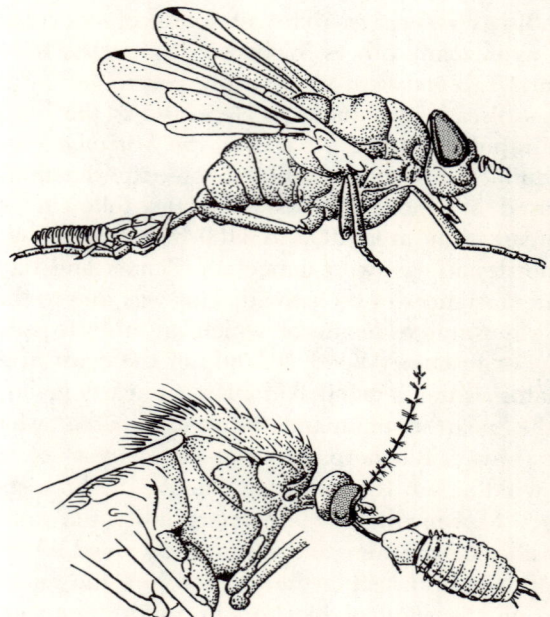

Figure 2-11 Phoresy.
Above: Pregnant female pseudoscorpion gets a ride on the hind leg of a fly. *Below*: A feather louse (Mallophaga) holds on to the snout of a mosquito (the snout's tip is visible on the extreme right). *After Schaller 1960.*

vides transportation as well as food. These beetles feed on monkey feces and travel on the anal region of the host, ready to alight when the new food supply is deposited (Osche 1963). Similarly, pseudoscorpions are often passengers on flies (Figure 2–11) and other flying insects that visit birds' nest and similar places abounding in organic debris. *Chelifer cancroides* L. [Cheliferidae, Arachnida] is the best known of these pseudoscorpions and is nearly cosmopolitan in its distribution. Another species, *Larca lata* (Hansen), is found in the nests of the black redstart *Phoenicurus ochruros* L. [Turdinae, Aves] and is suspected to have spread recently to England as a passenger on this bird, also a new arrival in the British Isles (Ressl 1963). This last example, though not yet fully investigated, would best show the role of phoresy in dispersal. Phoretic mites, pseudoscorpions, insects, and certain crabs and fishes use the live carrier mainly to reach their food source and only incidentally to spread. These forms are all characterized by having impaired locomotion and agility, and by the use of local, spatially isolated or restricted, and often ephemeral habitats. The phoretic relation is an obligatory characteristic of the passenger. The passenger has, as in the case of *Meloe*, specially adapted organs and behavior patterns, and cannot travel by itself.

Accidental (facultative) biochore transportation is an important dispersal agent. It is not restricted to certain specialists; it works on the principle of chance, and is therefore difficult to demonstrate. No matter how convincing some observations and analyses may be, biogeographers will inevitably be skeptical—especially those who are not familiar with the particular animals and the observed relations upon which the foregoing principle is based. How often does an ornithologist come upon a bird with freshly caught, wriggling animal food in its beak? Less often, but still within every bird watcher's personal experience, the food is dropped from the beak of the frightened bird and escapes unharmed. Wolda (1963, details discussed above) noted that his experimentally released snails were preyed upon by thrushes, and that these birds transported the snails across the water barrier of a small brook surrounding his study area. This was, then, biochore transportation, though on a microgeographic scale. Another example involves birds and plant transportation; though the evidence is circumstantial, it is very impressive. Taylor (1954) analyzed the flora of the remote antarctic Macquarie Island, he found 35 species of vascular plants. Four species also occur in South America, approximately 6,500 km away. The majority of these plants have small, hooked, barbed, or bristled seed which easily becomes attached to the feet of the seabirds that breed there. Such seed has indeed been found on the legs of albatrosses, and it was noted that some of these seeds were foreign to Macquarie Island. Invertebrate eggs adhere just as easily to the legs and plumage of birds. Examples of the transport of seeds and of the eggs of terrestrial invertebrates are scarce in the literature, but biochore transport of aquatic organisms is well known (cf. Niethammer 1953b, Löffler 1963, Maguire 1963, Proctor & Malone 1965). Økland (1955) mentions that snail eggs, probably of a species of *Succinea* [Succineidae, Pulmonata], have been found on the leg of a duck shot in the Sahara Desert. The duck was found at a distance of over 150 km from the nearest water where the snail could have laid the eggs. The transport of mollusks by birds has been observed a few times and hypothesized upon more often (Kew 1893, Baker 1958). Baker mentions, among others, land snails of the genus *Vitrina* [Limacidae] in which there is no apparent difference between the western North American species and the form restricted to high altitudes in Hawaii. Baker suspects shorebirds as carriers. In this connection, it should be noted that arctic and subarctic shorebirds winter in the Hawaiian Islands; they occupy mountain meadows as well as coastal flats, and therefore could easily have carried the snail into the cooler altitudinal zones. Zimmerman (1948) observed white-tailed tropic birds, *Phaethon lepturus* Daudin [Phaethontidae], nesting in the high interior rain forest of Samoa under such circumstances (among dense foliage) that they could easily dislodge insects, snails, and parts of plants while beating with their wings. These birds are open-ocean feeders, and cyclonic winds may carry them away. But sea birds are fast travelers, even without typhoons; recent banding has revealed, for example, that one giant petrel *Macronectes giganteus* Gm. [Procellariidae, Aves], which had been banded as a nestling on Macquarie Island, was shot, four months after the banding date, about 13,000 km away on South Georgia Island (Taylor 1954). Another specimen traveled a direct distance of some 11,000 km in five weeks (Falla 1960). Therefore, we cannot easily dismiss long-distance chance dispersal by bird carriers across great ocean expanses. Larger land animals are also good accidental carriers, especially when shorter distances are involved.

In most of the examples we discussed so far (excepting for the dispersal of endoparasites), the dispersing agent carried the spreading individuals externally. Some authors call this *ectozoic* transportation. *Endozoic* transport enables the spread of hardy spores, seeds, and fruit of plants, as well as eggs of various crustaceans; even adult ostracods have passed unharmed through the alimentary canal of birds in experiments by Proctor *et al.* (1967). One might call another kind of dispersal *endophytic* when the carrier is any part or a disseminule of a plant. While this part is dispersed passively, it may carry leaf-mining, twig-, seed-, or fruit-boring animals.

Anthropochore Dispersal: Adventives and Introductions

Man originally carried his own bacterial, protozoan, helminth, and arthropod parasites, and provided biochore transportation for only these specialists. In the course of evolving civilization, however, first the domestic animals and then food and game animals, together with their parasites, began to be transported willfully by man wherever he migrated and settled. The passive dispersal of all kinds of lower animals which use our vegetable or animal foodstuffs as hiding, sheltering, feeding, or breeding microhabitats is as ancient an occurrence as is food gathering. The supply carriers themselves (for example, ships and boats) were able to carry even relatively large animals such as rats, mice, frogs, and land snails. With the advent of railways and, later, of automobiles, airplanes, and huge ships—all of which have enclosed air spaces—the possibilities for the passive transport of unwanted animals have materially increased. By the middle of the twentieth century, this passage of *adventives*—inadvertently introduced animals and plants—has reached such proportions that many countries have established special bureaus, staffed by scientific and technical personnel, to block the arrival of unwanted pests. However, in many instances the introduction of adventives was not observed until after the faunal investigation had begun (Figure 2–12). It is often difficult in such cases to decide whether a species is native to an area or is an adventive. A sort of zoogeographical detective work has to be undertaken, and the true nature of the adventive species can be revealed by one or more types of criteria (Lindroth 1957). The most reliable is the *historic* criterion: the actual documentation of an introduction or of a historic event—for example, the record of a voyage—or documents made at a port of entry. The *geographic* criterion is helpful because of the fact that too small, immature distribution areas or unnaturally disrupted ranges, with gaps where ecological or faunal reasons would not warrant such gaps, become evident during a scrutiny of the geographic distribution. If a species is restricted to disturbed ruderal areas or cultivated habitats, this *ecological* criterion suggests an introduction. Often the *natural history* and the habits of the suspected species reveal its nature—for example, when it is restricted in its food to a similarly introduced plant. Often the criterion is *taxonomic:* the subspecific identity of the introduced form reveals its place of origin.

During recent decades, conscious transplantation or introduction of foreign animals has also reached gigantic levels. A great variety of game and food animals, pets, and familiar wild animals that have recreational

50 The Ecology of Dispersal

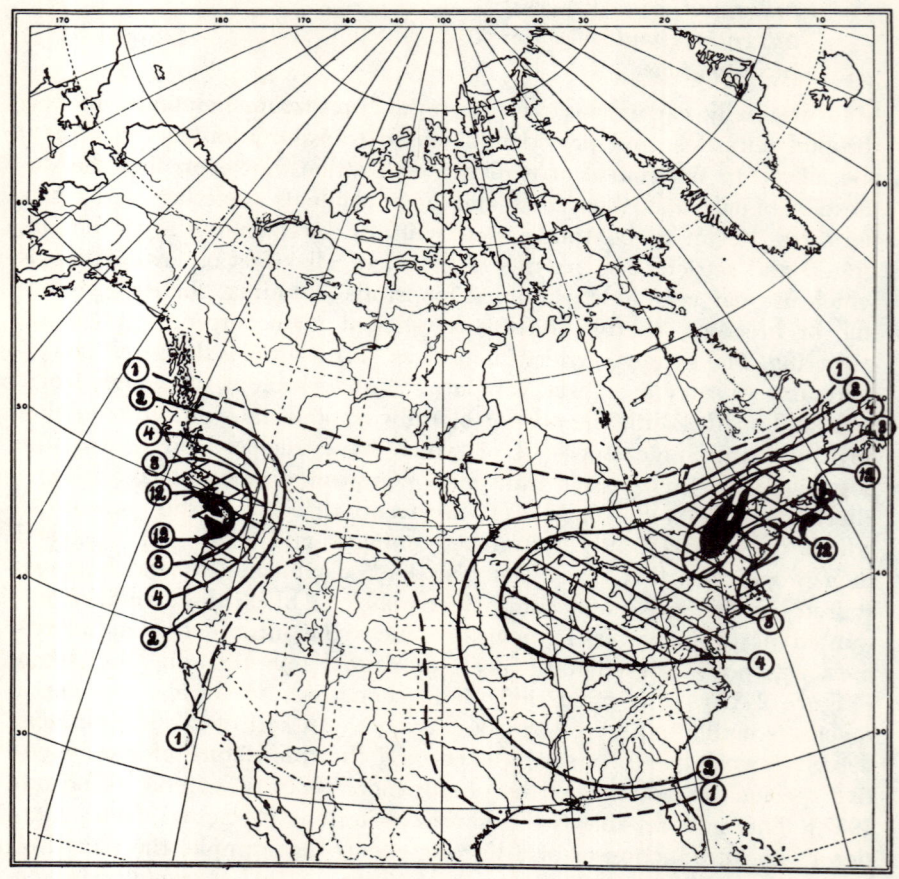

Figure 2-12 "Europeanization" of the North American carabid beetle fauna. The largest numbers of accidentally introduced species are found around the important harbors in the cool temperate climatic zones. From Lindroth 1957, (*Courtesy of C. H. Lindroth and Almqvist and Wiksell, Stockholm*).

value (or because of their aesthetic appeal or for sentimental reasons, or even through superstition) were transported in large numbers. During the second half of the nineteenth century, popular movements—naturalization societies, bird-protection clubs, and the like—or wealthy and influential individuals undertook these introductions. Later, government agencies, often known as fish and game departments, directed these introductions on a more or less scientific basis. Biological pesticides have emerged as a special category, developed for the purpose of exterminating or controlling native, introduced, or adventive predators and parasites—both plant and animal—which interfere with the yield of cultivated biota or which are otherwise harmful to human society.

Unfortunately, unwanted and accidental introductions sometimes result when live specimens are transported for scientific study. A case in point is the gypsy moth *Lymantria dispar* L. [Lymantriidae, Lepidopt.], a defoliator of deciduous trees in the Old World, which was imported to North America for study by Leopold Trouvelot at Medford, Massachusetts, in 1869. In that year, some caterpillars of the imported moths were accidentally released; within ten years a vast pest population developed which had spread over all of New England. Many predators and parasites of the gypsy moth have been introduced to the United States in an

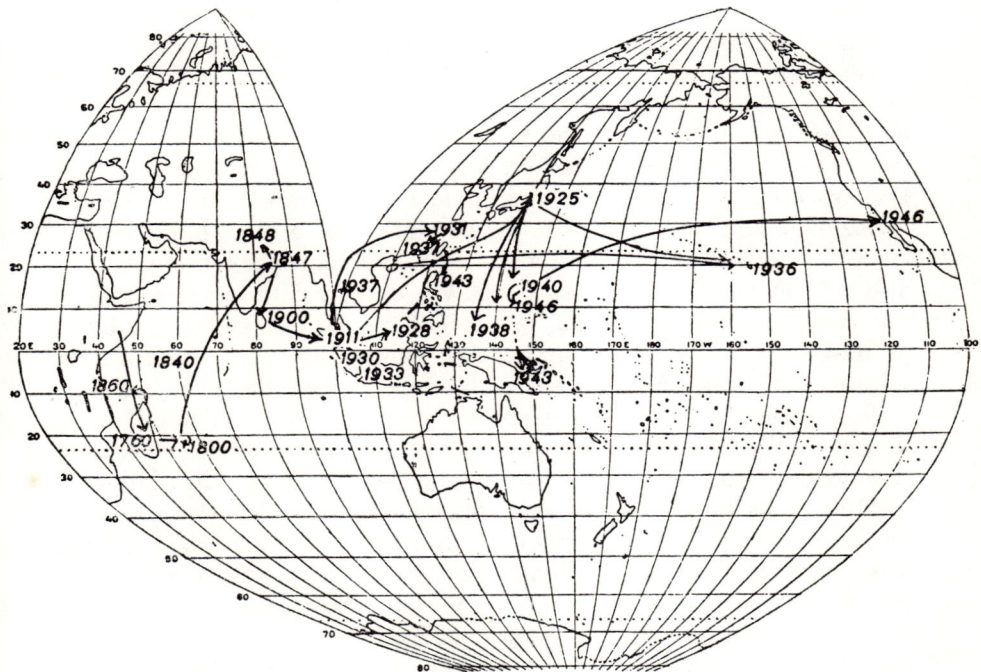

Figure 2-13 Anthropochore dispersal.
Routes of introduction of the giant African snail *Achatina fulica* Bowdich during about 185 years. Originally of East Africa, this land mollusk has been introduced to India, thence to other parts of the Orient and of the Pacific area. It is a serious pest in cultivated areas and in gardens. In the Hawaiian Islands, predatory snails have been introduced lately for its control, apparently with signs of success. Map from Økland 1955, after Brinkmann 1951. (*Courtesy of Aschehoug Company, Oslo*)

attempt to check its population increase by biological means (Brown and Sheals 1944).

The introduction of the giant African snail is a particularly illuminating example, because it is intentionally carried as a food source by certain peoples. It spread in this manner across the tropical regions of the Orient in about 150 years (Figure 2–13). Snails are not a common article of food in North America; yet, between 1948 and 1954 live specimens have been intercepted on sixty occasions by quarantine officers at American ports. During 1961 alone, mollusks of twenty-three species have been found in bags and parcels carried by American travelers aboard merchant ships on no less than 253 occasions (Hanna 1966). The muskrat, *Ondatra zibethica* (L.) is a widespread North American rodent which, because of its excellent fur, was introduced into Bohemia in 1905. It immediately became established there, and in subsequent years spread through all of Central Europe (Figure 3–22). This and other similar introductions are discussed and summarized in a short monograph by Elton (1958).

The faunal picture of those regions of the earth where civilization is old is thoroughly influenced by the long residence of adventive plants and animals. Adventives spread two ways. Military expeditions, warfares, tribal movements carry a mass of adventives across long distances. However, such events do not occur continuously. The spread of agriculture, civilization, and trade carries fewer immigrants at a time, but at a constant though slow pace. We find that this dual nature of introductions is also evident today. When a new highway opens a hitherto remote area, the trucks, trailers, and tractors of the building crew carry the first batch of unwanted immigrants. Later, when filling stations, shops, motels, schools, entertainment centers are developed along the new highway, civilization—and often even modern agriculture—becomes established. The new residents send and receive parcels, go on vacations, and receive commodities from their original residences—all of which furnish innumerable opportunities for unwanted faunal exchange. Much less is known of these adventives than of transcontinental introductions. The special eco-zoogeographical study of anthropochore dispersal and of the resulting distribution patterns is a neglected yet highly important aspect of zoogeographical research. Here are opportunities to study the ecology of dispersal, the broad-scale experiments in community reaction to intruding species, the short-term dispersal phenomena, and the adaptation of the species to the new favorable or hostile environments. Among the best means of studying niche requirements and competition phenomena is to concentrate our research on the ecology of introduced animals. This offers unlimited opportunities to provide new examples for the basic biological object of speciation studies.

CHANCE DISPERSAL

In addition to considering the external agents which help the passively dispersing organism, we must also examine the role of chance in initiating and governing dispersal. When we speak of chance in dispersal ecology, we do not mean that it is an entirely haphazard phenomenon which occurs without any predictable reason or against the laws that govern the environmental relationships of organisms. Chance should be especially disregarded in those situations in which conditions and causal relations of dispersal are unknown because the underlying facts and laws have not yet been adequately studied. The chance factor means, rather, that dispersal results from simultaneous effect of *independent* environmental variables. An equal number of potent or influential environmental factors may be independent of one another; and since most environments are complex and contain many such variables, there is a great variation in the outcomes of environmental encounters of the organisms, which can be predicted only on the basis of probability mathematics (see p. 374; also Darlington 1938*b*). Every environmental encounter (for example, an exposure to such a dispersal agent as a storm) results in a new state of the animal-environment complex. The number of possible states is large but finite, ranging from: no change of state, in which the animal weathers the storm unharmed and unaltered; and an extreme change of state, in which the storm carries the animal completely across a barrier. While we do not know and cannot predict in which state a given individual will eventually find itself, a number of storms acting upon a number of individuals will result in a scale of states, and the number of individual animals in each state will be statistically determinable. Our present interest is focused only on the extreme state in which the individual is successfully landed on the other side of the barrier. This attitude, or approach to the subject, was the basis of Simpson's (1952) reasoning—that in barrier crossing there can be neither absolutely insuperable barriers nor unlimited dispersal. In the consideration of any kind of barrier, one could predict that an infinitesimally small percentage of the potential crossers would be successful. If we were to combine this small number with the vast span of geological time, as compared to the lifetime of the organisms involved, every organism might be able to disperse by random aggregations of optimal conditions.

The nature of chance-governed success in dispersal makes it impossible to calculate its probability. The magnitude of the variables involved is in most cases inexactly known, especially since the initial variable would be the number of individuals of a population from which potential crossers would be recruited. The total number of individuals of a population has been estimated only for the most conspicuous and easily counted mammals and for birds of especially rare species. Yet chance-governed barrier crossing is to be expected from any widespread, abundant, and eurytopic species. Knowledge about their population size and turnover is negligible.

The probability of successful crossing and establishment further depends, as already indicated, on the time span during which such trials could be made; the probability that dispersal opportunities will be presented by the coincidence of several environmental factors (e.g., severe storms, favorable currents, or a ferrying animal undertaking a long-distance movement) will depend largely on the length of time-units multiplied by the number of individuals in every actual population exposed to these opportunities. Moreover, successful dispersal across barriers or uninhabited areas requires that the landing be at or near favorable habitats where a potential pioneer colony could be established.

It is not surprising that chance dispersal as a factor in historic zoogeography has had few advocates among workers in this field. Present or past discontinuous distributions can be attributed—in many cases with sufficient documentation—to bisection of formerly continuous ranges by a new barrier, to the extinction of connecting population, to the slowness of spread in the connecting unsettled area, and to the existence of a former connection across the present barrier. This reasoning applies especially to the explanations of intercontinental relations or to those between a continent and a large, separated island, such as Madagascar or New Zealand, the flora or fauna of which are related to those of the neighboring continent. One could almost classify zoogeographers and faunistically interested taxonomists into two camps: the bridge builders and continent joiners on the one hand, and the extinctionists on the other; yet the role of chance dispersal cannot be ignored.

Chance dispersal is more readily admitted by those biologists who study the unbalanced faunas of oceanic islands. "Unbalanced" means here that the proportions of the ecological life forms, or of the various taxonomic groups that make up these faunas, is different from those in comparable mainland areas; this is taken as an indication that some groups could not reach the island in question. It is curious that little study has been undertaken on chance dispersal through land areas. So-called ecological archipelagoes—isolated mountain tops, desert faunations surrounded by forested land or vice versa—could very fruitfully be studied with the methods island faunists use, and the degree of chance dispersal could be determined. We could, indeed, ascribe our reluctance to consider chance dispersal to those habitats partly to the fact that man himself is no longer permanently blocked by any kind of land barrier.

Simpson (1952) gives a statistical method for calculating the probability of chance dispersal for any given event for an assumed population size and for an assumed length of time. His calculation, however, is based on factual knowledge of the geological time available for the event to occur. Besides referring to his original essay, we shall note a few examples of distributions where research or observation clarified many of the basic ecological constellations that point toward such dispersals actually happening in the past or present. The staphylinid beetle *Micralymma marinum* Ström inhabits seaweed in the tidal zones. Its known distribution is amphiatlantic (Figure 4–54), which Lindroth (1931 and later) correlates with the course of the Gulf Stream, which has probably transported this flightless insect (Figure 2–14).

The same author (1957) made an exemplary study of the anthropochore dispersal of insects, especially beetles, across the Atlantic in the westward direction. His study will be discussed at some length because it gives us an insight into the work of the zoogeographer, and it will also aid in understanding factors of dispersal ecology other than the role of

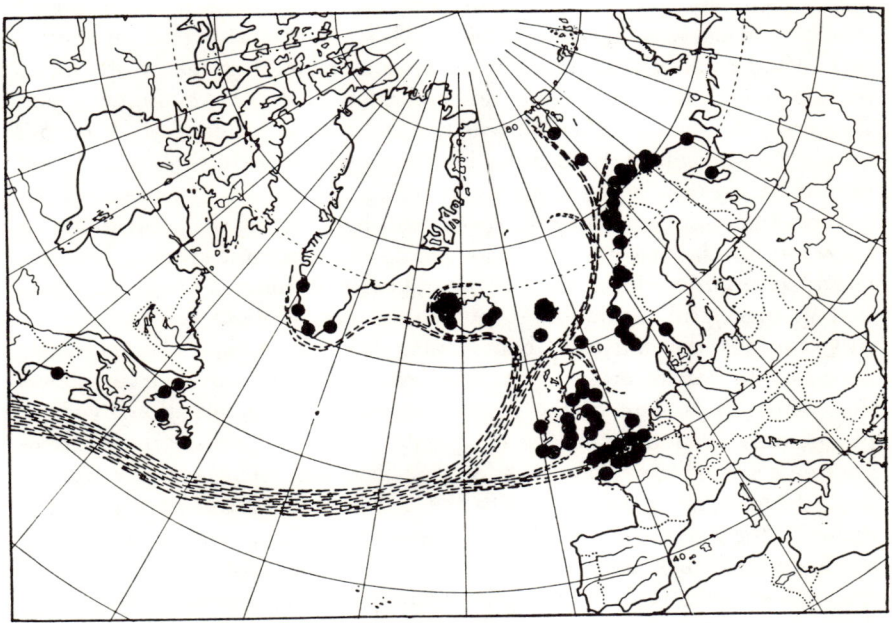

Figure 2-14 Hydrochore dispersal across an ocean.
Distribution of the staphylinid beetle *Micralymma marinum* Ström (*stimpsoni* Lec.) in relation to the Gulf Stream. From Lindroth 1957. (*Courtesy of C. H. Lindroth and Almqvist and Wiksell, Stockholm*)

chance. During a study of the then little-known beetle fauna of Newfoundland, Lindroth suspected that a number of species were European immigrants or adventives on the basis of the criteria discussed earlier in this chapter. Here we have, then, a number of animals—notably 67 species of Coleoptera—separated by a sea barrier from their main distribution area. Why were these and not other European species transported and introduced?

It is well known that fishing vessels sailed in ballast from Europe to the Newfoundland banks during the seventeenth, eighteenth and nineteenth centuries. Lindroth's historical study points out that most ships sailed from southern English ports in the spring. As Newfoundland had scarcely any inhabitants, these ships sailed empty except for the ballast which they discharged at or near ports of arrival. Shipmasters did not like to use sand for ballast, because it dries out and thus loses weight; consequently, they usually took loose, dry soil from near their port of embarkation. To prevent filling up its harbors, Newfoundland forced them to unload ballast on land. Thus, the ballast material was conveyed from land to land, and provided an easy means of ferrying the inhabitants of ground and soil.

Lindroth has shown that 90 percent of the accidentally introduced beetles live on waste-ground habitat. These insects also had an advantage when establishing themselves: areas surrounding the white men's harbors and habitations were not saturated with native animals, and therefore the emigrant could easily establish a beachhead. Lindroth visited the sites of the English harbors from which these ships sailed, and published a survey of the insects now living there. Only 75 percent of the collected beetles came from waste-ground habitats; of the individuals he collected, the largest number belonged to the emigrant species. He thus provided empirical proof for the theoretical reasoning of zoogeographers that in successful introduction, population size plays a role. Later, we will consider the intrinsic, behavioral, and tolerational characteristics necessary to barrier crossing. In these respects, Lindroth's survey shows that while 28 percent of the beetles at places of embarkation lived on plants, only 19 percent of the emigrants belonged to this group. There are two reasonable explanations for this: first, that such beetles were less likely to have been taken on with the soil ballast; and second, that—because of their tendency to move about and climb up and over any obstacle—those which *were* taken aboard were easily able to crawl out of the ballast, only to be blown overboard once they reached the windy deck of the ship. As for selection of the physical habitat, Lindroth found that although 25 percent of the nonemigrant beetle species lived on hygrophilic (wet) grounds, only 6 percent of the emigrants were hygrophilic species. Since only dry ballast was used in the vessels under discussion, it may be assumed that any numbers of a hygrophilic species that managed to get on board would have died either of desiccation or of the salt spray. This is because hygrophilic species usually have water-permeable cuticle and are extremely sensitive to dryness or to sea water.

The food factor has also been investigated. A monophagous species has little chance to find its particular food material during and after barrier crossing—and none of the adventive beetles were monophagous. In passive dispersal, such as ballast transport, great mobility is a handicap rather than an advantage, as we have just noted with respect to planticolous insects; indeed, 32 percent of the beetles that crossed the Atlantic were wingless or at least dimorphic. In this case, the wingless morphs had the better chance to cross, for these account for only 25 percent of the members of the surveyed fauna. Finally, reproductive conditions play an important role. All barrier crossers among the carabid beetles—22 species in all—hibernated as imagoes. The ballast was taken aboard in the early spring, and, upon arrival or even during the journey, these insects were ready to mate and found a colony of immigrants. Surviving individuals of those species which hibernate in larval condition had the disadvantage of not finding mates soon enough after they had completed the voyage and been cast up on the land. There were five parthenogenetic beetle species collected in southern England by Lindroth (see p. 128). All five had made a successful crossing.

Summarizing, then: to be judged especially fit to survive transport in ballast across the Atlantic and to becoming settled in North America, a terrestrial animal, e.g. an insect, should combine the following six properties: it lives on the ground (terricolous), has no pronounced moisture requirements (non-hygrophilous), prefers open ground of a waste-place character . . . is not dependent on special kind of food (polyphagous), is flightless, and has a parthenogenetic reproduction.

(Lindroth 1957, p. 209). The more of these six characteristics which occur in any one beetle species, the greater the probability that it has actually made the crossing. Seventy-three percent of the emigrants and, notably, only 46 percent of the nonemigrants exhibited a combination of three or more of these characteristics. This can be expressed in another way: of a total beetle fauna of 242 species, 129 were qualified crossers, and 38 percent of these actually succeeded in the crossing. Of the 113 less-qualified beetles, only 16 percent crossed the Atlantic. In this particular case, the functionally and ecologically better-qualified species had more than twice as much opportunity to utilize ballast transport than did the nonqualified species. Since Lindroth surveyed the beetles for only a few characteristics (the above-mentioned six plus population size), it is likely that in the selection of the less-qualified crossers and in the lack of success of some of the qualified ones, further intrinsic or extrinsic factors played the major role and narrowed the chances of successful crossing and the number of trials was not enough for eventual success. It must be emphasized that ballast transport was a highly artificial or—more objectively—an anthropogenic situation. Nature seldom provides rafts, storm winds, or other means of dispersal with such regularity and within so short a span of time. However, there can be little doubt that the selection of successful immigrants during geological time is based on similar principles—on time, take-off, ferrying, and establishment ecology.

TIME AS A FACTOR AFFECTING DISPERSAL ABILITIES OF ANIMALS

Earlier, we stressed the role of time needed for dispersing animals to chance upon a habitable area, and suggested that time, together with distance, constitutes a kind of barrier. We also considered time affecting the durability of barriers.

Before we discuss the intrinsic abilities of animals to disperse, we should stress that those attributes of animals which facilitate or inhibit their dispersal are also subject to dynamic changes.

The length of time that a species faces a barrier affects even the animal population itself. With the passage of time, genetic and adaptational changes may occur which make marginal populations better fitted for barrier crossing and hence improve the likelihood of dispersal across the barrier. This idea has been stressed by Reinig (1938), who theorized on the population-genetical consequences of marginal pioneering where the advancement is step by step and where chance-directed reshuffling and depauperization of the genetic material might occur in each generation. This "drift," however, affects the population nucleus in its success of colonization (*cf.* also Ehrlich 1958) rather than in its success in crossing the barrier, and will be treated in Chapter 3.

INTRINSIC ABILITIES OF DISPERSAL

The dichotomous treatment of dispersal factors separating extrinsic influences from the animal's own specific attributes has been suggested by several biologists, but it has been made axiomatic by Ekman (1922) and his Scandinavian followers. Many more aspects of the biology of animals have been brought into the focus of the environmental biologist's attention since that time, and we will endeavor to examine most of them —bearing in mind that every categorization is artificial and arbitrary.

These factors are usually parts of causal reaction chains that may or may not result in dispersal, and it is often difficult to determine which link of the chain was decisive in a given situation.

For convenience, we may group these factors as structural, functional (physiological or behavioral), population-genetic, and population-dynamic aspects. Each of these characteristics is marked by evolutionary adaptability and by individual, ontogenetic, and traditionally acquired variations.

Structural Attributes

Of major importance is the structure of the locomotor organs, of organs which further passive dispersal, and of those organs which aid the withstanding of environmental vicissitudes en route. Active locomotion is a basic characteristic of the animal world. Even sessile marine invertebrates are dispersed by active, swimming or floating larvae or eggs. The most inactive land animals are generally the soil inhabitants; if we consider that these may reproduce anywhere within their action radius (p. 75), *active* spread is possible only by slow penetration. Many flightless, creeping ground animals are in the same predicament. We learn from Oughton's (1948) study of land snakes in Ontario, Canada, that these traveled an estimated 900 miles (about 1,450 km) in 250 centuries —the time which elapsed between the present and the melting of the last (Wisconsin) ice sheet that once covered the whole province. The average speed was roughly 60 m per year. In England, a capture-release-recapture study of marked snails *Cepaea memoralis* (L.) showed an average annual displacement of only about 3 m per year (Goodhart 1962). If these snails actually spread by active locomotion at this rate, they could advance only 1 km in about 330 years; to extrapolate still further, they would be able to spread from Spain to Denmark in approximately half a million years.*

Though the evolution of locomotor organs is chiefly influenced by selection which achieves feeding or escape adaptations, there are some well-documented cases in which they became modified for furthering or hindering dispersal or where a morphic balance has been achieved especially in the structural adaptations of *wings*. Wigglesworth and Johnson (1963) and Johnson (1966) present a very convincing argument for the

---

*This kind of reasoning is perhaps misleading, because it is based upon the *average* snail's pace; the pioneers will be found among the fastest-crawling individuals, which would accomplish the journey in less than 100,000 years. Besides, as we observed on page 47, birds were definitely implicated as carriers, thus resulting in a far more rapid rate of transportation. There may also be other, occasional carriers for the snails. Therefore Oughton's speculation (900 miles in 2,500 years) may be closer to the truth than mine (900 miles in 500,000 years).

hypothesis that insect flight (and the accompanying structures) might have originated as a dispersal mechanism; in many insects, the sole or primary purpose of flight is dispersal from the birthplace or toward the place of reproduction. Johnson discusses "adaptive dispersal," in which even weak fliers participate or which may take place before the full flying power of the individual develops. In these cases, flight consists mainly of sustenance in the air; but the take-off and termination of flight are actively governed, and wings are absolutely necessary in these phases of aerial dispersal.

There has been much speculation, beginning with Darwin or earlier, about the direct adaptive use of secondary winglessness in *preventing* dispersal—e.g., on remote-island (birds, insects) or mountain-peak habitats (insects). The obvious explanation—the direct advantage of flightlessness in preventing the blowing away of the individual from the exposed habitats—has been challenged by Darlington (1943). He thinks that this is too simple an explanation and that wings may be lost, when flight has lost its usefulness, without the loss becoming harmful. One family of birds—the rails [Rallidae]—has often been used as an example. These ground birds live in good cover, and use flight principally as an escape mechanism; nevertheless, they migrate and spread on the wing. Island genera of rails tend to become flightless. Individuals that fly too readily and too often will be more exposed to the heavy winds that prevail in such places and will easily be drifted away, leaving only the structurally —as well as behaviorally—more sedentary genotype to multiply on the island. Flying winged individuals will also tend to migrate *if* they come from migratory ancestors—but rails of tropical continents are generally nonmigratory. Migration will expose their population to the same hazard that it does the population of migratory quail *Coturnix coturnix* L. [Phasianidae, Aves], introduced into the United States by game managers; these quail disappeared in the fall and never returned to the place of introduction. Insofar as the colonization of an island is considered with respect to maintenance of the island population, the faculty of flight can be deemed harmful to the degree that it might cause the removal of individuals of otherwise desirable genetic constitution. An additional selective factor is the lack of ground predators (mammals) on oceanic islands. The famous flightless rail of New Zealand *Notornis* is not subject to small-island selection, but enjoys, together with other flightless members of the New Zealand avifauna, a total freedom from mammalian predation; thus, no predator has acted as a selective agent of its flying ability. Reduction of an organ that has lost its adaptive significance may become an asset, depending on the energy or space budget of the organism so affected. The space can be utilized, and the material placed elsewhere (McCoy, *in litt.* 1967). Finally, for additional evidence we refer to the composite plants of the Pacific islands. Carlquist (1966a) shows that these plants lose their dispersability—the accessory structures of their seeds which facilitate wind dispersal—while they evolve to become sedentary island endemics.

Flightlessness in insects is often a polymorphic characteristic. *Polymorphism* is the phenomenon of discontinuous variation in a general population. Often the variation is multimodal, as it is in social insects (Kennedy 1961), or at least bimodal, as it is in the wing variation of many insects where winged (macropterous) and wingless forms or those with nonfunctional wings (brachypterous) appear in the same population. *Sequential polymorphism,* in which the different morphs represent development stages and each individual goes through these, is likewise a related, adaptive feature.

Lees (1961) studied the sequential polymorphism of the vetch aphid *Megoura viciae* Buckton, and found that the parthenogenetic summer populations are either winged or wingless. Such aphids kept under crowded conditions produced about 42 percent *winged* females, while those kept alone produced only wingless progeny. We can thus infer that producing a large number of flightless but parthenogenetically reproducing offspring is adaptive when a certain female *alone* occupies its habitat; the opposite circumstances would warrant the production of winged emigrants. Tischler (1963) found a similar phenomenon in an annual sequence of populations of the bug *Ischnodemus sabuleti* Fall [Lygaeidae, Hem.]. Macropterous forms appear every other year when the population is at peak densities. The macropterous females show an additional, functional adaptation. Their eggs mature later than those of the apterous females, and they are able to fly and disperse in a pregnant condition.

A taxonomic, ecological, and genetic study by Lindroth (1946, 1948, 1949) proves by experimental, circumstantial evidence that where macroptery and brachyptery occurs as polymorphism among insects, particularly among ground beetles [Carabidae], this mechanism greatly enhances dispersal (see Frontispiece). Whatever the original evolutionary impetus to evolving flightlessness, the winged forms have a definite advantage over the flightless ones in pioneering new habitats within or without the distribution area (Table 2). Lindroth noted the phenomenon of wing dimorphism during his collecting of carabid beetles. When he mapped the quantitative relations of the two morphs on the distribution maps of the Scandinavian area, some very revealing patterns emerged (Figures 2–15 and 2–16). Brachyptery is a dominant genetic trait of the beetles. When a population lived and thrived through many genera-

62   The Ecology of Dispersal

tions, totally or almost totally brachypterous populations were found. At the margin of distribution areas, or in regions which were suspect of having been colonized late during the short (between 7,000 and 10,000 years) postglacial period, the population of today is still totally or almost totally macropterous, because these pioneers were homozygous for the recessive allele, and brachyptery had to reappear as a new mutation or perhaps reach the area by cumbersome, passive dispersal. In either case,

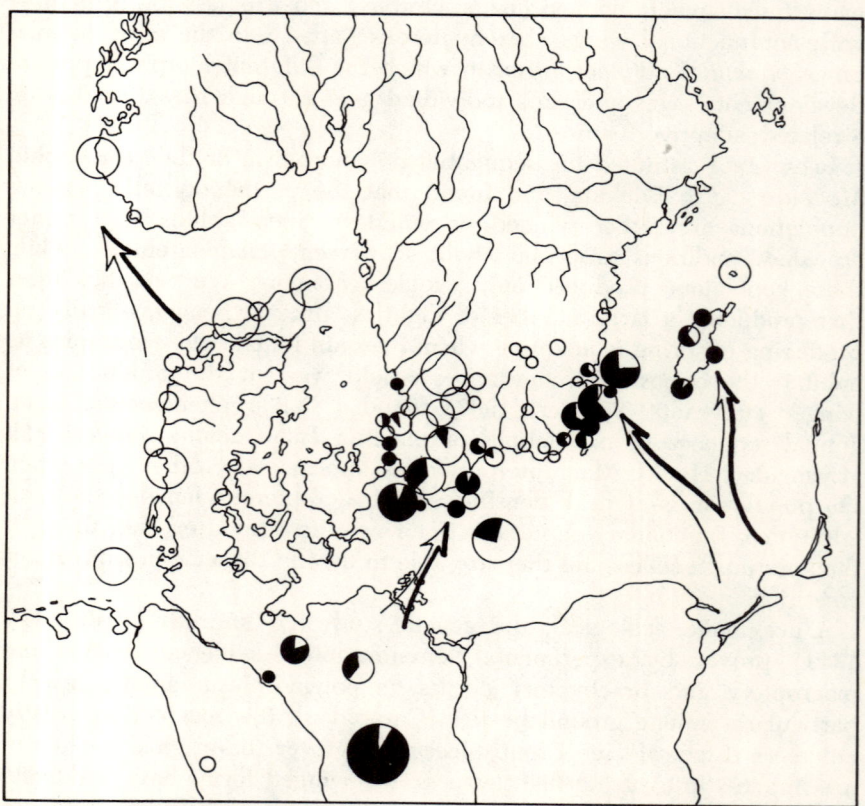

Figure 2-15   Wing dimorphism and distribution.

Distribution of wing forms of the dimorphically winged carabid beetle *Calathus mollis* Marsham in southern Scandinavia. (From Lindroth 1949.) Black circles and sectors indicate brachypterous individuals which cannot fly; open circles and sectors, macropterous individuals some or all of which can fly. The area of the circles is proportional to the number of individuals in each lot examined. The pattern of distribution reflects spread northward and northwestward by flight as indicated by arrows. (*Courtesy of C. H. Lindroth and the Royal Society of Gothenburgh, Sweden*)

Structural Attributes 63

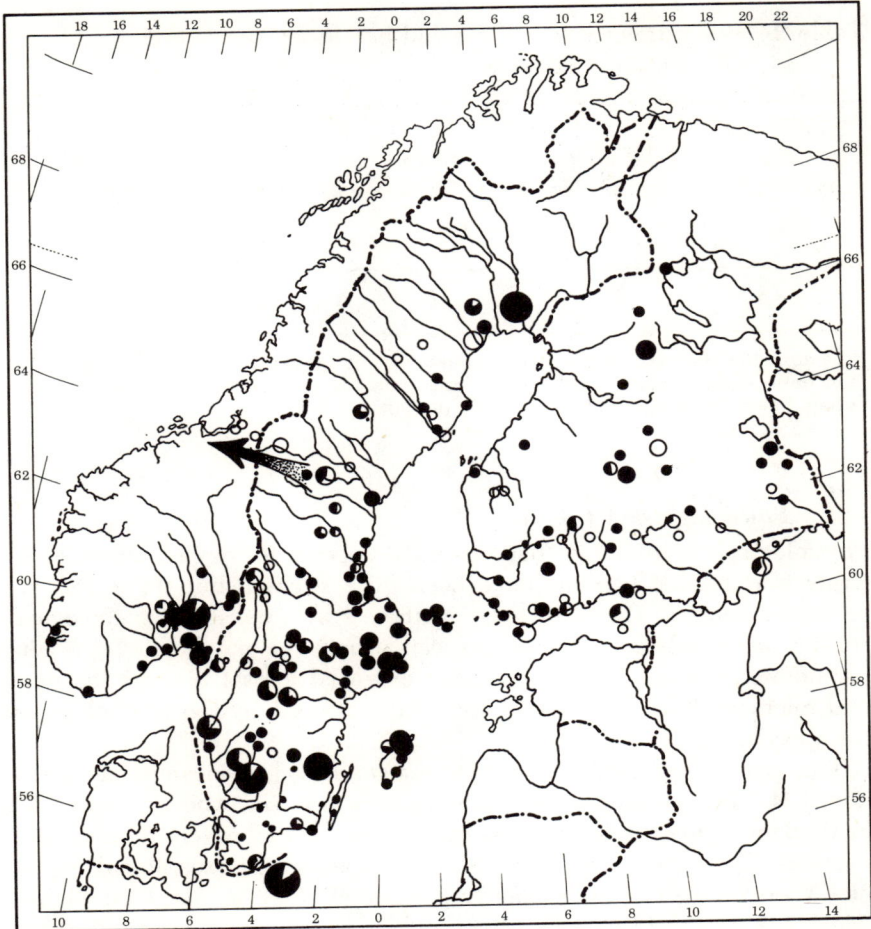

Figure 2-16 Analysis of wing dimorphism in the Fennoscandian distribution area of the carabid beetle *Pterostichus minor* Gyll., with the role of a forested pass emphasized.

Black circles and sectors indicate brachypterous (flightless) individuals; open circles, macropterous (mostly flying) individuals. The size of the circles indicates the size of the sample. Boundaries are biogeographic rather than actual political divisions. The species spread actively through a pass, in the subalpine birch region, to Norway, as indicated by the occurrence there, and through the pass, of only macropterous individuals in the samples. From Lindroth 1949, slightly modified.

the index of flying and nonflying individuals is an indicator of the direction of penetration and age of the immigrant species of the area.

Table 2 Wing dimorphism of carabid beetles on the Azores (From Lindroth, 1960.)

Number of islands occupied	Total number of species	Macropterous	Brachypterous	Species dimorphic[1]
1 − 2	15	5	9	1
3 − 8	17	16	1	0

Note that the majority of the macropterous, flying species are distributed on more than three islands, while the flightless, brachypterous species have a restricted distribution. [1] Both macropterous and brachypterous.

Physiological Attributes

Physiological adaptations that enable a species to spread across relatively inhospitable habitats have been little studied, but there are one or two examples. The knowledge gathered by experimental ecological studies or by tolerance physiology can usually be paralleled by a corresponding environmental factor that arrests dispersal beyond the general tolerance limit. Ultimately, this is determined by the physiology of the species.

The ground lackey *Malacosoma castrensis* L. [Lasiocampidae, Lepidopt.] is a European moth widespread in heaths and wooded areas; but in England, where it is a postglacial colonist, it exclusively inhabits salt marshes. These salt-marsh populations developed an exceptional means of dispersal: their eggs became salt-resistant, they are laid on debris, which the spring tides wash ashore and eventually leave stranded above the tideline. There, if the habitat is suitable, the hatchlings establish themselves (Ford 1964).

Behavioral Attributes

More widely discussed and of greater importance are the behavioral adaptations: the disposition of the animal to disperse. Birds have been especially well studied in this respect. Though basically flight specialists, birds do not by any means exhibit patterns of easy and haphazard dispersal. It was an accepted opinion among the early ornithologists that birds possess wings to bring them back to previous domains from which they have been accidentally displaced rather than to aid them in large-scale pioneering flights. An examination of the distribution of birds, even within a single continent, reveals the validity of this viewpoint, and strongly points toward the correlation of structural abilities and the behavioral utilization of these structures. Thus birds, while good flyers, have, as a rule, strong *philopatry* (the older German term *Ortstreue* is often used in English) i.e., the tendency or instinct consistently to return to their birthplaces or previous breeding grounds even after prolonged and regular juvenile wanderings or annual migratory flights. Thus, in the year-to-year biology of birds, long-range dispersal is the exception—almost in the realm of chance happenings.

On the other hand, environmental changes which invite range extensions are more quickly followed by birds than by any other terrestrial vertebrate, and here the advantage of flight structures, of behavior, and of the extended action radius (p. 75) is an important factor. Philopatry varies within the class of birds from species to species (Figure 2–17). It varies within a species, geographically as well as within one breeding population. It is likely that it has a genetically fixed background. Von Haartman (1949, 1960) evaluated the philopatry of a population of pied flycatchers *Ficedula hypoleuca* (Pallas) [Muscicapidae, Aves] in a long-term study in southern Finland. He found that while juvenile birds generally tended to choose a new domicile for their first breeding season, there were marked differences among the adult females. Many turned out to be of the nomadizing type, choosing a new locale each spring;

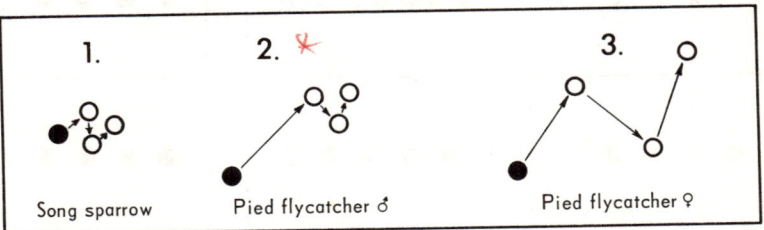

Figure 2-17 Three types of philopatry in passerine birds.
Dot = birthplace, Ring = breeding place
Type 1. Highly philopatric species—for example the song sparrow, *Melospiza melodia* (Wilson) [Fringillidae]. The first breeding place is in the immediate vicinity of the birthplace, as are subsequent breeding places (Geburtsortstreue). *Type 2.* The first breeding place is far from the birthplace, but the bird is philopatric during its later years (Nistortstreue)—for example, the male pied flycatcher *Ficedula hypoleuca* (Pallas) [Muscicapidae]. *Type 3.* "Nomadizing" species. The first breeding place is far from the birthplace, and subsequent breeding places are far from one another—for example, the female pied flycatcher, or the protonothary warbler *Protonotaria citrea* (Boddaert) [Parulidae]. After von Haartman 1949.

66 *The Ecology of Dispersal*

others were more philopatric, and remained faithful to their first choice of domicile year after year. Swedish birds behaved similarly, but studies in Germany, Holland, and Great Britain (cf. von Haartman 1960), have shown a generally high degree of philopatry in this species. The postulated historic explanation of these findings (Figure 2–18) emphasizes the significance of philopatry in range shifting and dispersal: (1) during the last glaciation, the flycatcher was confined to refugia in southern Europe. (2) In postglacial times, the favorable climatic belt gradually expanded northward. Nomadizing individuals—with low philopatry—found available habitat in, and extended the distribution area to, Central Europe. As habitat conditions grew more stabilized, the more philopatric individuals enjoyed an advantage in returning to their familiar, choice localities at the end of winter, and nomadizing thus declined as result of natural selection. (3) Most recently, Scandinavia became available, shedding its ice cover and becoming wooded, and was soon colonized by nomadizing Central European individuals. However, the species is still

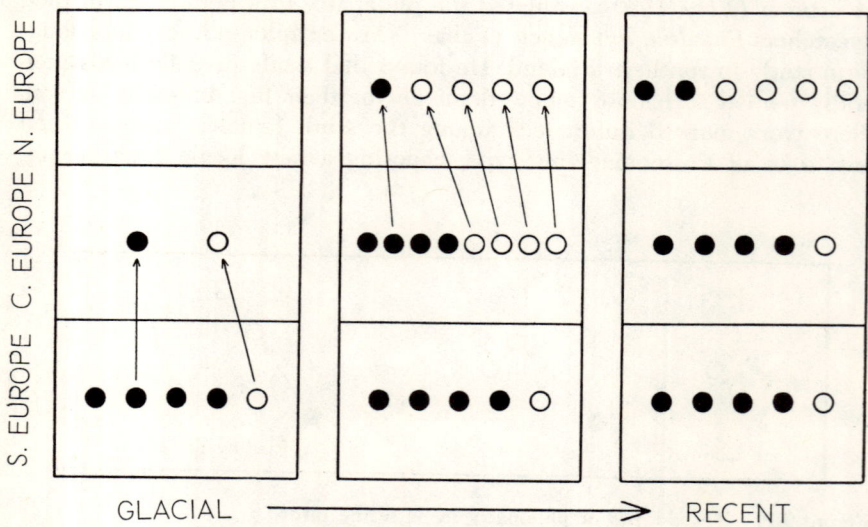

Figure 2-18 Diagram of the hypothetical origin of nonphilopatric bird populations in Northern Europe.

Dots represent philopatric, circles, nonphilopatric, nomadizing individuals. As the species expanded northward following the retreat of the Würm ice sheet (columns left to right, in time sequence) the nonphilopatric, nomadizing individuals had the greatest opportunity to pioneer the north. The longer an area is occupied, the more philopatric individuals accumulate on it, as in Central Europe (middle right) where they predominate. From von Haartman 1949, slightly modified. (*Courtesy L. von Haartman*)

relatively new there, and the habitat is scarcely better than marginal. Nomadizing, therefore, is advantageous, and is still present in a high degree.

Recent bird research also gives an insight into the behavioral mechanism involved in philopatry. German flycatcher fledglings, according to Löhrl's (1959) experiments, learn the environmental coordinates of the place where they spend their first fledgling weeks before the fall migration, and "home" there in the spring. The diagram map, with von Haartman's hypothesis, in Figure 2–18 is to be compared with Figure 2–15, which shows a structural polymorphism projected geographically. Figure 3–19 illustrates a functional-physiological characteristic. Philopatry, wing dimorphism, geographic parthenogenesis, and other examples give strong indication that *genetically determined polymorphic and functional mechanisms enable the species to disperse and to spread when and wherever it is advantageous for the maintenance of its population.*

Philopatry of comparable magnitude is exhibited also by mammals, but is less studied because of the nocturnal habits of most species; their success in return and in dispersal movements is diminished because of the more complicated ground environment, which birds can avoid by flying. Statistical studies of various rodent species (Dice and Howard 1951, French *et al.* 1968, and others) indicate that some individuals, as with birds, possess an innate trait of dispersing from the birth-place while most others remain. Studies of homing and philopatry have affirmed these characteristics in reptiles (including even sea turtles) and amphibians (Figure 2–19). Philopatry in fish is well known, though the main emphasis in fish studies—as also in ornithology—is on the homing aspect of the mechanisms involved. The salmon, to mention the most popular example among the fish (Hasler 1966), faithfully returns to the place where it was hatched.

Further down the evolutionary scale of land animals, where the behavioral aspects of adaptations are more rigid and stereotyped, fewer and fewer examples of philopatry are to be found. Such animals are increasingly bound by the physical and structural characteristics of the environment rather than by their own behavioral versatility. However, it must be borne in mind that studies of adaptive behavioral mechanisms are products of the present decade, and that only a start has been made. Mechanisms which are strongly reminiscent of the polymorphic behavioral mechanism resulting in philopatry or vagility can be discerned from the studies of Verhoeff (1938) in diplopods, Bates (1954) in mosquitoes, de Lattin (1960), and Ehrlich (1961) in butterflies. An interesting difference in activity—a behavioral characteristic presumably based on internal physiological attributes—is displayed by the tent caterpillars,

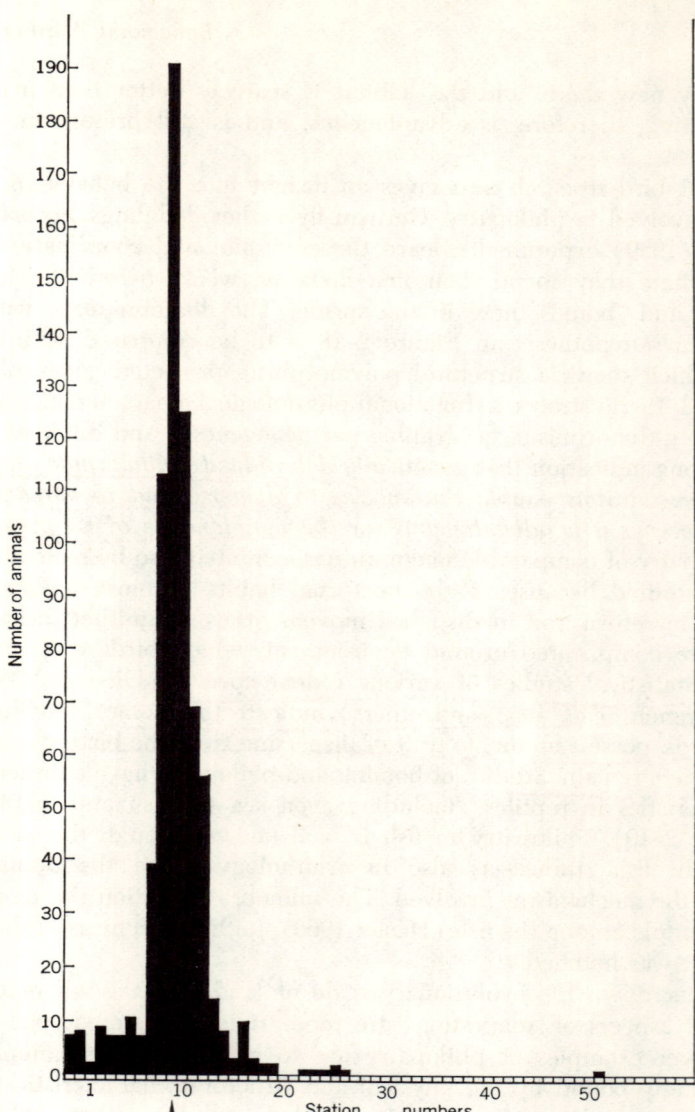

Figure 2-19 Philopatry.
The northern Californian newt *Taricha rivularis* (Twitty) [Salamandridae, Amphibia] migrates every year from the forested hillsides to stream pools several kilometers away, in order to mate in the water. A stream has been marked off into stations of about 46 m for a length of 2,653 m. From 1955 through 1960, all males (almost 2,000) were marked and released in station 9. The figure shows the location of 712 recaptures during the breeding season of 1964. The majority of captures were made at or near station 9. From *Of Scientists and Salamanders*, by Victor Chandler Twitty. W. H. Freeman and Company. Copyright © 1966.

of which the European form *Malacosoma neustria* (L.) [Lasiocampidae, Lepidopt.] and the western North American species *M. pluviale* [Dyar] have been investigated (Laux and Franz 1962, Wellington 1964). Among descendants of the same female there are agile individuals which are able to perform directed movements in response to environmental stimuli and which fly well as adults and undertake pioneering flights. Other individuals show different degree of sluggishness, and may in extreme cases fail to survive because of an inability to reach the food source from their tents. This characteristic seems to be an inheritable morphism, though its genetic background has not yet been fully studied.

Wellington found that the ratio of these morphs within a population varies with the fluctuations of the local climate as well as with the habitat features of the range. In forested areas, many of the pioneering actively flying individuals are stopped by tall trees and reproduce locally, leaving a larger percentage of vigorous, active descendants. They also pioneer the more open habitats where the climatic extremes are more prevalent. However, in favorable circumstances they are able to take advantage of the temporarily improved habitat, and establish their populations there. During deterioration of the climate, many of the pioneering populations also perish, but some survive and their descendants help recolonize the region. Thereby, this species is also presumable able to expand its total distributional area immediately following climatic or habitat improvements.

Wolda (1963) found in the much-studied polymorphic land snail *Cepaea nemoralis* (L.) that different morphs display a quantitative difference in their readiness to tackle the water barrier and in their success in crossing. "Emigration can be a selective factor as unbanded snails tended to leave the population more readily than did banded ones" (p. 411). This may be contrasted with our earlier discussion of flightlessness in birds.

An interesting, many-sided phenotypic condition of certain butterflies is discussed by Kaisila (1962) in his monograph on the immigration and expansion of Finnish Lepidoptera. When caterpillars of *Pyrrhia umbra* Hfn. [Noctuidae], for example, live under crowded conditions (as at peak densities or when the food plant is scarce and many females lay eggs on a single plant), the caterpillars assume a dark color, thereby revealing physiological changes caused by overcrowding. Presumably, their behavior is also influenced, for these caterpillars become expanding or wandering butterflies. The whole phenomenon is reminiscent of the sedentary and migratory phases of gregarious locusts.

Sociability is significant among other aspects of innate behavior patterns. Although a single fertilized female should be enough for successful spreading of bisexual species, in many cases the presence of the male or other social companion is necessary for the establishment of a pioneering population nucleus (cf. Chapter 3). The dispersal movement itself may be facilitated if performed by a group rather than by an individual. The positive or negative role of tradition and of versatile learning abilities is also of importance.

Natural History Attributes

All animals possess attributes which affect the whole organism or its reproduction, and thereby the growth of its population, but which are difficult to classify within the structural or functional categories discussed heretofore. These are the adaptive features one summarizes as traits of the *natural history* of a species—the number of eggs it lays, the timing of its life cycle, its feeding habits, habitat selection, and the like. In this instance, several functional and structural features act together but are equally important with respect to dispersal. For example, in comparing species of two phytophagous insect families, Gressitt (1956) concluded that those forms that burrow in dead twigs, logs, and dormant seeds or which mine into leaves show more passive dispersal in the South Sea islands than those which bore into living plants or feed on roots or leaves.

Population Attributes

Because the analysis of the ecological role of population genetic and dynamic factors is in its incipient stage, we can only conjecture upon how these factors influence dispersal. It is perhaps fair to surmise that many sudden changes in distribution areas—accomplished by lengthened dispersal of the individuals or by increased numbers of dispersing individuals—are initiated by variants or new genetic combinations which influence the structural or behavioral components of the dispersal potential. The sudden expansion in the 1930's of the collared dove *Streptopelia decaocto* Friv. [Columbidae, Aves] (see Figure 4–40) beyond a long-established stationary boundary is attributed, for want of a better explanation, to the sudden appearance of a nomadizing, expansive genotype (Mayr 1963). Because the success of pioneering usually depends on the numbers of individuals involved, population pressure and reproductive rate are important factors contributing to dispersal phenomena. The kind of population control which is established is also important. If a species exhibits emigration as a control measure in density regulation, pioneers will be available; if cannibalism is employed, there will be no accumulation of pioneers. The role of population density in dispersal and pioneering is twofold. It has been long suspected by students of range extensions due to climatic causes that where climate improves in marginal areas successful reproduction creates a population surplus which disperses, and that the individuals which pioneer beyond the limit of the range or on marginal areas achieve the range extention (Kalela 1944). In addition, the numerical relations of a large population favor the probability of chance dispersal as well. A third possible factor is that surplus individuals of a dense population are often obliged to settle on inferior, secondary habitats, which often border on the distribution area. The offspring of these individuals, already selectively bred by the harshness of their environment, might provide better pioneers into marginal areas. The study of these different ways by which overpopulation or crowding aids dispersal has hardly begun, and is well worth pursuing. Several biological mechanisms have already been explored. In locusts, crowding induces a phenotypical change in the developing larva which affects its morphology, physiology, and behavior (Uvarov 1928). The distinct "migratory" type, or "phase," (Key 1950) is gregarious, and undertakes long pioneering flights. Winged generations of many aphids arise as a result of crowding. Mites of the genus *Metatetranychus* [Tetranychidae, Acari] are induced by food shortages at high population densities to descend from trees on long threads that break off and from which the mites are carried away by the wind (Klomp 1964). In the highly philopatric population of song sparrows *Melospiza melodia* (Wilson) [Fringillidae, Aves], those individuals which were unable to secure a territory during years of population surplus emigrated and tried to breed in neighboring, inferior habitats (Tompa 1964).

Habitat improvement and population increase provide one kind of impetus for dispersal, and habitat deterioration or too sparse a population may induce another. Fire, flood, drought, cultivational change, and mass invasion by interacting biota cause drastic habitat alterations and thereby force many species to seek new domains (Figure 2–25).

Interplay of the Intrinsic Attributes

Many pioneering or spreading movements originate in some behavior pattern of the species which ordinarily serves as a spacing, mating, or feeding mechanism but which accidentally carries individuals beyond the limits of their range. Several zoologists have attempted to differentiate between the two phenomena: *dispersion within the range already occupied*, and *spreading beyond* its limits. Kalela (1940) considers ecological and regional distribution (ökologische und regionale Verbreitung) as two phases of the same process. Howard (1960) calls the first "ecological dispersal" and the second "innate dispersal." Andrewartha and Birch (1954) have likewise shown, in a convincing way, that near the limits of a range, especially when these are climatically determined, density fluctuations are correlated with fluctuations of the area limits, i.e., distribution equals abundance; but such an oversimplified statement does not aid our understanding of the phenomenon. We should rather interpret this axiom as meaning that in favorable periods the species becomes abundant, even at the terminal stations of its occurrence, and short-range pioneers then expand the breeding area. In adverse periods, the species is exterminated on the poor, marginal habitats which form the outskirts of its range, and thus the range shrinks; with prolonged periods of abundance, the range again expands (Figure 4–24). Thus, pioneering depends proximately upon abundance and ultimately upon improved conditions that affect the prime habitat as well as the marginal ones at or beyond the former area limits. It must be remembered that in these population phenomena there is also a genetic aspect (Baker and Stebbins 1964), which is still little understood. Pioneering individuals may increase in numbers in an increased terminal population; on the other hand, crowding may force normally philopatric individuals to leave because of the scarcity of unoccupied habitat, increased stress, or increased intraspecific aggression.

While Howard discusses exclusively the behavioral, innate aspect of the two kinds of dispersals, the foregoing treatment is based on the principle that structural, functional, or life history phenomena may be employed, either equally or simultaneously, to achieve the same evolutionary adaptations. Table 3 summarizes the positive and negative aspects of these attributes in maintaining adequate dispersion within the range as well as expansion when conditions permit it. Several of our examples lead us to suppose that polymorphic characters which simultaneously enable dispersion and expansion occur in populations. Such a dual functional role in a single biological system is not unique. We may, for example, consider the dual functional role of the integumentary system: preserving the internal homeostasis and simultaneously protecting the animal from harmful impacts that come from the environment. It remains to be proved whether the intrinsic dispersal capacity of animals is a by-product of the need to maintain adequate populations on the already occupied range, whether it developed for the purpose of utilizing new areas by expansion, or both. We favor the last alternative in most cases. A nonexpansive species may improve its habitat occupancy by specialization,* but will succumb when conditions change—and conditions do ultimately change. If evolutionists repeat the axiom that specialization is a blind alley which leads to eventual extinction, zoogeographers must

---

*Such as sexual dimorphism in size of the body or size of the feeding organs.

Table 3 Role of polymorphic dispersional systems in maintaining and expanding the distribution area

System \ Function \ Mode of adaptive role	1 Sedentary individuals	2 Mobile individuals	3 Nomadizing individuals (vagrants, wanderers, "accidentals," etc.)
	Offspring settles locally	Dispersion of offspring within the distribution area	Dispersal beyond limits of distribution area
Spatial	Right habitat at birth place readily available Stable habitat utilized constantly	Temporarily or newly available habitat away from birthplace located and utilized Varying or fluctuating habitat conditions within range easily exploited	Newly available habitat outside present distribution limits located by pioneers Distribution range shifts easily following shifts in habitat conditions
Qualitative	Locally adapted gene combination preserved and selected for	Mix the genotype Spreading locally arising mutants Maintain genetic homogeneity of species population	Pioneering qualities—e.g., nomadizing, accumulate in marginal, vulnerable areas and in bridgeheads of expansion
Quantitative	Maintain density level of local population Replace losses Increase population size when habitat, i.e., carrying capacity, improves	Relieve crowdedness of parental population Equalize population density by filling in elsewhere	Provide founders and nucleus populations on newly colonized areas Supply the replacement on marginal, temporarily inferior areas

Table 4 Some polymorphic dispersional systems

Examples discussed in chapters 2 and 3	Nature of morph or phase	Characteristic of morph maintaining local population	Characteristic of morph promoting expansion
Aphids[1] Carabid beetles, ants[2], termites[2]	Structural	Winglessness (aptery) Brachyptery (flightlessness)	Wingedness (alateness) Macroptery (fully winged, flying individuals)
Vertebrates: fish, newt, reptiles, birds, mammals	Behavioral	Philopatry	Vagility (nomadizing)
Western tent caterpillar	Physiological–behavioral	Sluggishness, slow metabolism	Activeness, fast metabolism
Migratory locusts[3], "expansive" butterflies[3]	Structural, physiological, and behavioral	Solitary phase Normally colored phase	Gregarious phase Dark gregarious caterpillars becoming expansive imagoes
Crustaceans: Notostraca[1], Daphnia[1], Aphids[1], Bugs[1]	Cytogenetic	Parthenogenetic (asexual) reproduction	Sexual reproduction
bagworm moth[4], beetles[4], millipedes[4], isopods[4], earthworms[4]	Cytogenetic	Sexual reproduction, diploidy	Parthenogenetic reproduction with polyploidy

NOTE:. [1]Morphs are cyclically alternating generations. [2]Only the sexual forms are winged. [3]Phases are phenotypic. [4]Morphs are sequential generations.

uphold the axiom that a nonexpansive species will succumb much earlier than an expansive one.

Table 4 summarizes some of the polyphasic dispersional systems from the examples given above and from those discussed in the ecology of reproduction of settling animals (Chapter 3). The presence of at least four types of such adaptive mechanisms points to the importance of the spatial, qualitative, and quantitative advantages of population maintenance systems where the adaptive features do not conserve a homeostasis but rather prepare for inevitable changes in time through space. As this phenomenon has not yet been recognized in its totality, and therefore has not been studied systematically, it is likely that future research will contribute many more examples. We may call the phenomenon *dispersional polymorphism.*

MODES OF DISPERSAL

Dispersal can be achieved by single individuals or by pairs crossing barriers if they are biologically capable of founding a new population. In addition to crossing barriers, such *long-distance pioneers* are often able to speed expansion across an expanded, generally suitable habitat (Figure 2–20A, C). The explosive dispersal of the collared dove has already been discussed (p. 70; see also Figure 4–40). During its expansion in Central Europe, it was once found 800 km north of the main front, in southern Finland (Merikallio 1958). The "accidental visitor" of the faunist (see Figure 4–3) represents a prospective long-range pioneer, but its chances as an individual settler are very slim.

The expansion of a population is usually achieved by a number of individuals, or by groups dispersing simultaneously, penetrating a short distance beyond the previous limit of the species. This *slow penetration* is perhaps the most important mode of dispersal (Figure 2–20A, B). Where a species area is not limited by an abrupt change of habitat, the limit fluctuates from year to year and the offspring of the border-inhabiting population steadily probes the neighboring territory by dispersing just beyond the limits.

Since animals do not move about at random, but each remains in a more or less fixed domicile (often called "home range"), one can theoretically consider this domicile as a circle within which the animal lives and carries out its activities. Let us call the radius of this circle the *action radius* of the animal. The animal may reproduce anywhere within its action circle (domicile) where it finds a suitable place for this function. Each of its progeny will have an action circle of roughly the same size. If the reproduction happened in the center of the action circle of the parent, the offspring will occupy the same area. If reproduction happened at the extreme limit of the action circle of the parent, the area occupied by the offspring will be different from that of the parent. The difference will be the length of one action radius (Figure 2–21). The domicile of

76   *The Ecology of Dispersal*

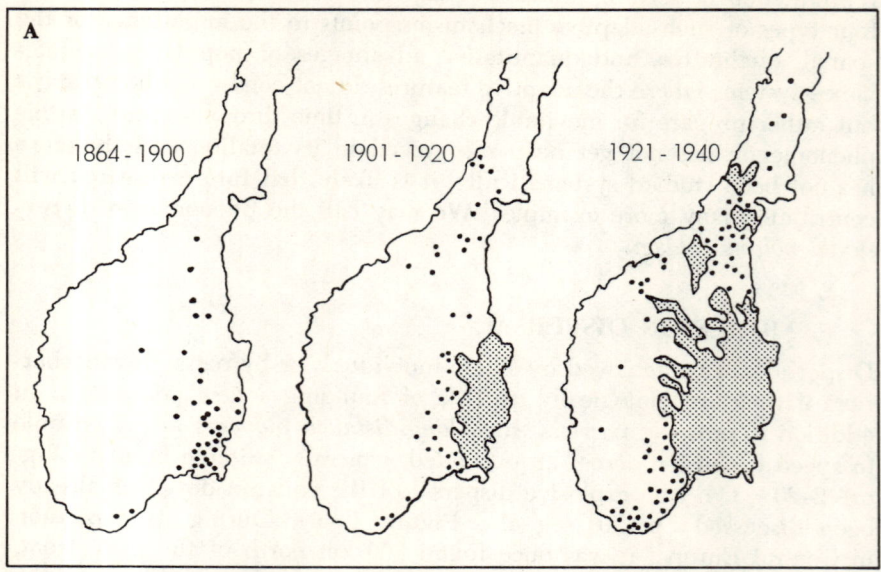

Figure 2-20   Expansion by pioneering.
A. Expansion of the roe deer *Capreolus capreolus* (L.) [Cervidae, Mamm.] in Norway during the periods 1864–1900, 1901–20, and 1921–40. Dots indicate single observations of wandering, pioneering animals; stippling, range of localized, breeding population. After Olstad 1943.
B. Dynamics of the range of roe deer in Sweden. Lines and dates indicate northern limits: 1750–1830, retreat; 1830–1955, expansion. After Curry-Lindahl, 1957.
C. Pioneering of Finland by roe deer 1909–53. Crosshatching indicates range of regular, breeding populations; stippling, area of occasional occurrence. Open circles show observations between 1909 and 1949; dots, from 1950 to 1953. Arrows indicate suggested routes of pioneering; dotted lines with arrows, doubtful suggestions of routes. Note that the animal observed on the central Baltic coast of Finland could have come through the ice from Sweden or around the Bay of Bothnia. After Siivonen 1953.
D. Map of Fennoscandia showing areas covered by maps A, B, and C.

the progeny will be on the average one-half action radius distant from the domicile of the parent individual. If this hypothetical species has no means of pioneering by emigrants, it may expand by the randomness of reproductive locals within the domiciles of its border individuals. The speed of such expansion will be, on the basis of the above argument, one-half action radius distance per generation.

The fringe-toed lizards of the genus *Uma* [Iguanidae, Squamata] live in the Sonora Desert of North America (see Figure 3–10). These animals never leave the shifting sand areas, but are found in the scattered sand

Modes of Dispersal 77

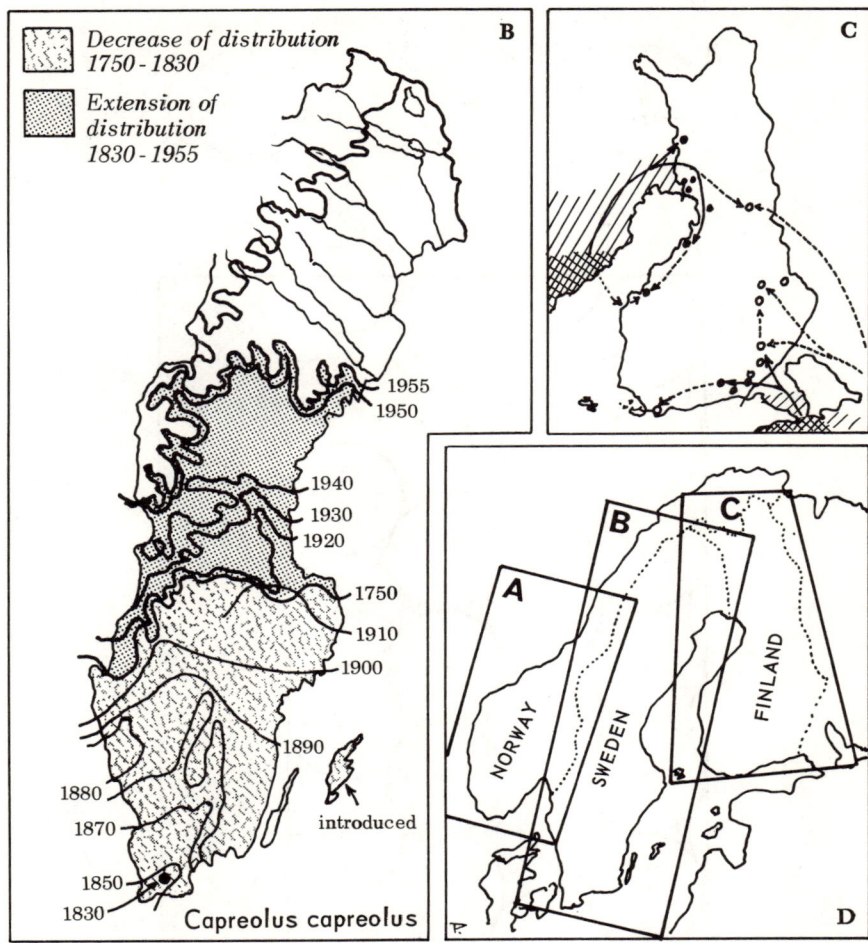

Figure 2-20 Continued.

dunes within a radius of several hundred km, wherever such habitat is available. K. S. Norris (1958) has shown that these sand dunes border ancient river beds which have transported the sand since Pleistocene pluvial times. Periodically, when the river beds are dry, piecemeal shifting of the dunes is resumed. *Uma* evidently traveled by slow penetration, following the slow, secular movement of the sand. The bag-worm moth *Solenobia triquetrella* F. R. [Psychidae, Lepidopt.] is a small, inconspicuous, wormlike European animal which hatches from the middle of May to early June and is ready to hibernate in September. It pupates in the

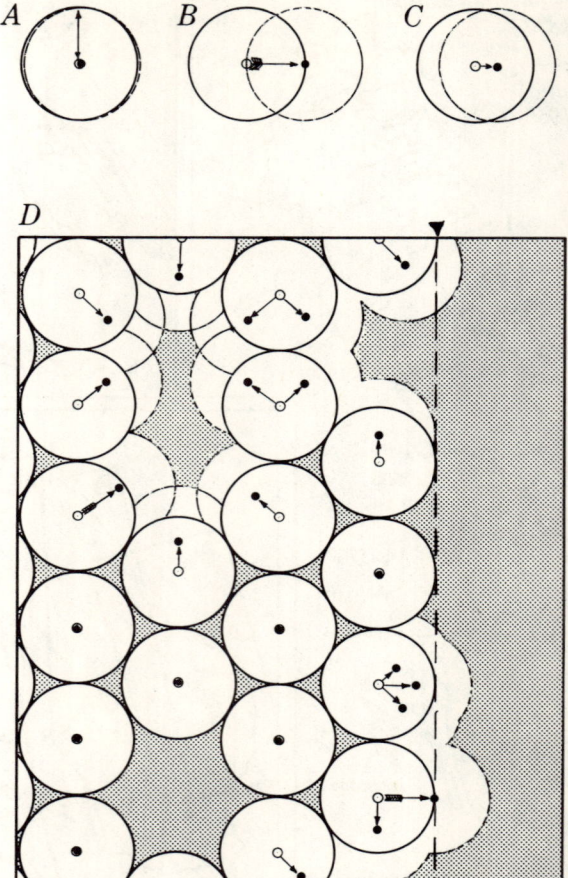

Figure 2-21 Expansion without dispersal mechanism: slow penetration.
A. An individual (o) has a circular home range; it dies after egg laying, and one of its offsprings replaces it. Sold circle denotes the *action circle* of the individual; double-headed arrow, its *action radius*. If the individual reproduces in the center (dot) of its action circle (or domicile, home range, and so on), the action circle of the offspring (dotted circle) will cover and replace that of the parent. This is one extreme of the reproductive locale.
B. This diagram shows the other extreme situation, when reproduction occurred at the periphery of the action circle and the offspring (dot) lives at a distance of one action radius from the parental domicile.
C. Average situation.
D. A diagrammatized population and its possible movements from one generation to another are shown among the same conditions as in A, B and C. (No dispersal mechanism; offspring generation replaces parental generation). Dotted line connect-

Modes of Dispersal 79

spring; when the imagoes emerge, the female appears as a strange, swollen larva—it is flightless, and does not even entirely leave the pupal bag as it lays its egg therein. Dispersal can occur only through the minute

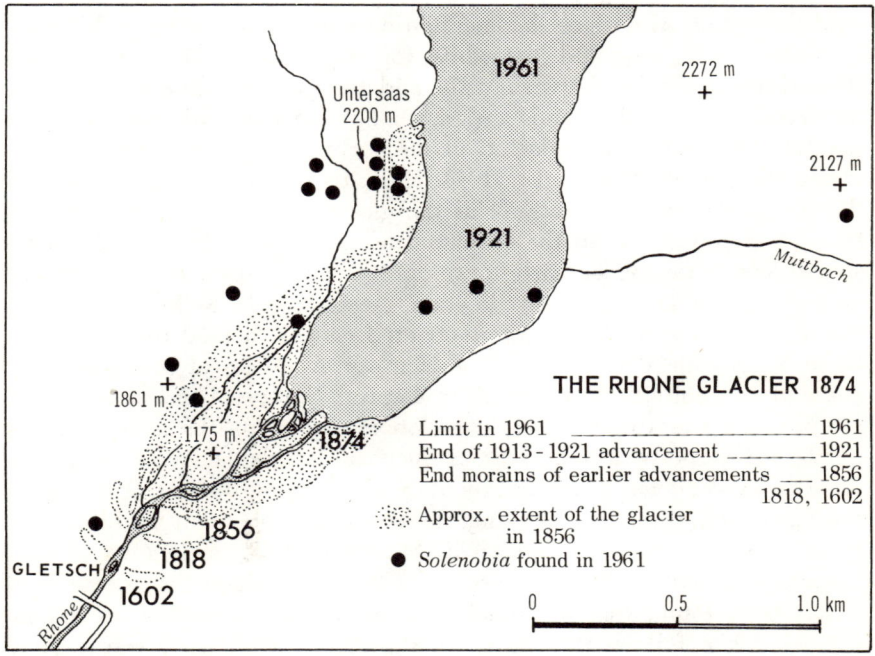

Figure 2-22 Expansion by slow penetration.
Retreat of the Rhone Glacier between 1874 and 1961 and the expansion of the bagworm moth *Solenobia triquetrella* F.R. The gray indicates the glacier as of 1874, after which it retreated as shown by the dates. The hatched area was still ice-covered in 1856. Dots indicate localities where the moth was found in 1961. After Seiler 1961.

---

ing wedges on margin indicates approximate limit of distribution. In a tightly settled area, such as at domicile A, population pressure keeps number and size of home ranges nearly constant. *Upper left*: Two individuals have died without reproduction, and neighboring offspring has filled in vacancies by slow penetration. *Lower left*: One idividual has died without reproduction, but no supplanting has yet occurred. *Upper right*: The situation is identical with that on upper left, but has resulted in an identation of the distribution border, as replacement can only come from one-half of the compass directions (180°) surrounding this place. *Lower right*: If the habitat conditions have improved, random penetration may cause a bulging of the distribution limit. Such an expansion is especially conspicuous where (feathered arrow) reproduction occurred near the periphery (in the manner sketched at B).

crawling movements of the feeding larvae and by the limited movements of the old larvae that are looking for suitable shelter in which to pupate. Yet this animal is common all through Switzerland, which was all but totally glaciated during the Würm (Wisconsin) period. Its microhabitat is not exposed to wind (it lives under stones, in crevices in the ground, or in the bark of trees). Therefore we must assume that, in spite of its extremely restricted locomotion, the moths spread by crawling and covered the whole Alps since deglaciation occurred. Painstaking field study by Seiler (1961) revealed that while, for example, the Rhone glacier retreated about 1.5 km between 1874 and the present (Figure 2–22), S. *triquetrella* followed it and settled, though sparsely, all the available habitat. The moth was collected in the early 1960's on the valley floor, where the glacier still stood in 1921. Here, then, is a reliable example of the speed of slow 'penetration': as the glacier retreated about 800 m in forty-seven years, so did the crawling moth advance at about the same speed. Water rafting or current was not a factor, because the advance was always in an upstream direction. According to Seiler, the erratic course of the rivers in the morain-covered valleys, helped the moth ford these rivers with "dry feet," and the 15 m—or even less—of yearly advance enabled it eventually to cover all of Central Europe. This example well corroborates Lindroth's (1949) calculation concerning the penetration of northern Europe by ground beetles since the last deglaciation. He postulated an average yearly advancement of 42 m during the past 9,500 years, during which these fast-running insects—many of them flightless—expanded into northern Scandinavia. Movements of glaciers have been studied in recent times at almost every locale in which they occur, and it has been found that the whole vegetation and faunation, as a continuous cover, follows the retreat of the glacier with early successional stages and pioneering forms coming first. In other habitats, the slowness of the changes—especially of the wooded vegetation and its inhabitants—obscures the penetration movements; in any case, their study is more recent. However, we can infer that this kind of individual and faunal movement is the most important among the modes of dispersal. We thus agree with Darlington that tropical rain forests of the Southern Hemisphere "evidently move as wholes at glacial rates" (1965, p. 135).

The mechanism of slow penetration is obviously identical with the mechanism of dispersion within the distribution area. At the present time we have no data to compare the effect of slow penetration to that of barrier crossing. Nevertheless, we are convinced that through longer—geological—time periods slow penetration is the more important means of dispersal.

In addition to single pioneers and the broad front advancement, mass exoduses are also known. Among these, *wanderings, migrations,* and *irruptions* may also lead to the establishment of new breeding stations, which can develop into new settlements. All are mass phenomena. Even if the participating individuals move singly and independently of each other, they nevertheless have common releasing and governing factors. "Wandering" usually means that individuals which move out of their regular range do not return. Yet the phenomenon has a certain regularity. Migration recurs annually, with the same individuals participating in the return movement as well. "Irruptions" are irregular wanderings; the species appears in areas in which it usually does not occur and propagate. It may or may not breed there, or it may return to its place of origin, depending on the vicissitudes it encounters. Nevertheless, an irruption may lead to the establishment of population nuclei, as do the other two forms of population movement. We should not place too much emphasis on the semantics of these phenomena, because it appears that they are neither chance-induced nor a hazardous undertaking but, rather, a dif-

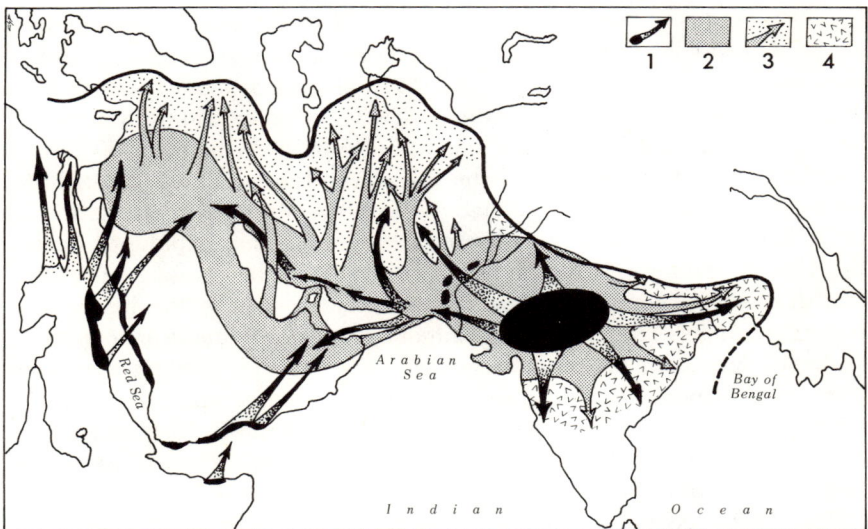

Figure 2-23 Dispersal by irruption.
Changes in the Asiatic distribution of desert locusts in connection with population increase. From Naumov 1955, after Zherbinovski 1952.
1. Zone of constantly inhabited areas which form the centers of periodic mass increase. 2. Zone of temporary increase in peak years. 3. Zone inhabited by wandering swarms temporarily during the summer months of peak years, when the locusts reproduce there. 4. Zone of farthest-advancing swarms, which are unable to reproduce there and die.

ferent form of evolutionary adaptation. By means of these phenomena, the species utilize habitats where living conditions are temporarily favorable but, which with annual or decennial regularity, become adverse (Figure 2–23). Otherwise, they relieve population pressure and help disperse the surplus as an expandable reserve. Although area expansion, as our exemples show, occurs occasionally or periodically through these mass movements, such movements more often provide opportunity for dispersion by passive carriers or actively by defective homing behavior patterns.

Many conspicuous mass phenomena of insect movements (usually called "migration") are really only wanderings, according to our terminology. Species of foreign origin appear in a certain season—usually when imagoes are most common—and this becomes a regular, annual event. Few if any of the individuals return to their place of origin, but if the area where their flight or drift terminates is suitable, they will try to reproduce. There are too many examples to discuss here; excellent summaries are available (Fraenkel 1932, Williams 1957). The spectacular flight of the monarch butterfly *Danaus p. plexippus* (L.) [Danaidae, Lepidopt.] to California, Texas, and Florida from its north-temperate breeding range is a real migration in the sense of the definition given above. Urquhardt (1960) reports recoveries of marked monarch butterflies, one specimen of which covered 2,112 km in not more than 46 days. The British Isles, especially southern England, enjoy almost yearly the influx of Mediterranean visitors among butterflies (Williams 1930). Such wanderings of pioneers stop in the fall at the time of heavy frost; however, on the Åland Islands, climatic improvements resulted in their colonization by the butterfly *Vanessa io* L. [Nymphalidae, Lepidopt.] during the 1920's (Palmén 1949a). It is no coincidence that among butterflies and dragonflies, good flyers and known wanderers have the largest distribution range. The migratory dragonfly *Libellula quadrimaculata* L. [Libellulidae, Odonata] is Holarctic. Another species, *Pantala flavescens* Fab., is pantropic, including the mid-Pacific islands, and has been observed actively flying in calm weather; so it may combine drift with wandering (Fraenkel 1932), as do the beetles and bugs spreading across southern Finland (Palmén l. c.).

Though insects move around in search of food like most other animals, the experience of entomologists shows that long-distance "migrations" usually end with reproduction and serve dispersal. Johnson (1966) gives a clear summary of insect dispersal by flight, and distinguishes three types of "migratory" movements: (1) Short-lived adults of locusts, termites, thrips, aphids, butterflies, and other insects emigrate and oviposit at the end of their journey before they die. This flight is clearly dispersive. (2)

Other short-lived adults (e.g., melolonthid beetles, many Odonata) undertake a flight but later return to, or near, the birthplace to oviposit and then die. This flight may or may not be dispersive—we postulate that it certainly must be. (3) The long-lived migrants fly to their hibernating or estivating locale and then return to lay eggs (this is a type of flight which even a vertebrate biologist calls "migratory.") They exhibit a typical dispersive flight. Homing, dispersion, philopatry, and vagility in insects have not been studied to any great extent. Such studies might find more parallelism with bird migration than simple spatial relations and the timing of the flight.

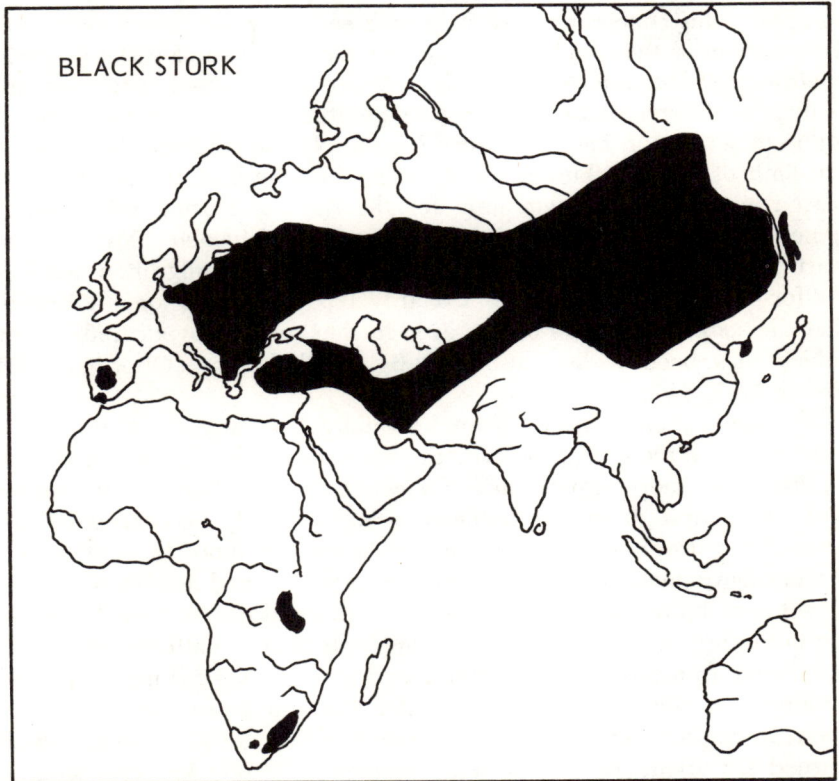

Figure 2-24 The role of migration in dispersal.
The distribution area of the black stork *Ciconia nigra* (L.) [Ciconiidae, Aves]. Colonization in several parts of Africa seems to be very recent—since 1900—by wintering Eurasian birds that remained to breed. After Voous 1960.

For birds, the role of migration in dispersal is threefold. First, migrating individuals and flocks are on the move, exposed to drifting agencies and chance. Already mentioned are the fieldfares which colonized Greenland as a result of migrational drift, and there are other examples as well. Second, some migratory individuals may fail to start a return flight; such lingering adults may be found at known winter grounds of northern birds in the southern Holarctic or the tropics. In a few cases, these individuals breed, like the tundra bird—the white-fronted goose *Anser albifrons* (Scopoli) [Anatidae, Aves]—on the salt marshes of Hungary (Keve and Udvardy 1951); in still fewer cases (Voous 1959), a breeding population becomes established (Figure 2–24). The third role of migration is the most important from the point of view of the zoogeographer: this concerns the settling to breed after the return migration. Termination of the return flight is under the influence of certain definite environmental conditions as well as internal, hormonal processes and philopatry. Correlation with spring temperature seems to be an important guide to settling in marginal areas (in the Northern Hemisphere, northern limit of the distribution area) where the length of the reproductive season approaches the minimum for the species. During cold springs, some migrants stop and breed farther south (Siivonen 1952); during warm, favorable springs, others expand, by prolonged migration, north of the former limit of their range (Udvardy 1956). There is some indication that the same mechanism occurs even farther south (Jovetić 1963). This subject has been widely discussed by ornithologists (Otterlind 1954, Fisher 1955, Swärdson 1957).

The mass phenomena of ground-dwelling land animals lead, for the most part, indirectly to dispersal: the great numbers exposed on strange habitats offer more opportunities for chance spread. Such mass wanderings, often through strange habitats or barriers, may be caused by rare natural catastrophes which one seldom witnesses more than once in a lifetime but which are frequent enough to be counted as a factor in dispersal geography. Almost everyone has read vivid descriptions by traveling naturalists of an earthquake tearing rafts from a battered coast, of a tidal wave transporting houses, cars, and other debris for miles in a few minutes, and the like. Naumov describes and illustrates such a phenomenon. In the dry summer of 1915, a giant forest fire in western Siberia burned for about 50 days over an area of 1,600,000 km^2. An expanse equal in size to Europe was enveloped in smoke, and insolation was decreased to 65 percent of normal, causing the ripening of crops to be delayed from 10 to 15 days. This forest fire caused many mass migrations of mammals; and numbers of squirrels were observed swimming large rivers such as the Jenissei and the Ob (Figure 2–25).

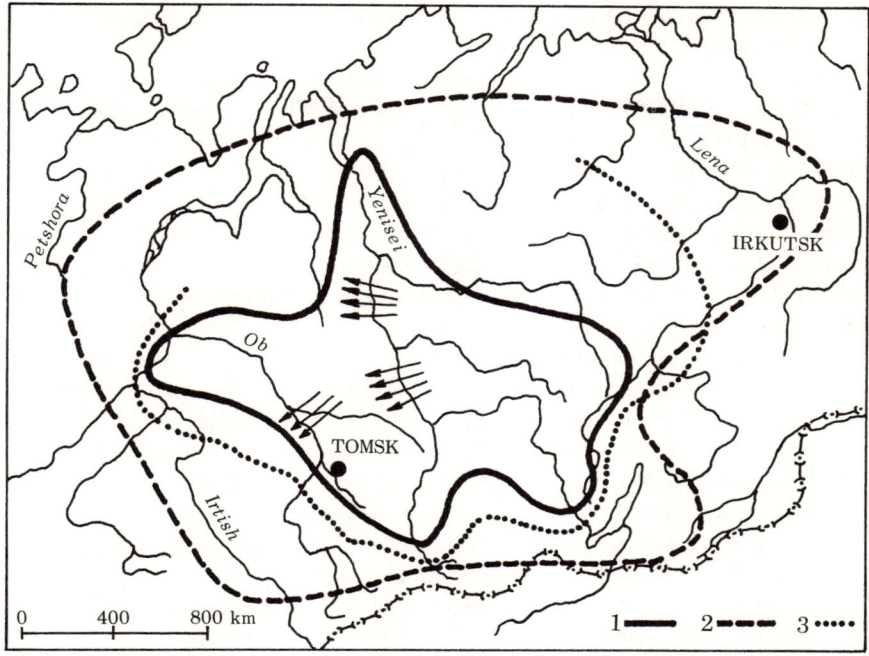

Figure 2-25 Dispersal by wandering.
Wandering squirrels were observed to swim across the large Siberian rivers Yenisei and Ob in 1915 at the points marked with arrows. After Naumov 1955. In this year, a huge forest fire burned over an area of six million km^2 (equal to the area of Europe) in Siberia. Wanderings of squirrels, bears, moose, and many other mammals were observed. 1. Border of the fire area. 2. Border of the danger area. 3. Area enveloped by espcially thick smoke. Swimming across rivers by squirrels is indicated by arrows.

DISPERSAL AND DISPERSAL ROUTES OF FAUNAS

In this chapter we have considered each major factor of dispersal and the ways in which individual animals may cross barriers and disperse. It remains now to consider the total effect of these factors on total faunas, with special regard to the importance of means of dispersal in historic zoogeography. Our starting point will be the present faunistic and faunational situation. We have no reason to suppose that the types of presently active barriers were not there in the past, and that animals had means of spreading other than those now known. The discussion which follows is based upon G. G. Simpson's scheme, as it was first expressed in 1940 and again in 1953 and 1965. Since Simpson, as a paleontologist, worked almost entirely with fossil evidence, resulting in faunal lists at the best, he discusses dispersal of faunas with little emphasis on the ecological relations of the species.

First, there may not be any apparent barriers between two neighboring faunas which are now in direct contact with one another. However, their definition as distinct faunas implies differences in their specific composition. The differences may have historic background—for example, when a former barrier has been removed and the faunas are able to mix. From the point of view of their faunation, they are identical in our example; and any existing difference is caused by the former separation, which facilitated different evolutionary forces and different immigrational processes. Thus, the present discrepancies may be due to lack of time for mixing, to the chance factor, or to the difference in the spreading disposition of certain members of these faunas. Finally, we may surmise that the action of biotic factors may also separate neighboring faunas—e.g., the slow adjustment of competitive situations between users of almost identical niches, or the especially aggressive predators or pathogens to which some prey or host do not fully adapt.

It is of course futile to look for identical faunas on any large geographic area, even though the ecological factors governing the existence of faunations are practically identical. The little we know about the composition of biotic communities tells us that not even the dominant or characteristic members are identical throughout the total area of a community, and that some of these are replaced by vicarious forms or are missing entirely. The composition of a community can be expressed only statistically, while the distribution of any one of its members is governed by the specific activity of a particular assemblance of present and past environmental factors. Thus the species composition of any faunation may also change in space even though the chief life forms remain the same.* One of the best-known examples was studied in the European deciduous forest belt, which has recently been separated by the last Würm glaciation but which now forms—or did at least form before anthropogenic changes—a mainly uniform faunational area. Yet the faunal list differs remarkably (particularly at the specific and subspecific level) in its eastern and western portions, mainly as a result of the past separation, which for many species still exists although the barrier itself is no longer there. Of the many avian examples, some of the most unequivocally vicarious species and subspecies are listed in Table 5.

The most common situation the zoogeographer finds in faunal or ecological analysis results from the interplay of historical background and

---

*The dynamic geography of communities will be discussed in more detail in Chapter 6.

Table 5 Species and subspecies pairs in the European avifauna which have vicarious distribution in related biotic communities. (Data from Stresemann 1919a and b, Salomonsen 1931, and Voous 1960.)

West European Species	East European Species
Picus viridis L.	*Picus canus* L.
Dryobates major (L.)	*Dryobates syriacus* (Hemprich & Ehrenberg)
Luscinia megarhynchos Brehm	*Luscinia luscinia* (L.)
Hyppolais polyglotta (Vieillot)	*Hyppolais icterina* (Vieillot)
Regulus ignicapillus (Temminck)	*Regulus regulus* (L.)
Ficedula hypoleuca (Pallas)	*Ficedula albicollis* (Temminck)
Certhia brachydactyla Brehm	*Certhia familiaris* L.
Emberiza cirlus L.	*Emberiza citrinella* L.
Western subspecies of	**Eastern subspecies of**
Aegithalos caudatus (L.)	*Aegithalos caudatus* (L.)
Sitta europea L.	*Sitta europea* L.
Garrulus glandarius L.	*Garrulus glandarius* L.
Corvus cornix L.	*Corvus cornix* L.
Pyrrhula pyrrhula (L.)	*Pyrrhula pyrrhula* (L.)

contemporary environmental differences. This is the case, for example, within the fairly uniform taiga (northern coniferous forest) belt in North America and northern Eurasia.

Faunal exchange is somewhat impaired when two areas are connected by a narrow *corridor,* even if this is ecologically suitable for both faunas. The narrowness itself acts as an impairing factor because of the limitation on the size of the population from which prospective pioneers would be recruited. Furthermore, the pioneers have great opportunities to stray in the wrong direction, because the advancing front must be narrow. For instance, the land mammal fauna of central France (exclusive of the bats) shows a similarity of only 84 percent on the species level of that of the comparable fauna of the similarly forested western Hungary, roughly on the same latitude but about 1,000 km farther east and interconnected by a forested corridor through Germany, Austria, and Bohemia. Of 44 species in central France, only 37 are common to Hungary, as had been calculated from the distribution data presented by van den Brink (1958). (To minimize ecological differences, the alpine species of

France and the prairie mammals in Hungary have been excluded from consideration here.) The longer the interconnecting area, the more time it takes to mix two faunas. In this instance, historic changes have also to be taken into consideration. The deciduous forest avifauna of Europe (84 species altogether, according to Stegmann 1938) has 25 species (30 percent) in common with the deciduous forest bird fauna of China: Of these, 13 species have continuous distribution throughout the deciduous woodland corridor along the taiga-steppe ecotone (Figure 2-26). However, it is likely that not all of them used the present corridor, as it was not at its present location when these related faunas were established before the Pleistocene glaciations.

The corridor offers conditions that are suitable only for certain elements of the fauna; it acts as a filter, and the sieving effect follows rec-

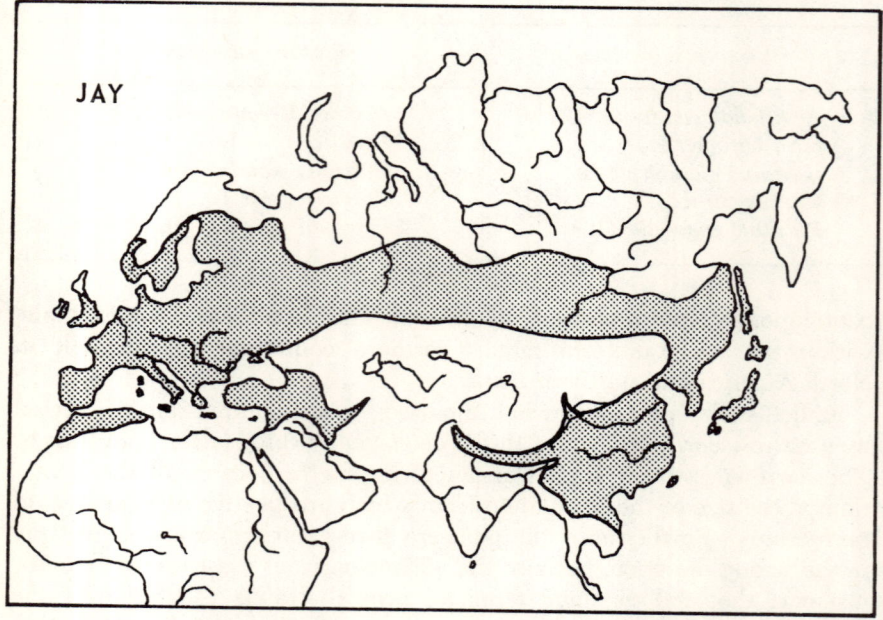

Figure 2-26 Faunal corridor.
The distribution area of the jay *Garrulus glandarius* (L.) [Corvidae, Aves]. After Voous 1960. The western part of the area in Europe and the eastern part in the Far East largely coincides with the area of the deciduous forest biome. The central, Asiatic, connecting area signifies the faunal corridor between the taiga in the north and the open steppe in the south. Deciduous woodland occurs along this zone, and members of the deciduous forest fauna could spread along it during interglacials and in recent time.

ognizable rules. If the two ends of the corridor are ecologically different, the filtering effect and the sum total of faunal migration will often be greater in one direction. Ecological barriers enabling transfiltration, corridors, and land bridges across the mighty sea barrier—all are *filter routes* (Figure 2–27).

A *sweepstakes route* may lead across a strong and lasting barrier; despite the low odds, chance dispersal throughout time allows some animals to cross it. According to Simpson,

The implication of the term is that as in a lottery or sweepstakes the odds against winning are enormous but nevertheless someone does win. . . . Where a corridor or filter exists, most or all of the groups adaptively capable of following the route will do so rather promptly. With a sweepstakes route, probabilities are so much against all groups that crossing may be long delayed. What groups do, in fact, cross, which do so first, and when they do so seem to be largely matters of chance, determined almost at random. (1953a, p. 24.)

We may add that, in addition to the chance factor, the ecological capabilities of each animal group also play a role. Thus, certain sweepstakes crossings nevertheless follow strict ecological rules.

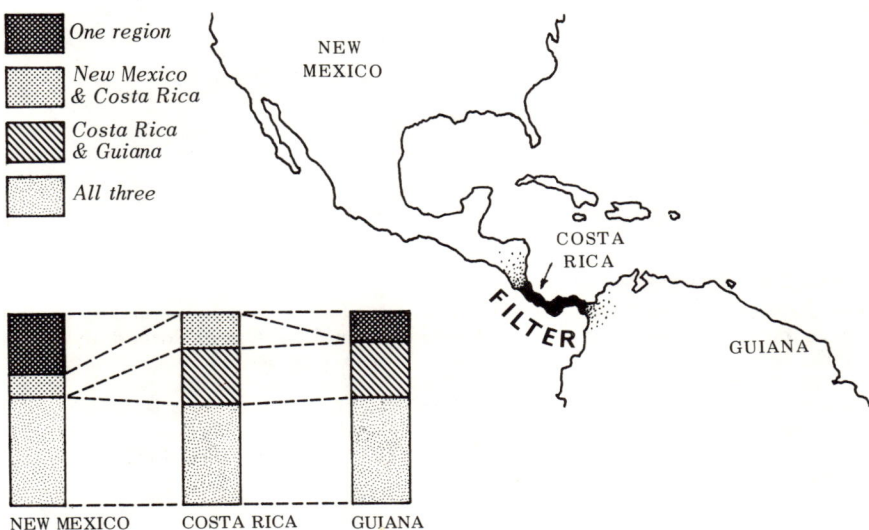

Figure 2-27 Filter route.
A sample area in each of two continents and one on the connecting isthmus are compared to demonstrate the filter effect of the Isthmus of Panama upon the mammalian faunas. Some families (stippling) are entirely blocked from the filter bridge area; others have crossed only part of the way. After Simpson *et al.* 1965.

90 *The Ecology of Dispersal*

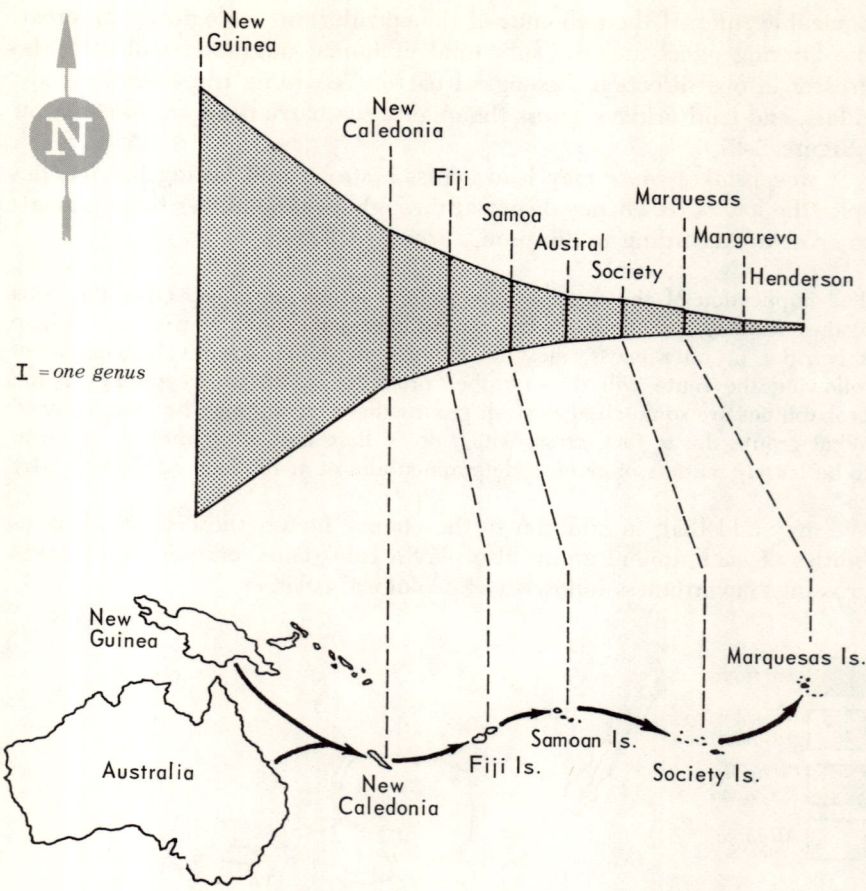

Figure 2-28 Sweepstakes route.
The progressively smaller number of genera (decreasing diversity of fauna) of the weevil family [Cryptorhynchinae, Coleoptera] from New Guinea toward the mid-Pacific islands. The height of the vertical bars in proportional to the number of genera at each place. If one turns the chart on its side so that the point is downward, it will appear as a great funnel fitted with graded filters; this represents the ocean barrier, excluding, by the chance and distance factors, progressively more and more genera from west to east. After Simpson *et al.* 1965, following Zimmerman 1948.

A typical corridor functions at the present time across the Eurasian continent and connects the deciduous forest area of Central Europe with that of the Chinese plains at the southern edge of the taiga belt across Siberia. Another corridor is formed by the Isthmus of Panama between tropical Central America and South America; it replaces a former sweepstakes route which was composed of island stepping-stones during the major part of the Tertiary. A narrow corridor connects the northern and southern Mediterranean shores via the Lebanese and Palestinian coasts, with the desert acting as a substantial barrier immediately inland.

The Sahara and other deserts and high mountain chains such as the Canadian Rockies, the Andes, and the Himalayas allow only sweepstakes dispersal. In island chains, each island acts as a stepping-stone for the expanding faunas but, at the same time, serves also as a convenient check for the zoogeographer; it shows that the filtering effect increases with distance from the source—the number of successful crossers diminishes, and the transported faunas take on a more haphazard composition (Figure 2–28).

CONCLUSIONS

Dispersal, in the wide sense, means shifting of domicile. Two kinds of spatial shifts have different biological meaning for the species. Dispersion *within* the distribution area means spacing of individuals, a concept of population ecology. Dispersal, in the stricter sense, spreads individuals *outside* the limits of the range of the species and expands the area of population—this is the zoogeographical concept.

Dispersal is a biological function, and its outcome depends on the resultants of furthering and hindering factors which may be either extrinsic or intrinsic.

A descriptive survey of the external conditions of dispersal reveals that certain environmental features and conditions bar dispersal but that the life of these features and conditions is finite. Barriers act in time; but, in terms of time span long enough for evolutionary adaptations to appear, they may be short-lived. Barriers may change before the species circumvents them or adapts to their presence.

Barriers are overcome by actively moving animals; special means of locomotion may have evolved in many species for the purpose of dispersion and dispersal.

Passive dispersal by carriers is a regular phenomenon, and is often especially adapted to by small land animals. Although the carrying agents seem to have an irregular nature, actually their incidence, when assessed for a long time span, is regular and predictable. Viewing the individual we find that *chance* will determine its opportunity for being passively dispersed, and chance will also determine its success in crossing a barrier. Viewing a population during a length of time, chance is seen to be replaced by an increasing degree of *probability*. The probability of dispersal is influenced by external and internal conditions, as pointed out above, and the process is demonstrable by examples of adventives.

The intrinsic attributes are subject to evolutionary adaptations. Survival depends on successful dispersal to new areas when the old domicile of the population ceases to be habitable. This applies to passive as well as to active dispersal, and may be especially important in successional biotic communities which are short-lived in any one area.

Several biological mechanisms which further dispersion are—like other adaptive features of the species—polymorphic. They simultaneously serve habitat occupation, dispersion, and expansion. The use of the collective term *dispersional polymorphism* is suggested.

Among modes of dispersal, all locomotor patterns seem to be employed, whether at the individual, social, group, or population level, and whether or not purposive or chance-governed processes are involved. The importance of slow expansion paralleling time-governed habitat changes is stressed.

Faunas are abstractions—synthetic lists of animals living on certain areas. Therefore the dispersal of faunas is as much an artifact as is the dispersal of genera or other, higher taxa. However, spatially and temporally different faunal lists may be usefully compared and the means of faunal dispersal profitably discussed.

Two main conclusions emerge from this chapter.

Because every species population has to live somewhere, and because habitats shift geographically and change ecologically through time, those species which have avoided extinction did so by evolving means and mechanisms of dispersal or by adapting already existing mechanisms which served other locomotive functions (e.g., dispersion mechanisms).

In the light of these means of dispersal and locomotion, one would expect wholesale crossing of barriers and practically unlimited dispersal throughout the available land area. However this is not so. Since the reason for this paradoxical situation is apparently not to be found in the dispersal mechanism, we must seek for explanation elsewhere. Limitations of expansion will be discussed throughout the following chapters.

3 / The Ecology of Colonization

> "La généralisation dogmatique est la mort du progrès biologique."—(Baron G. J. von Fejérváry)

The two disciplines taxonomy and ecology have so close and mutual a relationship with biogeography that it is often difficult to separate the one from the other. Taxonomy uses the geographical distribution of plants and animals as one of its chief criteria. Ecology leans on distributional knowledge to a still greater extent. This chapter will consider the use of ecological knowledge in zoogeography and its bearing on distribution and dispersal.

Here we shall discuss particular questions concerning the factors that enable a species to colonize a particular area. These questions pertain to the spatial relations of animals and therefore belong in the realm of ecology. In the system of biological topics outlined in the Introduction (p. 3), 10 of the 16 major fields contain ecological subjects. These fields are covered in such treatises on ecology as Bodenheimer 1938, Elton 1946, Dice 1952, Andrewartha and Birch 1954, Lack 1954, Naumov 1955, Tischler 1955, Macfadyen 1957, Balogh 1958, Odum 1959, Andrewartha 1961, Kendeigh 1961, Slobodkin 1961, Schwerdtfeger 1963, Kühnelt 1965, and Tischler 1965. Hesse (1924, translated and revised 1937 and 1951) and Elton (1958) treated ecological zoogeography. Several different approaches can of course be made to the study of the ecology of colonization.

An analytical approach assesses the major factors that may encourage or prohibit colonization. This is known as the *autecological* approach. The intrinsic capacities of the species must also be assessed: how these capacities enable the species to cope with environmental situations, by virtue of inherited tolerances and requirements—that is, by its adaptive range, for which Hesse (1924) uses the term *ecological valency*, and this term has priority in ecological studies. Also of importance is how the adaptive range may change as a result of learning or of other, non-inherited modifications in structure, physiology, behavior, and life history. However, zoogeographers will usually concentrate on the ecology of colonization which occurs—at least after the outset—by populations rather than by individuals. A completely different set of norms will apply to a breeding population; adaptive ranges characterize the individual rather than the group. Furthermore, the population, because of its numbers, has a profound impact not only on the biotic but also on the physical environment, and this impact may in the long run influence the success of colonization.

An indirect but often more useful approach is to assess the general environment (or the major habitat types) in which the species in question is known to be able to live, and to compare these with the general habitats of the new area. This is known as *habitat ecology*, which is really a geographical aspect of natural history.

Finally, there is a temporal aspect to ecological analysis. The *time factor*, combined with periodic changes of the physical environment, is a further prerequisite to successful establishment. The circadian, seasonal, annual, and other cycles of the environment must fit the rhythmicity in the life of the species: the alternation of sleep-wakefulness, foraging and other activities; the breeding cycle, seasonal migrations, and the like.

The examples which illuminate the principles of settling can be drawn from several sources. (a) Actual observations can be made of individual animals crossing a barrier and widening the distributional range of the species by settling on a new area. There are very few examples of this nature. (b) The process of settling in pioneer communities can be demonstrated in instances in which *vacant habitat* has been created either by such natural forces as the secular elevation of land (see p. 316) and the filling of tidelands; or as a result of such man-induced forces resulting from fire, erosion, blasting, or atomic explosion; (c) In localities in which groups of settlers have been actively introduced or unconsciously carried by man and his agents. (At first glance, the last-mentioned example may appear to be artificial, but on further consideration it will be seen to be otherwise. For man, a most effective and frequent carrier, is a natural agent. His introductions are, for our purpose, grand-scale experiments from which we learn much about the ecology of settling.) (d) Because examples from all the above-mentioned sources are extremely rare, one often resorts—as we shall do in this chapter—to inferences drawn from general ecological observations.

After a species reaches a new area, either by barrier crossing or by slow penetration, it can establish a nucleus of a new distributional area or an extension of the old one under the following conditions:

The colonist must *survive* on the area long enough to leave offspring.

Therefore the area must be suitable not only for existence of the settler but also for its *reproduction*.

The offspring now forms an *incipient population*, the existence of which necessitates more delicate and complex prerequisites of the environment than that of the original settler.

A kind of relativity applies even to animal-environment relations. An animal population will have as much impact on its environment as the environment will on the animals. Since both the environment and the species population are adaptable, *mutual adjustments* must follow.

Expansion will follow only if the new members of a local faunation become "accepted" in, or adapted to, its optimal environment. The available expanse of habitat will be saturated, and related and spatially available habitats may then be penetrated. This expansion thus leads to the beginning of a new cycle of range movement.

EXTERNAL FACTORS LIMITING SURVIVAL

The survival of an individual animal will be possible only if environmental conditions do not exceed its range of tolerance and if its minimum requirements are satisfied. This principle applies also to a colonizing population. As early as 1911, Shelford pointed out that the same environmental factors govern local and geographic distribution, viz. his Law of Tolerance* maintains that success of reproduction depends on the qualitative and quantitative completeness of the complex of conditions surrounding the animal. The failure of a species to breed in a particular habitat may be due to the deficiency or excess of anyone factor. This concept is close to the application of Liebig's law of the minimum to ecology of existence. Thienemann (1942, 1950) expressed Liebig's law as follows. "Die Entfaltung einer Art in Biotop wird durch denjenigen

*Botanist R. Good's (1931) theory of tolerance is a different concept. While Shelford emphasizes the tolerance of single factors, Good relates the entire range of tolerance to external conditions. "The tolerance of a species is a specific character subject to the laws and processes of organic evolution in the same way as its morphological characters, but the two are not necessarily linked." ... (Good 1964, p. 417.)

Faktor bestimmt, der dem Entwicklungsstadium mit kleinster ökologischer Valenz im Minimum zur Verfügung steht" (1942 p. 321): The establishment of a species in a habitat will depend on that factor which is presented in minimum amount to the developmental stage which is most sensitive. These factors and conditions are commonly called *limiting factors* or conditions. The study of their effect on the organism forms the nucleus of the ecological research spurred by Hesse, Shelford, and others from the 1910's onward. Ernst Haeckel, in 1866, was probably the

Table 6 Factors and conditions of ecology of colonization

Extrinsic Factors : Analytic Approach	*Extrinsic Factors : Synthetic Approach*
Weather Food Space and Shelter Other Animals & Plants Chance	Climatic and Soil Zonation Plant Formations Communities and Ecosystems Habitat Niche Environmental Resistance
Time Timing of Environmental Factor Duration of Effect	*Time* Circadian Periodicities in the Habitat Seasonal Fluctuations Successional & Evolutionary Processes
Intrinsic Factors : Analytic Approach	*Intrinsic Factors : Synthetic Approach*
Range of Tolerances and Requirements (Euryecy and Stenoecy of Individuals) Acclimations and Other Phenotypic Changes Structural, Physiological, Behavioral Adaptations	Extent of the Distributional Area Size of Population Structure of Population (especially eury-, amphi-, and stenotopy) Genetics of Population Interactions of the Population and Its Environment
In Time Longevity Seasonal Aspects of Life History Mobility in Time	*In Time* Population Dynamics Expansion, Retraction, Extinction Evolutionary Changes of the Populations

first to outline these limiting factors systematically. The following is a translation of Haeckel's table.

Inorganic requirements of existence:
Physical and *chemical* properties of the station
Climate: light, heat, humidity and electric properties of the atmosphere
Inorganic food
Properties of the *water* and *soil*

Organic requirements: All relations of the organism to other organisms
"Friends" and *"Foes"* that further or hinder existence
Organic food

The major environmental factors are listed in Table 6. The tolerances and requirements of an animal and the ecological terminology respecting the distributional aspect of these relations are illustrated in Figures 3-1, 3-2, 3-3, and 3-4.

It is generally recognized that a limiting factor may be greatly influenced in its action by other, simultaneously acting, modifying agents. It is therefore difficult to pinpoint a single limiting factor other than by in-

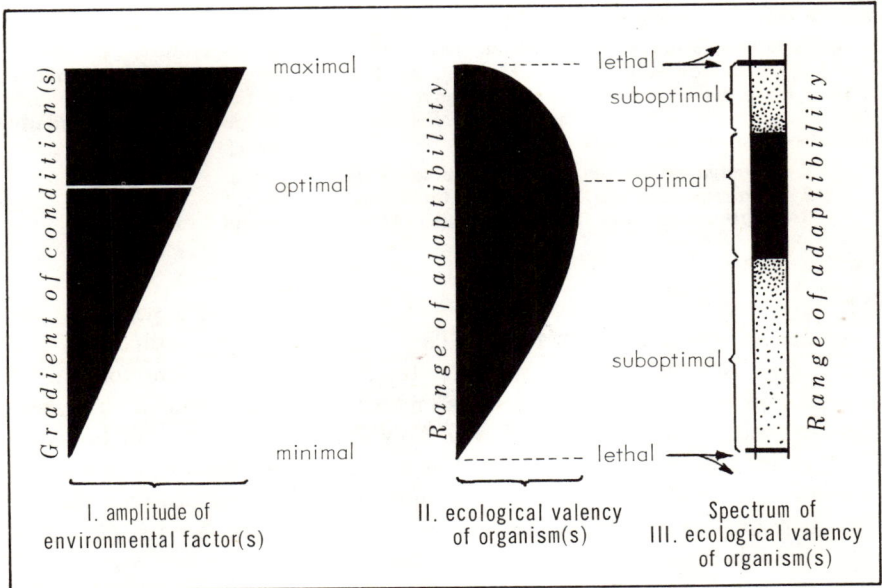

Figure 3-1 The concepts of tolerances and requirements I.
Amplitude of environmental factors and of adaptability (or acclimatization) of individual organisms. Parts I and II modified from Thienemann 1950, after Lenz, 1931.

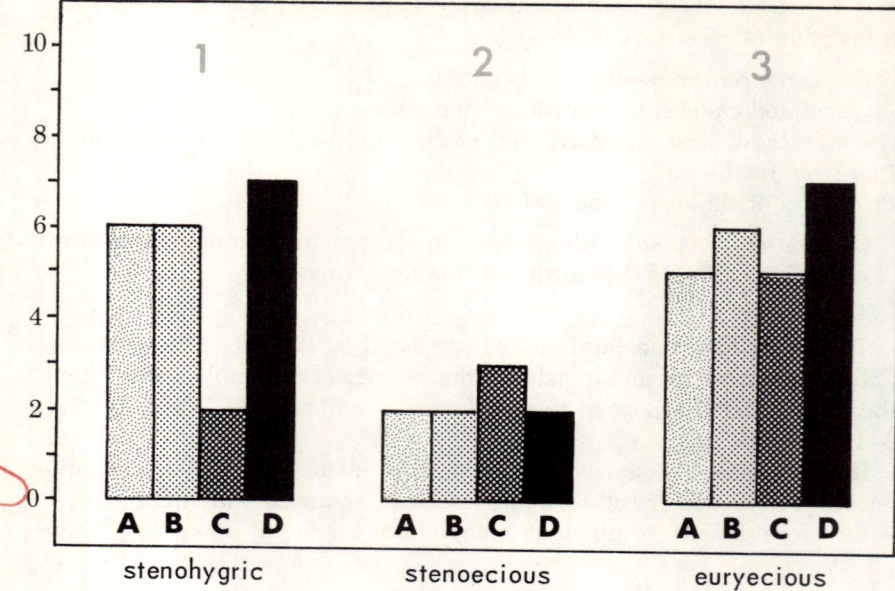

Figure 3-2 The concepts of tolerances and requirements II.
Amplitude of selection of environment, with respect to (A) temperature, (B) acidity [pH], (C) soil humidity, and (D) light, by three muskeg insects from Finland. (After Krogerus 1939.) The available range of the above factors was scaled arbitrarily on a 0–10 scale, and the width of the chosen sector indicated by a similarly scaled column on the graph.

1 = *Dictya umbrarum* L. [Sciomyzidae, Dipt.]   stenohygric
2 = *Boletina borealis* Zett. [Mycetophilidae Dipt.]   stenoecious
3 = *Bathysmaphorus reuteri* J. Sahlb. [Rhynch. Hemipt.]   euryecious

ference—that is, by the coincidence of distribution limits with the occurrence of the minimum factor. The botanist Rübel emphasized the Law of Substitution (1935), which states that any external factor can be substituted by another as long as its impact on the organism is the same. The organism is not concerned with how its requirements will be satisfied, as long as they are satisfied. For a hygrophilic organism, requiring a great amount of water, this can be supplied in the form of precipitation, fog dripping, or atmospheric humidity. Then too, low air temperatures can also substitute for moisture, for evaporation loss is much less in cold than in warm air. Rübel points out that the deciduous forests of Great Britain thrive with between 600 and 1,000 mm yearly precipitation in a fairly cool Atlantic climate; on the other hand, the closely related deciduous forests of the eastern United States require between 1,100 and

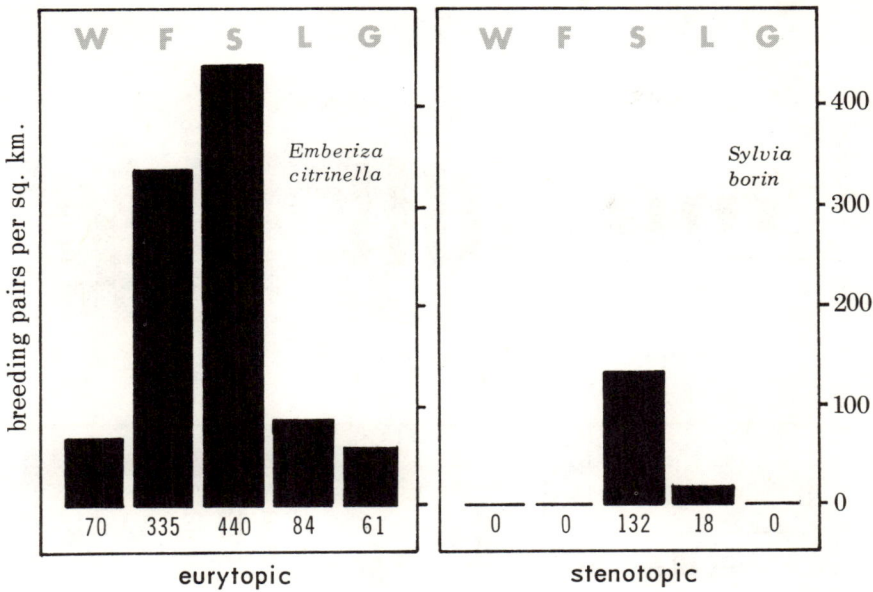

Figure 3-3 The concepts of tolerances and requirements III.
Amplitude of selection of environment with respect to various general habitats. Density of two breeding bird species in a gradient of five habitats at Tihany, Hungary in 1947. (From Udvardy 1947.) The optimal densities of both the yellowhammer (*left*) and the garden warbler (*right*) were in the same type of habitat.

    W = open oak woodland
    F = dense stand of mixed deciduous forest with understory
    S = open scrub forest
    L = low thorny scrub, pastured
    G = grassland on hilltop and slopes with some browsed thorny scrub

1,500 mm precipitation because of the much higher summer temperatures! Similarly, in the moist maritime regions of Ireland, 500 mm precipitation is enough for the growth of hygrophilic plants, while the same amount of precipitation in the continental, hot plains of the state of Colorado only supports xerophilic plants. Again, wind substitutes for aridity, because the wind dries out soil and plants alike and because evaporation is rapid in a windy climate. Auer (1951) studied in great detail this effect of "foehn" type, warm, downhill winds upon the vegetation of the Patagonian pampas. Syroechkovskij (1959) and Gressitt (1963) observed its effect on birds and insects in Antarctica. The structure of sandy soils has an effect similar to that of dryness in wet climates; but, paradoxically, in dry regions, sand can act as a relatively moist habitat. A group of small

100 *Ecology of Colonization*

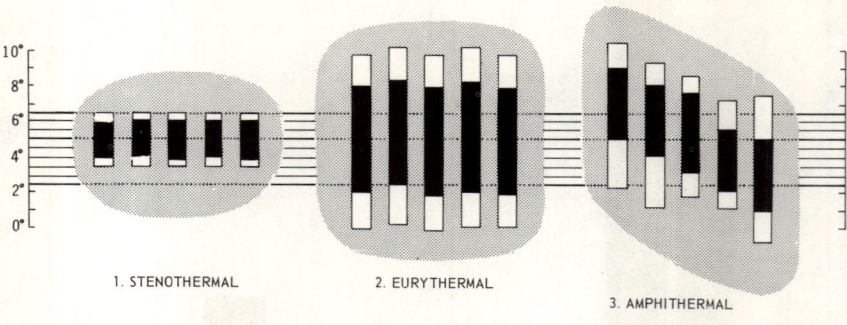

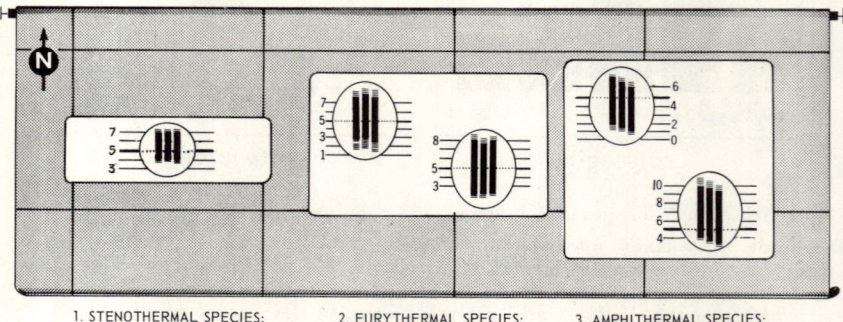

Figure 3-4 The three types of ecological valency and their effect on distribution.
Above. Three hypothetical species were sampled (five individuals of each) and their thermal tolerance illustrated on a spectrum similar to that in Figure 3-1, III. These live in the same temperature zone, which is optimal around 5° for each species, but in three different ways. Species 1 represents the classic concept of stenothermy. Species 2 illustrates the classic concept of eurythermy. Species 3 would also classify, according to the prevailing terminology, as eurythermal, but has a different tolerance structure, which we call *amphithermy*.
Below. Distribution ranges of the same three species are shown on a map, with population samples exhibiting tolerance spectra as above.

Species 1 is distributed along a narrow belt of optimal thermal conditions, isoclimatically, as an effect of stenothermy.

Species 2 is spread less narrowly, and as the spectra in the north and in the south indicate, adaptation to a wider thermal range is uniform. This population exhibits eurythermy.

Species 3 has large variability, and thus adapts easily to a much wider range of local thermal conditions, forming either clines or subspecies. This kind of valency is called amphithermy.

Condition 3 must be common in nature, and therefore the terms "amphiecious" and "amphitopic" may be introduced to denote general adaptability and wide habitat occupancy based on selection.

rodents, the pocket gophers (Geomyidae) are subterranean dwellers (like the better-known, insectivorous mole) in the western and southwestern United States. They are thoroughly adapted to the underground existence, and come to the surface only when they push out excess earth from their burrows. The California zoologist J. Grinnell pointed out that one species, *Thomomys monticola* J. A. Allen, which lives in the subalpine zone of the Sierra Nevada, extends its feeding tunnels into the deep winter snow, "thereby reaching in safety stems of plants about the ground surface!" (Grinnell, 1943, p. 178.) Here snow substitutes for soil.

I do not intend to present a complete compendium of the aspect of ecology which deals with limiting factors. Instead, I shall choose a few examples which show that such factors indeed restrict spread and, presumably also hinder successful settling of colonizing individuals or groups.

Light

Light itself may be such a limiting factor for organisms that have a definite light requirement. Petersen (1954) suspects that the cloudy, foggy summer climate of Scotland sustains only those species of butterflies that are adapted to conditions of low light intensity. Iceland, with similar or even duller summer weather, is devoid of butterflies, except for some casual, non-breeding immigrants. Most moths of the family Noctuidae, on the other hand, fly only in twilight. Rensch (1950), quoting Ryberg, gives an explanation for the northern limits of noctuid-feeding bats in Sweden. It appears that during the midsummer weeks temperature conditions are above the minimal requirements for bats. However, during this period many noctuid moths do not fly, since there is so much light at night. In late summer, with the advent of twilight, and when the moths are in flight, the cold limits the activities of the bats. In this case, photoperiod seems to limit the activity of the moth and, indirectly, the existence of the bat.

The Russian ecologist Kalabukhov (1938) has studied the light requirements of certain rodents, and found that in an experimental situation (using light gradients) the wood mouse *Apodemus sylvaticus* L. [Muridae, Mamm.] avoided the light more than did *A. flavicollis* (Melch.), and that ground squirrels (*Spermophilus*, Sciuridae, Mamm.) chose the brightest sunlight. He speculated that *A. sylvaticus* lives in more open woods than *A. flavicollis* and thus, by being more strictly nocturnal, may avoid certain predators. In such cases, we are dealing with very complex adaptations where light intensity *per se* may not be a limiting factor but where it is correlated with some other environmental factor exercising a selection pressure on the organism—i.e., light acts as a releasing stimulus (environmental trigger) which enables the animal to avoid unfavorable conditions.

Most animals are active under strictly fixed light conditions. Seibert (1951) noted the daily departure and arrival time of five heron species which roosted in the vicinity of his observation post. One species, the green heron *Butorides virescens* L. [Ardeidae, Aves], starts its feeding day half an hour earlier than do the others and also goes to roost later, at lower light intensities. This species arrives earliest in the spring, and is distributed considerably farther north than are the other herons. We may conjecture that its more crepuscular habits enable it to have a longer feeding day, thereby giving it sufficient energy for an early nesting, even in the north, where the favorable season is short. In this instance, the particular light requirements of the green heron enable it to extend its range into northern latitudes. The relationship between the bird and its environment may be illustrated as follows:

Particular light requirements ⟶ longer daily photoperiod ⟶ Increased length of feeding day ⟶ Earlier spring arrival ⟶ Expansion into northern areas with short favorable season but longer twilight.

Temperature

The zonal distribution of temperature, together with seasonal extremes, is expressed principally in the temperate and boreal zones of both hemispheres. The limiting effect of temperature has therefore received much attention from ecologists in the Northern Temperate Zone. Direct or secondary temperature limitations are evident when distributional limits are compared with seasonal mean isotherms, even though the direct controlling effect of temperature has not often been substantiated by experiments. Therefore a word of warning against overgeneralization is in place. Mail and Salt (1933), experimenting with the Colorado potato beetle *Leptinotarsa decemlineata* Say [Chrysomelidae], found that a temperature of $-7°$ C killed more than half of the beetles within a few hours and that $-12°$ C was quickly lethal for all of them. This expansive species could not spread north beyond the 54th parallel in Alberta, Canada, presumably because the temperature of the top layer of the soil, where the beetles hibernate, usually drops each winter to the lethal point. The northern distribution limit of the crested lark *Galerida cristata* (L.) [Alaudidae, Aves] coincides in Russia with the isometric line indicating a duration of 140 days for the winter snow cover. This lark does not migrate in winter; if we take the length of the snowy period as an indication of the severity of the winter, we can draw the inference, with Stantchinsky (1927), that winter climate presents an impassable barrier to the crested lark. During recent decades, with increasingly warmer and shorter winters, this lark extended its breeding area northward (Kalela 1950), thus corroborating Stantchinsky's earlier inference. Figure 3–5 illustrates the conditions of the crested lark. For comparison— and to illustrate the fact that one may find many correlations between the distribution of certain animals and different aspects of the winter climate—the relationship of snow cover to the northern limit of the roe

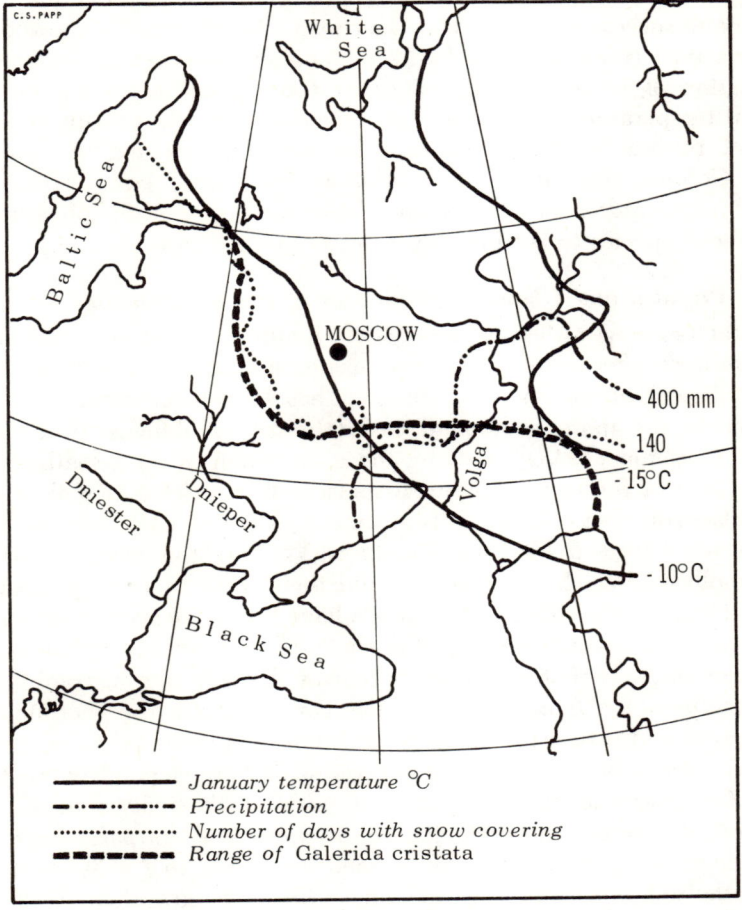

Figure 3-5 Environmental factors limiting distribution.
Northern distribution limit of the crested lark *Galerida cristata* (L.) in Eastern Europe and the duration of winter snow cover. After Stantchinsky 1927.

deer and that of the number of days with frost to the northern limit of certain bats are also presented (Figures 3-6, 3-7).

On the other hand, a species adapted to a relatively cool environment may be prevented by too high an environmental temperature from extending its range southward in the Northern Hemisphere and northward in the Southern Hemisphere. Kendeigh (1934, 1964) pointed out, in an experimental work on thermal requirements of the house wren *Troglodytes aedon* Vieill. [Troglodytidae, Aves], that the southern limit of this bird coincides with temperatures so high that the bird exerts so much energy in cooling itself instead of in nesting that its reproduction efforts become unsuccessful. Later, von Haartman (1956) actually showed that when a bird is subjected to high ambient temperatures it neglects the incubation of its eggs. Yeatter (1950) found that increasing environmental temperature before the start of incubation affected the hatching rate of pheasants (*Phasianus* sp.) but not the hatching of bobwhite quail *Colinus virginianus* (L.) [Phasianidae, Aves] eggs. Thus, high spring temperatures may be responsible for the failure of the pheasant, an introduced species to establish itself in southern North America.

Physical and Chemical Properties of the Environment

Such factors may—either as immediately limiting factors or by combined and indirect effects—block the colonizing individuals and thereby the establishment of the species. This fact has long been recognized with respect to the more soil-bound invertebrates (Holdhaus 1911, Hesse 1924). Oughton (1948), studying the distribution of land snails in Ontario, Canada, found that eighteen species are restricted to soils overlying Palaeozoic rocks. These soils are much richer in lime than are those which were formed by Precambrian rocks. Oughton stresses that for land mollusks, which use lime to build their shells, a lack of $CaCO_3$ in the soil can be a limiting factor. Snails have been observed to eat empty shells or even devour pieces of the shells of their live companions. Yet Oughton suspects that the lime content of the soil acts in combination with temperature. This fact has been borne out by the thorough experimental and observational work of Lindroth (1949, 1953) on certain ground beetles, which, like the Ontario snails, are dependent on limestone for their occurrence in at least parts of their distribution area. Lindroth first tested the preference of 15 species of *Harpalus* beetles in multiple-choice experiments. All seemed insensible to $CaCO_3$ or to the pH of the soil; nor could they distinguish limestone gravel from siliceous gravel. However, further testing of their thermal requirements revealed that the "calciphilic" species are, in reality, either strongly thermophilic or xerophilic or both. In further experiments, Lindroth compared the

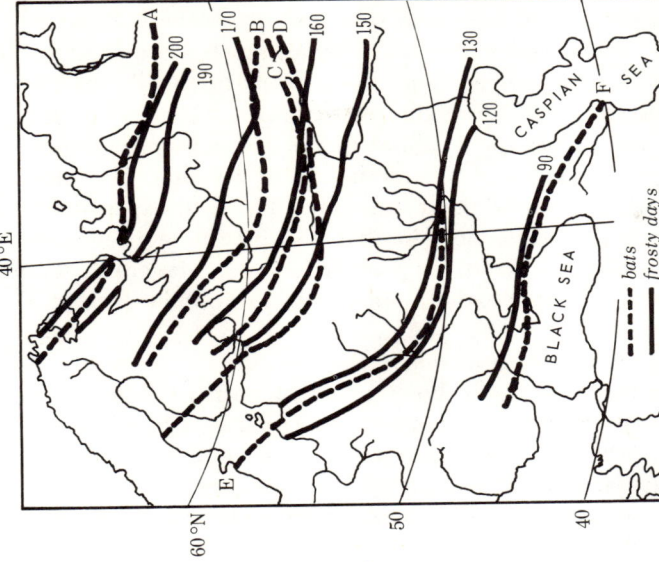

Figure 3-6 Environmental factors limiting distribution. Northern limit of the roe deer *Capreolus capreolus* L. in Eastern Europe and the duration of winter snow cover. (After Naumov 1955, from Formosov 1946.) The dotted line denotes the former northern limit of the roe deer; arrows, the direction of recent expansion (Cf. Fig. 2-20).

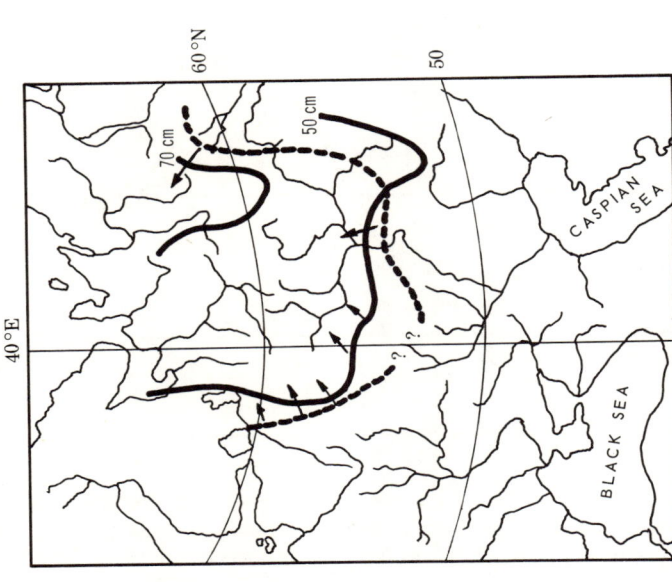

Figure 3-7 Environmental factors limiting distribution. Northern distribution limits of six species of bats in Eastern Europe and the isolines of the number of frosty days in a year. From Naumov 1955, redrawn.) A. *Amblyotus nilssoni* (Keyserling & Blasius). B. *Myotis daubentonii* Kuhl. C. *Selysius nattereri* (Kuhl). D. *Myotis myotis* (Borkhausen). E. *Vespertilio murinus* L. (A–E all Vespertilionidae.) F. *Rhynolophus ferrum-equinum* (Schreber). [Rhinolophidae].

106   *Ecology of Colonization*

thermal qualities of Palaeozoic limestone rock and gravel with those of Precambrian granite, and found that the latter had much more rigid responses to external heating and cooling. As a corollary to all this experimental work, Lindroth measured the *microclimate,* i.e., the climate immediately above and in the soil (habitat of the *Harpalus* beetles) in two locations on opposite sides of a fault fissure in Sweden. One side of the fissure was granite, and the other limestone. The field experiments indicated an even greater difference, in that the limestone prevented both cold and hot extremes and thus provided a more favorable environment for the more stenotopic, thermophilic, and xerophilic *Harpalus* beetles (Figure 3–8).

Lindroth's experiments with ground beetles also provide evidence of how chemical properties of the soil directly influence distribution (Lindroth 1949). Two carabids, *Bembidion aeneum* Germar and *B. minimum* Fabricius, reacted positively to the presence of very small quantities of NaCl in an experimental test. These species, however, are not restricted to seashore habitats as are other, more halophilic forms, for they exist also in the interior of Sweden. Their inland distribution (Figure 3–9) coincides neatly with the distribution of Yoldia clay. This type of soil was deposited on the bottom of the brackish Yoldia Sea, south of the margin of the retreating ice, in a phase of the postglacial period when

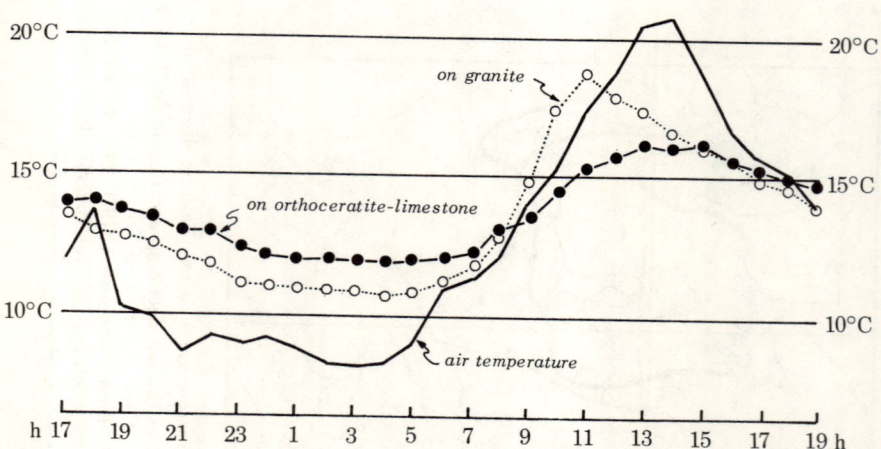

Figure 3-8 Physico-chemical properties of the substratum influencing microclimate. Temperature in C° from 1700 hours on June 4 until 1900 hours on June 5 on opposite sides of a fault fissure at Råttvik, Sweden. Note the temperating effect of limestone as opposed to that of granite. Redrawn, with the author's permission, from Lindroth 1949 and 1953.

Physical and Chemical Factors   107

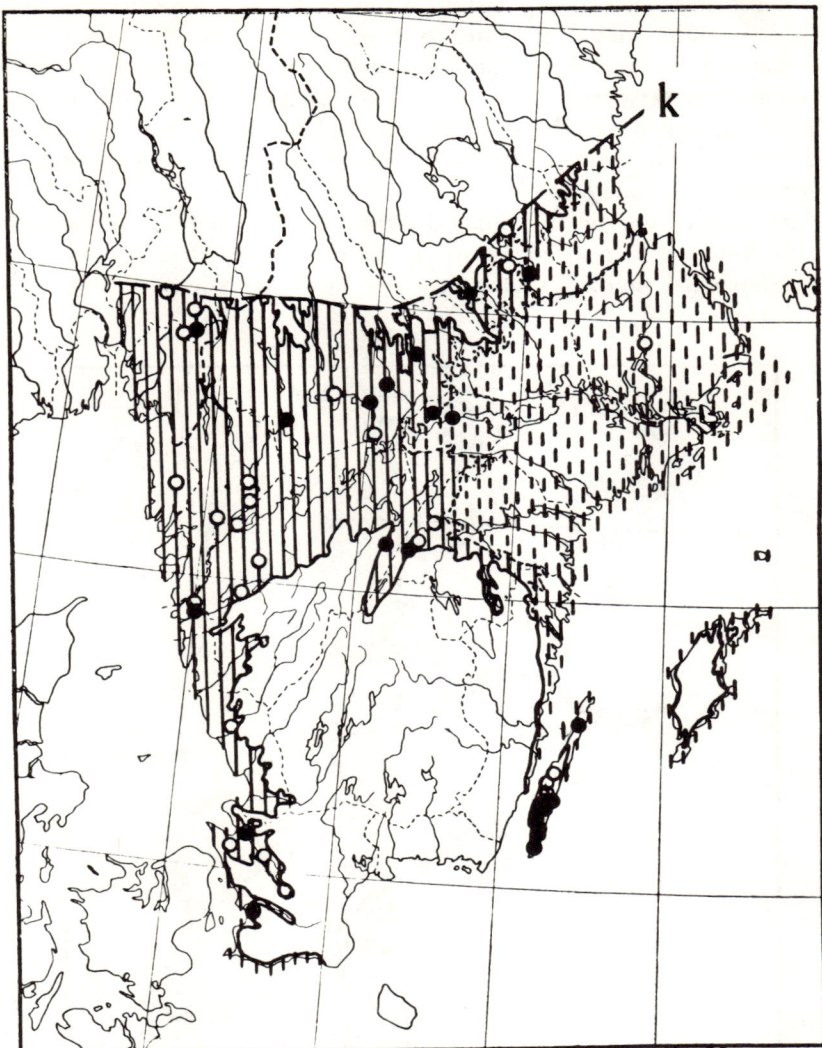

Figure 3-9   Soil limiting distribution.
Inland distribution in Scandinavia of the carabid beetle *Bembidion aeneum*. Circles (one specimen) and dots (several specimens collected) are confined almost entirely to the vertically hatched area of slightly saline soil on the left, and are largely excluded from the right (interrupted hatching) where the soil is not saline. About 9000 B.P., the glacial ice stood at line k; Yoldia Sea covered the lowlands of southern Sweden and deposited the salt, which, around 8000 B.P., a freshwater lake leached out from the eastern area shown. (Cf. Figure 6-8.) From Lindroth 1949 and 1953 (*Courtesy of C. H. Lindroth and the Swedish Natural Science Research Council*)

the Baltic Basin was connected with the North Sea across central Sweden. When the inland ice later retreated farther north, southwestern Scandinavia rose and the Baltic Basin, including the present southeastern Sweden, became a freshwater body—the Ancylus Lake of the geologist. This lake leached out the salt from the Yoldia clay; where the present soil lies on such leached elay formation, *B. aeneum* and *minimum* are conspicuously lacking.

The distribution of the iguanid lizard genus *Uma* in southwestern North America is restricted entirely to patches of desert covered by wind-deposited sand. A thorough study by K. S. Norris (1958) shows

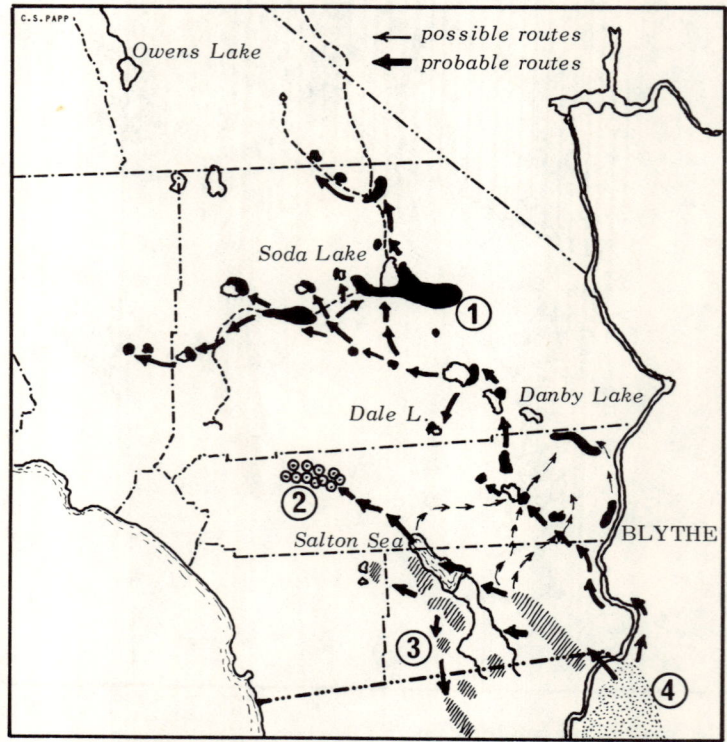

Figure 3-10 Distribution interrelated with a particular environment.
Distribution of the genus *Uma* [Iguanidae, Squamata] in western North America. The fringe-toed lizard is able to run on loose sand. Its patchy occurrence is tied to aeolian sand habitat. Observed dune movements are adequate to explain its dispersal during the Pluvial (Pleistocene) and postpluvial periods along the routes here suggested. (After K. S. Norris 1958.) 1. *U. scoparia* Cope. 2. *U. notata inornata* Cope. 3. *U. n. notata* Baird. 4. *U. n. rufopunctata* Cope.

that these fringe-toed lizards are adapted to this particular kind of environment. These sand dunes occur along desert water courses, lakes, and ocean beaches. Figure 3–10 illustrates the patchy, discontinuous occurrence of habitat and the distribution of this lizard.

Faunistic mapping, together with ecological field and/or experimental work, often pinpoint environmental factors likely to limit distribution. A further example, from Lindroth's zoogeographical work (1956a), illustrates this principle. Two carabid beetles, *Brachynus crepitans* L. and *Agonum dorsale* Pont., have almost identical northern limits in Sweden, and both live on dry meadows, presumably under nearly identical ecological conditions. Laboratory work on their tolerances (limiting intensity of factor) and requirements (preference in a gradient of controlled

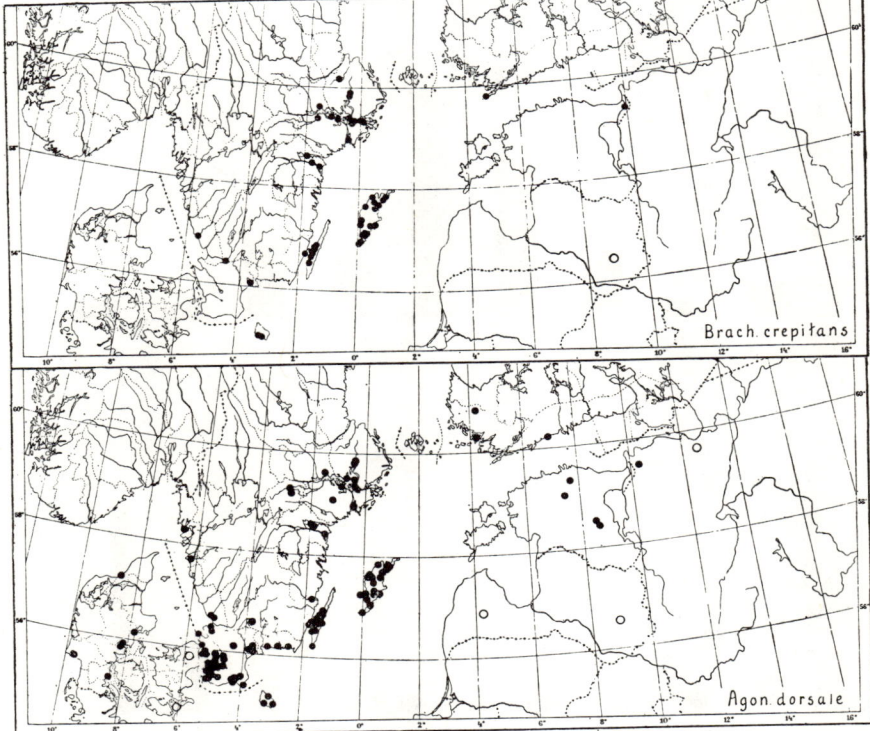

Figure 3-11 Limiting factors and distributional limits I.
Distribution of two carabid beetles in southern Sweden. They live on the same habitat and have almost identical northern limits. From Lindroth 1956a. (*Courtesy of C. H. Lindroth and the Swedish Natural Science Research Council*)

110   *Ecology of Colonization*

(a)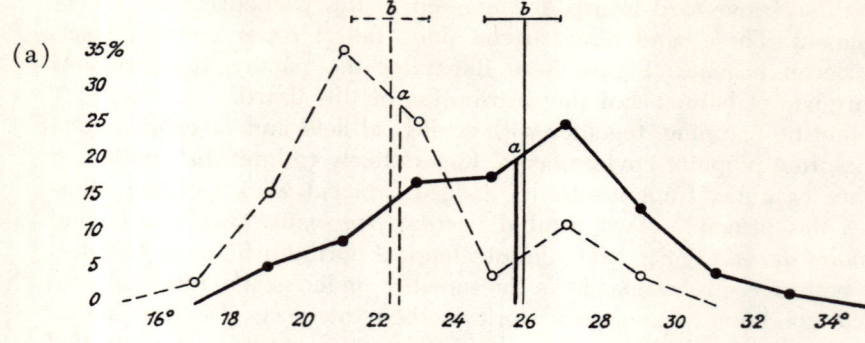

(b)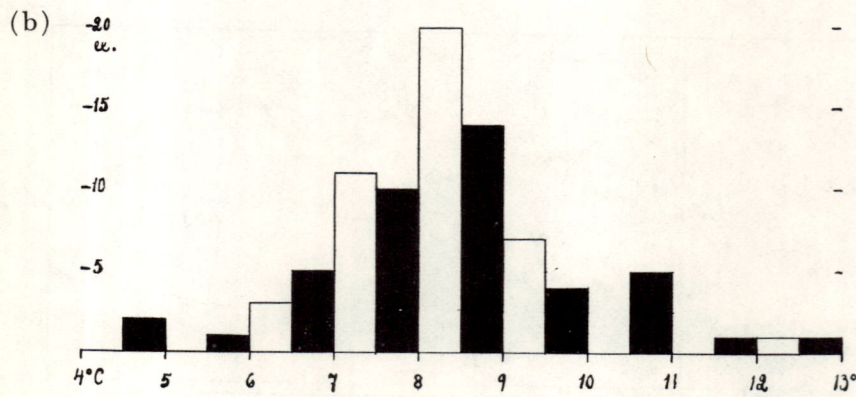

(c)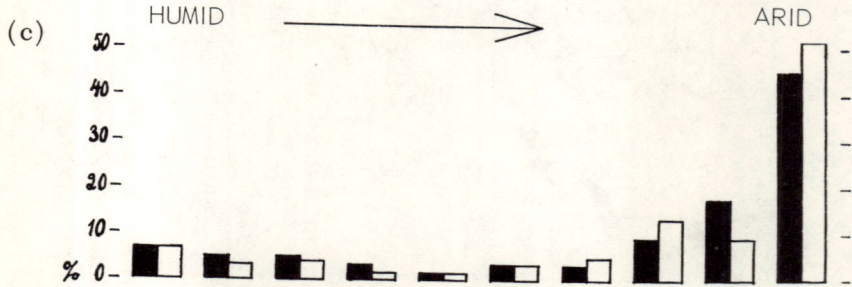

Figure 3-12 Limiting factors and distributional limits II.
Thermal (a), cold (b), and moisture (c) preference tests on *Brachynus crepitans* L. (black columns, solid line) and on *Agonum dorsale* Pont. [Carabidae, Coleopt.] (white columns, dotted line). (From Lindroth 1956a.) (See also Figure 3-11.) (*Courtesy of C. H. Lindroth and the Swedish Natural Science Research Council*)

environment) showed that cold tolerance and moisture preference were almost the same for both species. *Brachynus* exhibited higher thermal preference than did *Agonum*. The conclusion can be drawn that choice of habitat and limits of occurrence are determined by low temperatures rather than by the thermal optimum, and by optimal moisture rather than by danger of desiccation (Figures 3–11 and 3–12). Lindroth's monographs on adventitious (1957) and naturally dispersed (1963) insects of Newfoundland are exemplary treatments of the ecological factors influencing success of colonization.

THE PRINCIPLE OF THE INTERDEPENDENCE OF LIMITING FACTORS

The principle of interdependence or mutual moderation of limiting factors has not been precisely formulated, although it is well understood by ecologists and zoogeographers (Schwerdtfeger 1963, p. 429). The tolerance limit as well as the minimum requirements of an animal toward any one limiting factor, will be influenced by the presence and extent of other essential environmental factors. These may be called masking

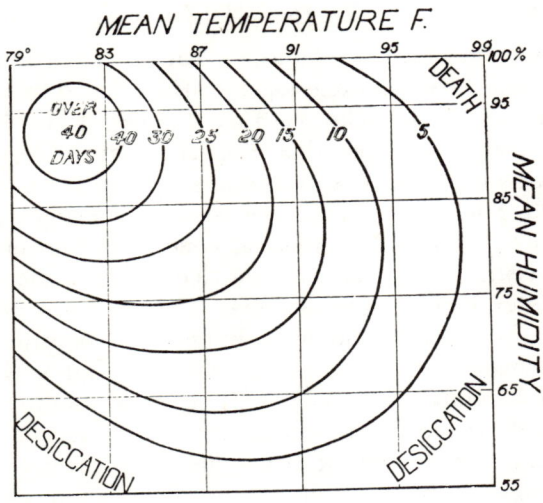

Figure 3-13 Interdependence of limiting factors.
Zones of maximum survival time of adult beetles of *Hoplocerambyx spinicornis* (Newman) at different combinations of temperature and relative humidity. From Linsley 1959, after Beeson and Bhatia 1939. (*Courtesy of Annual Reviews Inc., Palo Alto, California*)

factors when they prevent a limiting factor from operating on the organism to the extent that it would operate in their absence (Fry 1947). For example, Beeson and Bhatia (1939) show how experimental combinations of temperature and relative humidity reveal lethal limits of these conditions for the longhorn beetle *Hoplocerambyx spinicornis* (Newman) [Cerambycidae, Coleopt.]. Gradients of sublethal combinations of these factors influence longevity, but an optimum is reached in the vicinity of 27° C temperature and 88 to 98 percent relative humidity (Figure 3–13).

Wallgren (1954) experimented on the thermal tolerance of the yellow bunting *Emberiza citrinella* L. [Fringillidae, Aves]. He found that at high ambient temperatures the tolerance of the bunting depends on vapor pressure of the air, because this bird cools itself by evaporation. On the other hand, tolerance to the winter cold apparently depends on the length of the night during which the bird is without food and on the availability of food during the day. At night, when the temperatures are usually at their lowest, the birds are relatively well sheltered at their roosts; during the day, they are exposed while feeding. Thus, suitable shelter may mask the effect of low temperatures, and sufficient food may shorten the time of exposure to cold.

REGIONAL CHANGES IN ECOLOGICAL VALENCY

The study of ecological valency indicates that the range of tolerances and requirements may vary from one place to the other within the distributional area of the species. The Austrian ecologist Kühnelt (1943) was among those who pointed out that a species may be euryecious in its optimal habitat and stenoecious where one of the vital factors nears the minimum limit. Such observations are commonly made with respect to the temperature factor. Where thermal conditions become limiting, the animal still occurs in the habitat that—locally and microclimatically—offers it a better thermal environment. Thus, a species which, at its northern limit, behaves thermophilically and lives only in habitats with warm local climate, such as limestone soils, will tend in the south to occupy cooler habitats, such as bogs. Its thermal requirements may be the same, but these are found only in these restricted places. Near the geographic center of its area, it is not limited by temperature, and its other adaptations enable it to live in many different places (cf. also Warnecke 1934, Lindroth 1949, Thienemann 1950). The North American robin *Turdus migratorius* L. [Turdinae, Aves] occurs in southern British Columbia, at middle latitudes, at the forest edge and forest-woodland habitats from sea level to middle-altitudes. It is highly eurytopic at the lower altitudes; near sea level, it nests in five different habitat types (Horváth 1963). At the southern margin of its western North American distributional area, it rarely nests at low elevations, and is highly stenotopic (Grinnell and Miller 1944, Miller 1951). The yellow bunting *Emberiza citrinella* L. [Fringillidae], the blackcap warbler *Sylvia atricapilla* (L.) [Sylviidae], and several other birds reach their northern limits in Finland. These birds nest there only in certain southern kinds of habitats, in deciduous woods. Hungary lies in about the center of the distribution range of these songbirds; there they are eurytopic, living in a great variety of forested habitats and woodlands (Figure 3-3). In the Balkans, at their southern limits, they are mountain birds of the cool forests at high elevations. But all these examples indicate only that the vegetation zone in which they live has itself a climatically controlled distribution, and that increased altitude has an effect similar to that of increased northern or southern latitude both acting as proximate factors (p. 117) of existence. This observation was first recorded and elaborated by Alexander von Humboldt in 1807 and, regrettably enough, has not yet received sufficient attention by ecologists, geneticists, and zoogeographers.* Climatic zones are found at progressively higher land elevations as we approach the Equator. The parallel in the sea is the so-called *equatorial submergence*—in proceeding from subpolar toward equatorial areas (Figures 3-14, 3-15), the latitudinal temperature belts of sea water submerge from the surface to gradually greater depths. But the varying genetic structure of natural populations also provides examples in which truly regional differences in tolerance could be demonstrated—the "temperature races" of Dobzhansky (1935) and Timofeeff-Ressovsky (1940) (cf. Figure 3-4). These regional differences are in correlation with regionally different environmental conditions, and can thus be considered adaptive differences. Many insects—for example, the Lepidoptera—have a resting stage in their development, called the diapause, during which the caterpillar survives the adverse season in this inactive and resistant form. In temperate regions, diapause is photoperiodically triggered; the shortening of the late summer days induces the larva to diapause. Danilevskii (1965) showed by experiments that of four population samples of a moth taken along a south-north transect in Russia and kept under constant and identical temperatures, each population started its diapause at a constant day length, significantly different from that of the

---

*A similar phenomenon occurs when an animal is active during the warmest hours of the day in the north and during the coolest morning or evening hours in the south. In these areas, the thermal preference is about the same; but the preferred temperature range is available at different times, and the animal adapts its rhythmicity accordingly. Howden (1963) observed this phenomenon on scarabaeid beetles.

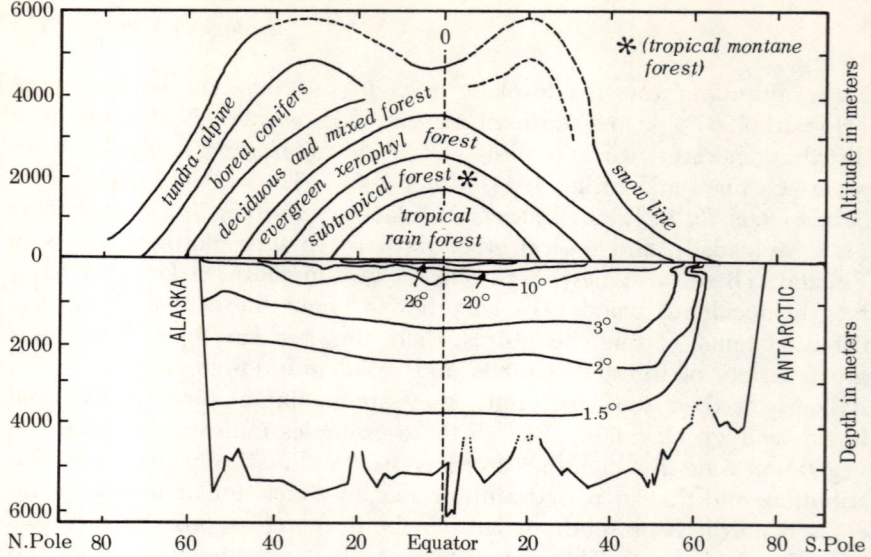

Figure 3-14 Relations of altitudinal and depth distribution with latitude I.

Above. Altitudinal zones of vegetation. After Walter 1964. Lack of seasonal temperature changes around 0° latitude obscures the similarity of the high altitudinal zones here to those at sea level at high latitudes; hence the saddle shape of the arctic-alpine zones.

Below. The equatorial submergence illustrated by depth zones of temperature in the Pacific. After Reid 1965.

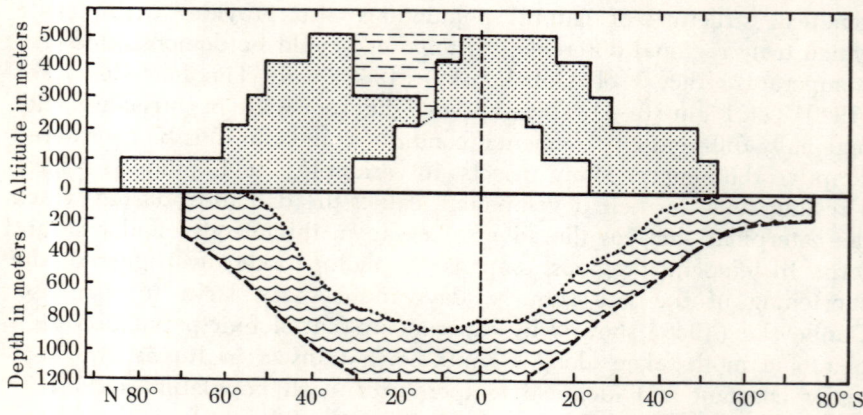

Figure 3-15 Relations of altitudinal and depth distribution with latitude II.

Upper scale. Altitudinal distribution of the butterfly genus *Colias* [Pieridae, Lepidopt.] in the New World. After Hovanitz 1958. Horizontal hatching indicates altitudinal zones where no mountain range of such height occurs.

Lower scale. Depth distribution of *Eukronia hamata* (Möbius) [Chaetognatha] in the Pacific. After Alvariño 1964.

others. The southernmost population needed the shortest days, while the northernmost sample diapaused even in nearly constant daylight—that is, in light which lasted from 20 to 24 hours of each day (Figure 3–16). The temperature requirements determining the onset of diapause were also zonally and geographically different in several butterfly species. The animals which exhibited such "physiological clines," or even physiological subspeciation, were morphologically uniform. Danilevskii (*l.c.*) points out that the same species possesses, in addition to these intraspecifically *variable* adaptations, remarkably *constant* environmental requirements, such as the temperature requirements of growth rate in larvae or the frost resistance of the diapausing stages, presumedly as responses to the relative uniformity of the corresponding environmental conditions.

Research in population ecology provides increasing evidence that ecological tolerances may also shift within a single population, depending on selective forces that vary from generation to generation (Wellington 1960, Laux 1962).

The results of these researches shed new light upon the adaptational structure of species and its geographical connotations. Expressions like

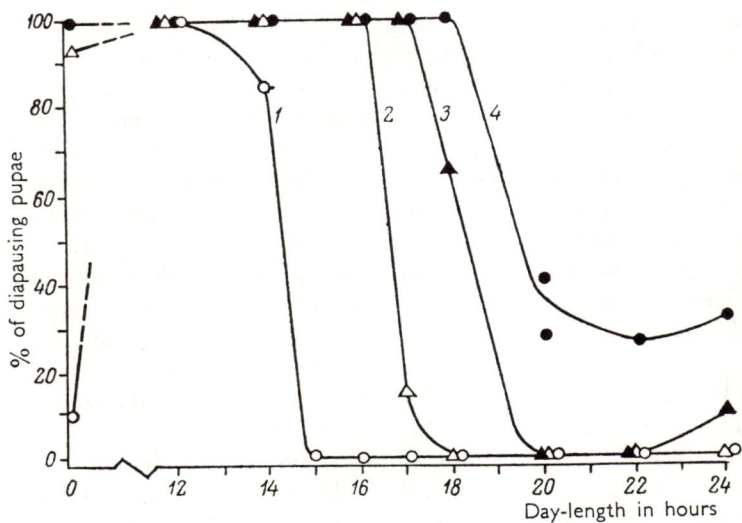

Figure 3-16 Regional variation of environmental requirements.
Geographic variation of environmental timing mechanism in *Acronycta rumicis* L. [Noctuidae, Lepidopt.]. Populations: 1. Abkhazian (43° N); 2. Belgorod (50° N); 3. Vitebak (55° N); 4. Leningrad (60° N). From Danilevskii 1965. (*Courtesy of Oliver and Boyd Ltd., Publishers, Edinburgh*)

stenotopic and *eurytopic* originated as empirical concepts of the statically thinking autecologists. We have already amended these concepts with a third kind of valency known as *amphitopy* (Figure 3–4). Yet these concepts have not yet been translated into fully satisfying terms of adaptational genetics (*cf.* Carson 1959). *The zoogeographer has therefore to use these terms with caution and care.*

THE INTRINSIC FACTORS OF EXISTENCE

A species' requirements, as opposed to its tolerances, indicate that the species shows *active preference* in a choice of several factors or intensities. The test of the salt preference of *Bembidion aeneum* (p. 106) is an example of this phenomenon. Ecologists have accumulated a large body of evidence showing that insects actively select certain combinations of habitat factors. Shelford (1911), who was among the first to demonstrate this fact with his beautiful treatise on tiger beetles, showed that the female searches at random until she finds the right environment for oviposition—in moist, warm, shaded, porous, and well-drained soil. The importance of habitat selection as an active, behavioral phenomenon was also stressed by Grinnell (1917a), Stantchinsky (1923), Dice (1931), Lack (1933, 1937), Miller (1942), and Errington (1943), among others, and is now the common knowledge of the zoologist.

Habitat selection is a behavioral mechanism, and as complicated as the better-known internal functional mechanisms such as the clotting of blood, thermoregulation, or respiration. When and how a functional mechanism is released will depend upon the predisposition of the organism and upon environmental clues. The animal will accept only a habitat which offers a combination of stimuli in a certain minimum intensity. The quality and quantity of necessary stimuli depends on inheritance and on the degree of internal motivation which is prerequisite of the search for proper habitat. A special type of learning, imprinting—specific learning during a restricted, usually early, life period of the animal—often plays a very important role in habitat selection. Without experimental proof, it is often difficult to decide which particular feature or factor of a habitat affects the animal's choice. It is even more difficult to pinpoint the minimum factor or factors which repel it. In England, the tree pipit *Anthus trivalis* L. [Motacillidae, Aves] inhabits cutover brushy areas or woodland edges. These habitats evidently provide—in addition to the general favorability of the physical environment which is well within the general breeding range of the bird—sheltered nest sites, feeding areas on grassy open patches, and song posts on high branches and treetops which the male require for patrolling and advertising his territory by song (Lack 1937). Based on his field observations, Lack had concluded that a high perch was an essential requisite for the pipit. He later found it necessary to alter his conclusion when he discovered that tree pipits could be found in an area devoid of trees but one in which there was an artificial song perch available—for example, a telegraph pole. The true, simplified, almost symbolic nature of the minimum requirement became apparent: the settling bird did not need a tree, as we recognize a tree—complete with trunk, crown, branches, leaves, and roots. Any habitat feature that was high enough to overlook the brushy territory and on which the bird could gain a foothold qualified as a song post.

The essential or minimum factors of the environment do not often serve as releasing stimuli (cues) of the habitat selection mechanism. These factors are often not perceivable, or they are not present, at the time a habitat is selected. Baker's widely used concept of ultimate and proximate factors* applies to this situation: the environmental releasers of settling are the proximate factors of the habitat selection mechanism. As an example, Klomp (1954) showed that early in the spring when the lapwing *Vanellus vanellus* (L.) [Charadriidae, Aves] arrives to breed in the Netherlands, it chooses barren areas which have the brown color of naked earth, and avoids green meadows. Another shorebird, the godwit *Limosa limosa* (L.) [Scolopacidae] behaves in the opposite way: it chooses green meadow habitats and avoids the barren, brown areas. The color of the ground is the most apparent criterion for bird and bird watcher (both being endowed with color vision), and it is likely that color is the chief proximate factor in godwit and lapwing habitat selection. The biological reason, however, manifests itself when the clutches of lapwings and godwits hatch and the chicks start to hunt for food in the meadows and fields. By then, the brown areas grow short grass, and the lapwing chicks crisscross them freely. The fields which had a good verdure early in the spring grow up in tall grass and tussocks, impenetrable for the lapwing. The godwit chick has longer legs; these and several other anatomical adaptations allow it to move with ease. Thus, an ultimate reason for the selection of different nesting habitat by these birds is the different adaptation of their locomotor organs.

*Baker (1938) analyzed the timing of the breeding season. Certain environmental factors ultimately determine the best time for raising young, e.g., food has to be most abundant when the young become fully grown and independent of the parents. This, and similar *ultimate factors* are the reasons for evolutionary adaptation. The bird adapts the breeding season to the food situation, but he will not know in the spring when to lay eggs unless some actual factor—for example, temperature or day length—is utilized to trigger the egg-laying behavior of the bird. Baker calls such an environmental releaser a *proximate factor*.

Although this example leads into the many other *intrinsic factors of colonization*—locomotor adaptations and the structural, physiological, and behavioral properties of the animal—we shall return to the analysis of habitat selection because it is a paramount intrinsic factor and because it exemplifies the behavioral mechanisms that are also used in the choice of food and feeding site, mating, competition, and practically all other aspects of the life of an animal. Such mechanisms have certain innate elements which assure that all members of a species behave basically the same way. The learned part of the behavior increases an animal's capacity for adaptation to given surroundings. The black-headed gull *Larus ridibundus* L. [Laridae, Aves] recently settled on the southern coasts of Finland. Originally a colonial marsh bird of old reed beds, it established its first colonies in freshwater marshes and reed-covered bays of the Baltic Sea. But as soon as the reedy marshes became overpopulated, these gulls started to build on neighboring sea rocks as well, an unusual site for marsh inhabitants. Faithfulness to the colony overcame the traditional choice of nesting habitat. But the choice of nest material could not be altered; the choice is made apparently in an innate way, and the gulls chose long, dangling, yellow reed sticks for nest building (Ahlquist and Fabricius 1938). The huge yellow nests are conspicuous on the barren, grayish granite rocks, and are therefore subject to predation by crows. Palmgren (1949) commented that since the selection of habitat is learned but the selection of nesting material is innate, the shift into this new habitat cannot be successful until evolution brings about a change toward better concealment of the nest.

The releasing mechanism of a behavior pattern consists partly of the internal readiness or motivation to perform (age, hormonal, and nervous elements play the chief roles internally) and partly of the summation of certain important environmental stimuli. One or more of these stimuli may be absolute requisites—for example, a nesting hole for a wasp or bird of a hole-breeding species, or a pliable soil structure for a pocket gopher. Other requisites stimulate the behavioral mechanism (the Innate Releasing Mechanism [sensu Lorenz] of habitat selection) in an additive way, i.e., their effect acts in combination (cf. Tinbergen 1951). If the internal motivation is high, less external stimulation by these factors is needed—beyond the absolute requisites—to release the behavior patterns of settling on the area. This delicate balance of the releasing mechanism assures that in an emergency an inferior habitat will also be accepted, within certain limits. Colonists often meet this situation, as we have seen in the case of the black-headed gull. Bergman (1946) gives another interesting example. Late in the spring an arctic tern *Sterna paradisea* Pontopp. [Laridae, Aves] alighted in the middle of a busy thoroughfare in Helsinki, where it started to perform the stances that indicate, to bird watcher and female tern alike, his choice of a spot for nesting. This occurred on a bridge connecting two parts of the coastal city. The shiny, gray asphalt of the bridge surrounded by the blue-green water presented the necessary minimum of external stimuli for an apparently highly motivated prospective settler.

Schneider (1952) observed the habitat selection of the June beetle *Melolontha melolontha* L. [Scarabaeidae, Coleopt.], and found that the beetles, on emerging in the spring after hibernation, first fly in vague circles; later they fly toward the highest part of the horizon—for example, a tall stand of conifers (Figure 3–17). When they reach this and it does not present further stimuli corresponding to their proper feeding habitat (deciduous or mixed woods), they circle low again and approach the nearest silhouette. If the latter is mixed or deciduous wood, they alight and start to feed on the leaves.

Certain insects possess a refined sensory mechanism which enables them to find the proper place for oviposition. Well-known examples are flesh flies, fruit flies, and carcass-eating beetles; in others, an *element of chance* also plays a role.

The tick *Ixodes ricinus* (L.) [Ixodidae, Acarina] changes domicile several times during its lifetime, and most of these changes are hazardous for the tick because of a lack of active habitat selection (Milne 1950). After the tick larva is successful in obtaining a meal of blood from a sheep which happened to pick it up, the gorged animal drops to the ground wherever the sheep happens to be. The tick then burrows in

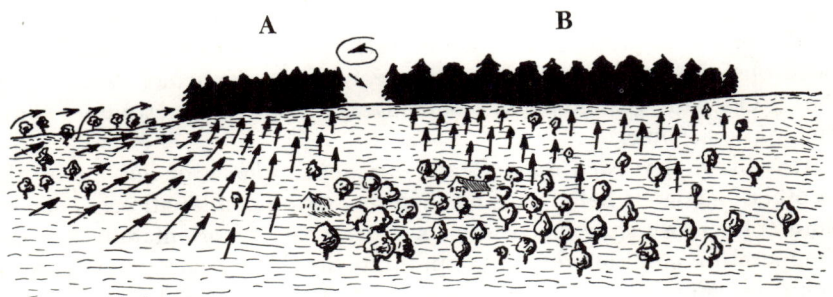

Figure 3-17 Habitat selection mechanisms.
Oriented flight of newly hatched *Melolontha melolontha* L. [Scarabaeidae, Coleopt.]. The june beetles crawl out from the ground (left) and aim first at the highest feature on the horizon (A), a dark spruce forest. There they circle above its highest point, and then dive deep and reorient toward the mixed forest (B). After Schneider 1952.

the surface layer of the soil or duff and spends the following year there. Next year, the tick climbs atop the vegetation and waits to be picked up again; if this happens, the whole process is repeated. In the following year, the nymph tries to obtain its third meal and to mate on the sheep. The mated female again drops to the ground, this time to develop her eggs. At all times, chance determines whether the insect is dropped in a moist, sheltered place where it can survive.

Wellington (1964) has studied the dispersional mechanism of the western tent caterpillar *Malacosoma pluviale* (Dyar) [Lasiocampidae, Lepidopt.]. The pregnant female moth undertakes a flight before egg-laying. If it encounters tall trees on its way, it alights and lays on them; otherwise, it will continue to fly until overcome by fatigue. The dispersion from a heavily infested area depends on—in addition to other, intrinsic factors—the height of obstructing vegetation.

On the other hand, Ford (1964) describes a mosaic of two distinct habitats on one of the small Scilly Islands off the southwest tip of England. The meadow brown butterfly *Maniola jurtina* L. [Satyridae, Lepidopt.] lives on the brushy meadow. Genecological studies showed that small populations, isolated by unsuitable lawnlike, turfy pasture, acquired a different genetic constitution. This isolation was achieved by avoidance behavior.

One could see the butterflies setting out in numbers over the two "lawns" from either end and, finding them continuously inhospitable, turning back as over the sea, after about ten yards; in the middle none were to be seen. Indeed here as elsewhere we find that 100 yards or less of unsuitable terrain proves almost a complete barrier to them, though they will constantly traverse as great a distance in a few minutes within one of their colonies. (Ford 1964, p. 60.)

Macan (1961) reports that active selection of oviposition habitat has been observed in four amphibious insect orders: Ephemeroptera (mayflies), Odonata (dragonflies), Culicidae of Diptera (mosquitoes), and Heteroptera (bugs).

It is likely that *each animal exerts an active choice of habitat* by this or other analogous behavioral mechanisms, using combinations of several different sensory stimuli that are its main means of orientation in space. Temperature and humidity are important stimuli for insects. Temperature influences migratory birds. Insects use visual (Tinbergen 1951), chemical and visual (von Frisch 1951), or chemical and tactile (Francia (1965)) stimuli. Fish use olfactory, tactile, and visual stimuli (Tinbergen 1951, Fabricius and Gustafson 1958). Birds mainly use visual (Palmgren 1938,

Bergman 1946, Udvardy 1956), kinaesthetic (Palmgren 1936), and acoustic stimuli. Mammals use olfactory, acoustic, visual, and tactile stimuli (McCabe and Blanchard 1950, Harris 1952).

Since knowledge of these behavioral aspects of settling ecology is of relatively recent origin, and since the subject is not yet universally studied, we do not have extensive information on their incidence. Active selection of habitat necessitates, to a certain extent, dispersal and, in this case at least, limited locomotion to find the proper microhabitat. Most animals possess both the locomotive properties and the basic behavioral reactions (kineses, taxes, internal releasing mechanisms, and so on), which are the components of a habitat selection mechanism. With respect to most plants, the depositing of seed or spores in a habitat suitable for its establishment and growth is usually the result of chance. Chance also determines the selection of suitable habitat for progenies of sessile (aquatic) invertebrates, when these are carried for long distances by currents. In such cases, massive reproduction ensures the survival of the species. Vertebrates, on the other end of the evolutionary scale, all possess habitat selection mechanisms; the number of their offspring is very limited, and they could ill afford to lose them in the wrong habitat (the aquatic and amphibious classes generally lay a large number of eggs but give no parental care). The roles of learning and tradition make their habitat selection especially adaptive. Land arthropods apparently have various habitat selection mechanisms, which keep pace with their locomotor abilities. Even the slow land mollusks and worms are known to select the microhabitat into which they lay their eggs, though it is usually not far from the birthplace of the parent.

The *timing of life processes* is another intrinsic characteristic of a species and also a result of adaptive evolutionary processes. For a colonist, general adaptations are more useful than specialized ones. These may fit many kinds of situations encountered by the pioneers. Certain earthworms are excellent colonizers; Omodeo (1951) attributes this partly to the fact that in these species the onset of an inactive torpid period is released by environmental stimuli. In this way they easily adapt their life cycle to the seasonal conditions of any new environment. In other earthworms, the onset of torpidity follows an endogeneous rhythm, which is specifically and rigidly adapted to the conditions prevailing in their ancestral range. These earthworms are poor colonizers, for they cannot adapt to the seasonal changes in the new areas to which they might disperse.

Many other attributes of a species play important roles in settling, since a limiting factor does not influence the animal in a strictly mechanical or physical way. The adverse or limiting factors are put into play against an organism, a sensitive and reacting body. Since the effect of each factor changes in space and time, the animal is exposed to it in a microhabitat, surrounded by a microclimate, and is often able to avoid the detrimental impact either by immediate and innate response or by a response modified by learning, i.e., by responsive behavior. Since in a bisexually reproducing population no two individuals are exactly alike, we can postulate that the reaction of no two individuals will be the same to the impact of the limiting factors. This question will be dealt with in further detail in the subsequent discussion of the ecology of settling populations. It suffices now to state that *the impact of every environmental factor depends on the structural and functional responses and on the range of adaptiveness of such responses in the organism.*

REPRODUCTION OF THE SETTLING ANIMAL

An animal may eke out a living in an inferior habitat but not be able to reproduce there because of its special requirements for mating, oviposition or of its growing offspring. This fact is very important with founders. Plant geography shows that the degree of habitat suitability can be grouped according to the stage of development a seed reaches in them: (1) it does not germinate at all in a totally unsuitable habitat; (2) it germinates but dies or barely survives temporarily in an unfavorable habitat; (3) it grows to healthy adulthood but does not bloom; (4) it blooms but produces no mature seed. In habitats (3) and (4), individuals or stands can be established which could even reproduce vegetatively; in (5) the most favorable habitat, the seedling grows and produces mature seed.

For a long time I thought that these habitat categories would not apply to mobile animals, for they could move out of an unfavorable habitat. Yet the same or similar habitat categories are occupied not only by pioneering animals but also by those individuals or breeding assemblages that are widening the distributional area as a consequence of crowding at their place of birth. These animals are often not able to move far in search of optimal habitat and are forced to settle down under quite adverse conditions. Recent research has clarified the role of inferior and optimal breeding habitats in land vertebrates. For example, Carrick (1963) studied the Australian magpie *Gymnorhina tibicen* (Latham) [Cracticidae, Aves] between 1955 and 1962. This bird lives on open savanna woodland, and forms social territorial groups of from 2 to 10 individuals. The group defends its territory as a team; many groups, consisting of various age classes, are compelled to live on inferior territories where trees, needed for shelter, roosting, and nesting, are scarce. The habitats, as well as the kinds of group, form a graded series. Carrick was able to distinguish the following main categories.

Choice habitat, with adequate or surplus amount of requisites, was occupied by *permanent groups*. These did most of the successful breeding.

Inferior habitat, deficient in either cover (too few trees but plenty of grassland for feeding) or in feeding area (open woodland with inadequate pasture for feeding at all seasons), was occupied by *marginal groups* of adults. These attempted to breed but rarely succeeded in raising their young.

Separate feeding areas in the open and nesting-roosting areas among trees were maintained by *mobile groups* during the breeding season—that is, by adult animals which could not secure better habitat at the onset of the reproductive cycle. These groups occasionally bred but always failed, usually at an early stage.

Still more inferior habitat—feeding and roosting areas widely separated, with the latter unsuitable for nesting—was held by *open groups* of nonterritorial, nonbreeding adults.

Wide-open pastures, subject to occasional freezing, were occupied by slightly nomadic *flocks* of birds of all ages that roosted in woods. Even this inferior habitat contained enough food; but in adverse weather the flock was compelled to feed in wet areas, where the birds became diseased by aspergillosis, a fungus of the lung. A contagious disease also killed large numbers of these birds but did not cause a single death among the territorial individuals, owing to lack of contact. This example demonstrates that where the choice habitat is fully occupied, socially inferior individuals are forced to live in inferior habitats where mortality is high and where breeding is either unsuccessful or cannot be attempted because of some environmental deficiency. The individuals or social groups seek to improve their status and move on into a better habitat as soon as it becomes available.

The temporary occupation of habitats not entirely suitable may be an important means of carrying migratory insects, fish, birds, and mammals through a time during which the breeding habitat is temporarily adverse. It may also help colonists survive until they find a more suitable place for permanent settling and for breeding. Where this is a recurrent phenomenon, and where the temporary occupants cannot be classified as migrants (since they hardly ever return to their place of origin), the phenomenon has a certain significance. These animals will have an impact on the ecosystem, for once conditions improve, the establishment of breeders may be facilitated by the previous presence of these temporary visitors (see further discussion in Chapter 4). Some well-known examples are the irruptions of locusts and butterflies, which carry swarms of adult insects beyond their ecologically suitable range; these expatriates may even breed, but they cannot establish permanent populations.

Reproductive and Cytogenetic Systems that Facilitate Colonizing of New Areas

The first condition of successful reproduction involves an intrinsic factor: the pioneer or *pioneers must be able to propagate themselves*. In the most widespread reproductive system—sexual *reproduction*—there must occur the simultaneous dispersal or settling of at least one male and one female or the settling of a pregnant female if the sexes are separate. In *asexual reproduction* and *apomixis*, one individual alone is enough to found a new colony or population. Asexual *vegetative reproduction* and sexual but monoecious and self-fertilizing reproduction greatly enhance pioneering in plants (Stebbins 1950, Baker 1953). Many forms of vegetative reproduction are known among the lower, aquatic phyla of animals. Asexual reproduction—that is, *parthenogenesis*—is rare among animals but occurs in every major phylum, including the vertebrates; it is most widely known for its benefits in evolution as well as in rapid multiplication in seasonally suitable areas (Suomalainen 1950, Mayr 1963). *Hermaphroditism* is also known among terrestrial animals—for example, in mollusks and olygochete earthworms. In this condition, self-fertilization is suspected but not universally accepted as having been proved; therefore, hermaphroditism alone does not differ from bisexual reproduction in its distributional consequences.

It is evident that any form of asexual reproduction must have an advantage for the dissemination of the species. If other conditions of dispersal and settling are found, single individuals are able to start colonies and rapidly colonize new areas. Each successful offspring of a parthenogenetic individual is likewise a female, and each individual thus gives rise to twice as many potential reproducers as in a bisexual form. Then too, pioneers are usually scarce, and in this instance an individual is able to colonize alone. Hermaphroditic *and* self-fertilizing animals share these advantages.

The cytogenetic condition *polyploidy* (multiplication of the basic chromosome number) often is the result of parthenogenetic reproduction. As in plants, where polyploidy is a common phenomenon (Löve and Löve 1943, Stebbins 1950), polyploid animal forms often have a distribution area different from that of the diploid form. Zoologists, as well as the above-mentioned botanists, believe (Suomalainen 1962) that this is the result of the broader ecological tolerances and requirements of the polyploid form. Vandel (1960 and earlier) found that the triploid, giant parthenogenetic females of the land isopod *Trichoniscus pusillus*

Brandt [Trichoniscidae] are distributed to habitats intolerable to the diploid form. Lindroth (1954) proved experimentally that this is true of the likewise parthenogenetic and polyploid weevil *Otiorrhynchus dubius* Ström [Curculionidae, Coleopt.].

Seiler (1961) made a monumental study of the parthenogenetic conditions in the bagworm moth *Solenobia triquetrella* F.R. [Psychidae, Lepidopt.]. This insect occurs at certain places in Switzerland and the Austrian Alps, in bisexual form. It also occurs at other localities, usually close to the range of the bisexual colonies, where diploid parthenogenetic populations are found. The rest of Central Europe is populated by a third, tetraploid parthenogenetic variety. The genetical survey showed that diploid parthenogenetic females rarely occur in a bisexual population—that is, even if a female is not fertilized by a male, she will still lay eggs, which are usually either not viable or will produce caterpillars which die shortly after hatching. In rare instances, however, one of these caterpillars may mature into a parthenogenetic female moth and become the founder of a diploid parthenogenetic population. This may result—also rarely—in a tetraploid individual. In a dense or crowded bisexual population, the chances are that every female will be fertilized soon after hatching, and it is not likely that parthenogenesis will occur naturally. Seiler and his collaborators made a thorough collection from all parts of Switzerland and raised large numbers of this moth to ascertain their cytogenetical constitution. Mapping the distributions, and comparing them with the well-known position of the late Pleistocene ice sheets that almost completely covered the Swiss Alps, Seiler found that the bisexual forms live on or near high peaks that once stood out, as *nunataks*, above the Würm (Wisconsin) ice field. The other form—diploid parthenogenetic—prevails in the lower mountains and on the foothills that first emerged after the regression of the glaciers. The valleys, which were the last areas to be freed of their glacial coverings, have pure tetraploid populations—the form that arises last in the experimentally proved time sequence. In the southern foothill country, the populations are almost pure tetraploids except for a small relict area on the Italian border which is on an unglaciated nunatak of the ice age. Seiler infers that, as the climate improved and more and more area became available, the slowly spreading (Figure 2–22) pioneers formed ever thinning populations, establishing the conditions for extinction of the bisexual forms but favoring the chance occurrence of parthenogenesis. The present bisexual populations, according to this theory, are relicts that survived because their ancestors spread upward following the dwindling ice cape of the nunatak peaks; thus the populations became denser rather than thinner, and therefore sexual reproduction could survive (Figure 3–18). One would wish to

126 *Ecology of Colonization*

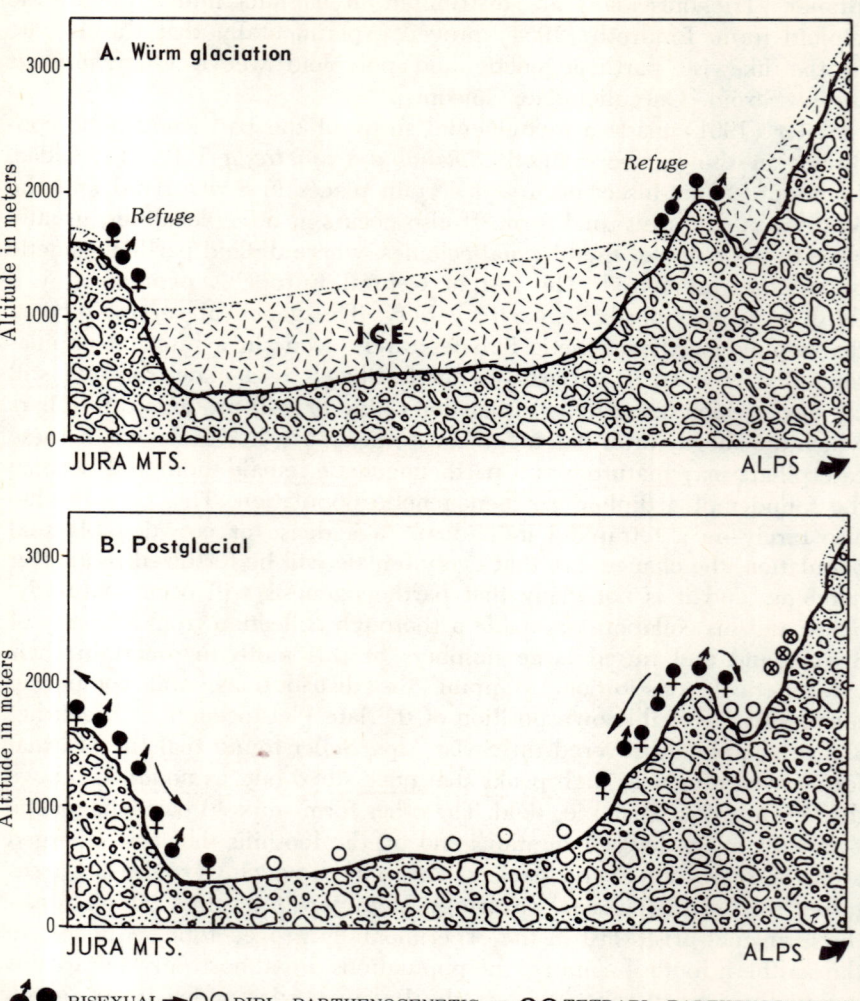

Figure 13-8   Geographic parthenogenesis I.
A. Schematic cross section (NW-SE) of northwestern Switzerland during the Würm glaciation with postulated *nunataks* serving as refuges for *Solenobia triquetrella*.
B. Postulated postglacial sequence of the origin (*below*) and expansion of geographic populations of *S. triquetrella* along the same transect. The Alps, to the right (west) of this transect, were totally glaciated, and today they contain tetraploid parthenogenetic populations.
Below B. Time sequence of appearance of experimental parthenogenetic populations of *S. triquetrella*.
From Seiler 1961 and Seiler *in litt.* 1964.

know more about the ecological—especially the population ecological—side of this fascinating insect.

The case of S. *triquetrella* is paralleled by that of many other organisms. There is an interesting incidence of parthenogenesis in Europe from the south to the more recently settled areas of the north, e.g., in millipedes (Vandel 1928, Palmén 1949*b*) and curculionid (Suomalainen 1947, 1953) and chrysomelid beetles (Suomalainen 1965). Vandel (1928), who first realized its significance in geographic distribution, called this reproductive condition *geographic parthenogenesis* (Figure 3–19). The study of adventives provides evidence that parthenogenetic forms (which are often polyploidic as well) have superior dispersal ability.

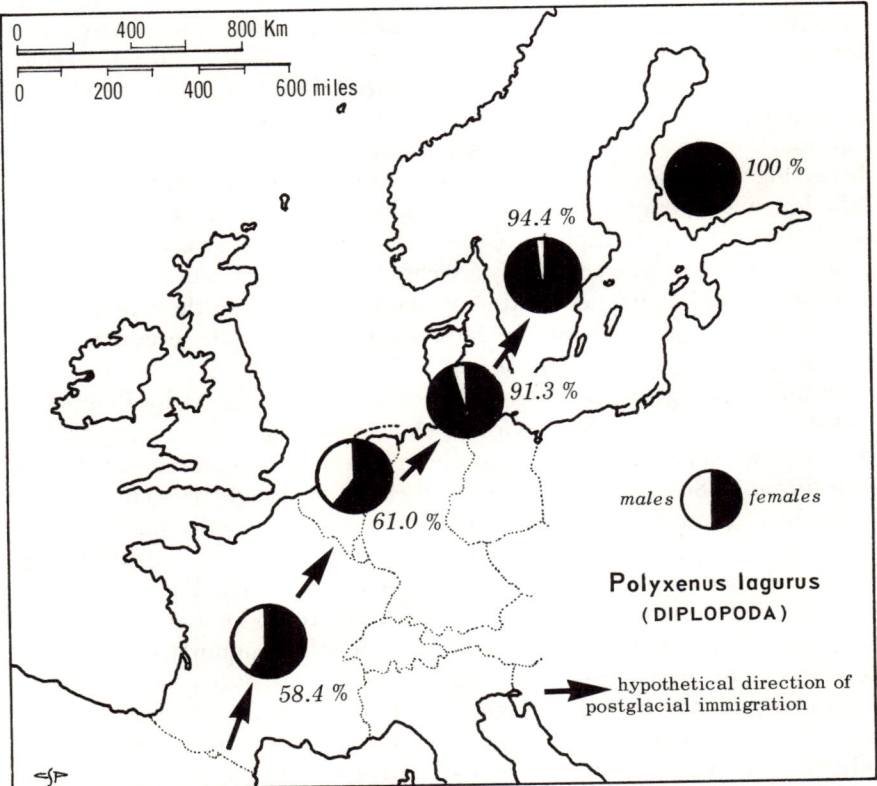

Figure 3-19 Geographic parthenogenesis II.
Percentage of females in collections of the millipede *Polyxenus lagurus* L. and postulated postglacial expansion routes to northwest Europe. From Palmén 1949*b*.

Lindroth (1957) investigated the biological characteristics of those English beetles that were introduced to North America in the ballast of sailing vessels. He found that all the parthenogenetic beetles that occur around the English ports became well established in the New World. Omodeo (1951) thinks that polyploidy, parthenogenetically originated, furthered the wide distribution of earthworms. Many terrestrial and widespread earthworms are suspected or proved adventives (Gates 1966); nevertheless, their reproductive system could be helpful in dispersal.

A special mode of reproduction—alternating bisexual and parthenogenetic reproduction in alternating or subsequent generations—is called *cyclic parthenogenesis* (Suomalainen 1962), and is known among the gall wasps (Hymenoptera) and the aphids. In the latter, as discussed in Chapter 2, this condition is paralleled by dimorphy of the wing, thus strengthening the conviction that the parthenogenetic reproductive system is important in rapid population increase as well as in pioneering.

There is a possibility, not yet investigated, that certain kinds of parthenogenetically reproducing animals may be able to revert to sexual reproduction. In such cases, as well as in those where the more slowly dispersing bisexual forms would eventually catch up with the established parthenogenetic pioneers,* an additional advantage may be postulated. The rapidly multiplying (asexual) pioneer forms may prepare the ground for these latecomers. Establishment in a new biotic community requires mutual adjustment and new adaptations of both the old and new members. This process can be facilitated by the early presence of a parthenogenetic pioneer population.

A similar change in the mode of reproduction was found by Longhurst (1954, 1955) in a freshwater Crustacean group, the Notostraca (Phyllopoda). The North African and southern European population of *Triops cancriformis* (Schaeffer) is bisexual; there is an irregular occurrence of males in Central Europe; hermaphroditism prevails in the north. Longhurst appears to have proved that hermaphrodites are self-fertilizing. A related species that lives on both sides of the North Pacific also has bisexual and hermaphroditic populations; this reproductive system may be termed *geographic hermaphroditism*. Though Longhurst is very cautious about his hypothesis of the advantages of this system in postglacial pioneering, it may fit well into the series of geographically polymorphic reproductive systems under discussion here and which are related to other polymorphic dispersional systems (Table 3, p. 73).

*Reminiscent of the flightless morphs that eventually replaced macropterous, flying pioneers in carabid beetles. See p. 61.

Other Intrinsic Conditions Influencing Pioneering

Intrinsic properties within the same category influence the success of pioneering bisexual animals. A single gravid or fertilized female may become the founder of a new colony if the circumstances of its life history enable it to travel while pregnant. However, this does not always occur; where the dispersing forms are not in a reproductive condition, chance must deposit both sexes in the same place at the same time. Wind-dispersed termites, for example, have little chance of colonizing because their mating takes place *after* the nuptial flight and thus the sexes have to travel together (Emerson 1952, 1955). In the Nearctic ant species *Pheidole sitarches* Wheeler [Formicidae, Hymenopt.], both habitat selection and mating are so specific as to make colonization very difficult. The males choose a special kind of habitat—barren ground amidst vegetation—above which they perform the swarm flight that attracts the females (Wilson 1959). The nuptial flight has a stereotyped pattern: the females fly through the swerving male swarm and are mounted in the air. However, the mounted pair copulates only on the ground beneath the swarm. Here the female remains and lays her eggs; she does not fly any more. Thus, the distribution area of the species can be advanced only over a short distance in each generation and settling by single founders is almost impossible.

Some behavior patterns concerned with host selection have been investigated for the wood-boring *Trypodendron lineatum* (Olivier), an ambrosia beetle of the family Scolytidae (Graham 1960, Francia 1965, Graham *in litt.* 1966). After hibernation in the forest duff, the insect emerges and takes to flight under the influence of skylight. The swallowing of air during flight appears to provide a mechanism which changes the relative threshold of response to light and host odor so that a response to the latter becomes manifest in the presence of the former. An anemotactic response prompted by host odor apparently guides the beetle to a host log where mating and tunneling in the wood take place. The necessity of the exact sequence—light, phototactic movement, overhead light, flight, air swallowed into stomach, olfactory stimulus, anemotactic response, alighting at the source of odor—has been verified experimentally. Dispersal capability in this example is an attribute of the adult animal before reproduction, and consists of a complicated behavioral mechanism which directs the animal to an environmental situation favorable to the propagation of progeny.

The timing and mode of reproduction are the intrinsic properties which might either limit or further settling. Where the favorable season is short, animals with slow growth rates are still able to become established if they can prolong their development over several seasons. It is a general characteristic of many insects that they pass several winters in the larval state in such climatic conditions. The moth *Dendrolimus pini* L. [Lasiocampidae, Lepidopt.] develops in Central Europe in one year, in northern Europe in two summers. The Mediterranean hawk-moth *Chaerocampa celerio* L. [Sphyngidae, Lepidopt.] produces from six to ten generations per year in North Africa, three in Jerusalem, and two in Nice. It does not occur farther north because the season is not long enough for development of one generation and because the species does not possess the adaptive characteristic which allows the postponement of its metamorphosis until the following summer (Bodenheimer 1930, fide Schwerdtfeger 1963). Land reptiles are able to shorten the required season by evolving viviparity—the eggs do not need the sun's heat to incubate but develop in the body of the female, and the young are born alive. Of the ten snakes and lizards which reach farthest north in Europe and in western North America (Darlington 1957), seven are viviparous (cf. also Neill 1964).

Food Requirements of the Young as a Condition of Settling

Often the place for oviposition or raising the young has very special requisites; thus reproduction is limited by different sets of factors from those that limit the existence of pioneering individuals. The special food requirements of many monophagous insect larvae are too well known to require documentation. Insects that are monophagous in some state of their life history are strictly limited to the distribution of a particular food, plant, or animal; other limiting factors further restrict their occurrence. Ross (1965) describes the difficulties some species encounter without a proper "host." The European satin moth *Stilpnotia salicis* (L.) [Lymantriidae, Lepidopt.] feeds on the Lombardy poplar *Populus italica* (Duroi) Moench [Salicidae]. Introduced Lombardy poplars thrive in the Pacific Northwest of North America, where the satin moth was accidentally introduced about 1922. In a few years the Lombardy poplar was nearly exterminated in the lower Fraser Valley of British Columbia, and the first very successful moth population dwindled. However, within a few years their numbers again rose to serious proportions, now threatening the cottonwood *Populus trichocarpa* T. & G. Laboratory work explained the adaptation phenomenon. In the absence of the original host, the moth lays its eggs on the closely related native tree. The first instar larvae feed on the new host but develop intestinal troubles and die in great numbers. The successive instars show decreased mortality; but the pupae develop much trouble, again resulting in high mortality.

Perhaps one of a thousand larvae reach pupation; but the few emerging new moths again lay on the new host, and the following generation of caterpillars develop vigorously, fully adapted to the new kind of food. This example is one of many in which man altered the distribution of the food plant and thereby interfered with the natural course of events. It also demonstrates the additional difficulties a pioneer species would encounter if the food organism, whether old or new, occurred more sparsely.

Special Habitat for the Young

In those species where reproduction takes place at special sites, the absence of such sites may be the factor that limits the reproduction of the pioneers. Cowles (1930, 1959) observed that the southern form of the Nile monitor lizard *Varanus niloticus* (L.) [Varanidae, Squamata] lays its eggs in a special place, in contrast to most larger tropical reptiles, which lay in warm sand. These monitor lizards scratch holes on the sides of huge termite hills of *Nasutitermes trinerviformis* (Holmgren) [Termitidae, Isopt.]. Once the eggs are laid, they are in perfect safety inside the termitarium: the termites surround the cavity with a new wall, and the eggs enjoy a favorable incubation humidity and temperature. The lizard hatchlings soften or dissolve the hard wall with the water remaining in the eggshell after hatching, and break through to the outside world. This species of lizard thus cannot reproduce without the termite hills. An analogous situation, noted by von Hagen (1938) and studied by Hardy (1963), exists between *Nasutitermes nigriceps* (Heldeman), a Nearctic relative of *N. trinerviformis,* and the orange-fronted parakeet *Aratinga canicularis* (L.) [Psittacidae, Aves]. This parakeet excavates its nesting cavity in the termitarium, which is built in the branches of a tree. The distribution of the bird does not exceed the range of the insect and even seems to be restricted by it (Hardy *in litt.*).

J. Marcus Buchanan (*in litt.*) has pointed out that the northern distributional limit in Mexico of two birds, the motmot *Momotus mexicanus* Swainson [Momotidae] and the green kingfisher *Chloroceryle americana* (Gmelin) [Alcedinidae], is limited in southeastern Sonora by the Río Mayo. On the north side of the river, the habitat does not change appreciably. Indeed, the kingfisher is known as an accidental winter visitor north to Arizona, but the creek beds there are gorges in hard limestone rock in which these birds cannot dig their nesting tunnels. This coincidence warrants further study.

The Combined Effect of Extrinsic Factors Influencing Reproduction and Thereby Colonization

The effect of environmental factors on reproduction are not always as simple as in the above-cited instances. One example will suffice: the effect of precipitation during the different seasons upon the survival and reproduction of grouse [Tetraonidae, Aves], studied by Siivonen (1962) and others in Finland and by Marcström (1956) in Sweden.

During the summer and early fall, the capercaillie *Tetrao urogallus* L., black grouse *Lyrurus tetrix* (L.), and the ptarmigan *Lagopus* sp. forage on the ground for insects, seeds, and greenery. Grit in their gizzard helps them to grind and digest this diet. When the late autumn snowfall covers the ground, the grouse flies up into the coniferous trees and subsists on a meager diet of needles, of which there are plenty. The ptarmigan picks the buds and end twigs of the willow and birch shrubs in its tundra and woodland habitat. For grinding this hard, fibrous foodstuff, the birds need a supply of large grit with which they usually fill their gizzard during the fall. In years when the first snow comes unseasonably early, they are not able to gather enough grit; and so they often perish by late winter, even though the weather may not have been exceptionally severe. Thus, a population decline follows winters with early snow.

A snow cover of too long a duration has an indirect effect on the reproduction of the grouse. Studies of the weight cycle showed that the hens reach their maximum weight just before egg laying, as is to be expected. Surplus weight indicates fat deposits, needed for the extra energy supply required in the production of eggs and for the hardships of the incubation period (after which they reach a low point of their body weight cycle). Browsing on needles during the short late-winter days is not very nutritious. Vitamin- and energy-rich food again becomes available only when the snow melts and the berries of the previous autumn emerge and when spring greenery appears. Capercaillie and black grouse both live in the Central European Alps and Scandinavia; in Austria, at 48° N latitude, they are able to fatten during two months between the thawing of the snow and egg laying. In southern Finland, at about 60° N, this period ordinarily lasts only one month. In northern Finland (66° N), there are only three weeks, since the season is shortened at both ends. The normal clutch size is correspondingly smaller in the north (Merikallio 1931). If the spring thaw is delayed or if new snow falls in March and April, there will be a catastrophic year for the grouse, with delayed and diminished clutches and underweight eggs and hens.

Cool, rainy summer days are also hazardous to the reproduction of the grouse. The capercaillie chick hatches with a large amount of the egg yolk (the embryonic food supply) still available. During the first couple of days of its independent life, the chick—not much different from the chicks of domestic poultry—explores its feeding environment. It possesses certain innate, inherited behavioral mechanisms whereby it pecks such small particles as gravel, grains of seed, tips of buds and grass, or even bits of eggshell or toenails of their clutch mates. Not all of these objects are edible; moreover, the edible particles vary according to season and locality within the distribution range of this grouse. The chick learns by trial and error what is edible. By the time the yolk supply is exhausted and the chick needs continuous foraging throughout the day, it has become an experienced feeder. If, however, those critical first days are cold and wet, the chicks huddle around the mother, who broods them most of the time; their thermoregulation is not perfected until much later. But, while thus brooded, their time is not spent in learning and apprenticeship in food getting; unexperienced and hungry, they are not able to feed quickly enough to gather enough energy for metabolism and growth. Easily chilled, they weaken and perish. These studies show that wet ground and rainy, cool summers limit the reproductive success of grouse and seriously weaken the whole population if such weather persists through several years (Figure 3–20).

This conclusion is strengthened when the thermal resistance and dis-

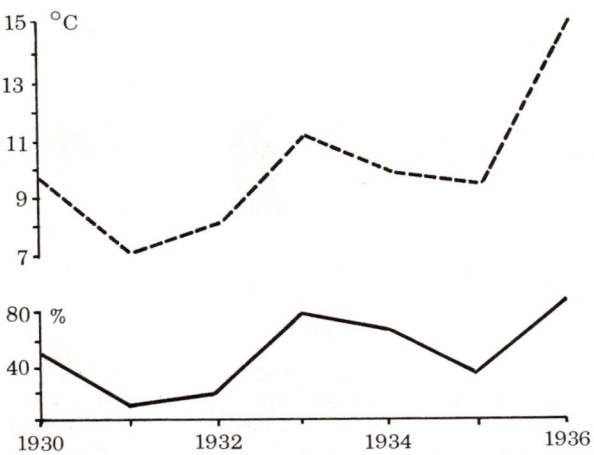

Figure 3-20 Temperature and reproductive success.
Relation of June temperatures and reproductive success of the capercaillie *Tetrao urogallus* L. *Solid line:* percentage of young grouse in the annual kill. *Stippled line:* average June temperatures. After Naumov 1955.

tribution range are compared, in the case of the *Carpodacus* finches of western North America (Salt 1952) and in the even more striking case of northern European ducks (Koskimies and Lahti 1964). Newly hatched ducklings of ten different species were tested for thermal resistance in southern Finland. The values so obtained are interpreted as an indication of cold-hardiness, and these cold-hardiness values are correlated with the latitudinal distribution of the species (Figure 3-21).

These examples remind us again of Liebig's Law of the Minimum as it actually affects distribution through the requirements of reproduction. Hesse (1924 (esp. p. 21), 1937) reformulated Liebig's law in the following way. "The continued presence of an animal in an environment depends on the developmental stage in which it has the least adaptability." The limiting factor—or the limiting combination of several factors—is different for each stage of the life cycle, but the stage which has the least adaptability is usually some phase of the reproduction and developmental process. It is here that the distribution limiting effect of the minimum factor or factors is felt.

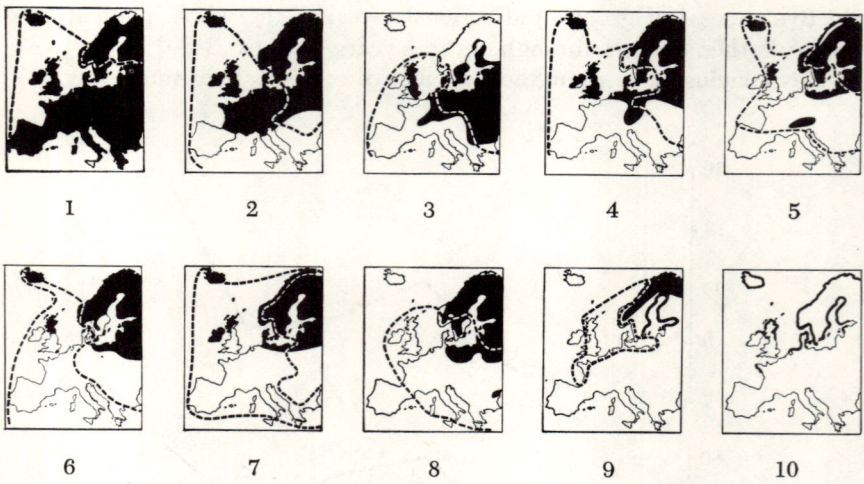

Figure 3-21 Cold limiting distribution through chick survival.

Distribution areas in Europe of ten species of ducks arranged in order of increasing cold-hardiness of one-day-old ducklings. Those species which exhibit maximum cold-resistance as ducklings have the most northerly distribution. (From Koskimies and Lahti, 1964.) Breeding area: black. Wintering area outlined by stippled line. 1. *Anas platyrhynchos* (L.). 2. *A. crecca* L. 3. *Aythya ferina* L. 4. *Aythya fuligula* (L.). 5. *Mergus merganser* (L.). 6. *Anas penelope* L. 7. *Mergus serrator* (L.). 8. *Bucephala clangula* (L.). 9. *Melanitta fusca* (L.). 10. *Somateria mollissima* (L.). Courtesy of J. Koskimies.

ECOLOGY OF SETTLING GROUPS

If a group of animals acted merely as an assemblage of individuals, the environment would not have a greater or a different impact on the group than it has upon the individuals comprising it. But it is a basic law of biology that *any natural group of individuals is more than the sum total of the individuals that make up the group.*

A group of mutually passive individuals, gathered merely by an environment which commonly attracts them, is called an aggregation. Because the activity of all members of an aggregation is concentrated in one place, it may considerably modify the substratum or medium and also affect other living beings. This environmental conditioning is a *mass effect*. Students of animal behavior find that in many instances the performance of an individual is enhanced by the presence of other, conspecific, individuals. This social stimulation and the other behavioral interactions between the members of a social group result in what may be called *group effect*, following Grassé's terminology (1946). A typical mass effect is the denudation of whole fields by swarms of migratory locusts, exposing the soil to such desiccation and wind erosion that it can support neither locusts nor other insects of the grassland habitat. An example of group effect occurs when less social animals, e.g., jays, which nest on individually defended territories, respond to an individual alarm and attack a prospecting mammalian predator with such vigor that it has little success in robbing any bird's nest in the vicinity. Further examples will be given in the discussion of population ecology.

Whether the colonists appear singly, in pairs, or in larger groups is clearly important. Groups may have a special advantage in starting a new colony.

ECOLOGY OF THE SETTLING POPULATION

In a wider sense, population is the unit of individuals of a species occurring in the same period of time. A special, restricted meaning is attributed to the word *"population"* in evolutionary and ecological studies; the number of individuals in a species—even the rare ones—is usually too large to serve as a convenient working unit. The local population (or deme), then, is a more closely knit group of individuals with the actual as well as the theoretical potential to interbreed. From the zoogeographical point of view, both the species and the local population are important; but our immediate concern here is the local population, consisting of the settler or settlers and their offspring. The population is a level of organization, and has its own group attributes; therefore the limiting factors affect a population and the individual in different ways. The most important attributes of a population are listed in Table 7. Rather than considering them systematically, we shall concentrate on only those attributes that have a special role in the establishment and spreading of a population. It is customary to stress the hazards encountered by a population, but factors favoring a population are equally important.

Size of the population is one of the most important attributes of a pioneering nucleus. All other intrinsic characteristics will depend to a greater or lesser degree on the number of founders. This may depend on chance, such as the size of a drifting, passively dispersing group or the success of one or several colonists. The smaller the population, the more is it subject to chance; the hazards of weather, catastrophe, interference, and other random factors of mortality can easily wipe out the founders.

We have discussed (p. 52) the spectacular spread of the muskrat in Central Europe. Now, more than sixty years after its introduction, the animal seems to be firmly established and still spreading (Niethammer 1963). The following circumstances contributed to its successful settling. There are such extrinsic factors as *favorable climate,* similar to that of its original home but without great extremes. Occasional heavy frosts or

Table 7   Population attributes

1. Size. (Absolute size, or density related to area unit.)
2. Spatial relations. The extent of the area and the important environmental parameters.
3. Structural attributes, such as:
    a. Genetic structure (which determines all adaptations, tolerances, variations, etc.).
    b. Age structure.
    c. Seasonality.
    d. Social structure (besides the physiological age, to know what proportion of the adults take part in breeding, etc.).
    e. Dispersion: internal structures within the area of the population: even dispersion, random dispersion, clusters, etc.
4. Functional (dynamic) characteristics (in service of maintenance, growth and expansion of the population).
    a. Turnover rates: rates of natality, longevity, mortality, immigration, emigration.
    b. Rate of growth or decline.
    c. Size fluctuations (seasonal, annual, etc.).
    d. Homeostatic regulatory mechanisms.
    e. Adaptational mechanisms.
5. Ecological attributes.
    a. Modifying effect on physical environment.
    b. Interactions with the biological environment.

floods may damage the population, but the losses are quickly replaced. *No disease* or *competitor* of the muskrat is known in Europe. It has no serious predator except man. In North America, it is heavily preyed upon by the mink *Mustela vison* Schreiber [Mustelidae, Mamm.]. As Errington (1943) showed, this predation mainly limits the number of surplus—mainly wandering and emigrating young animals. These surplus individuals are the important colonizers in the European spreading process.

Among the more important intrinsic advantages of the muskrat are its *very high reproductive rate*—an average of seven young per litter is cast three times a year, and the spring litter is mature by late summer—and the *low degree of philopatry* in the surplus animals. These undertake great migrations and easily find suitable habitats, where they settle. Others disperse inadvertently; for instance, floods, while drowning some litters, carry away other individuals to new locations.

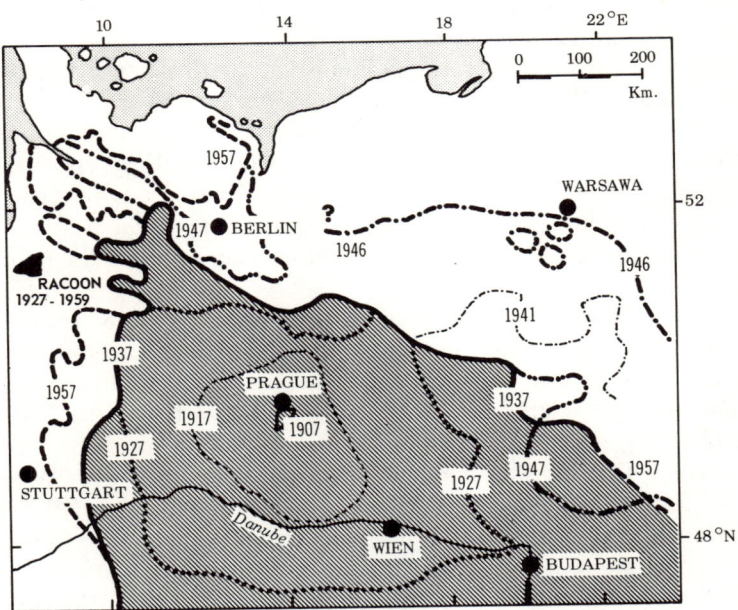

Figure 3-22 Population dynamics and expansion.
The spread of the muskrat (1905–57) and of one colony of raccoons (1927–59) in Central Europe. The muskrat's gain for each decade is indicated by a concentric line. The black triangle in northwestern Germany covers the 5,000 sq. km area occupied by the raccoon in 32 years; the gray area (1905–37) covers most of the 200,000 sq. km area the muskrat invaded in about the same time interval. Various sources, but chiefly after Niethammer 1963.

138   *Ecology of Colonization*

To illustrate the importance of certain aspects of population dynamics, the easy spread of the muskrat may be contrasted (Figure 3–22) with the slow penetration of the raccoon *Procyon lotor* (L.) [Procyonidae, Mamm.], both of which are introduced mammals of Central Europe. (Niethammer 1963). The raccoon has an average litter of four young each summer, and these mature in a year. This omnivorous raptor lives in wooded habitats bordering streams, rivers, and lakes. The climate in Germany is as favorable for it as for the muskrat. Moreover, it was protected until 1954. Yet its occurrence is spotty and restricted to the vicinity of the locations of introduction. It could not be said, though, that it failed to establish itself. Some six pairs were liberated in northwestern Hessen, in Germany, about 1927. By 1959, their descendents spread in approximately 32 years over an area of 5,000 km^2, and were estimated to number between five hundred and one thousand. The muskrat spread in the first thirty-two years after its introduction over an area of 200,000 km^2. There is a yearly kill of about 100,000 muskrats in Czechoslovakia alone, and so we can safely say that there are now many millions in Central Europe.

Two further attributes of a population are the *dispersal* and *dispersion* of the species. The dispersal habits of a species will affect the success of the founding group. If a particular species is very mobile and very nomadic, the animals may stray away from one another, and successful mating may not occur. The extraneous force that brought them to the area—for example, wind—may remove a number of the settlers again. Homing and philopatric behavior patterns may compel them to leave the area.

The introduction of game birds to North America provides many examples of the importance of *vagility* (the tendency of the animal to change its domicile in time) versus philopatry. A very successful introduction of the Asiatic chukar partridge *Alectoris graeca* (Meisner) [Phasianidae], a sedentary, nonmigratory bird, was made in dry areas of the West. On the other hand, several attempts, even with large numbers of founders, have failed to establish the Japanese quail *Coturnix coturnix* (L.) [Phasianidae]. This bird is migratory, and consequently possesses innate tendencies which enable it, within its natural range, to return from wintering areas. Outside this range, the migration pattern is disrupted and high mortality results. Most birds possess a pronounced degree of philopatry. The paradoxical statement has often been quoted (usually without citing its author): "Birds are conservative in keeping their own distribution area because they possess excellent means of transportation" (Stanchinsky 1923, p. 51). This serves to explain the difficulties encountered when attempts are made to induce artificial

Table 8 Success of introduction of mammals and birds to Europe and New Zealand.
(Data from Niethammer 1963 and Schilder 1956.)

	No. of Mammal Species			No. of Bird Species		
	Attempted	Successful	%	Attempted	Successful	%
Europe	47	32	68	85	13	15
New Zealand	44	26	59	130	24	18.5

settling. Both Elton (1958) and Baker and Stebbins (1965) discuss this subject in detail.

The study of extended—often large-scale or repeated—introduction attempts reveals fundamental differences between settling birds and settling mammals (and, by analogy, between various other groups) as Table 8 indicates. Here the data for Europe are slightly biased because they include the British Isles, to which many successful bird introductions were made. On the other hand, New Zealand is also an archipelago, although of much larger dimensions, and has no indigenous mammals. Most successfully introduced mammals occupy habitats which were disturbed or altered by man. Europe itself is of course most thoroughly altered. Basically, then, in spite of the hazards of ground locomotion and larger exposure to human and animal enemies, the mammals were much more successful settlers than were the birds.

Genetic constitution of the starting population.—Genetic drift—the random accumulation of nonadaptive genetic combinations—is very likely to occur in a small, isolated population. In other cases, the settlers are simply not heterozygous enough to allow gene recombinations likely to be adapted to the new environment. The "elimination" of genetic material (*sensu* Reinig 1938),* drift, and other genetic effects in small founder groups leads either to extinction or to speciation (see Mayr 1954 and 1963, Baker *et al.* 1964).

The size of a founding population can influence its genetic consti-

---

*Reinig assumed that expansion from the ecological optimum area, *e.g.*, from Pleistocene refuges—results from the gradual colonization of a few pioneers at a time. These can carry only a few alleles, and thus the new colony is poorer in gene combinations. With increasing distance, this genetic elimination also increases; if the habitat also changes gradually, eventually a limit will be reached at which the few remaining alleles are unable to produce adaptive combinations. This temporary static limit changes when new mutations cause a favorable change in the genetic constitution of the border population. Reinig's theory could apply to such cases as expansion through island chains, where pioneering is really carried out stepwise and not, as in most expansions on land, by slow penetration.

tution, which, in turn, determines certain structural and functional attributes and their ecological implications, which will not be discussed in detail in this book.

Whatever the intrinsically determined and genetically modified rate of increase of the population, its size will be further limited by two sets of components:

1. The first affects the *level of average density*—its commonness or rareness in relation to the area occupied. The chief environmental components—many of which are often interacting—into which Andrewartha and Birch (1954) grouped the limiting environmental factors, are as follows:

Weather, which either directly controls presence or absence, or limits the number of individuals through indirect, density-dependent action.

The quality, quantity, and accessibility of food material, including water, minerals, and trace elements.

The quality, quantity, and adequacy of shelter, including: the proper medium surrounding the animal; a substratum facilitating locomotion or rest; a shelter from weather, enemies, and interference; and a suitable place for reproduction.

Interactions of, by, and with, other animals and plants. In addition to comprising generally minor interactions—trampling, shading, obstacles to locomotion, disturbances to behavior—these mainly refer to such major phenomena as predation, parasitism, and competition. Because these are important in regulating population size and growth, and lead to mutual adaptation and evolution within a higher integration level (the ecosystem), they will receive special consideration.

Time. The time lapse after the initial penetration by colonists is extremely important in the settling and expansion process. Therefore the range of a settling population, as well as its density, is often determined by the amount of time the animals have at their disposal in their new locale. Time is of course considered in relation to speed of dispersal and dispersion.

The combined action of all these major environmental components also affects the *trend of the fluctuations* of the population. Changes in the severity of the limiting factor or factors may occasion either a population explosion or a decline, the latter eventually leading to local extinction.

2. The random or rhythmic *size fluctuations* that every population experiences may be correlated with periodic relaxation or strengthening of the components that govern the level of average density, or may be due to special, density-dependent regulatory mechanisms—either intrinsic (genetic, physiological, behavioral) or extrinsic (predation, parasitism, diseases, food shortage).

Since the population is the level of organization on which the evolutionary mechanism—as we know it now—operates, the impact of limiting or adverse environmental factors will show a population response different, in the group and throughout time, from that which may be experienced in observing or experimenting with single individuals. The mechanism of evolution consists of multiplication with genetic variation (mutation and recombination), environmental checks, and selection of the fittest. The result is usually not an all-or-none effect, but a slight change in the average characteristics of the group. "Adaptation to local conditions and evolutionary change are two aspects of the same genetic phenomenon" (Mayr 1963, p. 332). Similarly, Orians says (1962) that the general theory of ecology is natural selection. Although other biologists have dealt with the same idea (e.g., Hesse 1924, p. 7; Good 1931; Hesse *et al.* 1937, p. 27; Warnecke 1934; Timofeeff-Ressovsky 1940; Andrewartha and Birch 1954), Orians was the first to express the concept so succinctly.

The adaptational response of natural populations to the impact of the physical environment is amply shown, in drastic cases, by the failure of insecticides, after initial success, to wipe out populations entirely. Less drastic effects are countered by the slower, invisible population adjustments, which are detectable through detailed population ecological study. Boness (1953) showed that most insects living on meadows in northern Germany adapt their life cycles to the periodic mowing operations. Mayr (1963, p. 233) summarizes a wealth of evidence for the adaptive nature of geographic variation and concludes *inter alia,* that "every population of a species differs from all others genetically. . . . Geographic variation as a whole is adaptive. It adapts each population to the locality it occupies."

In the earlier discussion of the impact of the environment on the single animal and its reproductive requirements, stress was laid upon the limitations and perils caused by adverse environments. The present discussion, in contrast, is focused on the achievements of microevolutionary adaptations. One may wonder, then, why so few species are successful in colonizing new areas and why they do not establish locally adapted, yet versatile, widely spreading populations. This failure may be attributed to one or more of the following causes: the lack of time, the nature of the genetic makeup of the species, the nature of the limiting factors, and the spatial structure of the species. First, most expansions are slow and oscillating processes, and may go unnoticed. Second, the gene pool of each species population is limited. It contains coadapted genetic material in which each characteristic is controlled by the coordinated action of several genes and in which each gene is pleiotropic—that is, it affects several morphological and functional attributes of the animal. Successful rearrangement of the genetic makeup is rare, and successful incorporation of new and adaptive mutations is even rarer. Nevertheless, unless genetic recombination or mutation results in genotypes more or less well adapted to the new environment, the pioneer demes are rapidly eliminated. Third, many populations reach an absolute limit at a strong barrier; near the barrier they may flourish and the population may be dense. There are other situations in which living conditions become gradually more adverse and the population gradually dwindles to nothing (see Figure 4–24 and discussion there). Among such adverse conditions, a local population is often maintained only by the reinforcement of pioneers from more central and optimal habitats. The resulting inflow of genes from the main population—though probably strongly selected by the marginal environment—may prevent the fixing of locally adapted strains, if such strains occur. Fourth, where marginal populations have become established but demes are contiguous, unless the distance is very large between the central and these marginal populations, continuous gene inflow from the center populations prevents the locally better-adapted marginal populations from spearheading further expansion. Proof of this explanation comes from the genetic study of natural populations (Carson 1959, Birch 1960, Cook 1961, Mayr 1963; cf. also Baker *et al.* 1964), especially morphologically distinct subpopulations and geographic subspecies. In cases in which a species population is divisible into a number of distinct—and therefore presumably better-adapted locally—subspecific populations, it is possible to show that gene flow between these populations is limited either by distance or by other barriers. The barrier need not necessarily be a physical one; more often than not, distinct populations occupy ecologically different, although adjacent, areas, and ecological habitat selection keeps them apart and reduces contact to a minimum (for example, the mouse genus *Peromyscus;* see Blair 1953). But—to return to the original thesis—these marginal and distinct subpopulations are the potential new species of the future. If they sever all contact with the central populations, their independent evolution will soon produce local adaptations enabling the population to expand again. Species characteristics that act through the individual (such as ecological valency) and population attributes (such as sociability and group behavior or the number of surplus individuals available for pioneer reinforcement) are important factors which affect the spread of a species population, with or without subspecies formation. Another important factor is distance in relation to the locomotor and dispersal capacity of individuals or local populations.

Geographic subspeciation has long been recognized by morphological systematists, but its true nature and importance have been clarified only through work, beginning in the 1930's, by geneticists. The more subtle character differences within geographically distinct parts of a species area, such as clines (directional character gradients), polymorph ratios, ecotypes, and the like, are now commonly understood, although their ecological implications had been vaguely recognized earlier.* Thus, for example, Palmgren (1938, 1949) demonstrated that many pioneer species first show a rigid adherence (stenoecy) to the ecological conditions of their former homes. After they have increased and spread, they become euryecious i.e., able to occupy a wider range of environmental conditions. It is, however, possible that the pressure of increased population has forced some individuals to accept inferior habitats, and thus the change of adaptedness is not real (L. von Haartman *in litt.*). The mistle thrush *Turdus viscivorus* L. [Turdinae, Aves] is found throughout most of Germany. Local populations seem to utilize slightly different kinds of habitats. Peitzmeier (1942, 1950) discovered that the bird died out in Westphalia in the early 1930's. After 1941, the species reappeared there, but in a different habitat—in deciduous woodland, which the neighboring thrush populations of coniferous woods avoid. The newly occupied habitat resembles that of the plains of Flanders, where the mistle thrush is on the increase. Peitzmeier thinks it likely that the new settlers expanded to Westphalia from the southwest, from this suitably adapted French-Belgian population. Reinig (1938) also recognized the role of ecological versatility or—to adhere to Hesse's (1924) term—ecological valency in the expansion of a species' range. Reinig pointed out that a species which tolerates a wide range of temperatures may expand its distribution area centrifugally, because it would not likely encounter thermal conditions that would limit its expansion toward the north or the south. Those species which have only a limited range of thermal tolerance are able to live only within a narrow geographic zone where thermal conditions are favorable for them. These stenothermal species, if they expand, show expansion in a wedgelike formation—that is, they are restricted to a narrow belt which extends in an east-west direction (see Figure 3–4). Reinig calls such an expansion "isoclimatical," because it happens along a latitudinal zone of similar climate.

It is necessary to reiterate the importance of the time factor. Environmental conditions are not static, but fluctuate and change in time, and the expansion and retraction of area limits can thus alternate in time. Likewise, adaptational changes within the whole population of a species may in time take a course which enables population expansions. However, the role of geological time on evolution (cf. Simpson 1944, 1953*b*) and expansion is largely outside the realm of this discussion.

*Regarding plant populations, the development of these ideas followed a similar course; cf. Meusel (1943) and Baker (1959) on ecological valency and Stebbins (1950) for general references.

ENVIRONMENTAL RELATIONSHIPS OF THE POPULATION AND ITS AMALGAMATION INTO THE ECOSYSTEM

The individual must either enjoy, endure, or move away from the impact of the physical conditions imposed by the environment. The population, however, often has an active role in modifying these physical conditions so as to make them more beneficial; yet this conditioning of the environment is by no means a general phenomenon. Moreover, it is the mass effect (see p. 135) of individuals living in a limited space which is able to alter at least the local environment or the microenvironment. This alteration affects, positively or negatively, any population actually living in the area. Metabolically generated heat is usually lost by the individual to the environment, as is moisture that evaporates from the body in arid climates. These can be trapped even by single individuals provided they have properly insulated surroundings—for example, the nest of an incubating bird or the burrow or hole of a quiescent, resting, or hibernating vertebrate. However, this is the exception rather than the rule for individual animals. On the other hand, aggregations or social gatherings of individuals in a limited (and insulated) space actually magnify this effect; snakes, small mammals, and possibly even birds regularly use this means of creating a favorable local climate. Not only the physiological effect but also the mechanical impact of weather are modified by animal aggregations. Young ungulates in a herd formation are protected from hailstones. In social insects, several different means are known whereby termites, bees, ants, and tent caterpillars create a regulated medium in their dwellings. These means may also be employed by other animals, as in the egg-laying habits of the Nile monitor lizards. It is well known among European entomologists, that the chafer *Potosia cuprea* F. [Scarabaeidae, Coleopt.] lives in molding material under anthills, evidently without direct contact with the ants, and benefits from the heat generated by the ants. Earthworms alter the structure of the soil to their benefit, for in loosened soil their feeding conditions improve, and they also are able to alter the pH of acidic soils (Schwerdtfeger 1963). Colonies of sea birds alter the substratum by the amassing of guano (Hutchinson 1951). The swimming of a group of waterfowl keeps water surfaces from freezing and thus accessible for feeding, drinking, or safe roosting. Chemical and physical conditioning of the environment has hardly been studied as yet; it can be beneficial for groups of animals: for instance, when fish grow and thrive in water in which their conspecifics have been kept; when ungulates use the paths worn by their elders or ancestors; or when social caterpillars use the silk paths laid down by their more active companions (Wellington 1964). Social weaver birds congregate for communal nest building; other birds and also mammals such as prairie dogs, benefit from the nest-building or burrow digging activities of their companions or their elders. Some of these examples—in which special social population behavior occurs—indicate only that special evolutionary advantages were utilized by the coordination of group activities.

The impact of the living environment is twofold. In a general way, the biophags—predators, parasites, and pathogens—kill or weaken individuals. Competitors either tie up or occupy requisites—food, shelter, space, social status—which the settlers would have needed, or they directly and agressively hinder, by threat or force, the peaceful utilization of such requisities. Other adversaries interfere with the normal life of the settler: herds of such large animals as elephants trample down smaller animals or their nests and shelters; beavers flood valleys; caterpillars strip the trees of their leaves, exposing the nests of birds and small mammals to the elements and to plunderers. Because much of the earth is covered with a mat of vegetation, plants are dominant elements of animal habitats. This vegetation has its own local climate, seasonal changes of appearance, soil-forming and lake-filling capacity, and the like. Thus the plant is indirectly important to the settler. Vegetable feeders are directly influenced by the qualities of the food plants, especially where these differ from those to which a species is accustomed.

Two small birds live side by side in the mixed forests of southern Finland. The goldcrest (kinglet to the American ornithologist) *Regulus regulus* L. [Sylviinae, Aves] nests and feeds exclusively on conifers, especially spruce and pine. However, the willow tit (nearly identical with the black-capped chickadee of North America) *Parus montanus* Bald. [Paridae] feeds equally on birch, alder, spruce, and pine. Palmgren (1932) showed that a slight anatomical difference in the leg tendons prevents the goldcrest from hanging on the pensile twigs and cones of the birch tree, as does the tit when it feeds. Thus the goldcrest avoids birch trees entirely, and even its nest is woven amidst the hanging small twigs of the spruce bough. This bird lives also in England (where coniferous forests dwindled and died out during the Pleistocene glacials) and, as Lack (1937) remarked, it adapted there to the deciduous instead of spruce forests. The oak and other southern hardwood trees of the English forests do not have a pensile structure similar to that of the northern birches. Nevertheless, the goldcrest found a suitable place to hang its nest; instead of the spruce boughs, it uses the ivy *Hedera helix* L. [Araliaceae, Umbellales] vines that climb the trunks of old oaks. This example illustrates the importance of the right kind of plant as feeding ground and nest site and also the complicated nature of interrelations.

The length of a tendon, well adapted to a specific kind of locomotion, enables one species for, and prevents another from, utilizing a certain kind of forest habitat.

In particular, the living environment influences the population, because it either depresses its density or provides its density-regulatory mechanism. Intensive predation or parasitism, periodically recurring diseases, intensive interspecific competition, or other interactions may, without being dependent on the density of the settling population, influence the level of its density, i.e., the commonness or rareness of the settling population. The individuals removed or weakened by these agencies could have otherwise been instrumental in further expansion. Scarcity of food, either plant or animal, or the special requisites of an animal or vegetable nature also influence the density level.

Many ecologists theorize that food requirements must set the ultimate limit of the total animal population within an ecosystem. Theoretically, they are right. However, there are few instances known in which food supplies are exhausted by animal populations. Secondary requisites, such as nest site and shelter, are more often limiting. Still more often, the population is limited by enemies, i.e., biophages. There seems to be a common kind of mechanism—perhaps even more common than is known at present—by which a population is able to maintain a homeostasis. Such internal *regulatory mechanisms* as genetic fluctuation (Franz 1949, Chitty 1965, Wellington 1964), devices of social behavior (Wynne-Edwards 1962), birth control, or internally caused increases of mortality (stress diseases, cannibalism), and emigration (Tinbergen 1957)—all of which were formerly thought of as *intraspecific competition*—adjust population density to the requisites which the environment provides.

From our particular point of view of the general and particular biotic factors discussed above, the most important are those that influence not only the level of the settling population but also its further spread. Thus it is important to distinguish the factors that control population size at any level. Cannibalism or predation destroys the surplus individuals, while emigration or expansive pressure due to population increase compels these animals to become pioneers and potential new settlers.

A settling and expanding population also greatly influences its living environment. This influence is very conspicuous, because each animal feeds upon other living beings and is being fed upon, and these principal links with the living environment are usually much more forceful than other, rather subtle, periodic interactions. Interrelations with the living environment are not only harmful; they can also be beneficial, as in such forms as cooperation, mutualism, and the like.

The populations of all biota in a particular habitat are connected by these relationships. Biologists first recognized these conditions in vegetation; later, they found these between plants and animals, and defined the plant association (Kerner 1863) and the biotic community (Moebius 1877). A more recent view recognizes that plants and animals—particularly the edaphon or soil-building organisms—alter the physical environment or biotope of the community (Friederichs 1927, Tansley 1935, Sukatshev 1960). These components form a self-contained unit—the *ecosystem*, the highest complexity level in nature (Table 9). The ecosystem is a natural and relatively stable combination of various plants, animals, and protists on a certain physical habitat, and has particular attributes other than those of its member organisms. Its main functional characteristic is the utilization of the sun's energy and the chemical components of the habitat by the community. The members of the ecosystem are coadapted: some bind the energy, some utilize the living or dead tissues of fellow members; and, others decompose the surplus waste, thereby liberating energy and material for a new cycle. The physiognomical structure of an ecosystem depends on the producers—that is, the plants—which, in turn, are governed by the physical environment and by the competitive utilization of the available space. It appears that the func-

Table 9 Certain characteristics of ecosystems contrasted with those of individuals.

Individual	Ecosystem
1. Parts differentiate from primordia	1. Composition of already existing, complete parts
2. Components exist only as parts of the whole	2. Parts not dependent on the whole entity but are able to exist in other entities and are exchangeable
3. Harmony of functions is achieved by coordination	3. Harmony of functions achieved by antagonistic role of the parts
4. Lost parts are replaced qualitatively from within	4. Lost parts are replaceable only qualitatively from outside
5. The entity is delimited by internal conditions	5. The entity is delimited by external conditions
6. The corresponding parts of similar entities are mainly homologous	6. The corresponding parts of similar entities are mainly analogous
7. Physical environment is generally not part of the system	7. Physical environment is strongly modified by, and coadapted with, the system
8. Life span of individual limited; generations subject to microevolutionary changes	8. life span of ecosystems is geological; subject to steady but slow macroevolutionary changes

LEGEND: No's. 1–6 were modified after Tischler (1963), No's. 7–8 were added by the author.

tional complexity of the ecosystem depends on both its age* and its opportunity to evolve a balanced homeostatic system of energy turnover between the member populations. These members are exchangeable, and new members can replace lost ones, if they fit, or wedge themselves, into the functional chains. Analogous evolution of geographically separated ecosystems is a related attribute. These two characteristics are quite different from those of the lower levels of organismic complexity, and are based on the evolutionary history of ecosystems. The units of complexity levels of the living world (Figure 1–1) have progressively increased the length of life. Individuals live in terms of years or decades, possibly hundreds of years (with due exceptions at the upper and lower limits). Populations live from decades and centuries to as much as 10^6 or 10^7 years (based on the average ages of species). If we would consider major adaptive reorganization rather than speciation as the end of the life of a species, we would raise this limit by another exponential. Ecosystems may live as long as 10^7 or 10^8 years, although these are vague estimates, especially at their upper limits. However, since the age of the ecosystem seems to exceed the average age of the species population, it is evident that species in time can be replaced within the ecosystem. Further, since the ecosystem is dominated by the life form of the chief producer (plant) members, the evolutionary dominance of these plant life forms (manifested by a few homologous or analogous phyletic lines) is the cornerstone of the life of the ecosystem.**

Now that we have given perspective to the ecosystem, it is possible to assess the difficulties of a particular population of animals in trying to establish itself in an ecosystem unit. This is, as we said, a homeostatic unit in which the population levels and the oscillations of its members are coadapted and correlated, especially in older, "mature" ecosystems (for details, see Ross 1962, Margalef 1963, Pimentel 1963). An intruder has little chance of becoming numerous, because it lacks most of the adaptations for fitting into this functional system; it may, however, enjoy temporary success, scoring advantages due to the mutual lack of adaptations. If its innate capacity allows it to expand rapidly and if at the same time it is versatile enough, a rapid selection-adaptation process may help carry it through until the ecosystem adapts to its presence. This latter process is much more complicated, in that it consists of the adaptation processes of several community members that are directly or indirectly affected by the impact of the newcomer. For this stage of the pioneering process, ecologists use such expressions as "finding an

---

*In both senses: its evolutionary age, and its place in an ecological succession. A climax community has more members and more complicated food webs than a developmental succession, e.g., immediately following a forest fire.
**See discussion of life forms on p. 250.

empty niche," "competition," and "environmental resistance."* By "niche" —which we understand in a figurative rather than in a concrete or spatial sense, we mean the role of the particular animal or plant in the community which it fills by virtue of its tolerances and requirements (Hutchinson 1957) and its special combination of structural, functional, and population biological adaptations. Each animal species has its own niche, which is not exactly duplicated by any other related or unrelated form. By and large, however, functional counterparts in related ecosystems are said to fill the same niche; where such a counterpart is missing, the niche is considered to be empty (cf. Udvardy 1957, 1959). But partial overlapping of niches among established and pioneering species— such as might result when a limited requisite is claimed by both—may lead to interspecific competition; yet it is easier to assume such competition than to prove it. Food is thought by many (Lack 1954) to be a requisite that limits population size. However, in numerous theoretically food-limited biophags (Hairston, Smith, and Slobodkin 1960), behavioral or genetic density-controlling mechanisms (see above, p. 146) prevent depletion of the food supply (Pimentel 1961). Most of the proved cases of competition, at least among vertebrate animals, show either an extension of the behavioral density-controlling mechanisms to potential competitors (interspecific territorialism) or that the mutually sought resource was a nonexpandable, minimum requisite (e.g., nesting shelter) and that the competition affected members of otherwise widely different niches. In this interpretation of the enviromental niche, some of its parameters include relative safety from predators, parasites, and diseases as well as freedom from many other environmental pressures. Therefore "niche" is only another way of expressing the place secured by the species within the food web and other functional webs of the ecosystem; each species in each ecosystem has a particular, nonduplicated niche. A niche is never literally empty, for it is nonexistent until filled; nevertheless, there exists a potentiality of further exploitation of the ecosystem. The cycling of energy may take many courses, through many different organisms distributed among the three main energy cycling, or "trophic," levels of the community membership. Although the amount of energy available for each locality is limited and dependent upon the locality's position relative to the sun (latitude) and on the globe (climate), the utilization—and perhaps even the production—of energy is very incomplete (cf. Odum 1964). There is plenty of room in the ecosystem for

*Grinnell (1917b) used "niches," Elton stressed this terminology (1927) and also "competition" (1946)—see Udvardy 1951. Chapman (1931) introduced the concept of "environmental resistance." Elton's (1966) new relevant discussion of these phenomena is based on the long-term study of a community which showed remarkable stability in its species composition in spite of being "subjected to . . . ceaseless bombardment by species from outside" (p. 365).

more energy consumers. Ecosystem evolution is far from complete. Biotic complexity in the old, tropical ecosystems is still increasing, for speciation continues to occur and a corresponding degree of extinction has not been noticed. The temperate and subpolar ecosystems were depleted by the recent, drastic glaciations, and have only recently begun their renewed evolution (Dobzhansky 1950, Dunbar 1960). In conclusion, an intruder in a fairly complicated ecosystem meets resistance, and the success of settling will depend on the outcome of the mutual adaptation processes. Each ecosystem evolves toward more stable homeostasis, and its settling success may depend on the role the new species assumes in the functional structure of that ecosystem.

CONCLUSIONS

Keeping in mind the motto which heads this chapter, we conclude with the following cautious generalizations.

Since the genotypic and phenotypic adaptations of each individual animal are limited, certain essential or influential environmental components may limit the occurrence of individual animals in a locality. (Shelford's law of tolerance.)

Several essential or influential environmental components may have the same response in an animal, and therefore may act in substitution. (Rübel's law of substitution.)

Since each individual animal is a coadapted responsive system, and the environment is also a composite and variable system, a limiting factor seldom acts alone; its action is usually influenced by the simultaneous action of other environmental factors. (Schwerdtfeger's laws of compensation and moderation.)

Even if the average ecological potential (the sum total of adaptive features) of the individuals remains nearly the same throughout the distribution area, the complex of environmental components may change; therefore the apparent ecological valency (stenotopy and eurytopy) of the species may be different in different parts of the distributional area. These ecological valency attributes are mostly employed in a local and static sense. (Law of relativity of ecological valency.)

On the other hand, the adaptive characteristics that make up the ecological potential are variable and subject to selection. Therefore, they may change within the distribution area of the species: *clinally* (altitudinally or latitudinally or following some other, extrinsic or intrinsic gradient); *centrifugally* (central population versus marginal populations); relative to *population density* (large independent population versus small population, replenished and influenced to various degrees by its larger neighbors); and in mozaiclike fashion, with more *isolation* between geographical sub-population. (Law of amphiecy or amphitopy.)

Of the many adaptations of individual animals, the most important ones from the point of view of settling (and even of survival in general) are the faculty of *locomotion* and the faculty of active *habitat selection*.

The most critical stage of the life history of most animals is some phase of its *reproduction;* this circumstance is particularly important for settling individuals or social units. (Thienemann's law of tolerance in the critical stage.)

In certain circumstances, asexual means of reproduction may become important vehicles of successful settling in animals which are able to switch or change to such forms of reproduction. (Geographic parthenogenesis and the like.)

Any natural intraspecific group of animals is more than the sum total of its individual components; it has special *group attributes*, and these play an important role in environmental relations as well as in settling.

The most important group attributes, influencing settling, establishment, and expansion local populations, are: dynamic characteristics affecting *size* (optimum density, rate of increase, disposal of the surplus individuals); *genetic structure*, especially adaptability; and *time* needed to develop homeostasis (with regard to size and adaptations), within a new environment.

A population is part of an ecosystem, a higher organismic level of nature. Populations influence both the physical environmental components and the other biota that form the ecosystem. A new, intruding population upsets the homeostatic conditions of the ecosystem, just as the latter has its detrimental impact on the settling population. The survival of the population will thus depend on the outcome of a mutual adaptation process.

The establishment and continuing existence of a founding population may be endangered by environmental reactions which would, at its old domicile, have had only slight impact on that population. The main functional attribute of ecosystems is their energy cycling; therefore, success of a settling population ultimately depends on its success, often by chance, to find its place in the food chains of the ecosystem.

Ecosystems are subject to adaptive evolution and periodic destruction or extinction. We surmise from what we know about the energy metabolism of certain ecosystems that they are by no means saturated; other ecosystems are definitely in primitive conditions, owing to the drastic impact of Pleistocene events. Therefore, there is much room for existing as well as for evolving populations to penetrate new ecosystems, notwithstanding the fact that such phenomena seem to be rare by our human time measure.

4 / Areography: The Study of The Distribution Area

> *Forma, figura, locus, tempus, stirps, patria, nomen*
> *Haec ea sunt septem, quod non habet unus et alter.*
>
> Medieval philosophy*

The medieval concept of the uniqueness of the individual was expressed in this hexameter. In addition to name, ancestry, form and shape, and existence in time, there were two other attributes—location and homeland—and these had to do with the individual's position and relationship in space. From the vantage point of the zoogeographer, this concept may be aptly applied to the biological species. For just as the structural, functional, genetic, and phylogenetic elements all manifest themselves in individual animals, so the dispersional and existential factors result in the distributional range of each animal species. The biologist interested in synthetical aspects must piece together his findings and discover the operating principles through the study of individuals by observation and experiment. The zoogeographer also builds comparative and causal research on the elementary units of study—i.e., the ranges.

The importance of the range is infrequently evident in faunal studies whose major focus is taxonomic or ecological. Zoogeography is used chiefly as evidence in these studies. Works describing geographic or ecologic entities and, indeed, many local zoogeographic studies do not usually consider the total range of the biota concerned. The paleobiologist, in considering a fossil, is compelled to erect most of his postulates on the basis of a fragmentary knowledge of the geography contemporary to his ancient subjects, but he uses much circumstantial evidence to

*As quoted by Jeuken (1953).

overcome this handicap. The student of contemporary zoogeography does not need to be so handicapped, for a knowledge of the worldwide distribution of animals widens his horizon and permits his understanding of even the most local distributional riddles. The following discussion will suggest how much information can be derived from the particular study of distribution areas.

First, a word about terminology. The study of geographic distribution in zoology is almost the same as the study of *faunistics;* this discipline aims at assembling facts about the occurrence of animals in certain geographic areas. The situation is different in botany: there are *two* plant geographies! Because the influence of external, ecological factors on plant distribution is quite evident, and because visible plant formations enabled botanists to develop distributional thinking along *vegetational* as well as *floristic* lines, they felt rather early the need to consider total distribution areas in floristical analysis as well as in vegetational study. These two approaches led to the advancement of *areography*—the study of distributional area by analytic as well as by synthetic methods—in botany. *Floristic* area analysis had its beginnings in de Candolle's (1855), Hooker's (see Turrill 1953) and Engler's (1899) basic writings (zoogeographers had then just made their first approaches toward regional animal geography). The value derived from studying total distribution areas in comparative *vegetational* analyses further justified the use of this approach by biogeographers. Thus, plant areography has an impressive background; and when we try to introduce areographic concepts into animal studies, we may benefit from comparisons with plant areography. We could pay homage to this older sister discipline by adopting its terminology when appropriate. We should consider areography as the basic branch of zoogeography. Areography deals with the study, analysis, and comparison of the distribution areas of animal species as well as of the composite areas of higher taxa. As such, it will guide the researcher toward synthetic approaches which will, in turn, aid him in clarifying the principles of animal distribution. Among the first principles will be those causally determining the features of *distribution areas*. The term "range," synonymous with "distributional area" in the zoological vernacular, will be used sparingly hereafter, especially when something less than the entire area of a species is under discussion.

DISTRIBUTION AREAS AND THEIR MAPPING

Most land plants are firmly rooted and immobile, and so their total area may be precisely described. This is not the case with highly mobile animals. Diurnal, seasonal, or annual shifts in animal activities take place,

154 *Areography: The Study of the Distribution Area*

and are reflected in the *temporary* areas that certain populations periodically occupy. Therefore, we need to clarify whether or not the distributional area would also include these temporary habitations. If so, the range of the arctic tern would encompass most of the seas and coastal areas from the vicinity of the North Pole to the Antarctic coast, for this bird nests on the rocks of the Arctic coast and winters on the coasts of the Southern Hemisphere. In many respects the most important

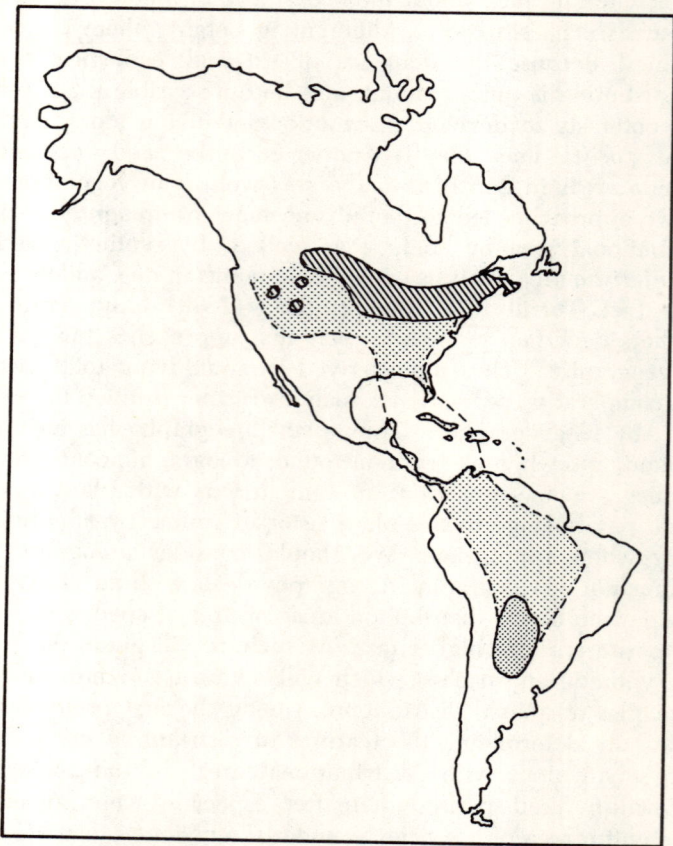

Figure 4-1 Seasonal occupation of distribution areas I.
Distribution area of a migratory bird, the bobolink *Dolichonyx oryzivorus* (L.) [Icteridae, Aves]. The North American *breeding area* (diagonally hatched) is occupied during the summer. The wintering area (*winter range*, dotted), on the pampas of South America, is utilized from the late fall through the northern winter. The *migration area* (vertically hatched) is traversed successively from north to south in the autumn and from south to north in the spring. After Lincoln 1939.

area is that on which successful propagation of the species regularly occurs. The distributional area of the animal therefore usually coincides with the *breeding* area (breeding range, reproduction area). The areas of *nonreproductive existence* are of two kinds. The first is the periodically but regularly inhabited area. This must always be designated with such special adjectives as "winter range," "migration area," "irrup-

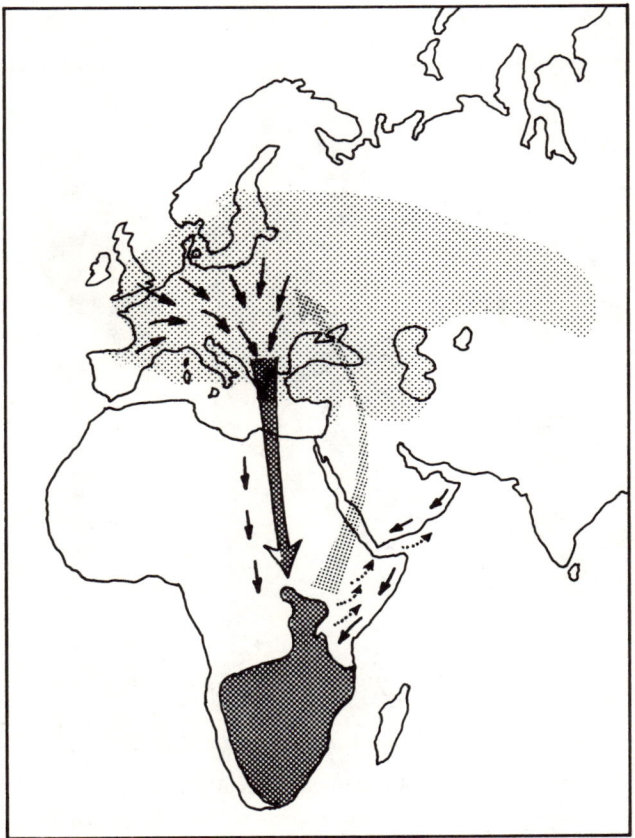

Figure 4-1 Continued
Distribution area of the red-backed shrike *Lanius collurio* L. [Laniidae, Aves]. Heavy line encircles the breeding area. Most European and South Asian shrikes migrate in the fall through Greece and through the Great Rift Valley of Africa to the savanna country in the southern part of Africa, which comprises the wintering area. The spring migration proceeds northward slightly to the east of the autumn route. Thus, in addition to the nesting and wintering areas, there are two areas on which the shrike is transient for a few weeks annually. Modified from Dorst 1956, after Verheyen 1951.

156 *Areography: The Study of the Distribution Area*

tion area," or the like. Although attention is usually given to breeding areas alone, these periodically occupied nonreproductive areas should not be neglected, for they may provide important stimuli for evolutionary adaptations of the animals (Figures 4–1, 2, 3, 4). Another kind of temporarily occupied area consists of places where the species occurs—and where it may even be found with a certain regularity—but which are unsuitable for its prolonged existence, let alone reproduction. Ekman

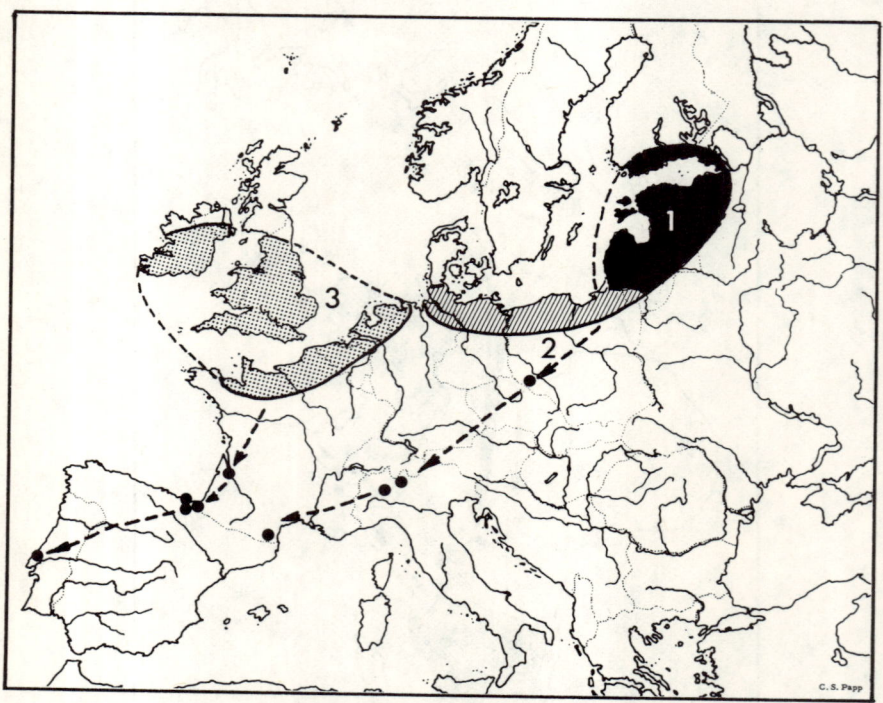

Figure 4-2 Seasonal occupation of distribution areas II.
Seasonal distribution areas of the starling *Sturnus vulgaris* L. [Sturnidae, Aves] population of the east Baltic area, based on banding and recovery data. 1. Breeding area utilized during the spring and early summer. 2. Intermediate area, where starlings spend the summer after the young are fledged. 3. Winter quarters.—Dots show recoveries of banded individuals (young birds of the year) from outside the normal migratory routes of the population. Arrows indicate the likely migratory routes of these juveniles, which bring them into wintering areas of other, more southern, starling populations. These individuals may not return to their birthplace, but may settle anywhere on their migration route. They show how expansion and/or mixing of genotypes occurs by these less philopatric pioneers. After Dorst, from Krätzig 1936.

(1953) calls these areas *sterile expatriation areas* (Figures 4–5 and 4–6).

For example, plankton is regularly carried by ocean currents into waters of unsuitable temperature where these organisms gradually die. This is true of arctic pelagic tunicates, notably *Oikopleura vanthoffeni* Lohm. and *O. labradoriensis* Lohm. [Appendicularia, Tunicata] in the waters surrounding Newfoundland, where the Labrador Current which carried them south mixes with the warmer Gulf Stream (Udvardy 1954). Wind-borne insects suffer a similar fate on the snowfields of alpine and arctic regions, as do pioneering locusts and grasshoppers (Figures 2–23, 4–6) in ecologically unsuitable regions (p. 82). However, these examples are in fact not different from the less conspicuous ones of *pioneering dispersal* (Figure 4–7), which must occur with at least similar frequency. These are seldom treated synthetically, and thus their frequency and importance are rarely evident in the faunistic literature. The

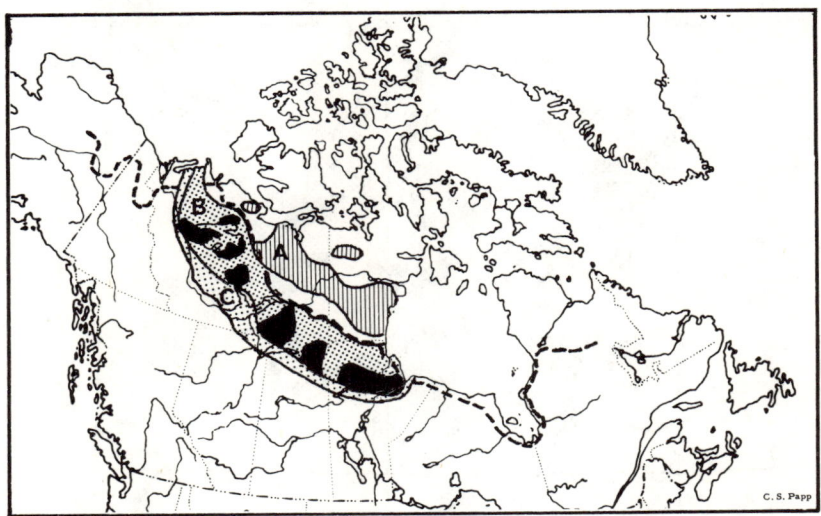

Figure 4-3 Seasonal occupation of distribution areas III.
Distribution area of the Barren Ground caribou *Rangifer tarandus arctica* (Richardson) [Cervidae, Mamm.] on the mainland of Canada as an example of a migratory mammal. The animals arrive at the summer range on the tundra in the spring and calve there (A). In late July or August, they migrate to the edge of the taiga forest, penetrate it during the fall, and have their winter range there (B, C). The northern limit of forest largely forms the southern edge of the summer distribution area of the caribou. Vertical hatching indicates the areas where herds had their largest summer concentrations in 1948 and 1949. Diagonal hatching indicates areas where the major herds spent the winters 1948/49 and 1949/50. Such areas of concentration (ranges of "herds") may shift annually as population size and the condition of the pasture require. After Banfield 1954; and Kelsall 1957, 1963, and *in litt*.

158   *Areography: The Study of the Distribution Area*

locations of *extralimital occurrence* in these cases cannot be consolidated into a regular area like the sterile expatriation areas, because there is no geographic limit—i.e., the limit varies with the density of the parental population (Figure 4–24), degree of vagility, nature of the barrier which forms the limit of the regular distribution area, and chance. Altogether, the frequency of incidence of potential pioneers ("drift visitors," "irregular visitors" of the faunist) varies with the reciprocal of the distance, and is influenced by the factors enumerated above.

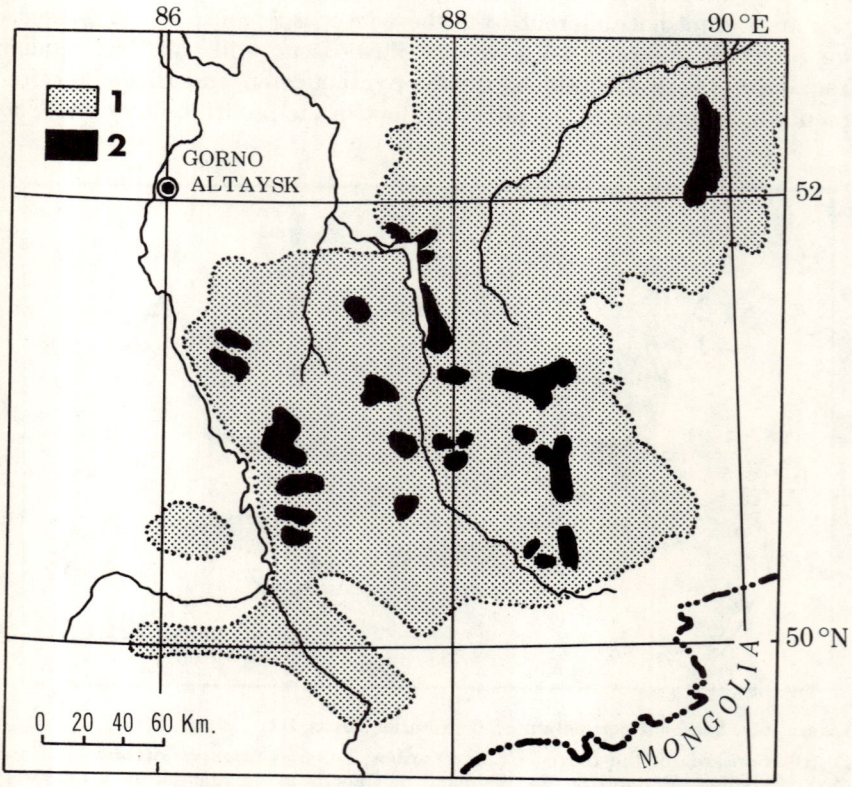

Figure 4-4   Seasonal occupation of distribution areas IV.
Distribution of the red deer *Cervus elaphus* L. [Cervidae, Mamm.] in the Altai Game Reservation of the USSR in 1936. This district comprises only a small part of the distribution area of the species.
The area occupied in summer is dotted; Areas of winter concentration are in black. We note that the summer distribution is largely continuous and that the wintering area is of a dispersed type. Seasonal migration occurs on a much smaller scale than with the Barren Ground caribou. From Naumov 1955, after Dmitriev 1938.

The basis of all areographic work is complete knowledge of the breeding distribution of a species of animal, and this knowledge must be acquired by field work. It would be hairsplitting to differentiate between faunistic and areographic field work, since the data on occurrence serve both the accurate mapping of the area of a species and the accomplishment of check lists of faunas in geographic areas. Yet the faunist tries to complete the total knowledge of a *geographic* area or region; for our present consideration, the total knowledge of *distributional* areas, species by species, is necessary.

The details of each distribution are needed for analytical as well as

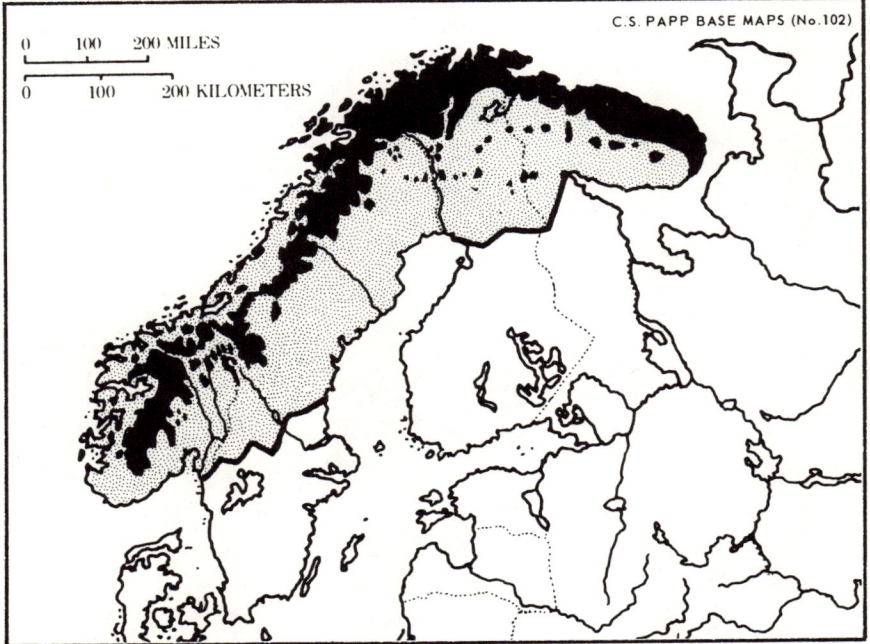

Figure 4-5 Sterile expatriation area.
Distribution area of the lemming *Lemmus lemmus* L. [Cricetidae, Mamm.] in Scandinavia. (After Ekman 1922.) This cyclic animal periodically shows mass increase, and therefore the distribution area fluctuates from year to year. Three categories of distribution can be distinguished. (Cf. Kalela 1949.) The area of permanent breeding distribution is black (alpine tundras and birch wood region). To the south and east of this, isolated black dots indicate isolated mountains (fells) which are inhabited and where reproduction also occur in years of peak numbers and distribution. The heavy line indicates the extreme limits of the *sterile expatriation area* where the species is found wandering but not reproducing in peak years.

for comparative zoogeographic work, and can be obtained only by mapping the area of a species. Mapping is a basic tool of the zoogeographer, just as microscopic slides are needed to see details of ultrastructure, models required for anatomical work, and preserved specimens for taxonomy.

The distribution of most animal species is known only in a superficial way (Figures 4-8, 4-9). Most faunistic data were not collected with special attention to completeness of single ranges; therefore, the zoogeographer often has to use caution when he outlines ranges. The most commonly encountered difficulty is that faunistic data tend to emphasize the extremes of occurrence. Usually we do not know enough of the homogeneity of distribution within a given area (Figures 4-10, 4-11); this is why the majority of distribution maps are outline maps. Actually, the mapping of single documented habitations by using dots is more accurate, and makes possible the outlining of the areas of coherent dis-

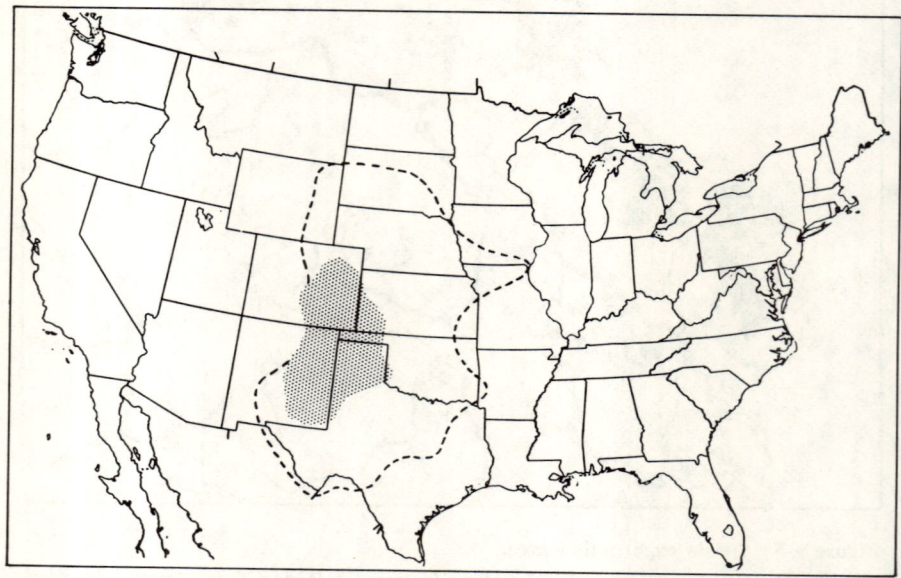

Figure 4-6 Sterile expatriation area as area of irruption.

Distribution area of the high plains grasshopper (*Dissosteira longipennis* (Thomas) [Acrididae, Orthopt.]. The dark central area indicates the extent of the breeding area. Broken line encloses the sterile expatriation area, to which adult swarms disperse but do not breed there. It may also be called the area of irruption as its occupancy is regular but not annual. From Herbert H. Ross, *A Synthesis of Evolutionary Theory*. © 1962. Courtesy of Prentice-Hall Inc., Englewood Cliffs, N.J.

tribution (Figures 4–12, 4–13). It would be highly desirable, if seldom possible, to indicate the density of the population as well (Figure 4–14). When the occurrence is very spotty, or is not well known, the use of single dots or other symbols is essential (Figure 4–15); too many kinds of symbols, on the other hand, make a distribution map confusing (Figure 4–16).

The great amounts of data necessary for the accurate mapping of distributions, together with their geographical and ecological details, require organized research, mechanized facilities, and the employment of a uniform system. The Distribution Maps Scheme of the British Floral Survey in Cambridge, England, provides an excellent example of this approach, and represents a highly commendable, pioneering attempt in the direction of creating uniform distribution maps. (Figures, 4–17, 4–18, 4–19,

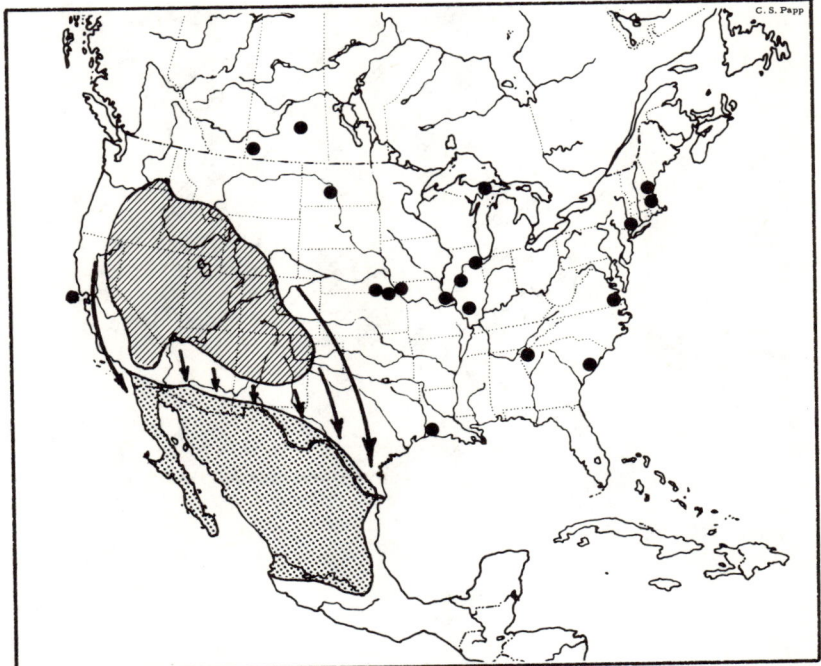

Figure 4-7 Extralimital occurrences.
The distribution area and extralimital occurrences of the green-tailed towhee. *Chlorura chlorura* (Audubon) [Fringillidae]. The approximate breeding area is hatched; the wintering area is dotted; on the area covered by the arrows, the bird is a regular transient and may stay over a mild winter. Dots mark the extralimital occurrences, all observed outside the breeding season during the past eighty years.

162 *Areography: The Study of the Distribution Area*

and 4–20). The collection and depositing of floristical data have been carried out in England over a long period of time, but not until 1950 was the work on the new Scheme begun. Between 1950 and 1954, with the assistance of collaborators in many parts of the country, a uniformly organized survey successfully collected data from earlier publications, from public and private collections, and, where necessary, from the field.

Figure 4-8

The relative exploration of beetles (Coleoptera) in the Fennoscandian area of Europe, according to Lindroth's subjective estimation. Black = well, squared = tolerably well, striated = badly, white = not at all investigated up to 1947. From Lindroth 1949. (*Courtesy of C. H. Lindroth and the Royal Society of Gothenburgh, Sweden*)

Distribution Areas and their Mapping 163

The data of occurrence for each species were then put on punch cards, the locality of occurrence in each case being based on a 10 km square grid system. After mechanized sorting and storing, tabulating machines printed the rough distribution maps and conspicuously indicated areas for which further data were needed. This was done by renewed field research and eventually, the *Atlas of the British Flora,* published in 1962, containing accurate distribution maps for every species was completed (Figure 4–17).

Modern taxonomic monographs are also a great help to the zoogeographer assembling distributional areas, for they usually contain detailed maps of localities where documentary specimens or series have been collected and of areas where some or all subspecies are missing (Figures 4–21, 4–22).

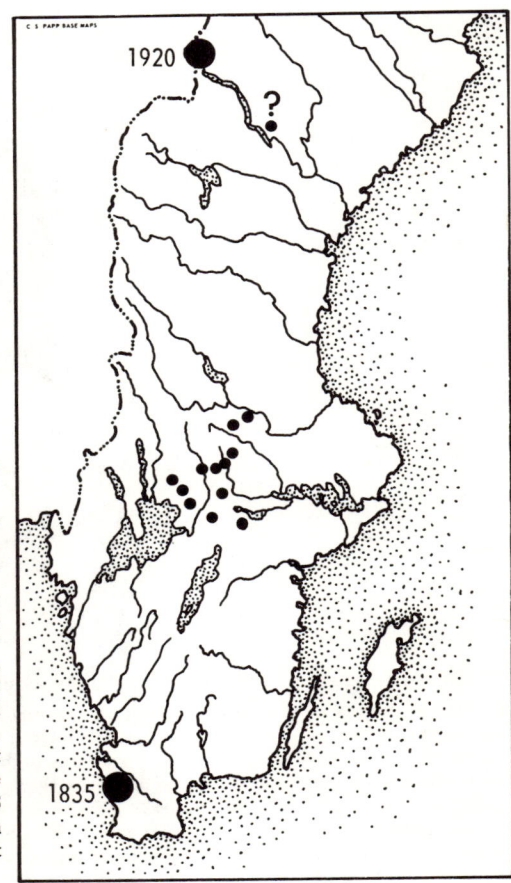

Figure 4-9 Incompletely known distribution.

The Swedish distribution of the birch mouse *Sicista betulina* Pall. [Dipodidae, Mamm.]. One occurrence, in the south of the country, dates from 1835. In the late 1910's, the birch mouse was found in central Sweden (1920 dot on map), and Ekman (1922) considered it to be a relict from the postglacial warm period (Boreal Period). Curry-Lindahl (1957) points out that most of the central occurrence data were collected from 1953 through 1956, and that the Swedish distribution of the species is not yet known completely.

164 *Areography: The Study of the Distribution Area*

The limit of a well-documented area should connect those extreme breeding stations which are not farther from one another than the distance that can regularly be covered by normally dispersing individuals of the species (Figure 4–23). Thus, knowledge of the ecology of dispersal is essential for dealing with distribution areas and their limits. The ecology of existence—the role of factors discussed in the preceding chapter—should also be kept in mind when establishing area limits, whether on the extremes of occurrence or within the general distribution area. For

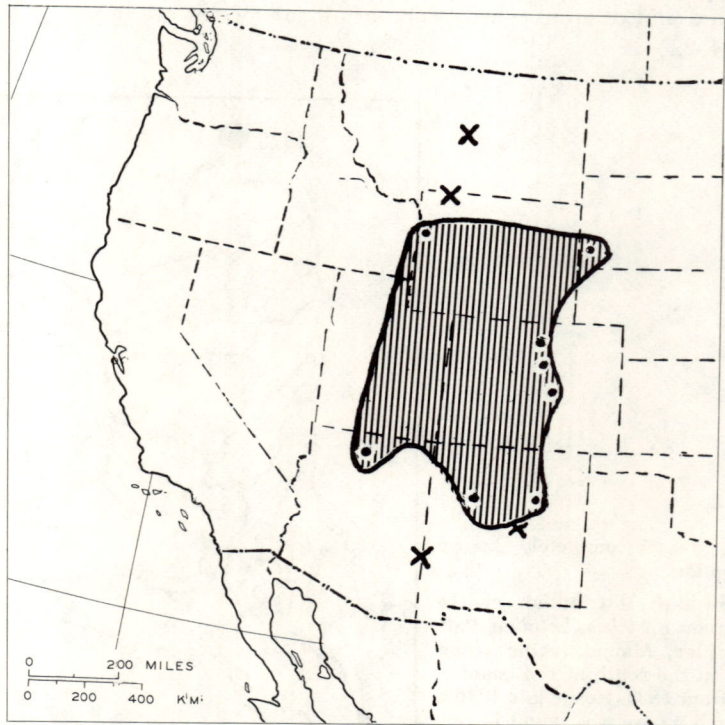

Figure 4-10 Mapping of distribution by emphasizing the circumference of the area. Distribution area of the dwarf shrew *Sorex nanus* Merriam [Soricidae, Mamm.] in the Western United States (hatched area). (After Hall and Kelson 1959.) In 1959, there were few known occurrences, and the eight extremes marked with dots were considered terminal and so mapped. Since then, Bradshaw (1961) has found it in Arizona; and Hoffman and Taber (1960), at two Montana localities (crosses on map). Hoffman and Taber say: "It seems likely that *S. nanus* is of wider distribution . . . since . . . almost every new record of the species has resulted in a significant extension of its known range (p. 234).

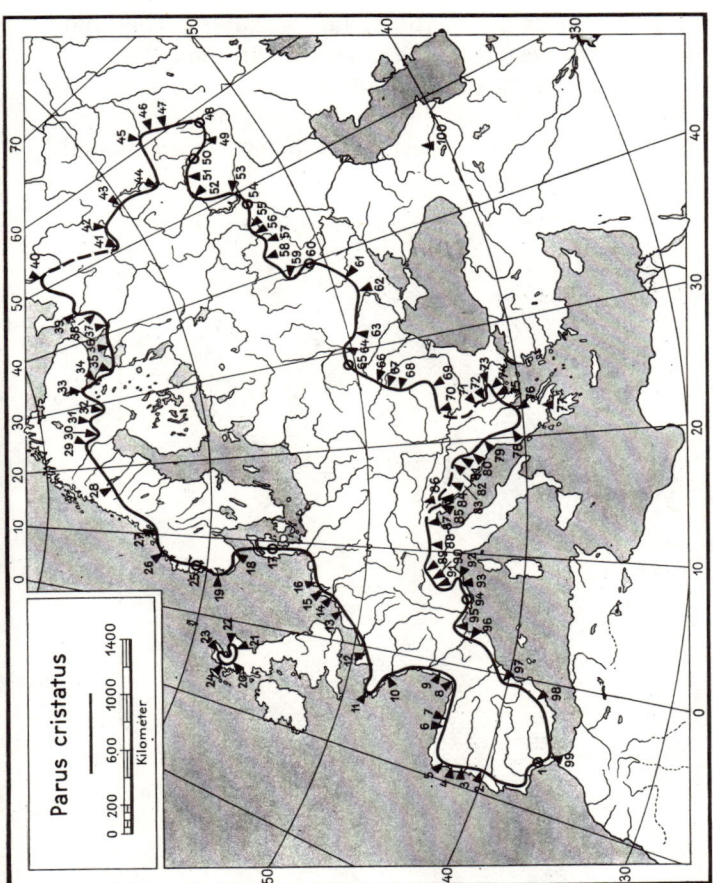

Figure 4-11 Mapping the circumference of the area.

Breeding distribution of the crested tit *Parus cristatus* L. [Paridae, Aves]. Each numbered wedge points at a documented breeding locality. Circles denote a larger area where its breeding has been known. These extremes of breeding occurrence are connected with the solid line surrounding its known breeding area. The limits of occurrence were weighed critically; "extralimital" or suspected but not documented breeding is excluded. Thus, the limits are as accurate as the available faunistic data allowed, but the internal structure of the area is not documented. Large expanses within the shown boundaries are not inhabited by the crested tit—for example, the Hungarian Plains or the northwestern European beech-forested expanses. (From Stresemann and Portenko 1967, *Atlas der Verbreitung palaearktischer Vögel*, map by Mauersberger and Stephan.) (*Courtesy of Akademie-Verlag, Berlin*)

166 *Areography: The Study of the Distribution Area*

the majority of animals a continuous distributional area ends with achievement of a temporary equilibrium based on the sum total of the spreading potential of the species (composed of its tolerances, requirements, and mobility) and the "resistance of the environment" (Figure 4–24). In some cases, the spreading potential of the species is sufficient to overcome the resistance of the environment, even though the speed of spreading is so slow that potentially available areas are not yet occupied. We believe that such occurrences are relatively rare in nature. The limit of a particular species is the result of the presence of some limiting factor. An introduced species is the best example: the raccoon has a low spreading potential compared with the muskrat in Europe (Figure 3–22). When the area is limited by the occurrence of an environmental

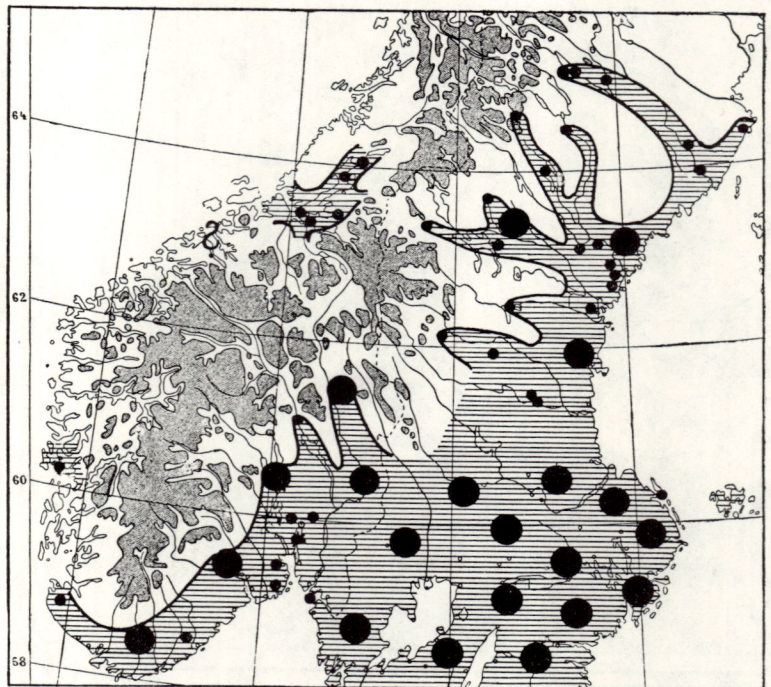

Figure 4-12 Accurate distribution maps.
The distribution of the wood pigeon *Columba palumbus* L. [Columbidae, Aves] in central Scandinavia, near its northern limit, as it was known in the early 1920's. Large circles indicate more or less regular occurrence in the area; dots, single records of importance for the distribution picture. Limits, where known, are drawn by continuous line. From Ekman 1922. (*Courtesy of Albert Bonniers Publishers, Stockholm*)

limiting factor, the distributional border usually follows the year-to-year geographic fluctuation of the limiting factor. If the limit is controlled by the slowness of the spreading process, it will also vary from year to year. *All distribution areas* are flexible with passing time. Therefore we ought to use contemporary data in range mapping, unless the particular animals are noted for slow spreading or their areas are limited by relatively stable barriers.

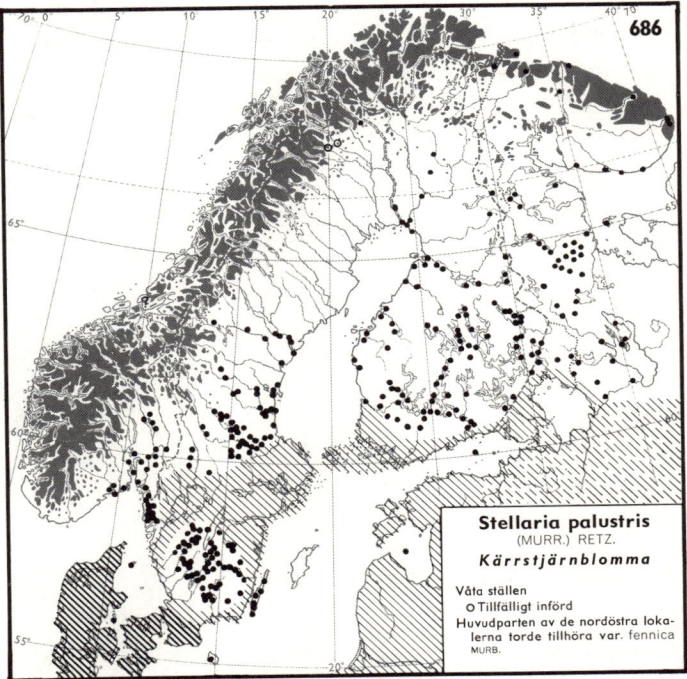

Figure 4-13   Accurate distribution maps.
The distribution of the chickweed *Stellaria palustris* (Murr.) Retz. (Caryophyllaceae) in Northwestern Europe as an example of very accurate and meticulous mapping work done on over 1800 plant species, from Hultén's *Atlas of the Distribution of Vascular Plants in Northwest Europe* (1950). Circles denote casual introduction by man. Each dot indicates a single locality where the species has been found. Each dot has a diameter corresponding to 16 km on the ground. If the finds are closer to one another than 16 km, the dots are confluent on the map. If this happens on a larger area, it is hatched. Fine hatching indicates that the species is not common, though the finds are closer than 16 km. Thick hatching indicates commonness. Dotted lines indicate that fewer data than usual have served as basis for frequency estimates, i.e., that the area is poorly explored. Note the similarity of the distribution area in southern Norway and Sweden with that of the wood pigeon (Figure 4-12). (*Courtesy of E. Hultén and A. B. Kartografiska Institutet, Stockholm*)

168 Areography: The Study of the Distribution Area

These principles of range limits and their mapping apply also, by and large, to plants. In plants, however, a fairly continuous boundary, at least within a largely uniform ecological area, means that the species is expanding and has not yet reached its limit of expansion. Fringed area limits and many large discontinuities usually indicate that the species is on the retreat but is still utilizing locally favorable areas (Cain 1944). These principles can be applied only to animals which are extremely slow in dispersal and when their area gains are comparable to the annual increment made by the radially spreading diaspores of plants.

In actively moving animals, which are able to choose their own pre-

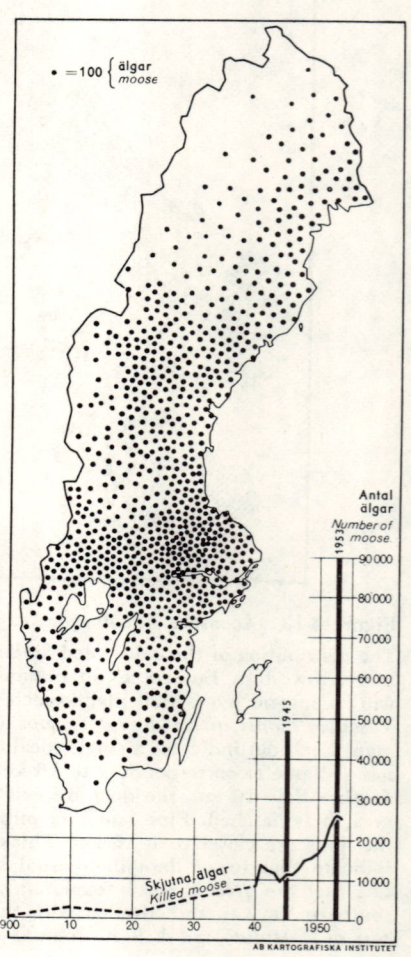

Figure 4-14 Mapping frequency of distribution.

Distribution of the elk (=moose) *Alces alces* L. [Cervidae Mamm.] in Sweden in 1953. Each dot represents one hundred animals; the estimated total was 90,000. The northwestern part of the country is above tree line; in the extreme south, coniferous forests are scarce and the land is intensively cultivated. Before the beginning of the nineteenth century, the elk was distributed over the same area but declined catastrophically and became limited to a rather small area in central Sweden. Through total protection for two decades, the species started to spread again after 1835 and has continuously increased subsequently. From Curry-Lindahl 1957. (*Courtesy of K. Curry-Lindahl*)

ferred environment, the situation can be different. A good example would be their reaction to environmental temperature. Ectotherm animals, such as insects, react to an increasingly cold environment by slowing down their reactions, finally lapsing into torpor long before reaching their lethal limit of cold. Endotherm vertebrates, on the contrary, react to cold by an increased muscular activity which often manifests itself in vigorous locomotion. A shelter or a more favorable local habitat in which the individuals can survive is thus reached with greater ease. The northern or altitudinal limit of a range where cold is the limiting factor is thus likely to be a more even line in the case of ectotherm animals. The opposite case, where warmth is the limiting factor, also merits consideration. The *optimal* range of temperature for activity is close to the upper, *pessimal*

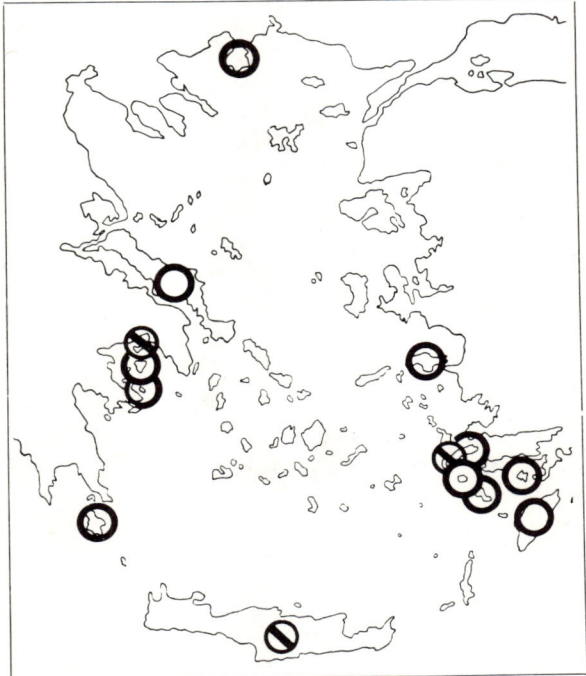

Figure 4-15   Use of symbols in mapping distribution.
Distribution of two rophalocere butterflies [Lepidopt.], *Pseudochazara anthelea* Hb. (circle with bar) and *Hipparchia allionii* Geyer (circle), on the Aegean islands. (from Bernardi 1961.) The single marks of occurrence on the islands are clearly distinct and visible; the occurrence of the two species are well separable. *(Courtesy of Centre National des Recherches Scientifiques, Paris)*

170 *Areography: The Study of the Distribution Area*

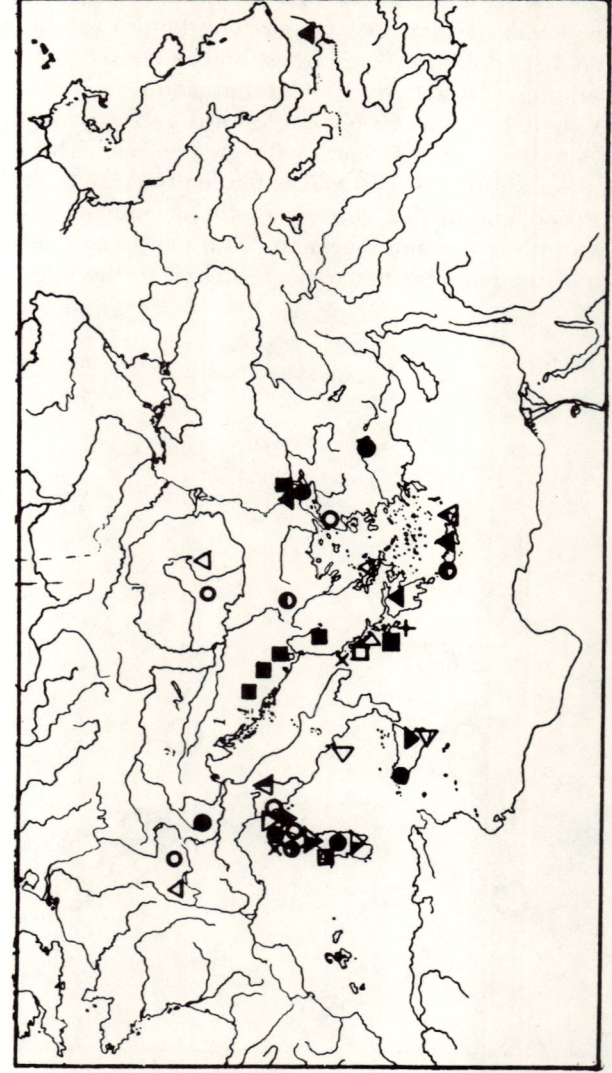

Figure 4-16 Use of symbols in mapping distribution. Distribution of triclads [Tricladida, Turbellaria] on the Mediterranean islands. (From Codreanu 1961.) There are actually sixteen different symbols, a question mark, and numbers 1–4 on this map attempting to illustrate twenty-one species. The effect is that of a picture puzzle. (*Courtesy of Centre National des Recherches Scientifiques, Paris*)

Distribution Areas and their Mapping 171

thermal limit in ectotherms, and thus many of these find locally cooler refuges when actively moving about. Lindroth (1956b), who originated this theory, observes that heat-relicts in Scandinavia (see p. 213) are

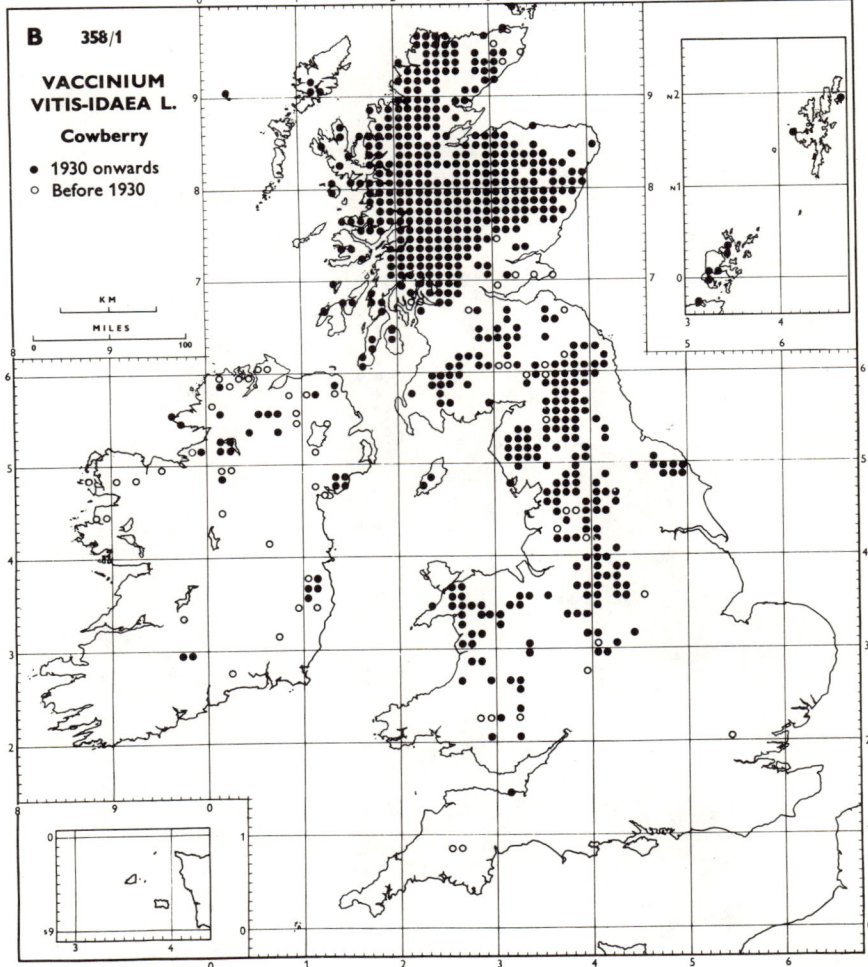

Figure 4-17 Use of grid systems in distribution mapping.
Dot map produced mechanically by the tabulator of the Distribution Maps Scheme of the British Floral Survey. From Perring and Walters 1962. (*By permission of the Botanical Society of the British Isles and Thomas Nelson and Sons Ltd., taken from their Atlas of the British Flora*)

172   *Areography: The Study of the Distribution Area*

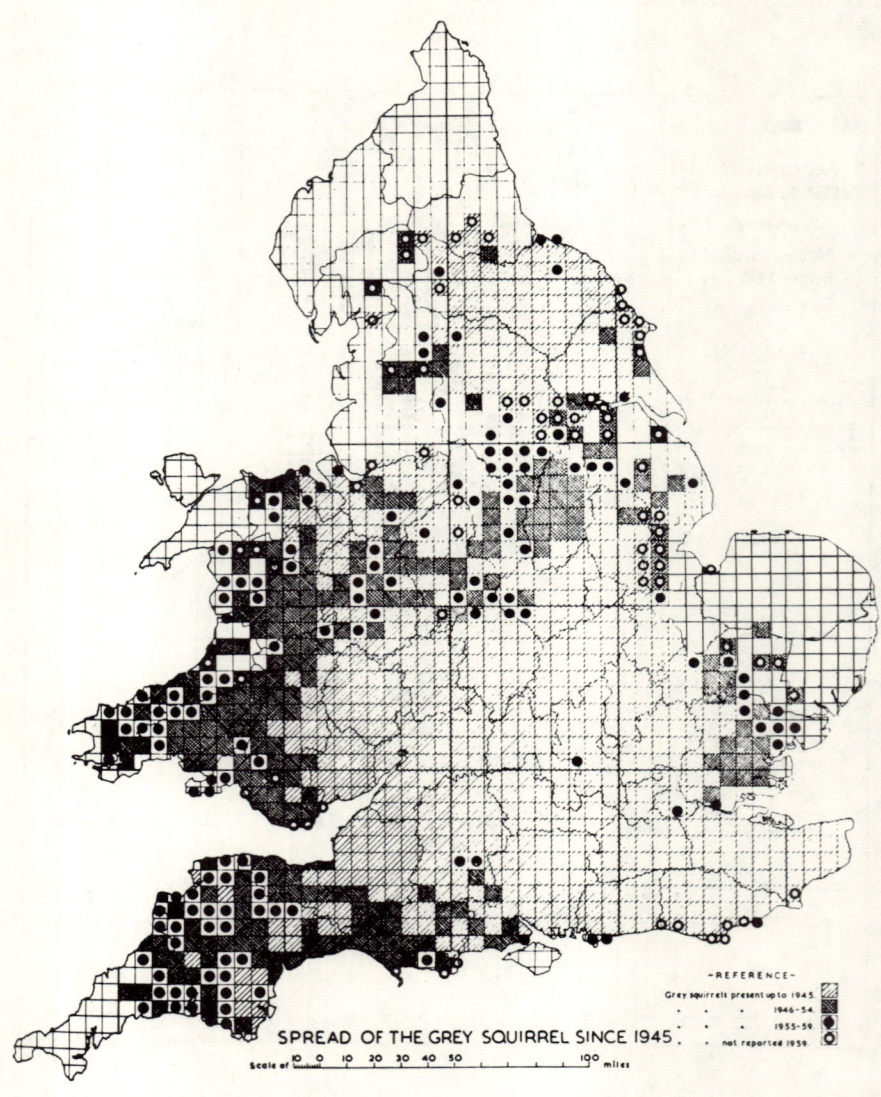

Figure 4-18   Use of grid systems in distribution mapping.
Distribution of the gray squirrel *Sciurus carolinensis* Gmelin, [Sciuridae, Mamm.], showing extension of the range in the British Isles between 1946 and 1959, using the 10 km square National Grid. From Lloyd 1962. (*By permission of the Comptroller of H. M. Stationery Office, British Crown Copyright*)

Distribution Areas and their Mapping 173

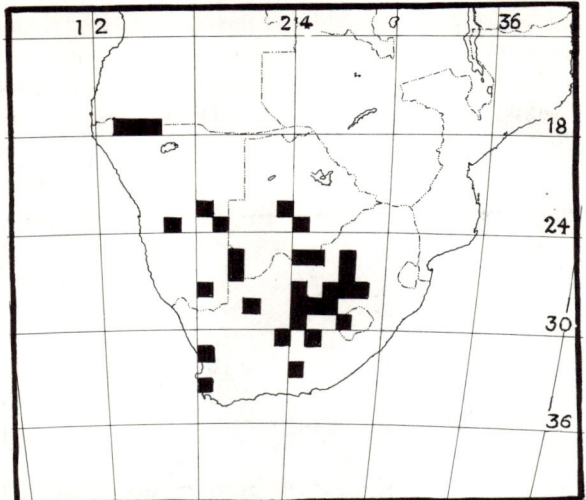

Figure 4-19 Use of grid systems in distribution mapping.
Degree square map showing the distribution of the mouse *Malacothrix typica* Wagner [Muridae, Rodentia, Mamm.] in South Africa. After Davis 1962, brought up to date by Davis 1969 (*in litt.*). (*Courtesy of D. H. S. Davis*)

restricted to wide areas of uniformly favorable climate. An exception is the endotherm southern birch mouse *Sicista subtilis* Pall. [Dipodidae, Mamm.], which, like southern plants, lives on isolated, widely scattered, relict habitats with warmer local climates (Figure 4–9). Lindroth enu-

Figure 4-20 Use of grid systems in distribution mapping.
Distribution of the nightingale *Luscinia luscinia* L. [Turdinae, Aves] (open circles) and the thrush nightingale *L. megarhynchos* C. L. Brehm (dots) in north-central Europe. Occurrence in each quadrangle is indicated by the corresponding symbol. The blackened area of the circle indicates the proportion of the occurrence of the two species where they overlap. From Schilder 1956. (*Courtesy of G. Fischer Verlag, Jena*)

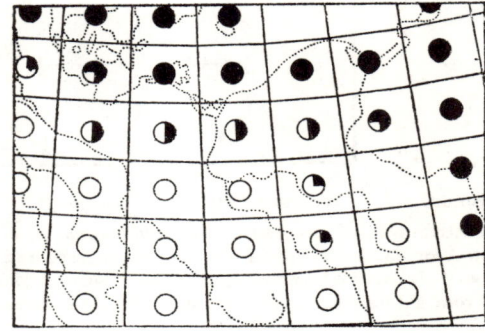

174 *Areography: The Study of the Distribution Area*

merates several species of carabid beetles that have fringed southern limits in Germany. Occurrence in scattered peat bogs of the more northern type is also well known among the ectotherms of southeastern North America. In endotherm vertebrates, temperature tolerance is combined

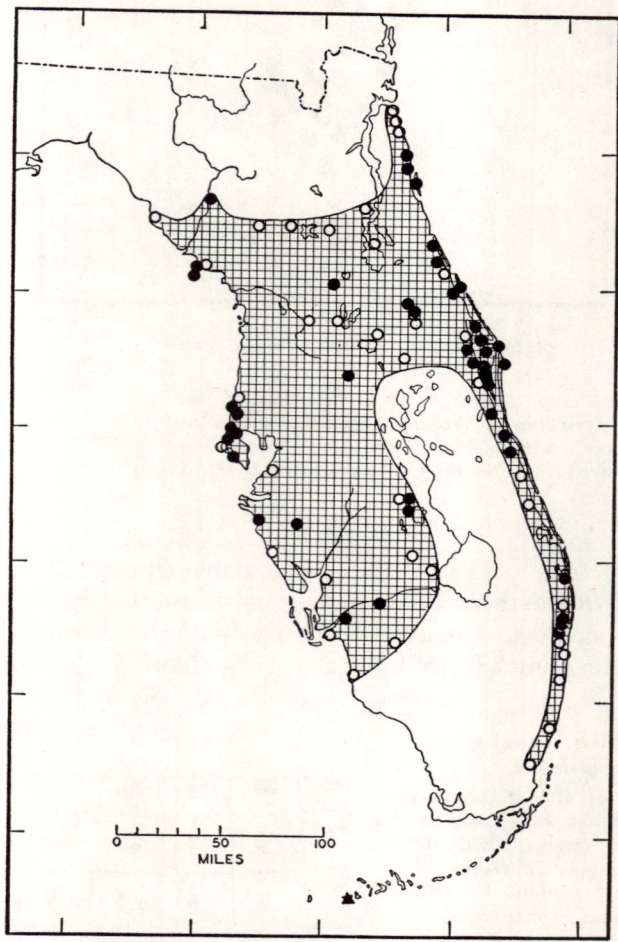

Figure 4-21 Accurate distribution mapping.
Distribution of the scrub jay *Aphelocoma c. coerulescens* (Bosc.) [Corvidae, Aves] in Florida. Dots indicate localities represented by specimens; circles, literature reports; triangles, vagrant specimens. Figure 4-22 is based on detail maps like this one. From Pitelka 1951. (*Courtesy of F. Pitelka and the University of California Press*)

with water balance, therefore the situation in these animals is not as simple as in the case of beetles or butterflies; in vertebrates, there is also a very pronounced role of active, behavioral, habitat selection. Schmidt (1950), for example, thought that the northern limit of the five-lined skink *Eumeces fasciatus* (L.) [Scincidae, Rept.] is temperature-controlled in the northern United States. Yet isolated areas of favorable local climate are populated, especially during exceptionally warm summers, and the limit is strongly fringed. Thus, much survey work would be desirable in order to test Lindroth's ideas respecting vertebrate area limits.

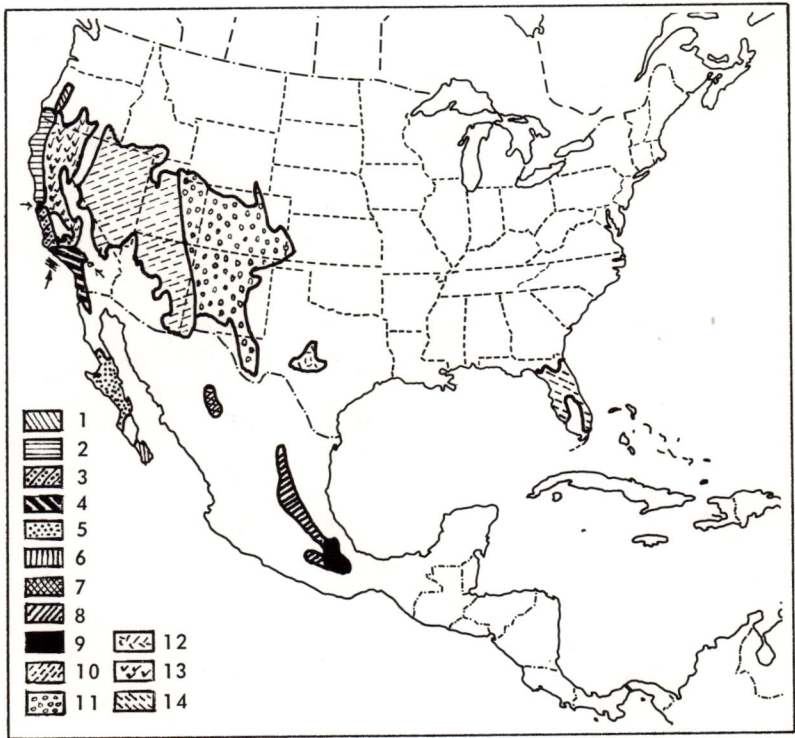

Figure 4-22 Accurate distribution mapping.
Total distribution area of the scrub jay after maps (see Figure 4-21) in Pitelka's (1951) taxonomic monograph. Over a thousand localities where specimens have been collected are on the original maps. Over 4,800 specimens have been studied from 745 localities; 18 subspecies are shown on the map; southern Mexican occurrence is still incompletely known.

176 Areography: The Study of the Distribution Area

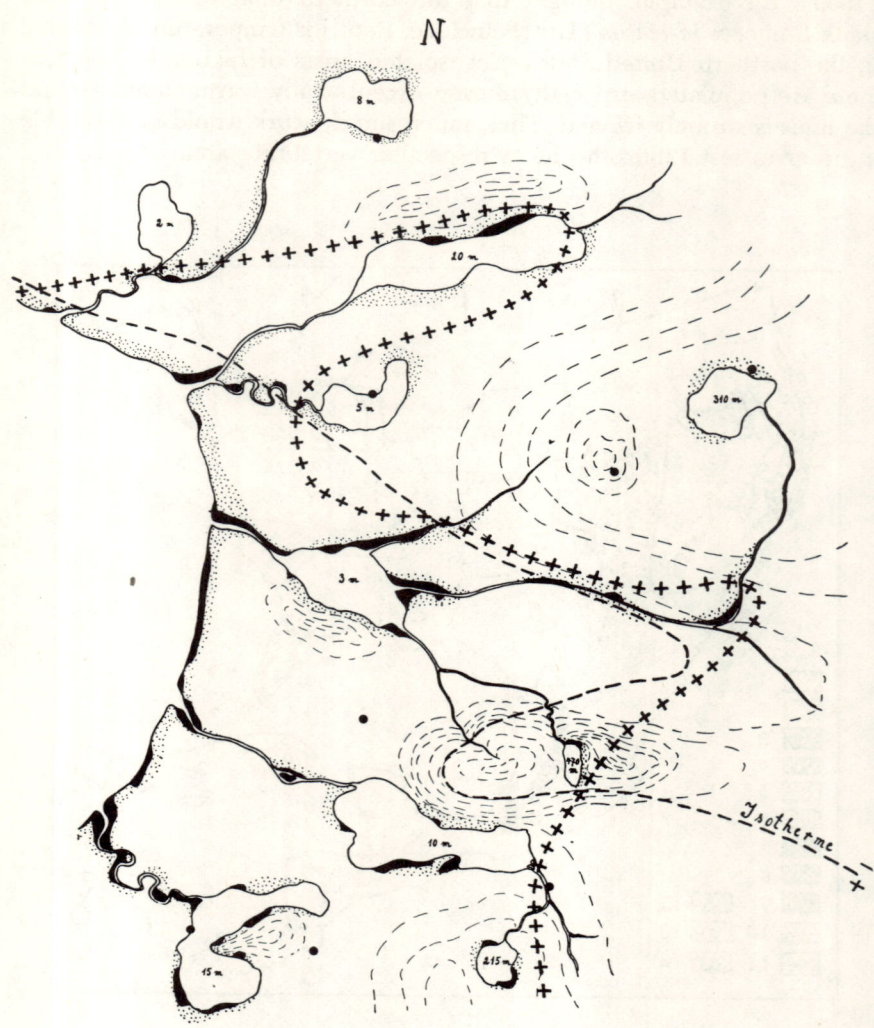

Figure 4-23 Area limits.
Northern limit of a hypothetical temporary "area" of an insect restricted to shores and banks of a certain type. Black areas indicate populations; black circle, single specimen; dotted areas, suitable habitat; crossed line, the zoogeographical "limit of the area." From Lindroth 1949. (*Courtesy of C. H. Lindroth and the Royal Society of Gothenburgh, Sweden*)

The Shape of the Area

In addition to the theories cited above, the history of a species also greatly affects its distribution area; the reverse is also true, in that distribution areas supply clues for the paleobiologist. For this reason, the general shape of an area is entirely relevant to the present discussion.

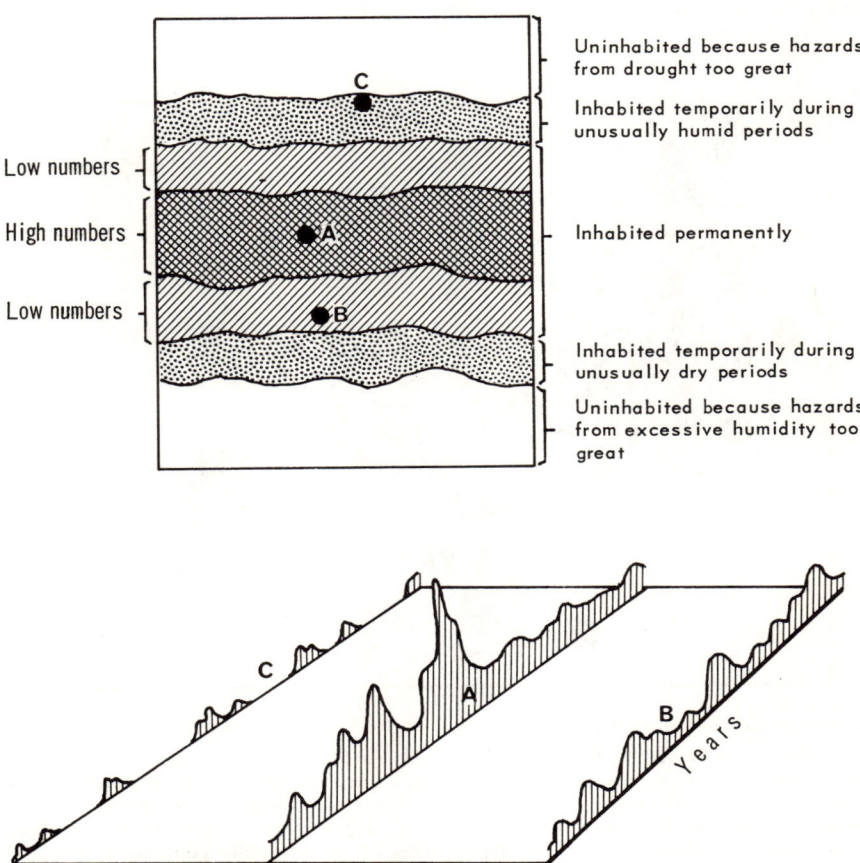

Figure 4-24 Area limits.
Andrewartha and Birch's models showing the relation of distribution to habitat distribution and population size. *Above:* Zonal habitats and population density. *Below:* Grasshopper populations in the localities marked on the upper map during many years. The actual animal upon which the model is based is *Austroicetes cruciata* (Sauss.) [Acrididae, Orthopt.]. Redrawn from Andrewartha and Birch 1954.

178 *Areography: The Study of the Distribution Area*

Areas can be continuous or discontinuous, and the latter may be treated as disjunct or disperse areas (Figures 4–25, 4–26). A distributional area is *continuous* if within it no suitable habitat remains unsettled—that is, there are no gaps larger than can be covered by normally dispersing individuals of a species (Figure 4–7). This definition represents an ideal situation; but in nature only approximations of the ideal can be found, because no habitat is uniform over a great expanse. Then too, the local density of every animal population fluctuates and approaches zero under adverse conditions in at least the most inferior part

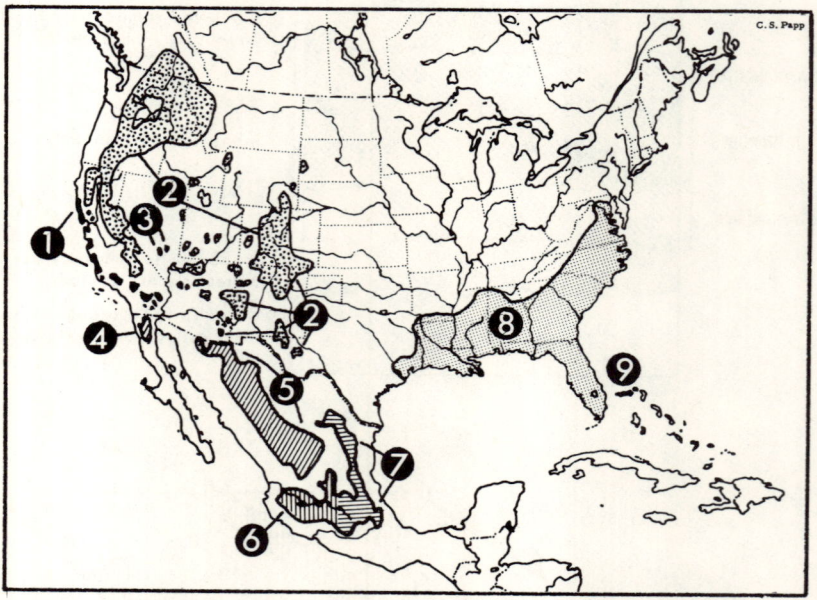

Figure 4-25 Types of distribution area.
Distribution area of the brown-headed nuthatch *Sitta pusilla* Latham in southeastern, and of the pygmy nuthatch, *S. pygmaea* Vigors [Sittidae, Aves] in southwestern, North America. (After K. A. Norris 1958.) These two species closely resemble each other in morphology, behavior, and ecological requirements. They are *vicars*. Their speciation is evidently caused by the splitting of an originally continuous range in continuous pine woods during a Pleistocene glaciation into disjunct western and eastern areas. Both species are sedentary and philopatric; yet the distribution areas have striking structural differences, owing to the different geographic configuration of the habitat in the west and in the east. *Pygmaea* has an extremely *disperse area*, with many small population isolates in the desert-encircled southwestern mountains; seven are recognized as subspecies (1–7). *Pusilla* has a *continuous area* occupied by a uniform population (8), except for the isolated small island colony on Grand Bahama Island (9), which is subspecifically distinct.

of its habitat (Figure 4–24). Another circumstance—that, in practice, only a water barrier can result in sharp area boundaries—is responsible for the fact that most continuous areas show some fringing of their outline. When two populations of great vagility are separated by an unin-

Figure 4-26 Disjunct distribution area.
Distribution of the Holarctic nuthatch *Sitta canadensis* L. [Sittidae, Aves]. The North American population has a large continuous area, and is not differentiated into further subspecies. It has been found fossil in the late Pleistocene of Southern California, indicating that it had crossed to North America at an earlier date. In Europe, it has a small relict distribution area on the Mediterranean island of Corsica. There is a fairly continuous part area in Asia Minor and the southern Caucasus and another disjunct part area in northern China and Korea. (Data from Voous 1960.) Recent behavioral observations by Löhrl (1960–61) revealed that the Corsican and North American forms would not mate, and have differences in calls that warrant specific status. Therefore Vaurie (1959) considers the four populations specifically distinct rather than one polytypic-polytopic species.

180   *Areography: The Study of the Distribution Area*

habited region, the discontinuity is more apparent than real, for, due to their vagility, genetic continuity is maintained (Schmidt 1950). Circumpolar species of ducks, for example, maintain contact across the Atlantic as well as across the Pacific (Figure 4–27).

Continuity in most instances is most often based on conjectural evidence. If the distribution of a habitat is known to be confluent, and if a species is known to occur at fairly regular densities, it is assumed, even in the absence of faunistic data, that the species occurs in connecting areas as well. While in a majority of cases the zoogeographer has to rely upon this kind of evidence, it is nevertheless evidence which must be used with extreme care, and findings based on it must be expressed clearly to keep from misleading other investigators. Taxonomists scrutinizing specimen localities or early faunistic accounts become painfully aware of erroneous data of this sort. On the other hand, findings suggestive of scattered and disconnected occurrence very often indicate the

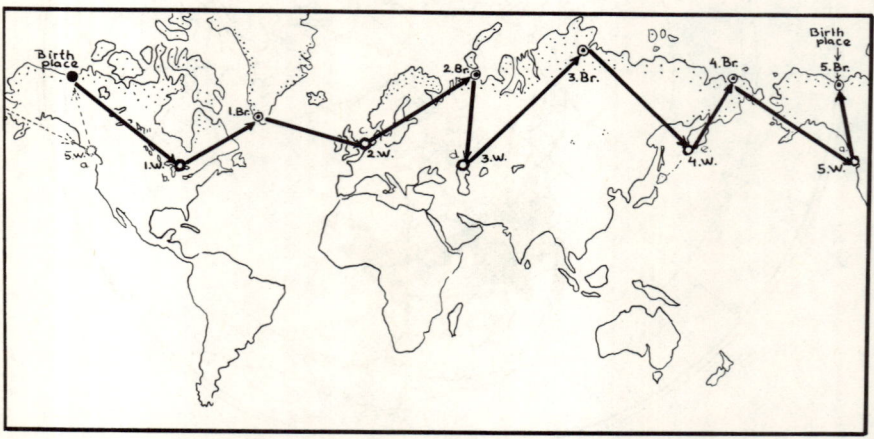

Figure 4-27   Apparent discontinuity of the area.
The long-tailed duck or old squaw *Clangula hyemalis* (L.) [Anatidae, Aves] is circumpolar. The area is widely discontinuous geographically. On the map, the breeding area is dotted, and some of the wintering localities are marked with circles. Ducks from several parts of the breeding area may use the same wintering places; as they mate there, mixing occurs. (Cf. Salomonsen 1955.) Thus, the species population is continuous in spite of the apparent geographic discontinuity, and no subspeciation is present. A hypothetical male, hatched at the Mackenzie Delta in Arctic Canada (dot) may, in five consecutive years, breed and winter at ten different localities around the Northern Hemisphere (1.W. = first wintering place, 1.Br = first breeding place, and so on) since males always follow their mate during the spring migration to the breeding grounds, but leave these areas ahead of the females in the company of other males.

sparseness of field workers rather than of the animal in question. In faunistic as well as in paleogeographical work based on fossils, a large volume of discontinuities has been filled in as a result of intensified field research during the past few decades. In sum, the continuity of a range may be either real or apparent and both can be underestimated or overestimated.

Discontinuity of areas is characteristic of the majority of animal species, and the study of discontinuities and their causes is the chief source material of causal zoogeography. A *disjunct* distributional area consists of a few major areas widely separated in space; an area is said to be *disperse* when it is composed of many relatively small but distinctly separate subranges. Naturally, all kinds of transitional types occur, from strictly continuous to disjunct and disperse areas. A species is called *polytopic* if it consists of genetically separated populations, i.e., when its area is discontinuous and no regular crossing of intervening barriers occurs.

The reasons for areal discontinuities are always historical but may in addition be ecological. Most animal species which reproduce sexually originated on one small area; therefore, all these forms had a continuous area in the beginning. However, discontinuities may arise either by expansion across barriers or by retraction of formerly larger areas, while leaving behind some local populations which, in this way, become separated from the basic stock. Whatever the origin of the discontinuous areas, they may, at the time of observation, still be ecologically separated by barriers. Discontinuities may also be maintained because the spreading potential of the species is not large enough to fill the gap across an otherwise suitable habitat. A third reason for discontinuities may be based upon a fallacy—when the supposed discontinuous areas are inhabited by two unrelated animals (that are similar by convergent evolution) whose phylogeny has been misjudged. Such occurrences are usually characteristic of higher taxa and not of a discontinuous area of the same species (Figure 4-28). If any doubt arises about the specific identity of two congeneric populations that occupy geographically discontinuous areas, it is usually based on morphologic (sometimes behavioral or ecologic) differences. If such differences really exist, the discontinuity is proved by them. Specific differences could arise only by spatial separation during a considerable time. In judging such cases, the rule should be whether the disconnected populations (i.e., those which do not maintain genetic continuity) appear indistinguishable on the subspecific or specific level (Figure 4-25). If these discontinuous populations are specifically distinct, their further study leads to a zoogeographically founded consideration of their evolutionary history, in which

182 *Areography: The Study of the Distribution Area*

areographical skill would be a useful aid but no longer the center of interest.

The Structure of the Area

The shape of the area is a purely geographic concept, which becomes apparent after mapping of the distribution. Causal analysis of area shape, like every zoogeographic analysis, leads into ecological and historical considerations beyond the purely geographic. On the basis of the foregoing discussion, it follows that the shape of every area is greatly dependent on its internal structure; for certain wide geographic discon-

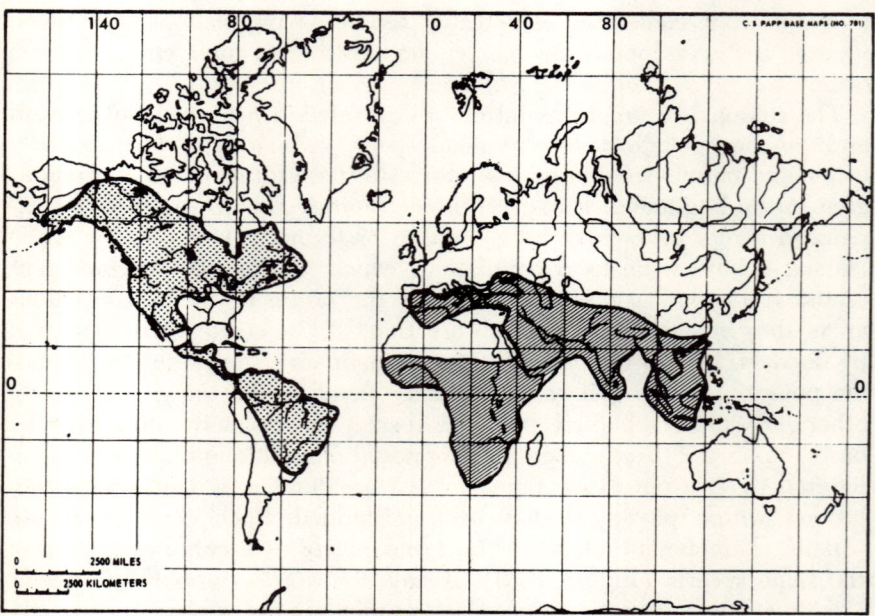

Figure 4-28 Fallacious discontinuity of a taxon.

Distribution areas of the Old World porcupines (superfamily Hystricoidea, Mamm., broken lines) and of the New World porcupines (superfamily Erethizontoidea, Mamm., solid lines). Both are after Bartholomew *et al.* 1911, slightly modified. The porcupines of both worlds are extremely similar in morphology and habits, and were formerly thought to represent discontinuous branches of the same suborder, Hystricomorpha, their distribution evidencing former continuity of the southern continents. Many modern researchers in mammalian paleontology (Wood 1950, Colbert 1955; but not Landry 1957) believe that they evolved convergently, from unrelated rodent stocks. The discontinuity, and historical arguments based upon it, is not real but fallacious; the animals involved are *pseudovicars*.

tinuities paradoxically within the limits of continuous occurrence are not true discontinuities. Dispersing individuals cross these and connect the genetic pool of the population on both sides of such an internal barrier. In other words, although the area itself is discontinuous at such a locality, the population is not.

Structural analysis of an area leads from geographical to microgeographical considerations, and these deal essentially with the chorology of the species—namely, the spatial occurrence of the individuals, with population structure in space.* The spatial structure of a population depends: upon *intrinsic factors*—viz., the specific systems of dispersion (territorial, home range system, colonial dispersion, and the like); and upon such *extrinsic factors* as the competitive utilization of environmental requisites, and the many other ecological and geographical features of the typical habitat of the species. Thus, many types of dispersion (Figure 4–29) and local distribution (Figure 4–30) occur.

The size of a total species population is another important but little-known structural characteristic of an area. Quantitative population studies and estimates are still in their infancy, with the exception of a few well-studied portions of such areas (see, for example, Figure 4–14) or minute, often dwindling populations of some large-bodied vertebrates whose numbers seldom exceed a million individuals. Few sources give a quantitative specification for the term "rare" as it applies to a species of beetle, land snail, or lizard.

Figure 4–31 presents a hypothetical example in which the role of environmental and historical factors upon the shape, structure, and dynamics of an area are illustrated. The distribution area of a monophagous, plant-eating insect, and the average level of population densities in any of its parts (A), is here controlled by the geographic distribution of the food plant, limiting climatic influence, and a very influential predatory organism (B). In the area where optimal physical and biological environments coincide, or, to use Rubtsov's (1937) expression, where conditions closely approach its "synecological optimum," we find the highest population densities. Elsewhere, the zonal effect of the climatic factor, as well as heavy predation, affects population densities within the range of an animal, thus bearing out the thesis of Andrewartha and Birch (1954) that "distribution equals abundance." Stated differently, where mortality is lower than natality, a dense population builds up and the limit of the distribution area is that line along which mortality (plus emigration) exceeds natality (plus immigration). With the passage of time, some of these environmental components may change in intensity

---

*This was discussed from another point of view in Chapter 3.

184 *Areography: The Study of the Distribution Area*

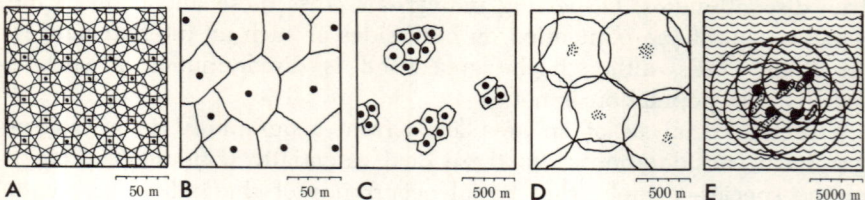

Figure 4-29 Structure of the area: some schemes of dispersion.

All these examples are different in their effect on the population and its ecology, genetics, and evolution. (B, D, and E after Wynne-Edwards 1962.) In schemes A, B, and D, the population (dots) is uniformly spaced; in A – D, the habitat is uniform. Action circle of individuals (home range or territory) is shown by circles or polygons.

A Random, even spread of individuals, with totally overlapping home ranges.

B Territorial spacing. The action radius of each individual (or pair, or family unit) does not overlap.

C Same as B (on diminished scale), but territories are clustered. In this particular instance, there is low density with unoccupied space between clusters, which indicates that either social attraction or low spreading potential caused the clustering.

D Home ranges of colonies, with slight overlap at the margins.

E Large colonies on the only available sites on islands (e.g., sea birds, seals). The home ranges—that is, feeding areas—overlap, and their sizes depend on the action radius of the individuals.

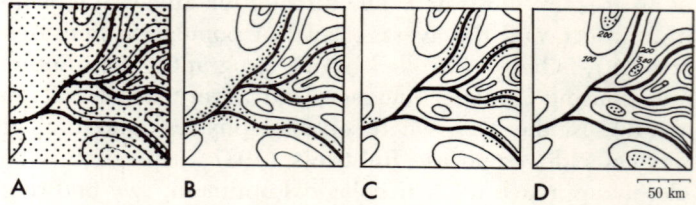

Figure 4-30 Structure of the area: some schemes of local distribution.

All these examples have different effects on the population and its ecology, genetics, and evolution. All available habitat is occupied. Dotting indicates population density; thick lines indicate rivers; thin lines follow contours at each 100 m of elevation. A and B after Timofeeff-Ressovsky (1940).

A Local distribution with varying density. All the area is occupied, but lowland habitats are inferior and have lower densities.

B Distribution in floodplain habitat, continuous population.

C Discontinuous local populations on particular habitat (cool streamside) of restricted occurrence.

D Zonal distribution with small discontinuous populations. Our theoretical species frequents only the vegetation zone which occurs at altitudes between 200 and 300 m.

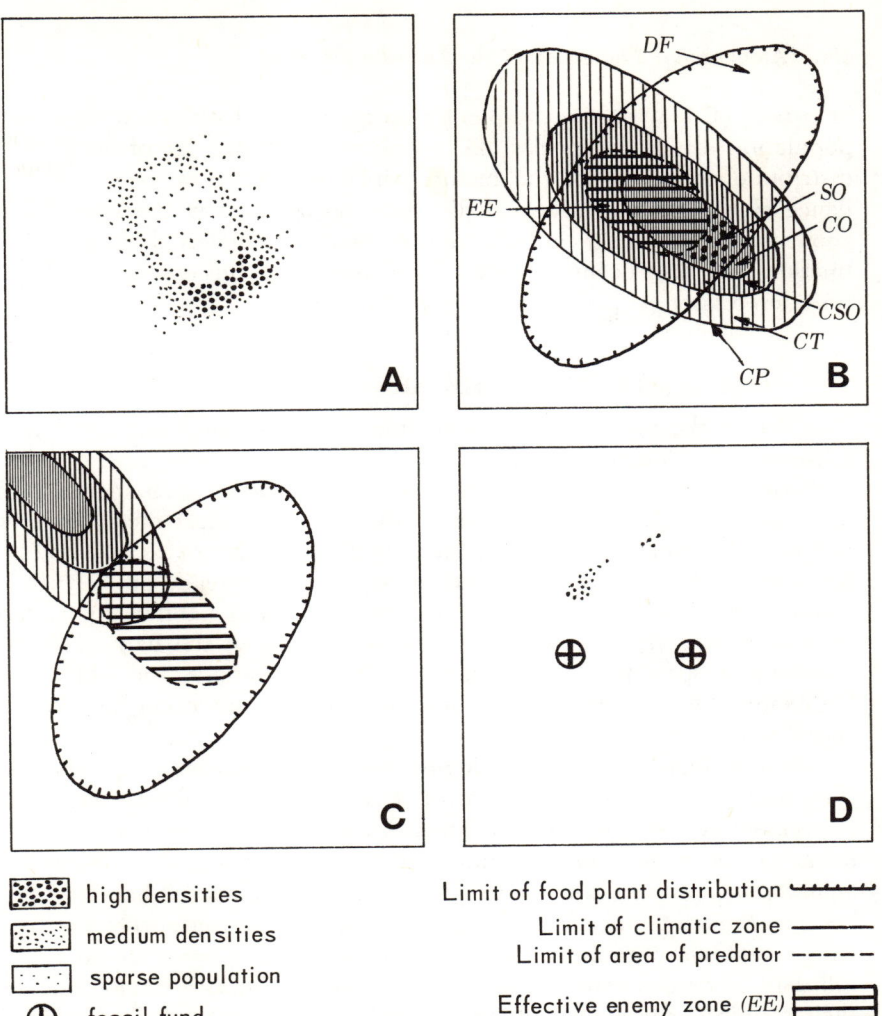

⬚	high densities
⬚	medium densities
⬚	sparse population
⊕	fossil fund

Limit of food plant distribution ⌣⌣⌣⌣
Limit of climatic zone ——————
Limit of area of predator − − − − −

Effective enemy zone (EE) ≡≡≡

Figure 4-31 The effects of ecological and historical factors on the area.

A Distribution area of a hypothetical monophagous animal, with three density zones indicated by increasing density and size of dots.

B Range of occurrence of three important ecological factors and of the *synecological optimum* of our hypothetical animal on the same area as A. CO = climatic optimum; CP = limit of climatically possible distribution; CSO = climatic suboptimum; CT = climatic tolerance zone; DF = area of food plant; EE = effective enemy zone; SO = synecological optimum. After Franz 1964, modified.

C Range of the same ecological factors at a later time on the same area. Climatic zones shifted toward the northwest, but the range of the food plant and the predator were not affected.

D The conditions depicted in C result in two small relict populations on disjunct, relict areas. All direct evidence of the past conditions is a few fossil funds of the animal.

or extent (C), with a consequent change in the distribution area and population size of the species (D). We have no knowledge of the former distribution area and of the factors which influenced it; our only evidence of the former, historical distribution (A) is now the knowledge about two fossil localities. Following further environmental changes, extinction may also occur. The historical aspects of this example will be considered later (p. 351).

THE ECOLOGY OF THE AREA

The size of the area depends on all factors of existence and spreading (discussed in Chapters 2 and 3).

These factors may be summarized as follows. Since each species had its origin in a given area, and since all except the most recent species had time for expansion—or even for retraction—the area size will reflect: the ability of the species to occupy many widespread, or only few, localized habitat types; the size or extent of these habitat types; the spreading capacity of the species; and the nature of the barriers to be overcome in spreading. These are all ecological relations, and, only after their effect on area size is taken into account, can attention be concentrated upon the historical factor.

When ecologists speak of ecological valency or plasticity of a species, they refer to the range or extent of its environmental limitations. It is convenient to use the prefixes "steno–" (indicating a narrow range of tolerance or requirements) and "eury–" (indicating the opposite). *Stenotopic* organisms are able to inhabit only certain special habitat types because their tolerance limits are narrow or their requirements are specific. *Eurytopic* organisms, on the contrary, are able to adapt to many different kinds of habitat because their tolerance limits are wide and/or their requirements are small. Stenotopy may also restrict distributional areas, whereas eurytopy allows wider spread. Widest spread is exhibited by *amphitopic* species (Figure 3–4).

As long as the required habitat is fairly evenly distributed, it is available for the animal. Thus, the geographic extent—the distribution and size of the habitat or habitats—is also an important factor influencing the size of the range and of the species.

The intrinsic ability of a species to spread is called its *vagility*, which of course includes morphological, behavioral, and other elements, such as suitable locomotor organs, willingness to enter unfamiliar ground, and readiness to accept strangers into a social unit. Philopatry and vagility are opposite tendencies in terms of spreading potential. A highly philopatric population will provide few pioneers; conversely, a vagile population has a low degree of philopatry.

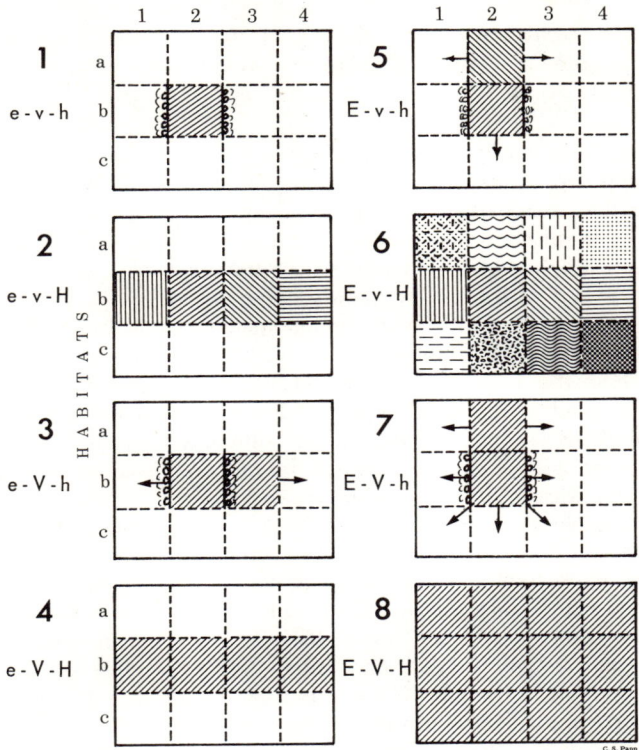

Figure 4-32 Models of the eight theoretical distribution area types resulting from the ecological relations of species 1–8.

Each model represents a theoretical continent within a major climatic zone, consisting of three habitat types (*a*, *b*, and *c*) and four geographical regions (1–4) separated by lesser barriers (dashes running vertically). The area of habitat *b* in region 2 is delimited by strong physical barriers which are represented by semicircles in models 1, 3, 5, and 7. Crosshatching and the like covers the distributional area; different crosshatching within a model indicates subspeciation. Since our models illustrate extreme cases, no subspeciation is indicated in models 4 and 8, though slight subspeciation is possible in both instances. As discussed in the text, only types 1, 2, 4, 6, and 8 seem to occur with more or less permanency (dynamic changes could not be considered on these models). However, types 3, 5, and 7 may exist as transient stages in new or rejuvenating areas; the arrows (small where mobility is restricted; large in strongly vagile species) indicate the transitional nature of these areas which would in time become types 4, 6, and 8, respectively.

e = restricted ecological valency = narrow range of adaptability = stenotopy
E = large ecological valency = wide range of adaptability = eurytopy
v = restricted mobility or vagility
V = high vagility or mobility
h = restricted habitat
H = widespread habitat
1,2,3,4 = geographic regions of a continent
a, b, c = habitats (zonally arranged, horizontally related habitats)

188 *Areography: The Study of the Distribution Area*

Though distributional areas are of a great many different types, and their synthetic treatment is a different matter, a certain grouping will elucidate the role of ecological factors in relation to area size. (Figure 4–32, Table 10).

1. There are stenotopic species with low vagility, on a local, restricted habitat type. Such "specialists" are geographically restricted, usually monotypic animals, and the small extent of their area does not provide much opportunity for developing locally different populations. The smallness of their habitat and their stenotopy might indicate that they are members of a dwindling faunation, possibly relicts of a formerly more widespread group.

2. If the habitat is broader, a stenotopic species would most likely also be widely distributed. Despite its restricted vagility, a long existence enables spreading. Over the long term, increased inbreeding and the potential for mutations may lead to the development of local adaptations causing intensive subspeciation. This is especially true if the habitat is separated by minor barriers which, together with the intrinsically induced isolation of each local population, would cause isolation necessary to incipient speciation. The widely distributed specialist is highly polytypic—that is, subspeciated.

3. For two reasons, a stenotopic, highly vagile form does not seem to manifest itself on a restricted area. Because of its vagility, a species of this kind would lose too many emigrants to afford such steady losses without any benefit to the population in still clinging to its habitat. Selection would sooner or later favor some stationary allele or combination, or vagility would result in expansion through the barriers. If there is such a type in existence, it would, in a very short time, change into model type 4.

4. The same combination of ecological species characteristics enabled the stenotopic, vagile species to spread over widely distributed habitats. Its vagility assures steady mixing of the local populations, even though they are far from one another and perhaps isolated by large expanses of unsuitable habitat. In widespread but monotypic or only slightly subspeciated "specialists," we usually find physical* and ethological** signs of high vagility. If their special habitat is spread across many climatic zones and covers several continents, these species best represent a kind of cosmopolitanism.

5. To find an eurytopic, slow-spreading species on a local restricted habitat would appear to be a paradox. Relict distributions are unlikely because of the eurytopy involved; therefore, such a type probably does not exist.

6. An eurytopic and slow-spreading animal is more likely to dwell on a widespread, major habitat type—e.g., a biome—where eurytopy enables it to utilize the locally varying conditions. The extent of the habitat and of the

---

*Such as the strong legs of an accomplished runner or hopper, the pointed wings of a migratory bird, and macroptery in insects.

**Such as the preference to fly or circle over forest canopy, the chasing of the young from their natal domicile by adults, the phototactic orientation in the flight of an emerging insect.

Table 10 North American vertebrates exemplifying some distribution area types.
(Area types are shown on Figure 4–32, and are explained in the text.)

Area type	Scientific Name	Common Name	Family, Class
1 e-v-h	*Mustela nigripes* (Audubon & Bachman)	Black-footed ferret	Mustelidae, Mamm.
	Vermivora luciae (Cooper)	Lucy's warbler	Parulidae, Aves
	Dendroica kirtlandii (Baird)	Kirtland's warbler	Parulidae, Aves
	Gopherus polyphenus (Daudin)	Gopher tortoise	Testudinidae, Rept.
	Bufo canorus Camp	Yosemite toad	Bufonidae, Amph.
2 e-v-H	*Microtus pennsylvanicus* (Ord)	Meadow vole	Cricetidae, Mamm.
	Carpodacus purpureus (Gmelin)	Purple finch	Fringillidae, Aves
	Chrysemys picta Schneider	Painted turtle	Emydidae, Rept.
	Ensatina eschscholtzi Gray	Eschscholtz's salamander	Plethodontidae, Amph.
4 e-V-H	*Gulo luscus* (L.)	Wolverine	Mustelidae, Mamm.
	Anas acuta L.	Pintail	Anatidae, Aves
	Aegolius funereus (L.)	Boreal owl	Strigidae, Aves
	Cnemidophorus sex-lineatus (L.)	Six-lined racerunner	Teidae, Rept.
	Rana sylvatica Le Conte	Wood frog	Ranidae, Amph.
6 E-v-H	*Peromyscus maniculatus* (Wagner)	Deer mouse	Cricetidae, Mamm.
	Melospiza melodia (Wilson)	Song sparrow	Fringillidae, Aves
	Thamnophis sirtalis L.	Common garter snake	Colubridae, Rept.
	Ambystoma tigrinum Green	Tiger salamander	Ambystomidae, Amph.
8 E-V-H	*Ursus arctos horribilis* Ord	Grizzly bear	Ursidae, Mamm.
	Pandion haliaetus (L.)	Osprey	Accipitridae, Aves
	Hirundo rustica L.	Barn swallow	Hirundinidae, Aves
	Rana catesbeiana Shaw	Bullfrog	Ranidae, Amph.

area of the species testifies to a certain geological age of the faunation generally and of the species studied. Polytypy is a necessary feature of such animals, and the most heavily subspeciated animals are among this type.

7. Eurytopic and mobile species would not be restricted to localized, special habitats; hence, this type is not likely to exist.

8. Eurytopic and especially mobile species are likely to include many different habitat types within their distribution area, although steady interchange of individuals and, through them, of genotypes, would not allow heavy subspeciation, i.e., restriction of their eurytopy by local adaptations. These species are the most genuine cosmopolitans, showing distribution across many climatic belts and major regions separated by generally active barriers.

Although types 3, 5, and 7 are paradoxical, they may indeed exist for a short time as developmental stages of types 4, 6, and 8. It must be kept in mind that these five distribution types are broad, though useful, generalizations, albeit several objections may be raised against their unrestricted validity. First, neither of the three criteria (ecological valency, vagility, habitat utilization) is well known for all parts of the distributional area for most animal species. Second, these categories presuppose that the participating species are distributed throughout the total available habitat, but this is not always so. Many species are now enlarging their ranges by slow penetration of quick expansion. Others are temporarily restricted in distribution by strong barriers (e.g., the shore of oceanic islands), and their spreading might thus be hindered. Third, eurytopy, stenotopy, and the different forms of vagility are not all-or-none types of species characters, since intergradations and morphisms frequently occur in this respect. It is well known that at least some species have different tolerances in different parts of their areas, i.e., they may be stenotopic here, eurytopic there, or stenotopic to a different degree in another area. Innate vagility—e.g., local variations in wing structure—and philopatry also differ locally. Clinal variation of characters and elimination *sensu* Reinig (see below) are also phenomena that make the pigeonholing of areas and species difficult. Finally, in comparing ecological species character with the geography of the habitat, a foundation is provided for consideration of the evolutionary history of a species or of the fauna, without which the picture is incomplete.

In ecological treatments of area types, we often find mention of ubiquitism and cosmopolitanism. It is best to apply these related terms to two different concepts.

A *ubiquitist* (from the Latin *ubique* "everywhere") is a highly euryecious* form which can live almost anywhere within a large type of

---

*This term, in its widest sense, is almost synonymous with "ubiquitous."

habitat—that is, one form may live in every kind of freshwater; another might exist in any kind of soil; and the like. Thus the term is purely ecological. A *cosmopolitan* (from the Greek, "inhabitant of the whole world") is widespread geographically, though true world-wide inhabitants (dwelling on all six geographic continents or in all world oceans) perhaps do not exist (Figure 4–33). This, then, is the geographical term. A cosmopolitan may be a ubiquist—i.e., widely eurytopic and euryecious —like many small aquatic invertebrates: protozoans, rotatorians, and crustaceans (belonging to type 8, above, of our ecogeographical classification of species). It may be a vagile specialist on a widespread habitat (type 4) but not at all ubiquitist. A ubiquitist may be restricted in geographic distribution by barriers, but utilizes a great number of different habitats—for example, many of the island animals.

The geographic extent of the climatic belts and continents explains that some "tropicopolitan," pantropic species are nearly cosmopolitan in their distribution. The term is more meaningful when applied to higher taxa, for they are per se more widely distributed (page 234). Handlirsch

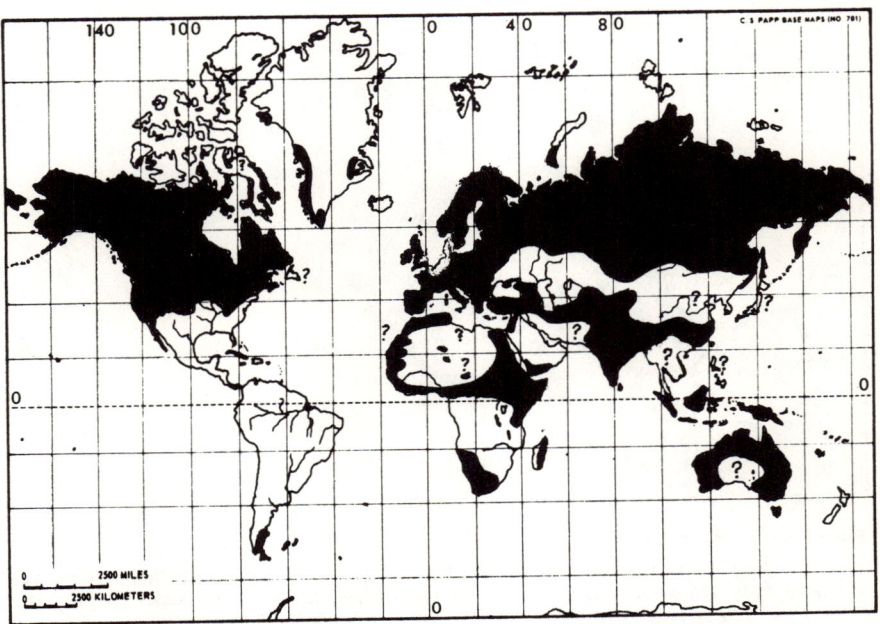

Figure 4-33 Cosmopolitan distribution.
The worldwide distribution of the peregrine falcon *Falco peregrinus* (Tunst.) [Falconidae, Aves]. After Voous 1960.

(1913) lists 116 insect genera which extend to six continents making 1.4% of 8,300 analyzed genera. Bartholomew et al. (1911) show four cosmopolitan freshwater mollusk families and two mammalian ones (Vespertilionidae and Muridae). Emerson (1955) finds only two cosmopolitan termite genera: *Kalotermes* and *Neotermes;* his criterion of cosmopolitanism is occurrence in each of the Sclater-Wallacean zoogeographical regions. Cosmopolitan plants are mostly vagile species living on unsettled habitats; most are aquatic or tropical. De Candolle (1855), who was probably the first to tally them, found only 19 species of higher plants which occurred on more than half of the earth's surface.

The concept of *vicarism* is used widely, by botanists and zoologists alike, when comparing certain types of distributional areas (Figure 4-34). Though its original meaning is not ecological, it is now often used in this sense and therefore will be considered in discussing the ecological connotations of an area. Moritz Wagner (1868) describes vicariant as two closely related species which are separated by a barrier but which otherwise live in geographically related areas, such as certain Carabid beetles on two opposite slopes of the Caucasus. Botanists also use this term—much more frequently than do zoogeographers—to define geo-

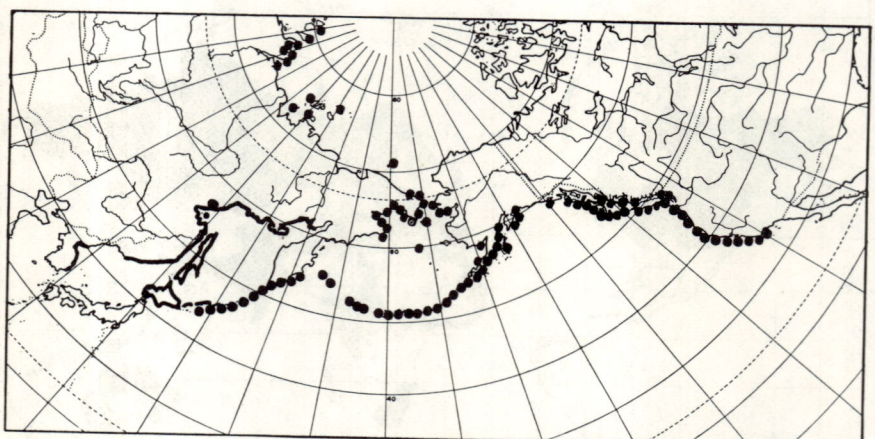

Figure 4-34 Vicarism (true vicars).

The breeding distribution of the North Pacific sea birds pigeon guillemot *Cepphus columba* Pall. [Alcidae, Aves], dots, and spectacled guillemot *Cepphus carbo* Pall., thick line. These two species are closely related, and have similar if not identical ecological requirements. They most likely speciated from common stock during an earlier Pleistocene glaciation, and are now allopatric except on one of the southernmost Kurile Islands, where their coastal, linear areas meet. After Udvardy 1963a, modified. (*Courtesy of Bishop Museum Press, Honolulu*)

graphically (and ecologically) allopatric pairs of species or higher taxa. Ecological and seasonal substitutes are also vicars as long as they remain taxonomically related. Cain (1944) calls ecological vicars species pairs which differ in ecological requirements in a complementary way, e.g., in the pH of their soil. He points out, however, that such differences may arise after geographic separation as part of the speciation process. Most kinds of subspecies are vicars, i.e., allopatric—by definition. Species belonging to the same section, tribe, subgenus, and related genera, subfamilies, and so on, may become sympatric and at the same time ecologically widely divergent in the course of their history. Those which really "vicariate" on their allopatric areas deserve the distinctive term *vicars*. The concept may apply even to communities. Many find that similar but geographically distant parts of the world have communities with similar

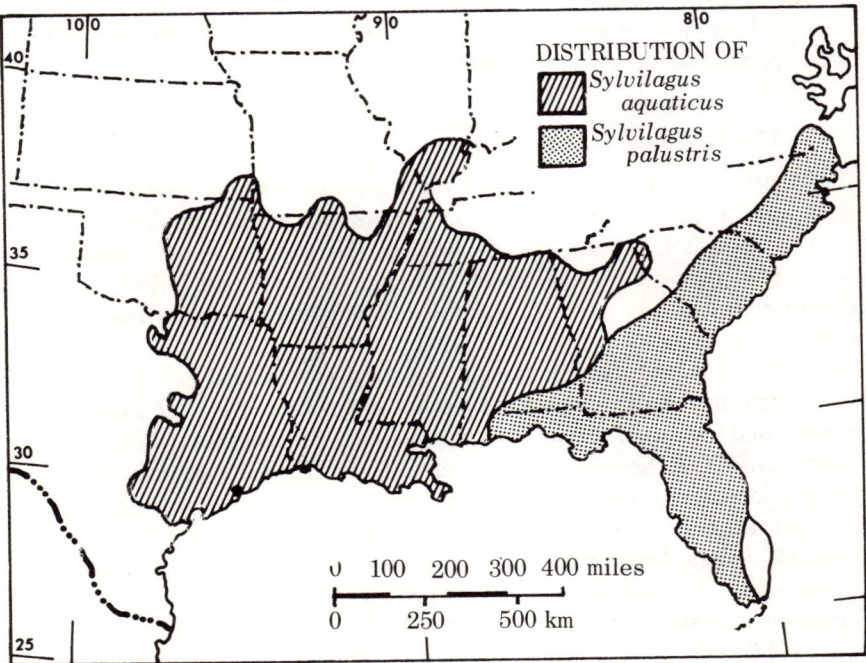

Figure 4-35 Ecological vicars.
Distribution of the swamp rabbit *Sylvilagus aquaticus* (Bachman) [Leporidae, Mamm.] and the marsh rabbit *S. palustris* (Bachman). (After Hall and Kelson 1959.) These species are members of the same subgenus but not the closest relatives, and they differ in size. They have allopatric, adjacent distribution areas, and are ecological substitutes in similar habitats.

structural patterns, consisting of related species, or species of the same life forms (cf. Poore 1962).

The *ecologically similar, distributionally allopatric, and phylogenetically unrelated ecological counterparts*, the "pseudovicars" of Vierhapper (1919), are converging analogues and not homologues, as are the true vicars (Figure 4-35). Figures 4-34 and 4-44 show the distribution of some true vicars. Voous (1959) points out that the horned lark *Eremophila alpestris* L. [Alaudidae, Aves] occurs in the same habitats of the New World as does the skylark *Alauda arvensis* L. in the Old World. Though Voous calls these birds ecogeographical substitutes, the vicar concept applies here because they are members of the same family. The frogs *Acris gryllus* (Le Conte) [Hylidae, Amph.] of southeastern North America and *Pseudopaludicola falcipes* (Hensel) [Leptodactylidae] of southeastern Brazil appear to occupy identical niches in similar habitats, although they belong to different families and live on different continents (Milstead 1961). They are superficially identical in shape, size, color, color pattern and variation, voice, and daily and seasonal activities. These frogs are pseudovicars, i.e., ecological counterparts.

The Synecology of the Area

The basic study of the area is geographic, and ecological connections are more appropriately attached to the species population itself. One point needs emphasis here—viz., the relation of the species area to that of the community (ecosystem) in which the species lives (see below, p. 235), for communities are also geographically existing entities, though of a different rank.

It was noted long ago that when major environmental conditions allow the occurrence of a widespread community, there is a change in the species making up the community although its members may be sufficiently alike for the structure to remain much the same. On the other hand, related communities may harbor the same species of plant or animal yet differ in some other important aspect. For instance, the western hemlock *Tsuga heterophylla* (Sarg.) occurs in the related rain-forest communities of coastal western North America (the Douglas-fir-hemlock-red-cedar forests of British Columbia, Washington, and Oregon; and the redwood forests of California and southern Oregon), though the climatic conditions of the summer are different in these areas.

Coincidence (overlap) of the species and community area depends chiefly on: the age and extent of the community; the distribution of the minimum factor or factors within the community which limit the occurrence of the species, i.e., the homogeneity of the habitat (biotope) factors; the age of the species; and the length of the association of the species with the community. These factors alone allow many different combinations, so that there is no rule for predicting coincidence. A species area may be smaller or larger than the area of the community of which it is a regular member. Yet the area of wide-ranging plant or animal species may be much larger than that of wide-ranging communities (Figures 4–36, 4–37). This follows from the understanding of eurytopy in wide-ranging species. These are adapted to related niches (environmental situations, widespread constellations of certain essential habitat elements) which occur across many different communities, while the vegetational basis of the community necessitates constellations of mainly physical factors, which are more limited geographically.

Overlap with community area may be complete but, as follows from the age criteria above, it might concern subspecific, specific, superspecific, or even generic boundaries (see Figures 4–61, 4–62).

HISTORY OF THE AREA

It is often worthwhile to scrutinize the morphological, ecological, and other characteristics of an area in the hope that these might provide clues to its history and, consequently, to the distributional history of the species. Biogeographers attempted many times in the past to find the principal correlations between an area and its history but, for the most part, they failed. Their hypotheses were widely known for a time, and stirred much controversy, interest, and research; but most of them were eventually discredited. They will be reviewed only briefly here, since they are readily available in the literature.

Several attempts have been made to determine the *center* of a geographic area—with the understanding that this helps to find the geographic origin of the species. Several criteria can be used to define a center within an area. The center of *density* is that part of an area with the most abundant population. The center of *maximum variation* refers to genetic structure. The center of *frequency* is that part of an area in which a species occupies the largest variety of habitats—i.e., where it is most euryecious. The currently distant parts of the area have been reached from the center of *disperal*. The center of *dispersal* is the point from which a species began its expansion into the current distribution area. It is not necessarily the center of *origin*, since the whole range might have shifted or might have started new expansion from small cores after a regressive stage in its history. These "relict" areas could have served as centers of *preservation*, helping the species to survive an adverse period of time. Diverse kinds of "centers" might be added indefinitely, according to the background and interest of the investigator. These could be based on some

196   *Areography: The Study of the Distribution Area*

specific characteristic which varies geographically according to the rate of its incidence and the environmental (selection) pressure. Another example would be the center of *efficiency*. Density alone may not mean the

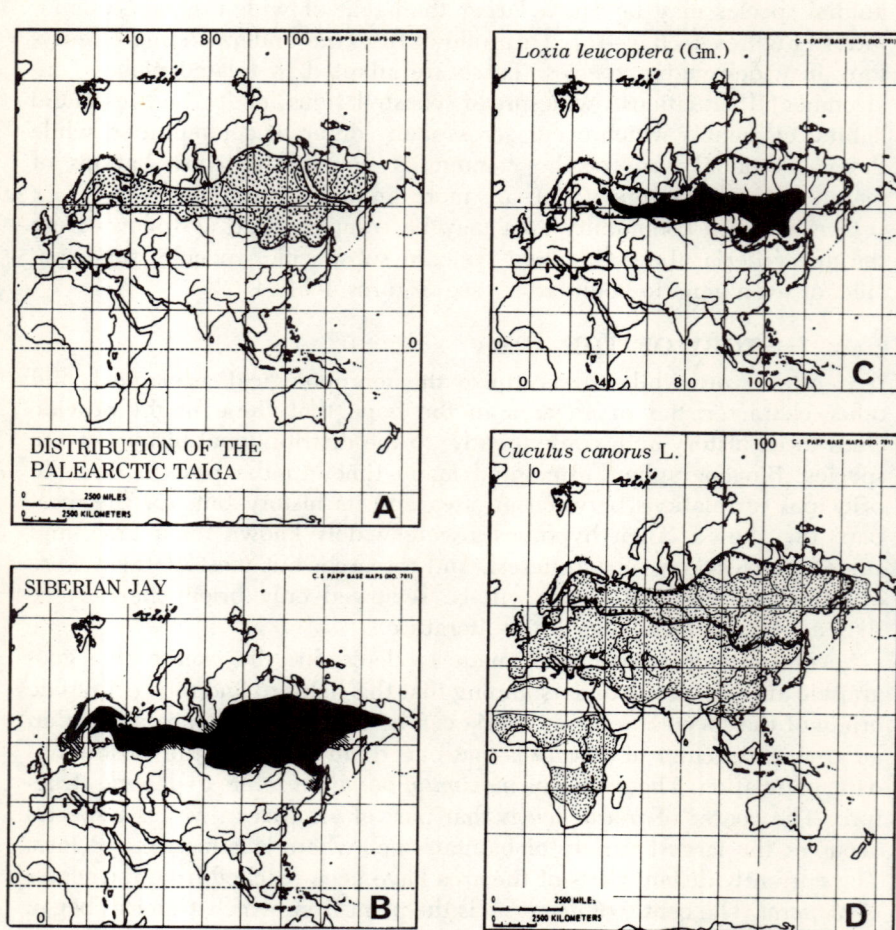

Figure 4-36   Correlation of species and community areas.
The Eurasian taiga belt (A) and some of its characteristic birds. (B). The Siberian jay *Perisoreus infaustus* (L.) [Corvidae], a bird stenotopic in, and closely associated with, the taiga. (C). The white-winged crossbill *Loxia leucoptera* (Gm.) [Fringillidae] has a much narrower distribution than the taiga (which is also shown on the map), and is restricted to a particular community type. (D). The cuckoo *Cuculus canorus* L. [Cuculidae] has a very wide distribution area which includes the taiga but also many other forested biomes. B, C, and D distributions based on Voous 1960.

same as the combination of longevity, body size, metabolic rate, action radius, and numbers within a generation per area. The center of efficiency should indicate population success. All these centers may be of assistance in speculating about the center of *origin,* but the latter cannot be found by geographic considerations alone, especially when the areas

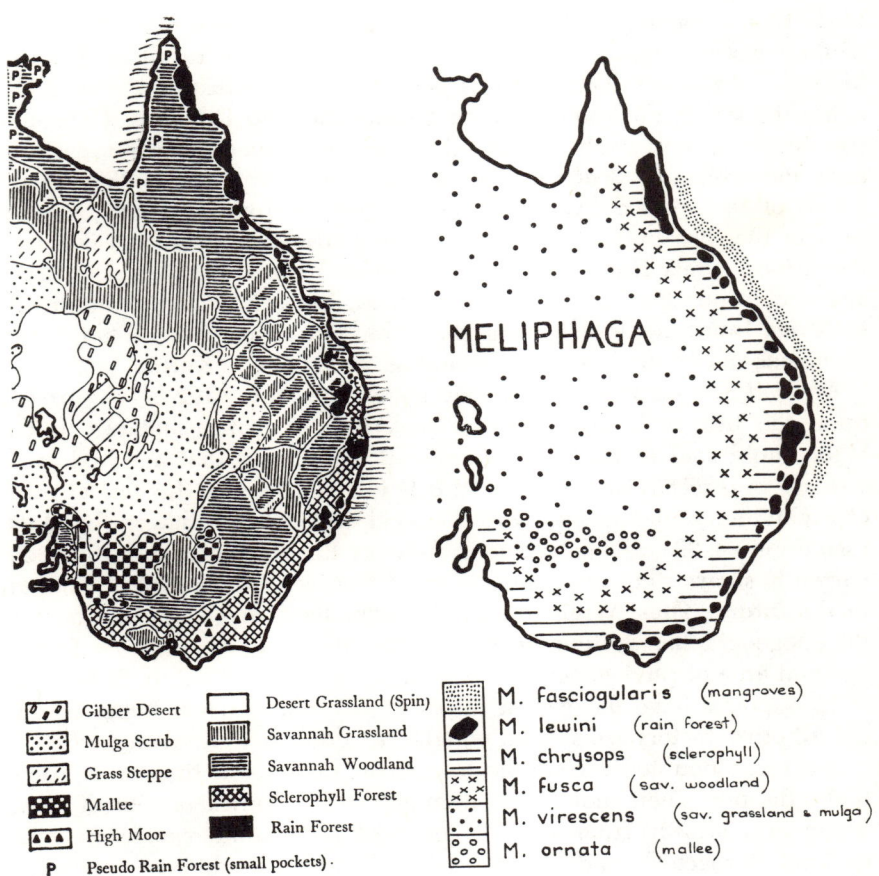

Figure 4-37 Correlation of species and community areas.
(*Left*) The distribution of the major vegetation formations in eastern Australia. From Keast 1962, after Prescott and Wood. (*Right*) The distribution of six species of birds from the genus *Meliphaga* (honeyeaters, Meliphagidae), somewhat diagrammatically. From Keast 1961. Left: (*Courtesy of the Royal Society of Victoria, Melbourne University Press, and G. W. Leeper*) Right: (*Courtesy of A. Keast and the Museum of Comparative Zoology, Harvard University*)

of related forms and of the nearest higher taxon are not simultaneously taken into consideration. Yet the most crucial argument against finding the center of origin within the present area of the species is that during the lifetime of all but the most recent species (which would differ depending on the kind of organism, but would probably be anywhere between a half a million and ten million years), very drastic climatic changes have resulted in large-scale shifting of climatic belts not only in the temperate and cold regions but even in the tropics. Thus it is likely that a majority of the present species do not even live at the site of their historic origin or that, if they do so, it is the result of chance. Many fossil records document this phenomenon. For instance, paleontologists think that the spotted hyena *Crocuta crocuta* Erxleben, [Hyaenidae, Mamm.] most likely originated in Asia; its suspected ancestor is the early Pleistocene *C. sivalensis* Falconer & Cautley from India. Pleistocene fossils of the spotted hyena are known from Europe to China, yet at present the distribution of this animal is restricted to Africa. Likewise, the striped hyena *Hyaena hyaena* L. is today mainly distributed in Asia, and occurs but sparsely in northern Africa. Yet it has been widespread in Africa from the early Pleistocene times, and an earlier African form, *H. namaquensis* Stromer, is suspected as its ancestor (Kurtén 1968).

Age and Area—a slogan of biogeographers during the 1920's—was based upon the title of a once famous book by the British botanist J. C. Willis, published in 1922. Willis' original theory was that age and area size are in direct relation. While this holds true for the theoretical case of expanding young species with identical ecology and uniform habitats (see Figure 4–32), no one can accurately predict whether a species which currently shows signs of regression or of halting will or will not expand in the future. Thus, Willis' theory does not have much application for the zoogeographer desirous of tracing the history of an area. As to the age and area of phylogenetic units (taxa), Emerson (1952, p. 224) says, "Older groups have the widest dispersal, all other factors being equal, but all other factors are seldom equal." His point is well taken.

Species undoubtedly continue to appear, exist for a time, and eventually die out. Their individual "lifetime" would consist of: "birth," i.e., origin in a locality from a small, incipient population; expansion until its limit is reached; and in time, with changing conditions, regression and extinction. An "offspring"—i.e., a new species that might have originated during any phase of the life of its "mother" species—may survive; but few if any species follow this ideal pattern. Expansion results from oscillations of the boundaries and not even simultaneously in all directions. Long periods of retreat may divide the area, only to be reunited in later, expansive phases. Instead of dying out, the species might gradually be changed into another, newer form within a substantial part of its area. Besides these and other vicissitudes of history, the size of the area will definitely show correlation with the power of dispersal and with the action radius of dispersing individuals.

It is difficult to find an example of a young, expanding species in an ecologically uniform area. Therefore we may consider related cases, such as (a) species that human transport helped across a major barrier and that are now in the process of spreading and apparently have not yet reached, at the time of mapping their area, either their ecological limits or a major, hindering barrier—for example, the muskrat in Central Europe (Figure 3–22); (b) the distribution of certain subspecies, or at least of individuals with some well-recognized and outstanding (usually morphological) characteristics. The vole *Microtus arvalis* (Pallas) [Cricetidae, Mamm.] reaches its northern limit in Central Europe in Denmark (Figure 4–38). Immediately south and east of this peninsula, the sampled vole populations show 85 to 90 percent incidence of a genetically controlled aberration of the third upper molar tooth (the lack of the fourth transverse ridge), the so-called simplex condition. Zimmermann (1935) examined around 4,200 skulls from this and surrounding areas.* The incidence of the simplex condition diminishes evenly, along concentric circles which center on the Denmark–West-Pomerania area. The river Elbe proves to be a barrier, but not insuperable. North of it, the simplex molar shows about 90 percent occurrence; on the south bank of the estuary, 50 percent. Looking at the map we notice that its eastern and western limits are uniform; but in the south, its spread is largely blocked by the hills of Central Germany. Yet voles are abundant in this area. Though the appearance and distributional history of this mutant are not known, it can be surmised, from Zimmermann's map, to have arisen somewhere within the area where it now predominates and thence spread in concentric circles. If this assumption is true, its area in the beginning of its spreading was presumably related to its age. However, later studies (Stein 1958; but see Zimmermann 1958) claim that this mutant has locally high incidence elsewhere within the species' area on habitats that are inferior and marginal to the species. The geographically marginal area of southern Denmark might provide increasing opportunities to the dominating occurrence of the simplex form. Thus, not even this single example is a convincing proof of the age-area relation.

The morphological and genetic structure of the species can be studied

---

*Most of the skulls were extracted from owl "pellets," the undigestible remains of prey that these birds regurgitate.

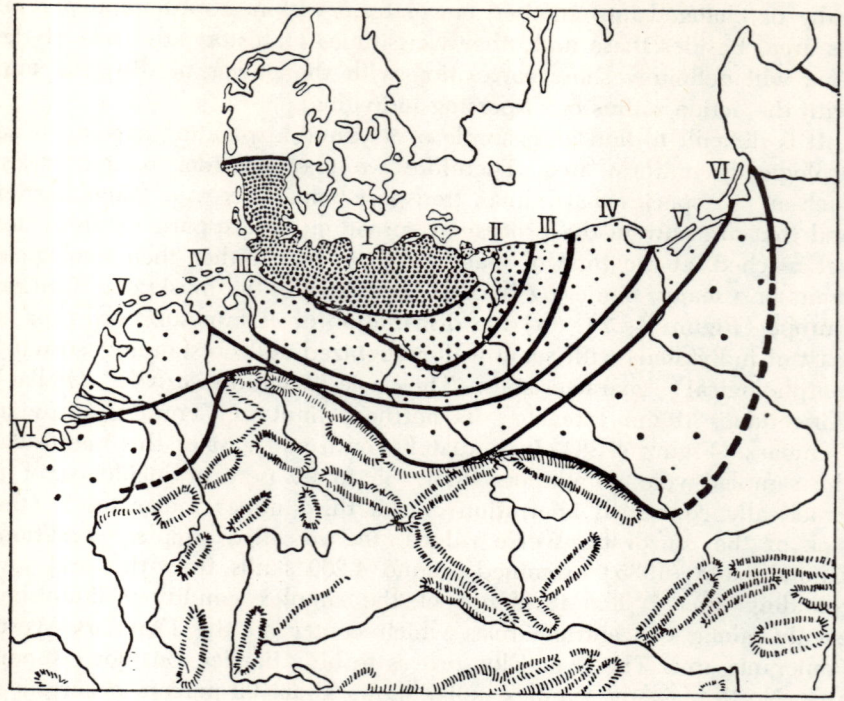

Figure 4-38 Distribution of a genetic mutation.

Incidence of the "simplex" condition (a morph characterized by an aberrant molar tooth ridge) in the northwestern European marginal population of the vole *Microtus arvalis* Pallas [Cricetidae, Mamm.]. Incidence zones (percentage of the measured sample within each zone): I = 85–90%, II = 65–70%, III = 50%, IV = 20–30%, V = 10–15%, VI = 5%, VII = less than 5%. From Zimmermann 1935.

from the geographic point of view, for these aspects are interrelated with the structure of the area. The species may consist of a uniform population or of contiguous, or intergrading, sharply delimited or geographically isolated part populations, of which many are named as subspecies (Figure 4–25). The degree of structural variation is most often correlated with present or past ecological diversity of the range and with the history of barriers. Thus, the number of subspecies is likely to be low on large, ecologically uniform areas; if subspecies are found—especially without apparent geographic or ecologic barriers—they may signify spread from some formerly isolated parts of the historic area of the species. They may not have had time to merge, or peculiarities of a hybrid belt

(Mayr 1942, 1963) or of some other, intrinsic, isolating factor may have kept them apart.

A morphologically uniform species may exhibit physiological, behavioral subspeciation or clinally changing character gradients within its distribution area. The modern genetical concept of species population visualizes the predominance on each part of the distribution area of a gene pool which best exploits the local conditions at that time. Earlier authors called these variants *biotypes* or *ecotypes*. Since the introduction of the *ecological valency* concept, it has been realized that stenotopic and eurytopic populations may differ not only in the range of individual tolerance but also in that within a stenotopic population there are only a limited number of biotypes while many an eurytopic population has a greater variability and therefore possesses greater potential for adaptation to diverse conditions (polytopy). Reinig (1938) draws a parallel between the shape of the geographic area and the genetic constitution of the species in postglacially reoccupied temperate regions. A species which has a large variety of biotypes, i.e., an euryecious species—spreads *centrifugally* (or *aclimatically*), for it possesses adequately adapted biotypes capable of expansion under many and diverse environmental conditions. A species with restricted variation, poor in biotypes, is able to spread only *isoclimatically*—it is able to fill in the limited habitats within the climatic belt to which its few biotypes are adapted. We find examples of both of Reinig's categories among the area types discussed above (Figure 4–32). Many North American authors use the term "zonal distribution" to "isoclimatic distribution." This usage originates in the concept of life zones (see Chapter 5), and is more acceptable if we restrict it to distributions occupying an *altitudinal zone*, which usually means that a single vegetational zone or belt is inhabited along a mountain chain (Figure 4–57).

Discontinuities of range may arise either by spreading over and across a barrier or by splitting a formerly continuous area. Barrier crossing is by far the more seldom used way of spread across a land barrier, especially for nonflying animals; it is very common in overcoming stretches of sea. All oceanic islands are populated by a selection of actively or passively dispersing animals. All continental islands that have been devastated by such catastrophes as volcanic eruption, glaciation, and flooding of the continental shelf have regained a relatively varied fauna within a geologically short time through dispersal across water. On the other hand, disjunctions of ranges within land areas can usually be explained by a climatic or tectonic change in the past: former habitat connection and range continuity have been broken by the intrusion of an area of

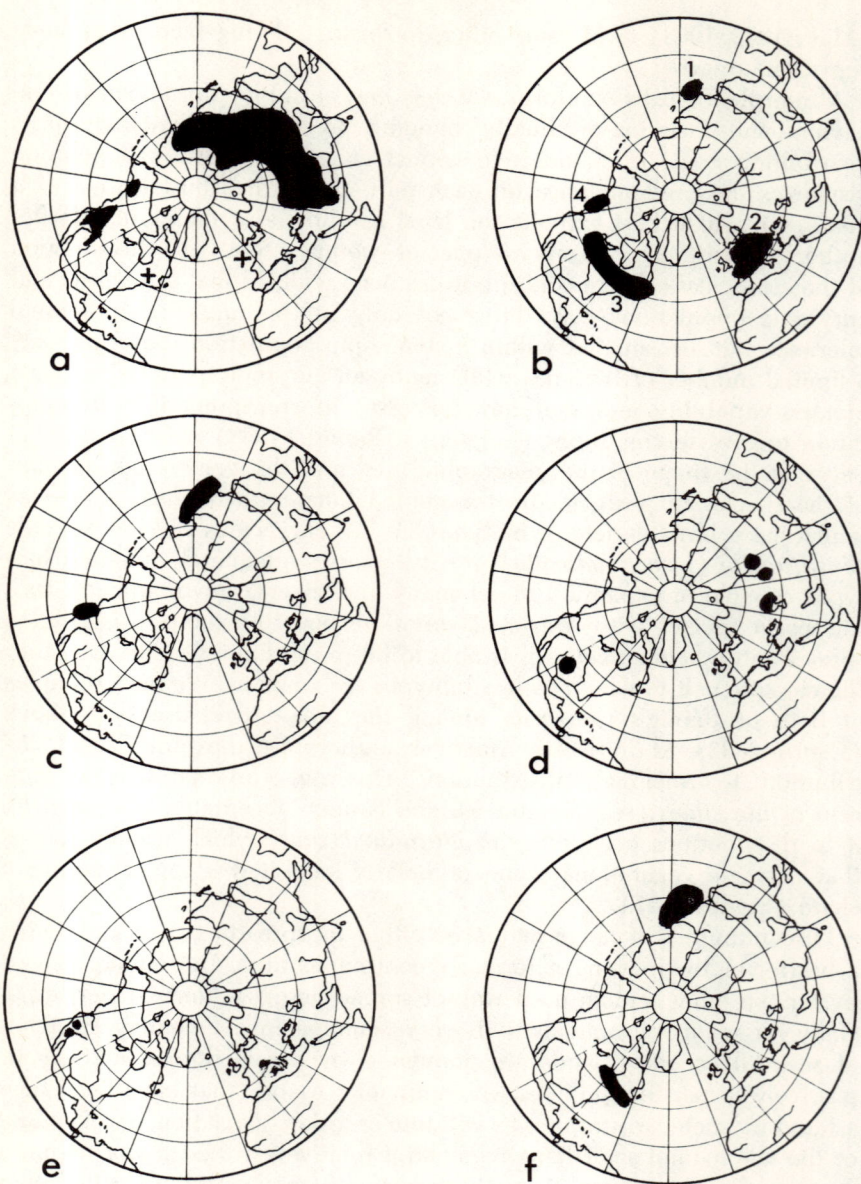

Figure 4-39 Some discontinuous distribution patterns in the Northern Hemisphere (in the Holarctic region).

In the following, we substitute the term Palearctic for temperate and arctic Eurasia; Nearctic for North America including Greenland; and Holarctic for the combination of the two regions.

unusual habitat. Contraction of a formerly wider area results in disjunctions and discontinuities. Many of these contractions of the past million years fall into well-known patterns, since the biogeographical history of the Pleistocene, especially of its later stage, is well authenticated in several geographical regions. The discontinuous patterns themselves provide first-class evidence to piece this history together; in a few instances, fossil finds also help (Figure 4–39).

Any area discontinuity, however far apart the part areas are, is suspected as evidence of former habitat continuity. This is true in the Northern temperate Hemisphere (Figures 4–22 and 4–26). Discontinuities between the southern continents are rarer, and since these are much older than those in the Northern Hemisphere, there are less clues to their history. They are, therefore, more subject to speculation as to whether land bridges or long-range dispersals were the connecting agents, or whether they are evidence of a continuity stemming from the remote past.

Discontinuities of areas within a geographically or ecologically limited region may occur as the result of barrier crossing by species which possess especially vigorous dispersal capacity and which live in habitat which is patchily distributed, such as many insects do; then too, there is the rare instance of a species which is currently in a state of vigorous expansion by pioneering groups or individuals, and which eventually will fill the intervening area. For example, the recently spreading collared dove *Streptopelia d. decaocto* Friv. [Columbidae, Aves] demonstrates such temporary discontinuities in Europe (Figure 4–40).

PATTERNS

(a) Central and eastern Palearctic, western Nearctic, fossil (x) in the western Palearctic and eastern Nearctic; therefore historic circumpolarity of this Holarctic genus is evidenced.
(b) Eastern and western Palearctic, eastern and western Nearctic.
(c) **Amphipacific disjunction:** Eastern Palearctic, western Nearctic.
(d) Central Palearctic, central Nearctic.
(e) Western Palearctic, western Nearctic.
(f) Eastern Palearctic, eastern Nearctic.

DISTRIBUTION AREAS

(a) The mammalian genus of pikas, *Ochotona* [Lagomorpha]. Data from Bartholomew et al. 1911, Broadbrooks 1965, Ognev 1940.
(b) Three closely related beetles: 1. *Catops sparcepunctatus* Jeann. [Catopidae, Coleopt.], 2. *C. subfuscus* Kelln., 3 and 4. *C. basilaris* Say. From Jeannel 1942.
(c) The beetle *Leptura obliterata* (Haldeman) [Cerambycidae, Coleopt.]. From Linsley 1963.
(d) The beetle *Blethisa eschscholzi* Zoubk. [Carabidae, Coleopt.]. From Lindroth 1957.
(e) The salamander genus *Hydromantes* [Plethodontidae, Amphibia]. (From Gorman 1964.)
(f) The beetle *Sachalinobia rugipennis* (Newman) [Cerambycidae, Coleopt.]. (From Linsley 1963)

204 *Areography: The Study of the Distribution Area*

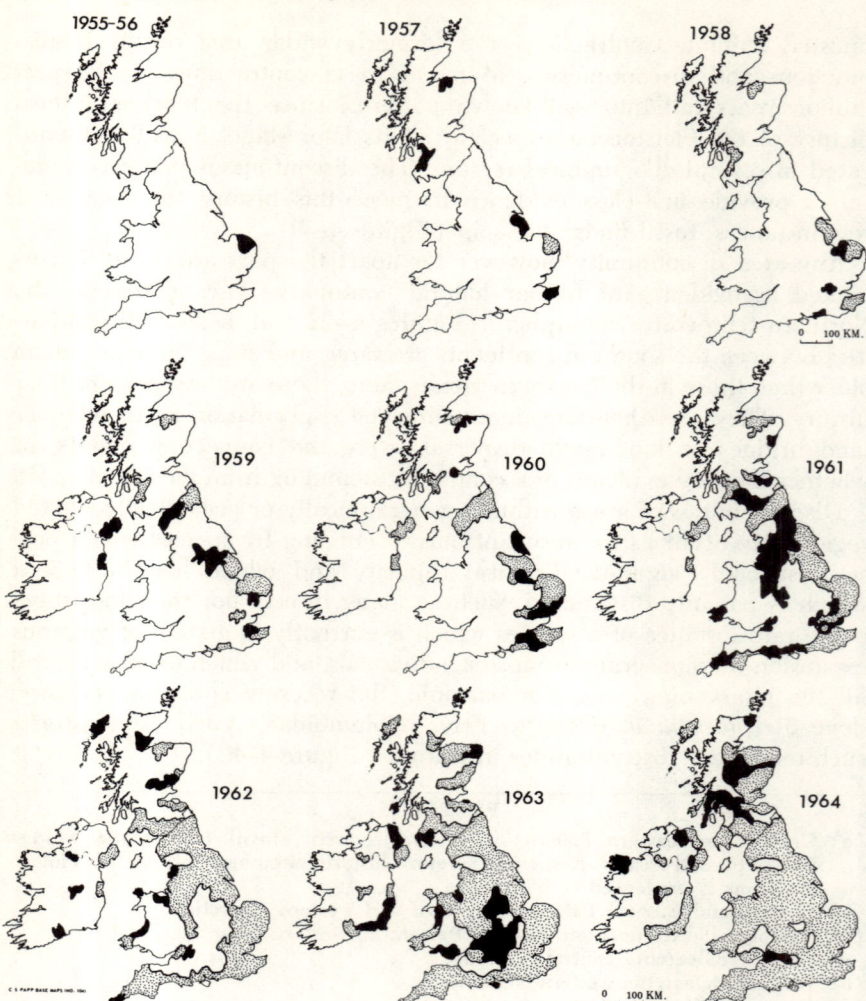

Figure 4-40 Discontinuous distribution due to spreading by long-distance pioneers. Breeding distribution of the collared dove *Streptopelia d. decaocto* Friv. [Columbidae, Aves] in the British Isles 1955–1964. The first birds were seen in 1952 in Lincolnshire, about 430 km from the nearest mainland breeding localities in Holland. Our series of maps (modified from Hudson 1965) shows (in black) the vice-counties where new breeding was observed and the areas (stippled) where breeding occurred in previous years as well. The colonization of Ireland shows the pattern especially clearly. In 1959, pioneering pairs arrive by long-distance flight and settle to breed; local breeding populations are established by 1960 and 1961; from these, some juvenile pairs (siblings) again fly away and establish new breeding units in 1962; meanwhile, the first established populations to expand, (1963), and so on.

The ten-year pattern shows that the whole island group contains suitable habitat for the colonists. Thus, the British Islands are part of the *potential area* of this species.

Relict Areas

The concept of *relictness* is most properly discussed in connection with the historic reasons for area discontinuities, because the zoogeographic term "relict" has mostly been applied to the smaller portions of discontinuous areas. A *relict* (from Latin "relinquere," "to leave behind") is something left behind after the disintegration or disappearance of the whole. The term is used in biology in two senses—phylogenetic relict and biogeographical relict.

Phylogenetic or *evolutionary relicts*, in their historic and taxonomic senses, are archaic forms still in existence while all other members of the major taxon to which they belong have died out during a past geological period. A better name for them would be "survivors," but it is difficult to eradicate the use of "relict" in this connotation (Figure 4-41).

A good example is the North American ruminant family of Antilocapridae (Artiodactyla, Mamm.). Members of this family were widespread in the late Tertiary and in the Quaternary of North America, and they underwent a spectacular adaptive radiation, analogous to that of the bovine antelopes of the Old World (Figure 4-42). However, they declined, and only one species—a survivor or phylogenetic relict—the prong-horn antelope *Antilocapra americana* (Ord) exists at the present time. Its distribution area largely covers that of the formerly numerous members of the family. The concepts of "survivor" and "relict" converge when one considers cases such as the celebrated lizard of New Zealand, the tuatara *Sphenodon punctatus* (Grey), which is the only survivor of the order Rhynchocephalia. In the early Mezozoic, this order was widespread on several continents; thus, as far as the area of the *order* is considered, the present New Zealand distribution is a relict distribution. The tuatara is itself a *survivor*.

A *biogeographical relict* is characterized by its area. "A species (or genus, etc.) is a relict in a region if it occurs there in isolation from its main center of distribution and if its presence can only be explained by the fact that it or its ancestral form was left behind under different natural conditions than exist at present" (Ekman 1915, restated 1953). If the species is extinct elsewhere, but otherwise corresponds to the above definition, it is still considered a relict (often called a "reductional relict"), and its total area is reduced to become a *relict area* (Figure 4-43). *A relict area is a part or total area which is left behind by regressing species from the past times when natural conditions had differed from those now existing around the relict area.* There are many partly conflicting definitions of a relict in the biogeographical literature. Ekman's definition excells in clarity, and ties the geographic connotation to the relict taxon.

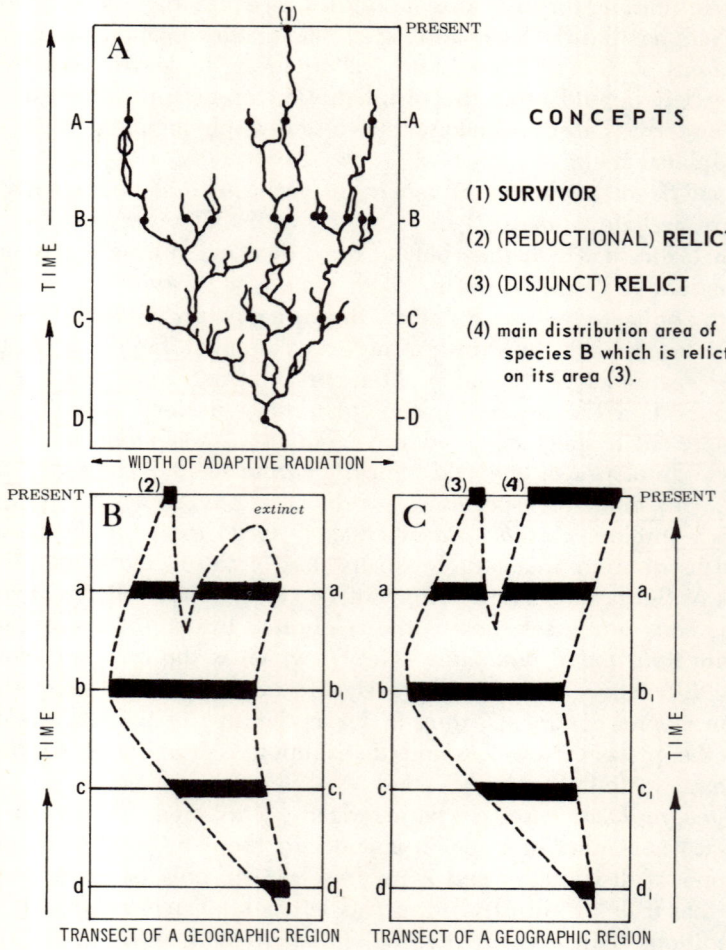

Figure 4-41 Relict concepts.

A Survivor (phylogenetical relict). Model of the evolution of a group through time. Dots along the lines A-A, B-B, and so on, indicate the species in existence at the particular time. At D time before present, one species—the ancestor—lived. C years before present, the taxon contained six species; B years ago, nine or ten species. Later, A years ago, this number diminished, by extinctions, to four. At present, only one species is found, and this is a *survivor*. This model is hypothetical, but reminds us the much less completely known evolutionary history of the family Antilocapridae. (See Fig. 4-42.)

B and C Geographic distribution of two hypothetical species in the same geographic region at the present time and historically at times when their distribution along the transects $a-a_1$, $b-b_1$, and so on, is shown by the thick lines. Species B is a *reductional relict;* species C has a disjunct area, the smaller part of which is a *relict area*.

Figure 4-42 Survivor concept.
The pronghorn antelope (1) is the only survivor of a family [Antilocapridae, Mamm.] which appeared in the late Tertiary of North America and exhibited a spectacular adaptive radiation convergent with the Old World antelopes [family Bovidae]. They all, except the pronghorn of the Western plains, died out at the end of the Pleistocene. Reconstructions from Frick 1937. (*Courtesy of the American Museum of Natural History, the pronghorn antelope, of A. S. Leopold and the University of California Press*)
1. Pronghorn antelope *Antilocapra americana* Ord. (From Leopold 1959)
2. Quentin's pronghorn *Stockoceros onusrosagris* (Roosevelt & Burden)
3. Hay's pronghorn *Hayoceros falkenbachi* Frick
4. Spiraled pronghorn *Ilingoceros alexandrae* Merriam
5. Osborn's pronghorn *Osbornoceros osborni* Frick
6. Merriam's pronglet *Merriamoceros coronatus* (Merriam)

The French Biogeographical Society devoted a session in 1947 to questions pertaining to relictness. It was agreed that the phylogenetic and biogeographic concepts should be separated, using the two forms which are used in French just as in English: A *relic* (French *relique*) is

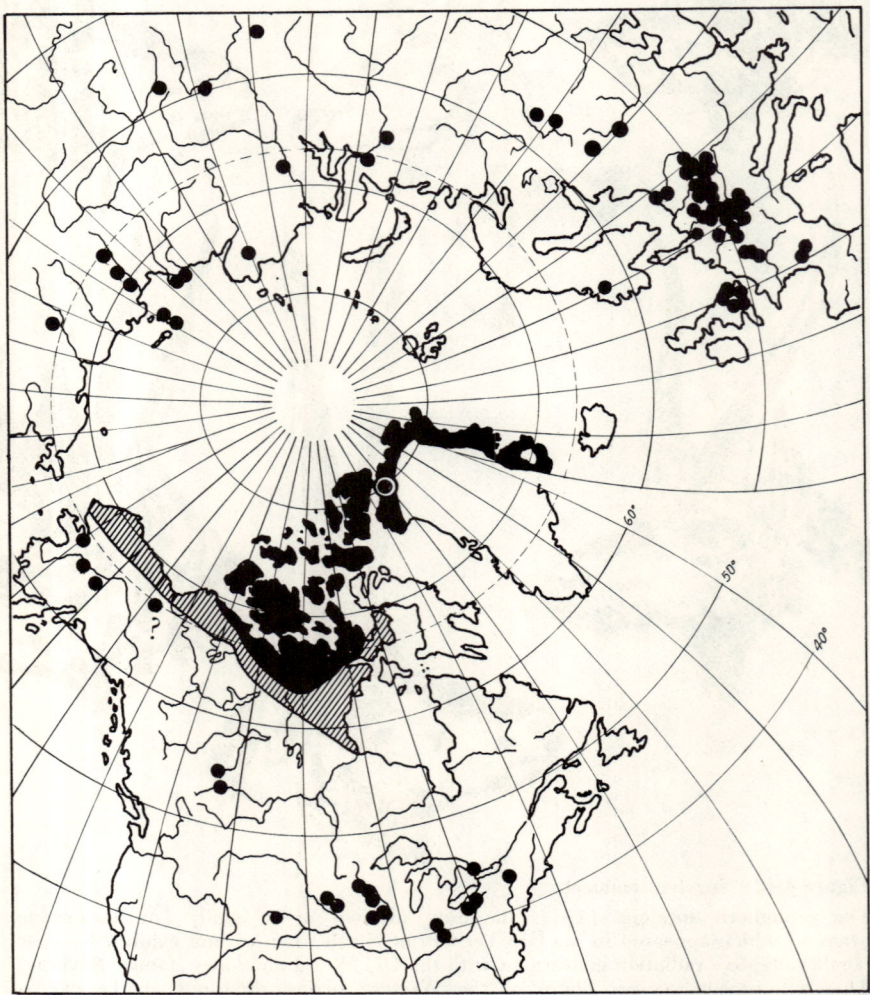

Figure 4-43 Reductional relict.
Present and past distribution of a reductional relict, the muskox, *Ovibos moschatus* (Zimmermann). [Bovidae, Mamm.] Dots indicate fossil localities; black, the present distribution area; hatched area, distribution during historic time. (Recent and North American fossil distribution from C. Ray *in litt.* 1966). Present distribution from Macpherson 1965. Old World data (some westernmost data refer to *O. fossilis*) from Ekman 1922.

a species surviving from an ancient lineage; it has a very restricted distribution area. A *relict* (French *relicte*) is a species isolated from its normal distribution area, most often as a result of climatic changes (Furon 1958). But because these two closely related words may easily be confused—especially when they will continue to be used by scientists of widely different background—it is perhaps safest to use *survivor* for the phylogenetic and *relict* for the biogeographical entities.

In further ecogeographic analysis of the relict area, we find, with Matvejev (1954) that there are *refugional relicts*. These are relict populations that live on an *ecological refuge*,* i.e., on an isolated habitat that has preserved the environmental conditions which were widespread in the past, before the refuge became isolated from the main ecological unit (ecosystem, faunation, community, etc.). The species is thus in harmony with its surroundings and with the members of the faunation and vegetation of the refuge, but the refuge itself is distinguished by its unique ecologic and geographic position in the region. Such refugional relict species are conservative relicts (Rylow 1921)—they are not able to adapt to the conditions prevailing outside the refuge. Hultén (1937) claims that certain groups of coastal plants are still confined to their glacial refugia because they are not plastic enough to break out from them; their variability became restricted during the vicissitudes of the past glaciation when new variants have been, or would have been, at a disadvantage during the singularly and uniformly harsh conditions. Though this theory seems genetically sound, it needs detailed testing on many animal species before it could be universally applied.

When a relict occurrence of an area is in an environment that is atypical for the species (as compared with other, more characteristic parts of its area, or at the time of its former large distribution), Matvejev's term *disharmonious relict* is applicable; relicts of this type are in disharmony with the ecological characteristics of the other organisms, or at least with a majority of the other organisms on that biotope. The zoogeographer has reason to suspect that such a relict is the most adaptable or the most euryecious remnant of an extinct faunation, for it is able to survive among the changed conditions by virtue of its wide tolerance range or its adaptability. Nevertheless, dwindling and doomed populations of conservative relict species are also well known, and we cannot find a norm to distinguish these from the more vital disharmonious relicts.

The Ozark Highlands in the United States form an ecological refuge with their locally cool climate and corresponding vegetation and habitats of more northern character than that of the surrounding areas (Figure

---

*Often the Latin form *refugium* is used, especially in the plural: *refugia*.

210 Areography: The Study of the Distribution Area

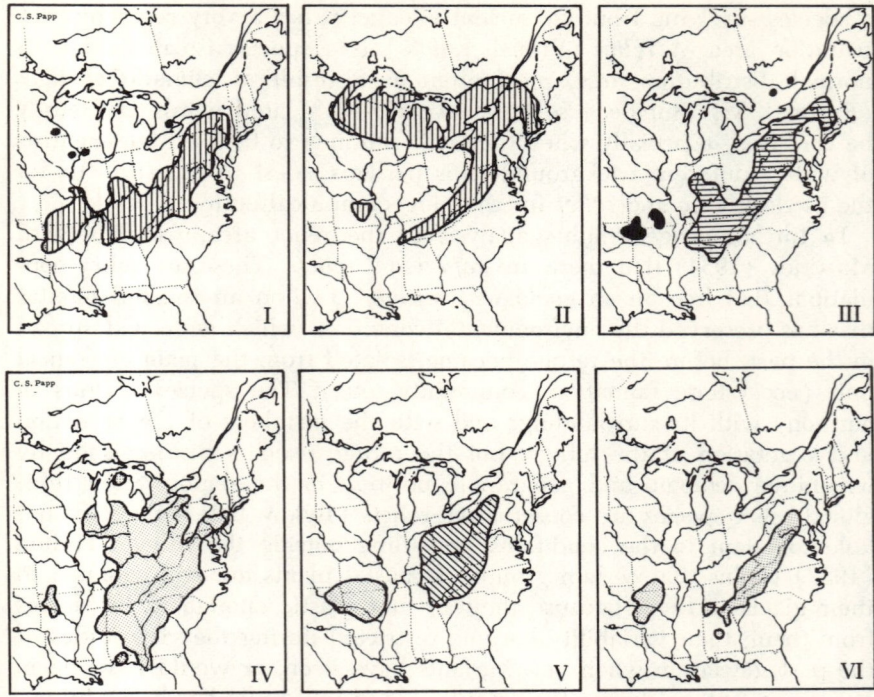

Figure 4-44 Refugional relicts.
In the south-central United States, the Interior Highlands form an ecological refuge. Their cooler environment preserved communities composed of many elements of a highland biota which has its main distribution in the Appalachian Mountains region and also farther north.

Distribution areas of four species of eastern North American winter stoneflies of the genus *Allocapnia* (Plecoptera). (After Ross 1965.) These areas are related, and the age and history of these animals are presumably very similar. The rate of area disjunction and of speciation presumably depends on intrinsic (tolerances, vagility, etc.) and extrinsic (distribution of the habitat) differences. Stoneflies live in fast, cool streams.

I The largely continuous area of *A. rickeri* Frison is ecologically limited. Note the constriction of the range where the Mississippi Embayment (south) meets the formerly glaciated area (north) and nearly separates the Ozark Highlands population.
II Disjunct distribution of *A. pygmaea* (Burmeister) with relict population in the *refuge* of the Ozark Plateau area.
III Similar distribution, but the two part areas harbor a species pair: *A. mohri* Ross & Ricker (*west*) and *A. recta* (Claassen) (*east*). As the smaller western population reached species rank presumably due to spatial isolation and local environmental differences, it can not be called a relict any more. The two species are true vicars.

In order to emphasize the refugional nature of the Interior Highlands, the distribution areas of a snake, a fish, and a land plant are presented, From Conant 1960. All these are refugional relicts.
IV Queen snake, *Natrix septemvittalta* (Say) Colubridae, Squamata.
V Streamline chub, *Hybopsis dissimilis* (Kirtland), Cyprinidae, Cypriniformes.
VI Black locust tree, *Robinia pseudoacacia* L., Papilionaceae, Leguminosa.

4–44). We find here refugional relicts, such as the red-backed salamander *Hemidactylium scutatum* (Schlegel) [Plethodontidae, Amph.] and the wood frog *Rana sylvatica* Le Conte [Ranidae, Amph.], according to Blair (1965). The winter stonefly *Allocapnia pygmaea* (Burmeister) [Capniidae, Plecopt.] is also a relict in this area, disjunct from its more northern and eastern (mountain) populations (Ross, 1965). More widespread but still restricted to this area is the stonefly *A. mohri* Ross and Ricker, disjunct from its eastern sister species or vicar, *A. recta* (Claassen).

A good example of a disharmonious relict is the pickerel frog *Rana palustris* Le Conte in the same region. South of its continuous distribution area, relict populations live in the states of Alabama and Mississippi in the immediate vicinity of caves. They survive the hot summers in the cool twilight zone of the caves and thus maintain their small relict populations in an ecologically strange environment (Blair 1965). Holdhaus (1954), on the contrary, stresses the role of caves in having preserved, during Pleistocene glaciations in the Alps region of Europe, relicts of southerly distributed species far north of their main distribution area. The caves provided a generally warmer environment than that of the region. Mesophil forests, which preserve moisture on their shady floors, dwindled to a few small refuges on the Hungarian plains during the past 150 years, and were mostly replaced by steppes. Forest-dwelling ferns and amphibians survive as refugional relicts in the remaining forest reserves and as disharmonious relicts in the cool, shady environment of large, deep wells in the dry, open grazing land.

In studying disjunct distributions, the following points of view are of great help in discerning the relict nature of the isolate:

(1) The area in question is effectively isolated from the main distribution area; thus its existence is not due to reinforcement from the latter by regular means of dispersal.

(2) There exists a barrier separating the part areas; viz., the separation is not due to temporary thinning of a connecting population or to time lag in establishing contact with pioneering centers. The barrier may be climatic, ecologic, or physical.

(3) There are indications that, in the past, habitat connections were uninterrupted between the relict area and the main distribution area.

(4) There is reason to surmise that the former distribution was uniform and continuous.

(5) Several species, as if forming a group, show similar disjunct distribution, or these (and other) species show similar distribution elsewhere in related circumstances. The first alternative reveals the refugional relict; the second may help in the case of disharmonious relicts.

We may apply the five principles set forth above to the distribution

212 *Areography: The Study of the Distribution Area*

in the state of Missouri of the winter stonefly *Allocapnia pygmaea* (Burmeister) (Figure 4–44).

(1) The isolation that the map shows is real; no connecting populations or pioneers have been found in this faunistically well-known area.

(2) Silty, sluggish waters or warm, sluggish streams are ecological barriers separating the part areas; winter stoneflies have very low vagility.

(3) During glacial advance, the streams between the Ozark Plateau area and the southeast part of the United States were presumably colder than they are today, for there is much evidence of boreal habitat belts in the Pleistocene of this area.

(4) Therefore, and because the populations on the Ozark Plateau and on the main range bear the closest morphological relationship, they may be thought as formerly geographically continuous areas and biologically continuous populations.

(5) Other species of stoneflies, amphibians, etc. (see p. 211) show similar distribution patterns.

In a discussion of relict distributions of the Pleistocene epoch, Deevey (1949) points out that a relict does not necessarily occupy exactly the same geographic area throughout the history of its discontinuous existence. The relict area may shift—for example, as a bog or lake becomes

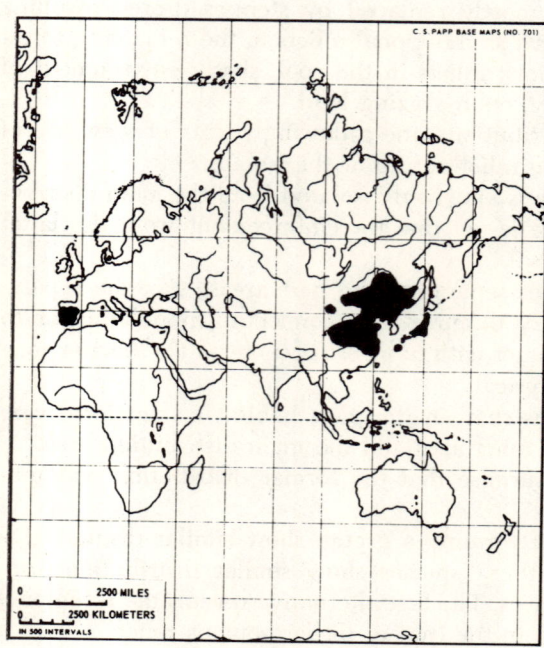

Figure 4-45 Tertiary relict. The azure-winged magpie *Cyanopica cyana* (Pallas) [Corvidae, Aves] can be considered a Tertiary relict on the Iberian Peninsula. Its habitat is Mediterranean woodland, but in the Far East it also occupies many other wooded habitats. It is evident that the now widely discontinuous part areas did not originate by barrier crossing; vicissitudes of the Pleistocene climatic changes wiped out the species from the connecting areas. After Voous 1960.

filled and its relict faunation finds a new foothold on another relict habitat, but not on the same relict geographic locality.

Several relict categories are well known from the biogeographical literature by such self-explanatory names as *Tertiary relicts, glacial relicts,* and *xerothermal* or *hypsothermal* or *heat relicts*. These names refer to relicts that survived general area changes since the Tertiary (i.e., from times before the Quaternary [Pleistocene] glaciations), since the Pliocene-Pleistocene glacial periods, since the postglacial dry and warm period, and so on (Figures 4–45 to 4–52).

Plant ecologists often talk about *relict communities,* which live on refuge habitats. Their *successional relict* is a community or a single species left behind after the whole vegetation changed from on successional stage to another. Another synthetic use of the relictness concept is, when analyzing floras or faunas of certain areas, and it becomes evident that a

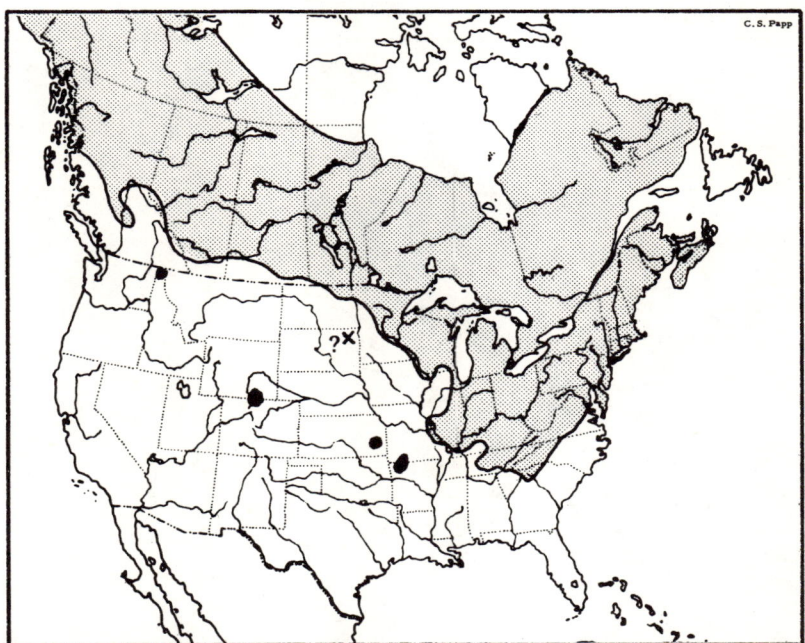

Figure 4-46 Pleistocene (glacial) relict.

Relict areas south of the main Boreal distribution of the wood frog *Rana sylvatica* Le Conte [Ranidae, Amph.]. All these relict areas are on refugia; the Rocky Mountain refugium, for example, also harbors a relict population of another Boreal amphibian, the Dakota toad *Bufo hemiophrys* Cope. (See discussion in Blair 1965.) Map after Conant 1958.

214 *Areography: The Study of the Distribution Area*

group of biota do not show ecologic and geographic-historic relations with the majority of the biota but are relicts. One may even talk about *relict faunas.* In other words, faunistic and floristic analysis may reveal relictness without knowing of the refuge character of the ecosystem that harbors them. These usages of the relict concept show the versatility of biogeographical approaches.

Finally to be considered is the great impact of human civilization on the distribution areas of animals. While adventive and introduced favored species widen their range, a multitude of species are dwindling. Their areas are shrinking; they are divided by strange and man-made habitats; and in many cases they are reduced to relict size with a doubtful future. Though these processes are historical, and though human impact on natural distribution areas was probably felt throughout the Pleistocene period, the explosive spread of civilization in the past two hundred years makes necessary the introduction of the concept of *secondary relicts.*

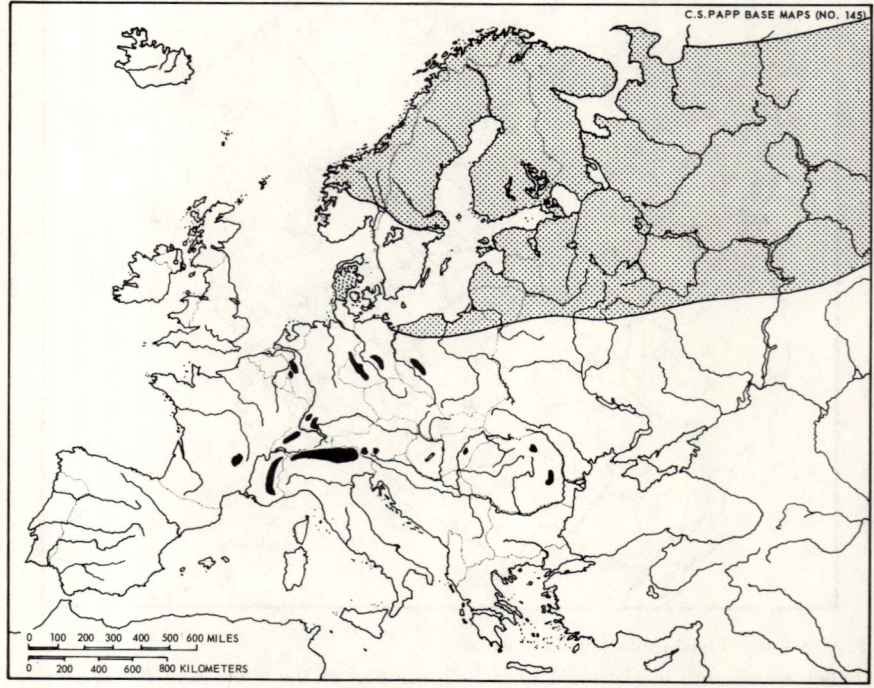

Figure 4-47   Boreo-alpine relict.
The distribution in Europe of the butterfly *Colias palaeno* L. [Rhopalocera, Lepid.]. After Jeannel 1942.

The areas of primary relicts dwindled through "natural" causes, including the impact of primitive man. In the case of secondary relicts, no other ecologic or historic factor was responsible except the recent impact of civilization (Figure 4–53). Outstanding examples are the North American buffalo *Bison bison* (L.) [Bovidae, Mamm.], now restricted to a small part of the natural range of one of its subspecies; the artificially disjunct area islets of the European beaver *Castor fiber* (L.); and the world areas of the rhinoceros. Lindroth (1949) uses, in a discussion of the relict concept, the term "anti-culture-relict," and mentions several Carabid beetle species on retreat or near extinction in Scandinavia because of habitat destruction and alteration by man.

Endemicity and Related Historic Concepts

A species is *endemic* to a geographic area if the specific distributional area is smaller than the geographic unit in consideration. Because most biogeographic regions cover a whole continent, and because many species of a fauna are continent-wide in their distribution, it is not often that a species is described as being endemic to a continent. It is more meaningful to restrict the concept to a subcontinent or an even smaller area. When the distribution of higher taxa or other groups is considered in a worldwide sense, such expressions as "the endemic North American

Figure 4-48 Boreo-montane distribution, boreo-montane relict.

Distribution area in the Palearctic of the circumboreal tree-toed woodpecker *Picoides tridactylus* (L.) [Picidae, Aves]. The Central European discontinuous areas are in the montane and subalpine coniferous forest zone of the mountains, and the species has boreo-montane distribution and is a boreo-montane relict. After Voous 1960.

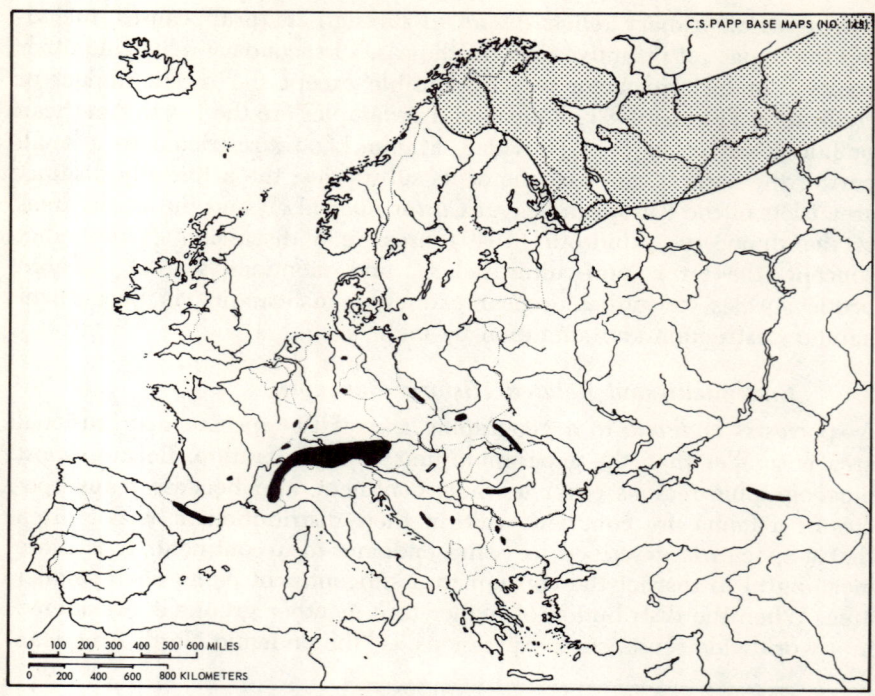

Figure 4-49 Arctic-alpine distribution.

European distribution of the circumboreal, Holarctic ground beetle *Amara erratica* Duft [Carabidae, Coleopt.]. After Holdhaus 1954. Holdhaus uses the term "boreo-alpine" for this kind of distribution with Pleistocene relict areas in the alpine zone of mountains in southern latitudes. As the beetle lives on the tundra and above tree line in the Scandinavian mountains and above tree line on alpine meadows near snow patches and fields in the Alps, we would call its distribution *"arctic-alpine,"* with Pleistocene relict areas as shown on the map.

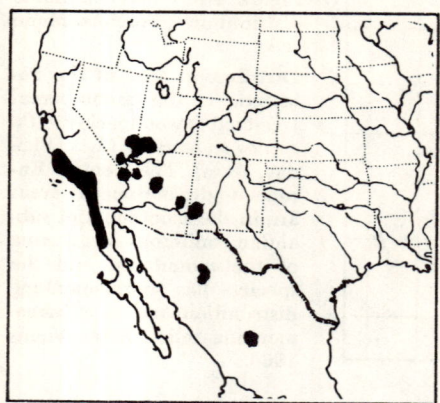

Figure 4-50 Pluvial relict.

Distribution of the southwestern toad *Bufo microscaphus* Cope [Bufonidae, Amphib.] in southwestern North America. After Stebbins 1966. Blair (1965) found only insignificant morphological differentiation in the disjunct inland populations; thus, a much greater, and more continuous range during the Pleistocene (Wisconsin) Pluvial is plausible.

snail fauna . . .". "Most marsupials are endemic to Australia," and the like, are more commonplace. The "reductional" relicts, i.e., those whose area has contracted and is now a single small relict area—are always endemic (by definition) of a small geographic region. The species which is relict only on some disjunct part of its area is evidently not endemic there, though it might be endemic considering its total range and a larger geographic unit. *Thus, not all relicts are endemics.* An endemic area may be a young area, in the process of expansion. *Endemism*—the concept of being endemic—is used in faunal analysis because the ratio of endemic versus widespread species depends on historic and ecological factors, which help to explain the dynamics of the fauna. In general, the more endemics an area contains, the older is the fauna of that area—if they are old endemics.

The concept of endemism is borrowed from medical science. A disease is said to be *endemic* if it is confined to a certain area, and *epidemic* if it is widespread. Yet this pair of terms do not reveal per se the origin of endemic species. The species that originated on the geographic area where it now lives, or on the area under consideration, is said to be *autochthonous* or *indigenous* there. If it has spread to other, unrelated area, it is *allochtonous* on those areas; the term *adventive* is almost synonymous with *allochtonous*, but implies that human activity helped in the spreading process.

Examples of this are provided by the distribution areas already illustrated in this chapter. The muskox (Figure 4–43) is endemic in the North American Arctic, for today it lives only there. It is not autochthonous there, for it most likely originated in the Old World (Kurtén 1964). The dwarf shrew (Figure 4–10) is endemic to the Great Basin area of North America. The bobolink (Figure 4–1) is a North American endemic bird. The winter stonefly *Allocapnia mohri* (Figure 4–44) is endemic to the Ozark Plateau; but, according to Ross' interpretation (1965), it is also autochthonous there, for he surmised that it speciated in that area. The collared dove (Figure 4–40) is allochthonous in Western Europe, because it is positively known that it immigrated only recently to that area.

GEOGRAPHY OF THE AREA

When the zoogeographer scrutinizes and compares distribution areas on the map and globe, he often uses a special terminology for certain commonly occurring area types. These terms usually refer to the geographic position and often also to the ecological nature of the area. Most of these terms refer to the Northern Hemisphere.

The names of the broad ecogeographical zones may be applied to the distribution areas. Though the meaning of these terms (Figure 4–54) is by and large clear to every biologist and geographer, it is difficult to

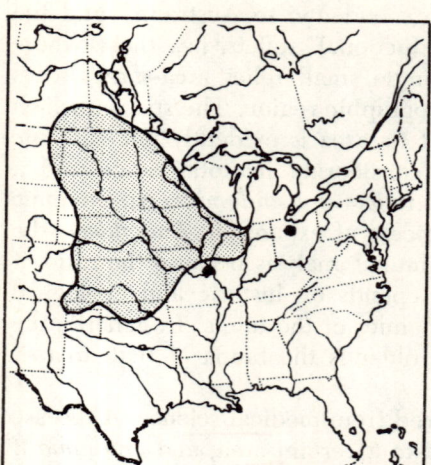 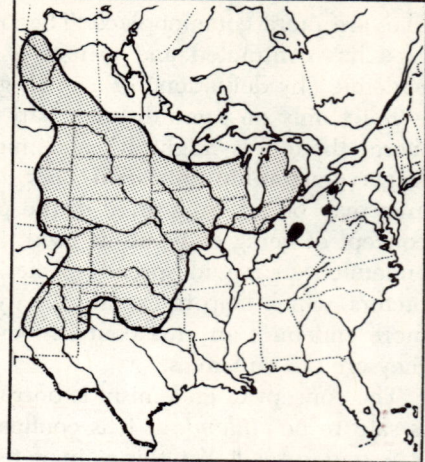

Figure 4-51   Xerotherm relicts I.
Xerotherm or heat relicts in eastern North America on an area northeast of the Mississippi valley where there was a "peninsula" of the prairies in the postglacial warm period, as the many relicts testify. *Left:* Plains garter snake *Thamnophis radix* (Baird & Girard) [Colubridae, Rept.]. *Right:* The ground squirrel *Citellus tridecemlineatus* (Mitchill) [Sciuridae, Mamm.]. (After Smith 1957.)

define them accurately because they are used so often and without definition in the literature. Their limits do not follow the latitudes as exactly as on our diagram, but are climatically controlled transitional zones meandering across continents and following certain currents in the ocean.

The prefixes *pan-* (the whole), *circum-* (around), *sub-* (under), and *amphi-* (on both sides) are often combined with these and with geographic names. Some of these common terms are explained below.

Circumpolar is a distribution area which encircles one of the poles; Arctic or perhaps even boreal areas may be embraced by this adjective. *Circumboreal* is also used; but because neither land nor water encircles the globe entirely at boreal latitudes, it is better to speak in terms of *panboreal* land or sea animals (Figure 4–55). Similarly, *pantropical* is the more frequently used term for a *disjunct* distribution with part areas in each of the four continents containing tropical regions (some include even the subtropical belts). Longitudinal disjunction of area is indicated by the terms *amphiamerican, amphipacific,* and *amphiatlantic; amphitropical* areas are disjunct, on either side of the Equator (Figure 4–56). However, if the latitudinal disjunction has resulted in two part areas, each nearer to the respective pole and farther away from the Equator, they are best described as *bipolar* areas.

Geography of the Area 219

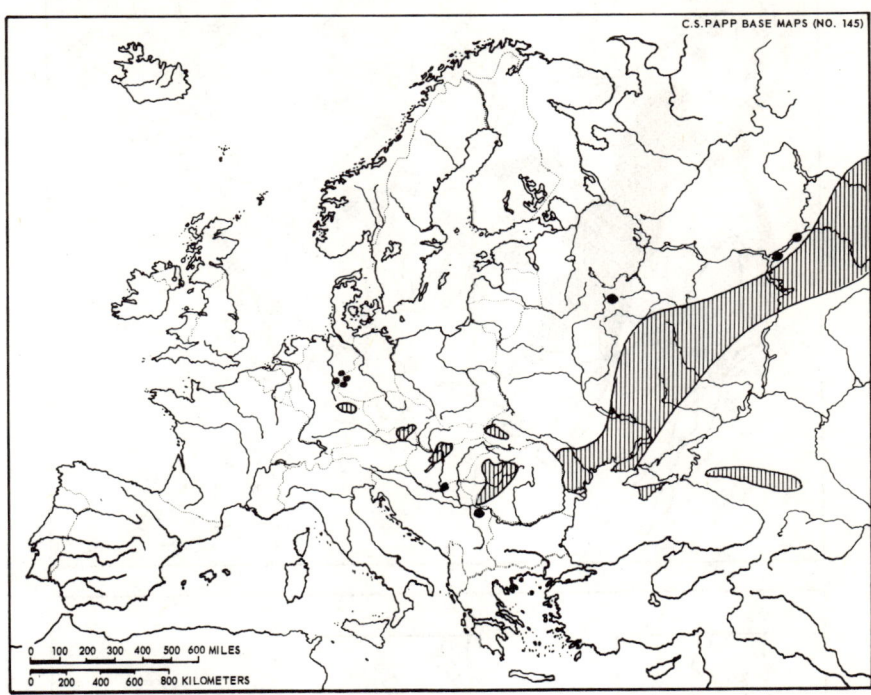

Figure 4-52 Xerotherm relicts II.
Distribution area of the feather grass *Stipa stenophylla* Czern. [Gramineae]. (After Meusel 1943.) Distribution areas of the Central European xerotherm (heat) relicts form a mirror image of their eastern North American counterparts (cf. Fig. 4-50). Isolated warm and dry localities within the zone of mesic forests harbor refuges of the elements of Pontic steppe flora and fauna. (Franz 1936).

Two important types of area discontinuities are well known and classified by northern zoogeographers. A distribution is *arctic-alpine* when the zonal, arctic area of the species has extensions along mountain chains running southward. Often the alpine part of the area is disconnected from the main arctic range by a barrier of intervening lowland. In these cases, the alpine area isolates are on glacial refuges, and the species is a *glacial* (Pleistocene) *relict* in that geographic region.*

Exactly the same explanation and reasoning apply to those boreal areas which have extensions along a mountain chain or discontinuous part

---

*In many instances, the reverse is true; alpine species came into contact, during the Pleistocene glaciations, with arctic habitats and became widespread in the Arctic, but still have discontonuous, alpine areas in more southern latitudes. There are no dogmatic rules in zoogeography—only rules and grouping of phenomena for convenience.

220 Areography: The Study of the Distribution Area

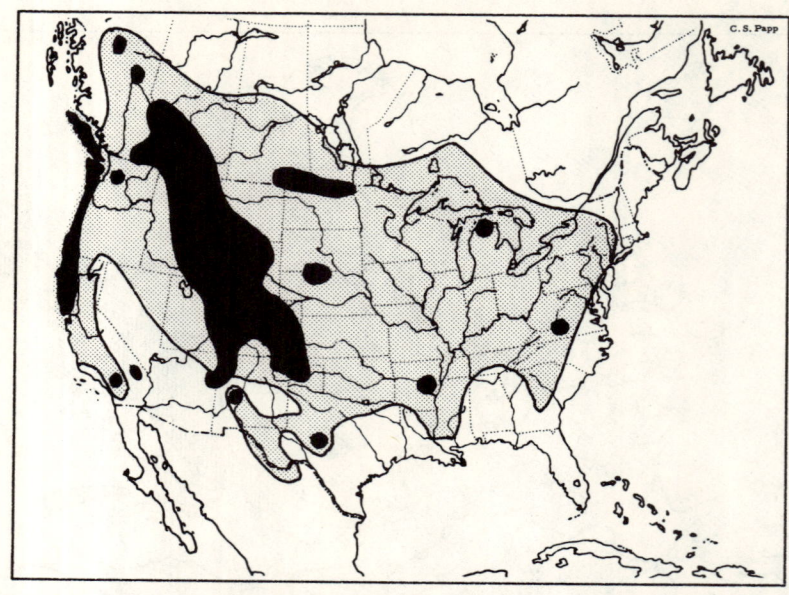

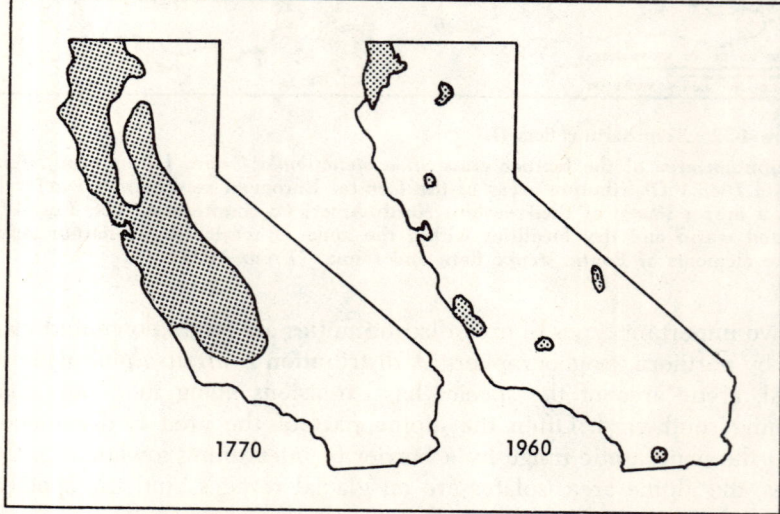

Figure 4-53 Secondary relict distribution.
Historic and present distribution of the wapiti *Cervus canadensis* Erxleben [Cervidae, Mamm.]. *Above:* Total historic distribution is outlined; present, *secondary relict* areas shown in black. Large black dots indicate reintroductions to areas where the natural population had already died out. (Data from Hall and Kelson 1958 and Burt and Grossenheider 1964.) Below: California distribution in 1770 and 1960. After Dasmann 1964.

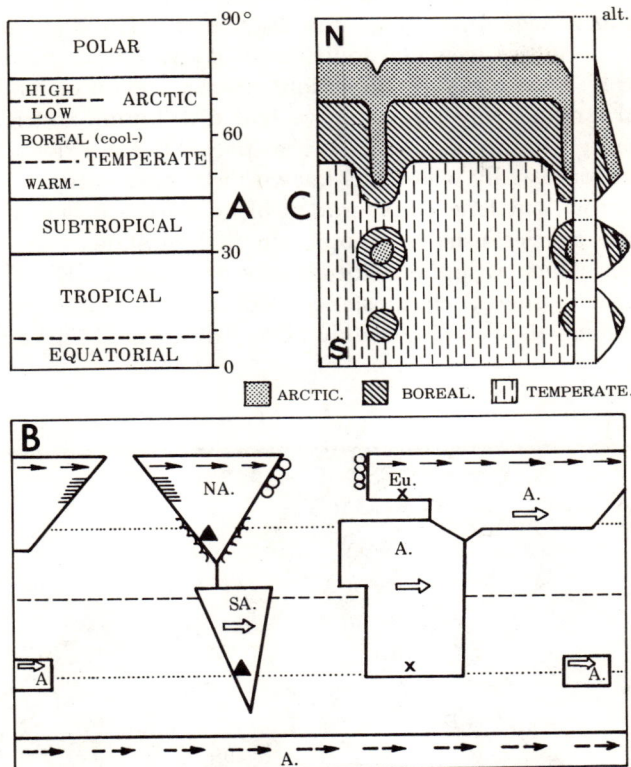

Figure 4-54 Illustrations of some zoogeographical terms.

A A scheme of ecogeographical zones as used in characterizing distribution areas.

B Some geographic terms used in characterizing distribution areas, illustrated on a schematic map of the earth.
Arrows = *circumpolar* distribution area in the Arctic (*Holarctic species*). Vertical stippling of disjunct area on both sides of the North Pacific = *Amphipacific* area. Circles on both sides of the Atlantic = *Amphiatlantic, linear* area. Triangles in north-tropical North America and south-tropical South America = *Amphitropical* area. Crosses in temperate Europe and subtropical Africa = *Bipolar* area. Semicircles on both coasts of tropical North America = *Amphiamerican* area. Wide arrows in all four tropical regions = *Pantropical* area. Broken arrows along the coast of Antarctica = *Circumpolar Antarctic* area.

C Two important distribution area types of the Holarctic (of the Northern Hemisphere). Dotted area = area of *Arctic-Alpine* species. Two isolated alps have disjunct, relict type of part areas of the species. Diagonally hatched area = *Boreomontane distribution* area. The main distribution area lies in the boreal forest zone, with outlyers along the mountain chains projecting southward. Two isolated part areas show zonal distribution of the species which is *relict* there. On the right, an altitudinal section of the area illustrates the *zonal* distribution of both species.

222 *Areography: The Study of the Distribution Area*

areas in mountainous areas south of the boreal belts of coniferous forests and bogs. The term used here is *boreomontane*.

Zonal and Linear Areas. Since the distribution area is composed of all geographic localities in which individuals of the species population live and mulitply, most areas, by definition, cover a section of land and have width and length. Only rarely do we find distribution restricted to one single locality (Figure 4–57). If such a species is not a dwindling relict, later collection usually widens its known distribution area.

Very narrow and long configuration of an area is called a *zonal area* because the species usually occupies in these instances a narrow ecological zone which is limited latitudinally or altitudinally (Figure 4–57, Figure 3–4). In special circumstances, one finds distributions which are *unidimensional*—they are *linear* only (Udvardy 1963a). This circumstance

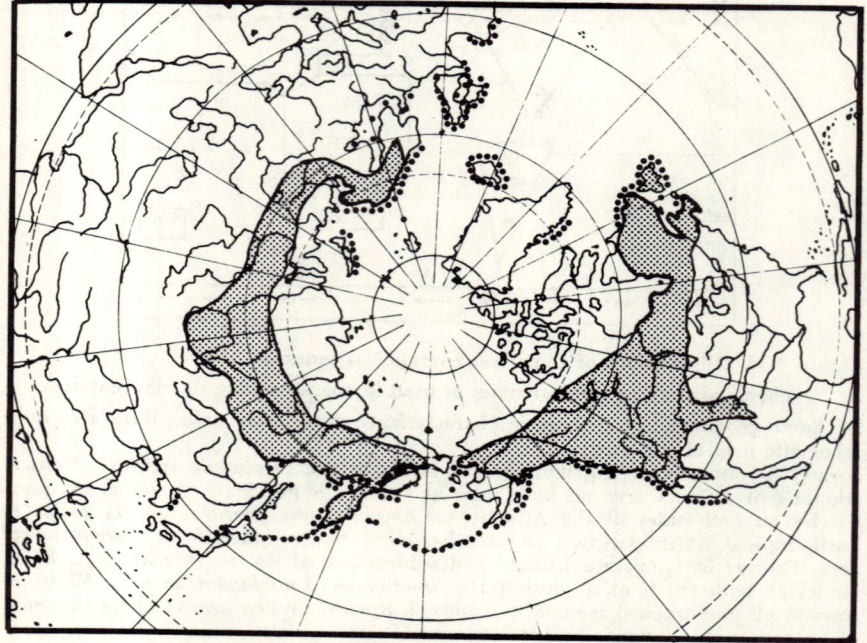

Figure 4-55 Panboreal distribution.

Distribution of two panboreal species of birds. The pine grosbeak *Pinicola enucleator* (L.)[Fringillidae] is a bird of the boreal coniferous forest belt (taiga belt) in both the Old and New Worlds. Its distribution range is disjunct, as the Bering Strait separates its two main parts. The distribution of the common murre (guillemot) *Uria aalge* (Pontopp.), [Alcidae] is disperse but panboreal; all boreal coasts are inhabited.

Geography of the Area 223

reflects of course linear configuration of the habitat of these animals. For example, on recent, relatively uneroded mountain chains, one would expect narrow altitudinal habitat zones with corresponding distribution areas (Figure 4–57). A further case of linear distribution is found when ecotone animals utilize the margin (which is of course linear) of two different habitats for their activities. However, their total distribution area is seldom linear, for ecotones delimit shifting zones of habitats, and these are regularly fringed or interdigitating (Figure 4–30B). The seashore is such a marginal area; along fairly straight coastlines, littoral and coastal forms exhibit *linear distribution* (Figure 4–34).

Causal analysis of linear, coastal distributions has an advantage over three-dimensional distribution in that, because of its longitudinal situation, we may confine our attention to two limits only—the polar and equatorial extremes. On the other hand, a distribution area on land may be limited by many different sets of factors in the different sectors of its margin. In a simplified example, where only gross climatic effects are considered, polar limits may be caused by cold; equatorial ones, by warmth; those east of an ocean, by humidity; further inland, behind a mountain chain, by aridity. The latter two may be coupled with either lack or excess of solar radiation. Between these main delimiting factors, any degree of combined effect may occur (Figure 3–13)—for example,

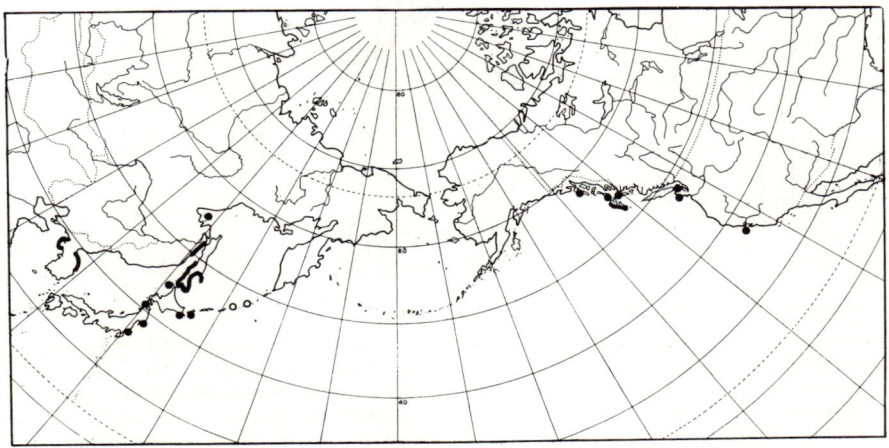

Figure 4-56 Amphipacific distribution.
Known distribution of the rhinoceros auklet *Cerorhynca monocerata* (Pall.) [Alcidae, Aves]. The two populations are conspecific; nor has any subspecific difference been found. From Udvardy 1963a. (*Courtesy of Bishop Museum Press, Honolulu*)

with the widespread, Palaearctic "Large White" butterfly *Pieris brassicae* L. [Pieridae, Lepid.]. Danilevskii (1965) concludes, that at the far northeastern end of its range in Siberia, the limiting factor for this butterfly is cold, for the frost resistance of the diapause stages is low. In Western

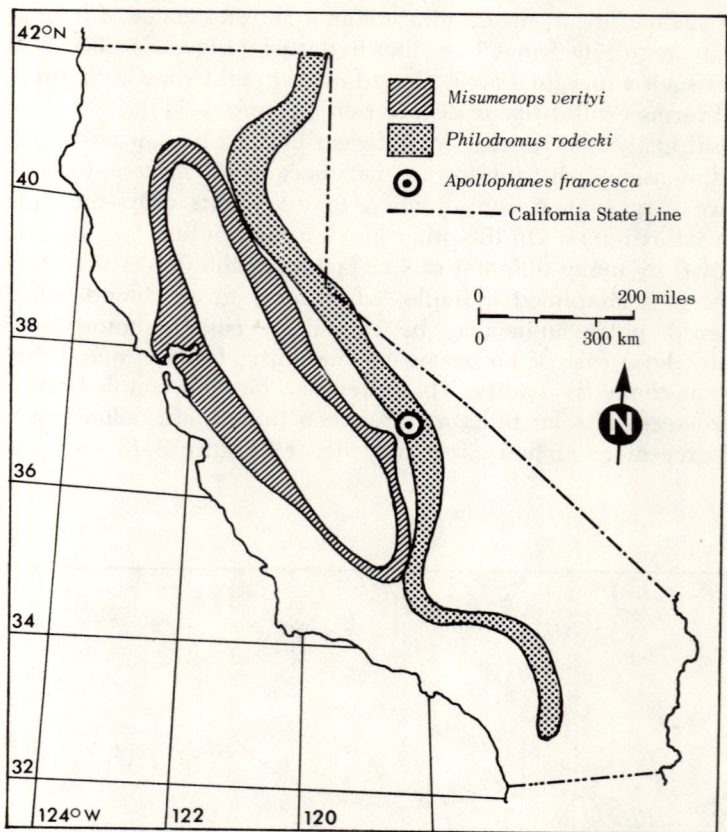

Figure 4-57 Some types of distribution areas.

Three species of crab spiders [Thomisidae, Araneida] after Schick 1965. *Misumenops verityi* has a zonal distribution similar to that of certain species of Manzanita bush (*Arctostaphylos* sp., Vacciniaceae), with which it is associated; they both reach the valley floor at the north end of the Central Valley of California. *P. rodecki* is mainly associated with coniferous forest; the connecting part of its range between the south end of the Sierra Nevada and an occurrence near the Mexican border is hypothetical and may be discontinuous (an occurrence that is not proved is hypothetical) (R. Schick, *in litt.* 1966.) *A. francesca* is known from only one locality; it is either extremely rare and has escaped the concentrated and continuous attention of Schick or it is a dwindling relict species.

Europe, in the regions of maritime climate, the winters are relatively mild. Here summer temperatures (namely the effective, accumulative temperatures necessary for the development of the caterpillars) seem to limit northward distribution. Moore (1952) studied the distribution and ecology of the mink frog *Rana septentrionalis* Baird [Ranidae, Amphibia] in northeastern North America, and demonstrated that the northern limit coincides with spring water temperatures which the developing embryo cannot tolerate. The physical conditions at the southern border do not limit the tadpoles of this frog; therefore, Moore surmises that predation by the abundant southern species, the larger bullfrog *Rana catesbeiana* Shaw, might be one of the factors limiting southern distribution.

DYNAMICS OF THE AREA

Most of the past zoogeographical thinking was twofold: the evolution and shifting of faunas in past geological ages were recognized, but the speed of the changes was underestimated; The present distributional picture, however, was looked upon as an unchangeable status quo, where man was the main culprit who altered the face of the earth and, within it, the distribution of its biota. This dualism hampered the development of contemporary, dynamic thinking in zoogeography just as it retarded the clarification of modern ideas about organic evolution. In actuality, an awareness of the dynamics of animal distribution is essential to an intelligent consideration of the basic elements—the distributional areas. Therefore, the basic steps in the documentation of physical, biotic, environmental, and internal causes of that dynamism are outlined as follows.

(1) The cosmic system is dynamic; it is evolving steadily.

(2) The crust of the earth, being part of the cosmic system, evolves; it is not settled, but is under the influence of tectonic and plutonic forces; it builds up and subsides; erosion of different kinds, together with sedimentation and the like, is steadily reshaping both the surface of land and the physiognomy of the sea bottom.

(3) The history of the earth's surface, as far as it can be traced, shows periodic changes in gross climate, globally as well as locally. The climatic periods break down into oscillations of different magnitude which can be measured from geological periods to a few millennia; other oscillations last a few centuries only; others are measurable in decades, and no two subsequent years have the same climate extremes anywhere on earth.

(4) Cosmic, geologic, and climatic events affect the environment in which biota live, ultimately causing varying and variable patterns of distribution.

(5) Partly as a result of changes in the physical environment, biota undergo progressive evolution. Species arise, expand, change, or, after regression, die out. The pattern of life appears to grow more complicated as time progresses.

Growing, changing, and dwindling distributional areas follow the evolutionary changes of the species which inhabit them.

(6) The life cycle of an area may repeat itself many times during the lifetime of the species, governed by chance coincidence of all the factors here enumerated.

(7) A species as an evolutionary unit lives a dynamic life; those of its characteristics which intrinsically determine the area size are dynamic components, and the changes in them are among the causes of area fluctuations.

(8) In spreading and settling, chance plays an important role. It is through chance that each interacting species influences the distribution area of other species. Once they happen to be together, harmful or beneficial effects may result. If the influence is not direct, it may come through the web of chain reactions in the community.

(9) Finally, co-adapted ecosystems—combinations of vegetational and faunational units—are somewhat more than the sum total of their parts. Thus, they affect one another's distribution and, through that, the distribution areas of their faunistic or floristic components, the species.

The area, then, is to be considered *a dynamic characteristic of the species*. It is subject to evolutionary changes in the future as it has been in the past. Any given area has a transitory status reflecting the historical and currently acting external and internal forces that by chance govern its existence. *The area is a biological attribute of the species;* biological zoogeography keeps this fact always in mind.

Because area size (and area limit) depends upon the three groups (physical, specific, supraspecific) of evolutionary factors, there is little predictability with respect to the dynamics of an area. Certain groupings are, however, possible, and may prove useful.

(1) *Constant areas.*—Where the geological-geographical nature of a region and the age and lack of variability of a species exemplify constancy over a long period of geological time, we may think in terms of areas of constant size—but only when backed by paleobiological evidence. Species with constant, small areas are few; species with constant, large areas may be numerous among the lower organisms, but they have not yet been recognized; such areas have not been studied.

(2) *Expanding areas.*—There are three main categories:

(a) A new, young species, evidently filling an empty niche is said to be in the expansive stage. Since species are forming all the time, this type of expanding area is probably common, although hard to recognize if the species has not yet shown its ecological versatility.

(b) A species may have just commenced to utilize new opportunities, either by overcoming a barrier or by altering its internal spreading potential. This type is common, and examples from the latest period of active zoogeographical research are numerous in certain well-known animal groups (Figure 4–40).

(c) A species may be in an expanding phase, penetrating into a habitat which improved as a result of climatic or tectonic changes. This kind of expansion is also well documented, especially in the northern temperate, the subarctic and the arctic zones (as a consequence of postglacial climatic improvements) and on deglaciated new habitats (Figures 2–20, 4–58).

(3) *Dislocation of areas.*—The whole area of a species can gradually change location by simultaneously expanding in one and contracting in the opposite direction, as when a climatic belt shifts. It is more difficult

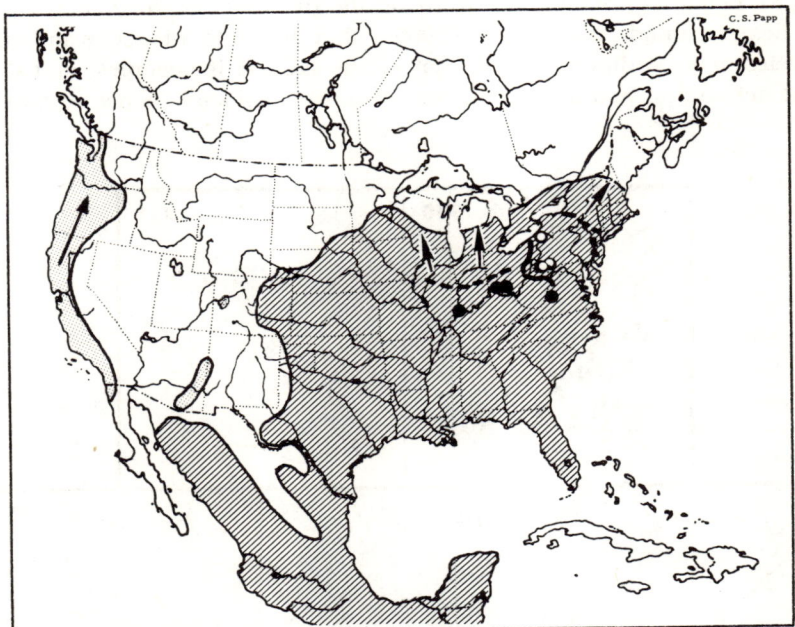

Figure 4-58 Dynamism of area.

Distribution area of the opossum *Didelphis marsupialis* L. [Didelphidae, Mamm.] in North America. Based on map by Hall and Kelson 1959. The dotted discontinuous areas in the west are secondary areas. The species there is an adventive, having been introduced and having spread north on its own during the most recent decades. In the Northeast, the species is now expanding. Thick line represents presumed northern limit around 1400–1600 A.D.; archeological sites from this period contain opossum remnants (dots) only to this area, not (circles) farther. Broken lines are approximating the northern limit of about 1900 A.D., and arrows show the present expansion. The map is from about 1957; since then, the limit has shifted farther north. (Archeological data from map in Guilday 1958.)

228 *Areography: The Study of the Distribution Area*

to document this theoretically common instance of area change, because many other controlling elements of topography and environment may obscure the movement or its speed on one or more borders of the area.

(4) *Contracting and retracting areas.*—These are always in the process of shrinking on some part of their boundary (Figure 4–59, 4–60). As a result, they may divide or totally disappear, resulting in extinction. Exemples are many, although the causes are only becoming evident by thorough analysis; it is often difficult to separate "natural" causes from mankind's detrimental effects. This is not to suggest that human influence is less "natural" than that of any other animal, but it surpasses all other causes in importance and, at the present time, undubitably influences practically all habitats on earth (see above: Secondary Relicts).

(5) *Potential area.*—When studying the dynamics of areas, it is useful to bear in mind, with Good (1965), that the rate of expansion of a species may be different from that of the habitat it occupies, and thus, that actual and potential areas do not always coincide. Since expansion

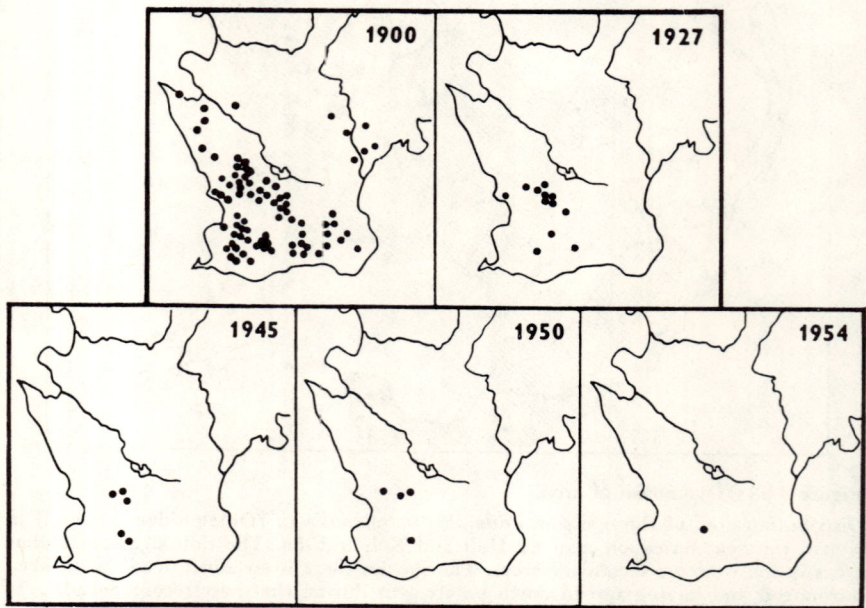

Figure 4-59   Extinction.

The decrease and extinction of the white stork *Ciconia ciconia* L. [Ciconiidae, Aves] in Sweden. Dots indicate breeding pairs. The two nestlings of the only 1954 nest died half-grown; there has been no breeding since 1954. After Curry-Lindahl 1961. (*Courtesy of K. Curry-Lindahl*)

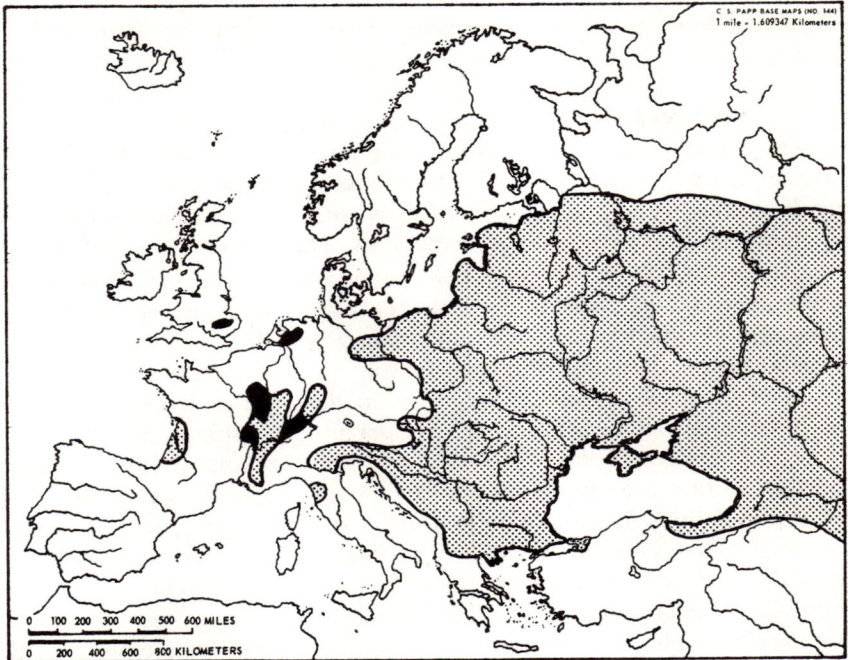

Figure 4-60 Extinction.
Extinction on part of the distribution area of the butterfly *Thersamonia* (*Chrysophanus*) *dispar* Haw. [Lycaenidae, Lepid.]. This species demonstrates dynamic changes, presumably as a consequence of climatic changes. Isolated relict populations in Western Europe remained after the main area regressed toward the east. These populations have been exposed to different ecological conditions and, as a result, they have developed different degrees of geographic variation. Some, such as those that lived in England and Holland, became extinct during the past hundred years (black areas on the map); the subspecies around the Alps is presently regressing. From de Lattin 1958, redrawn.

depends not only on the age of a species but also on its vagility, as well as on the nature of the barriers it encounters, it is evident that more often than not the potential area is not occupied. It is hard to find proof of what actually *is* the potential size of an area other than that indicated by occupation after natural or man-assisted crossing of a barrier (Figure 4–40), such as is witnessed in the instance of many introduced species of animals and plants.

Dynamism of the distribution area manifests itself in *dynamism of the area limit*. The following conditions may be recognized: *stable limit*—as e.g., a seacoast or other geologically long-lasting "absolute" barrier; *fluctuating limit*, a broader zone of marginal occurrence, within which the actual limit fluctuates from year to year; *advancing limit;* and *retreating limit*.

Area and Extinction

Faunal changes are the results of evolutionary processes, of which extinction is one negative phase. This phenomenon will be examined in detail in Chapter 6, but the present discussion of area dynamics leads to an immediate consideration of extinction. The extent of a distribution area, as previously explained, is determined by the range of essential and limiting ecological factors. (See the example in Figure 4–31 and the discussion on page 183). The model maps present a static picture of distribution, abundance, and the limiting or influential factors upon a hypothetical insect species. If changes in the ecological situation eventually cause the disappearance of the potentially optimal area (C), disjunct and sparse *relict* (D) populations may remain. In this example, the distribution area dwindles into two small disjunct parts; two small relict populations live under suboptimal conditions on *refuges*. This species might have become extinct before conditions reached directly lethal, critical levels, if its population became so sparse that random, periodic numerical oscillations has brought it to zero level e.g., because mating frequencies declined or accidental survival of inferior genotypes (drift!) occurred. On the other hand, the population may now show intensive local adaptations, resulting in subsequent expansion and, eventually, in *speciation*. If, however, the adaptional processes are slower than the rate of environmental change, *extinction* will be the inevitable result because the habitable area equals zero.

THE AREA OF HIGHER TAXA

The study of the geographic area of higher taxa is different from that of a species because the area of a higher taxon is an artificial composite area. Basically, all features of a species area apply also to a composite area; but the characteristics become progressively more generalized as we deal with higher taxa, and finally they become practically meaningless.

In order to grasp the great distinction between species areas and generic areas, it is necessary to keep in mind the fundamental differences between species and the other taxonomic ranks. The biological species is a population of related and similar individuals stemming from a common ancestry (for more complete definitions, see Mayr 1942, 1963); while the higher taxa are artifacts, i.e., abstractions. Higher ranks are absolutely necessary for a hierarchial classification; but when a taxon contains an excessive number of subunits, it becomes unmanageable, and the systematist subdivides it. Therefore, the different ranks of our classification have different value—as far as the evolutionary relationship of their members goes—depending on the total size of the group. For example, among the amniote vertebrates, the birds are generally considered a class; but vertebrate zoologists are aware that birds are no more than "glorified reptiles," with outstanding specializations for their flights, but that in most other respects they remain at the reptilian stage of development. What is called an avian subclass, a superorder, and order should be equivalent with the reptilian superorder, order, suborder or superfamily, respectively. Brundin (1965) believes that the family of chironomid midges is fully comparable to the class of birds, even as to the number of known species. Evolutionary scientists often wish that the major taxa would correspond to the evolutionary relationships of their members—which they do not. Phylogenetic relationships have developed throughout time, and the branchings are irregular; in geologic history these relationships are obscure, although paleomorphologists have done much to clarify them. In summary, the higher taxa are abstractions, compromises between the wishful thinking of the evolutionist and the practical need of the systematist and the nomenclator (cf. Voous 1964 for a highly pertinent discussion).

In view of all this, and since there is no uniform definition of the higher taxa, *one cannot properly attach a geographic criterion to a higher taxonomic entity.* The fact that this is done throughout the historic zoogeographical literature can be explained as follows. Much zoogeographical writing results from, or is a by-product of, the synthetizing work of specialists in the taxonomy of animal groups. These specialists catalogue the occurrences of taxa and, in doing so, find that the presence or absence of various taxa invites comment and interpretation. Evolutionists want to trace the geographic history of groups within their field of interest, and their use of distributional clues is here fully justified; they are obliged to use the accepted taxa because they have no other nomenclatural means at their disposal. But since not all zoogeographers are evolutionists or taxonomists, the general recommendation stands that *zoogeographical research should be based on the distribution of species* and on species areas.

Keeping these points in mind, we may cautiously proceed in discussing briefly the areas of higher taxa. Generally speaking, we may say that a genus is usually spread across a continent, and that a family is usually distributed over several continents. However, many genera have even wider distribution; but genera that belong to smaller geographic entities and families that are restricted to a single continent are in the minority (Figures 4–61, 4–62).

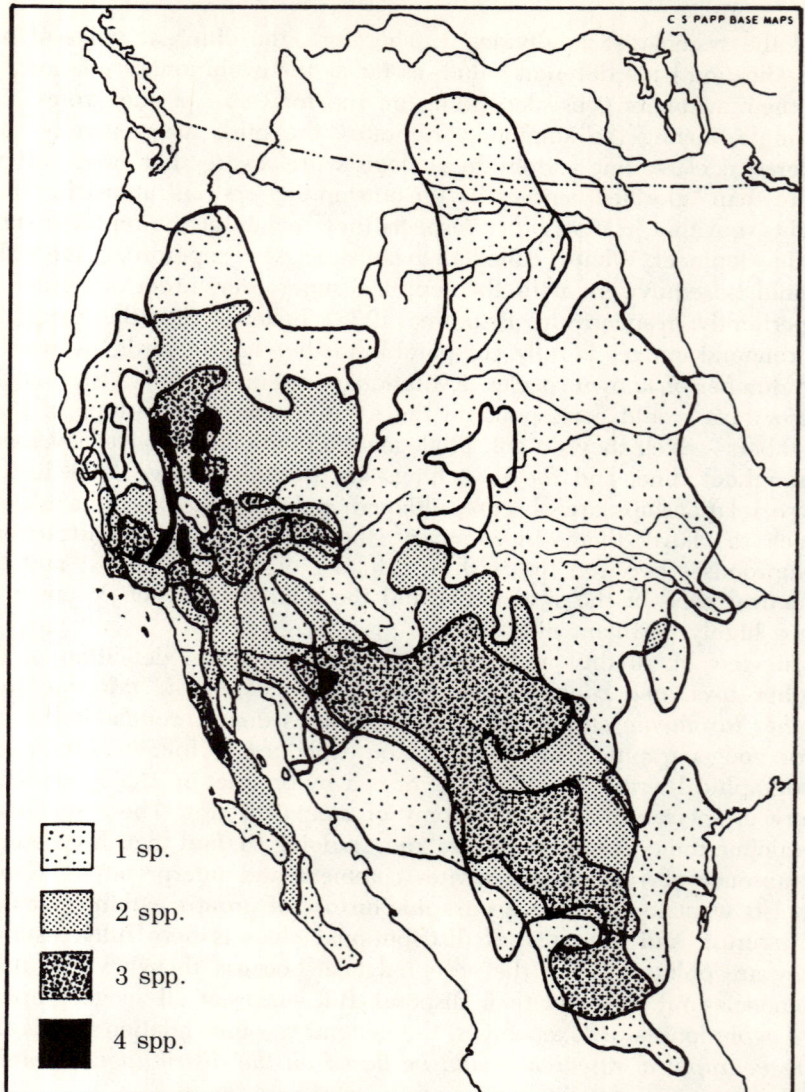

Figure 4-61 Generic area and center of diversity I.
Distribution of the North American endemic kangaroo rat genus *Dipodomys* Gray [Heteromyidae, Mamm.]. The areas of all thirteen species are superimposed. Two "centers" are thus shown: one in the southwestern part of the Great Basin; the other in the southern Rocky Mountains–Mexican Plateau area. Data from Hall and Kelson 1958. The essential environmental factors of kangaroo rats are found through several biomes of arid southwestern North America. This fact is reflected in the distribution of the genus. These factors are: arid or semiarid climate; proper soil drainage (for their burrows); light ground cover, also providing seeds for food, shelter, and dusting places. Dale 1939.

This wide distribution of the higher taxa is due, first of all, to their generally greater age, but it is due also to the ecological factor. Genera include closely related species, which are differentiated into related, major ecosystem entities and which are at least partially vicariating for one another. Their distribution reflects the variations of the ecosystem units within the same grossly limiting environmental elements.

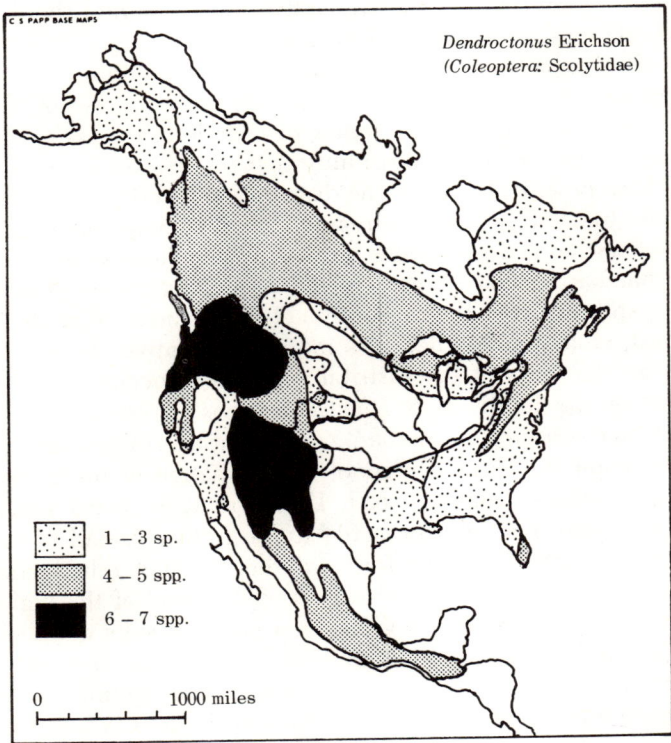

Figure 4-62 Generic area and center of diversity II.
Distribution of the bark beetle genus *Dendroctonus* Erichson [Scolytidae, Coleopt.] in North America. There are thirteen North American and one Eurasian species in this genus. The map shows the superimposed distribution areas of thirteen species. The North American area of the genus extends through all wooded plant formations which contain coniferous trees, on which the beetle larvae live. Eight species normally attack pine (*Pinus*) only; three, only spruce (*Picea*); one, only larch (*Larix*); and one, mainly Douglas fir and its relatives (*Pseudotsuga*). Thus, while many of the species show certain host specificity the genus is widespread. Two centers of species density in the west indicate the diversity of coniferous forests there, not necessarily the center of diversity or origin of the beetle genus. Data from Wood 1963.

Members of a family are genera that have diverged to fill related niches in many ecosystems. These ecosystems have little in common, but on their common properties are based the chief adaptive features of the members of the family. If we call the proper ecosystem taxon a biome, then we may say that families are spread across *unrelated* biomes. The emphasis is on the adjective, for the main structural element of the habitat, to which the family is adapted, occurs in many unrelated biomes. Structural and ecological characteristics of the genera in a family may be widely different, as are the limiting or essential environmental factors in the habitat. Many families show adaptive radiation, and thereby exhibit the interrelationship between structure, environment, distribution, and history—all of which are different aspects of evolution.

Geographic distribution—i.e., area, in the widest sense of the word—also plays a role in historical studies of orders and classes. The higher the taxon, the greater is the possibility that its distribution will be of the cosmopolitan type. If, however, the distribution is restricted and the taxon is endemic to some geographic region, this fact may serve as a clue to the age of unchanged conditions on that area; relictness of such a taxon reveals that barriers have acted in the past. Because emphasis of such studies is shifting from the currently acting factors toward those acting in the past, areal analysis becomes more comparative. A generic area is clarified on the basis of the distribution of the species areas within it; the range of the family shows concentrations of species or genera in one place and scarcity in another. Zoogeographers often use these findings to pinpoint the latest centers of differentiation of the taxon or even its center of origin (Figure 4–63). Discontinuities reveal relict species of the genus, and relict genera testify to the former spread of the family. Endemic and authochthonous forms, if they are not relicts, are usually in or near primary or secondary centers of dispersal of the higher taxon. The relatively late arrival of a group in a region results in allochthonous areas. Yet, at best, these conditions reveal only the present and the immediately past conditions of distribution. The argument against indiscriminate use of such "centers" as clues to the past is best exemplified by reading a statement by E. Truessart (1907) and a subsequent remark by Furon:

If there is one zoogeographic axiom which no one has yet thought of contradicting, it is this: as a general rule, that part of the globe in which at the present time a natural group of animals is represented by the largest number of species and by the largest variety of forms must be considered the center of its dispersal. . . . . Following this reasoning, Madagascar is undisputably the home of the lemurs. (Quoted by Furon 1958, pp. 40–41; my translation from the French original.)

Furon adds that fossil lemurs are now known everywhere in the world except Australia. The oldest (Paleocene) known fossil ancestors of this animal have been found in North America and in Europe (Colbert 1955).

COMMUNITY AREAS

Although ecologists are careful to point out that ecosystems consist of ecologically interrelated animal and plant species in a definite physical environment, little comprehensive work has been done on biotic communities beyond the elaboration of the concept. Botanists are far ahead in this area, and the study of plant communities and higher vegetational formations has progressed along geographical lines as well. At least one important school of plant geographers—geobotanists—maintains that a community is an entity comparable to a species and that its tangible manifestations are community individuals. The location of individual stands of the community combine to comprise the distribution area of

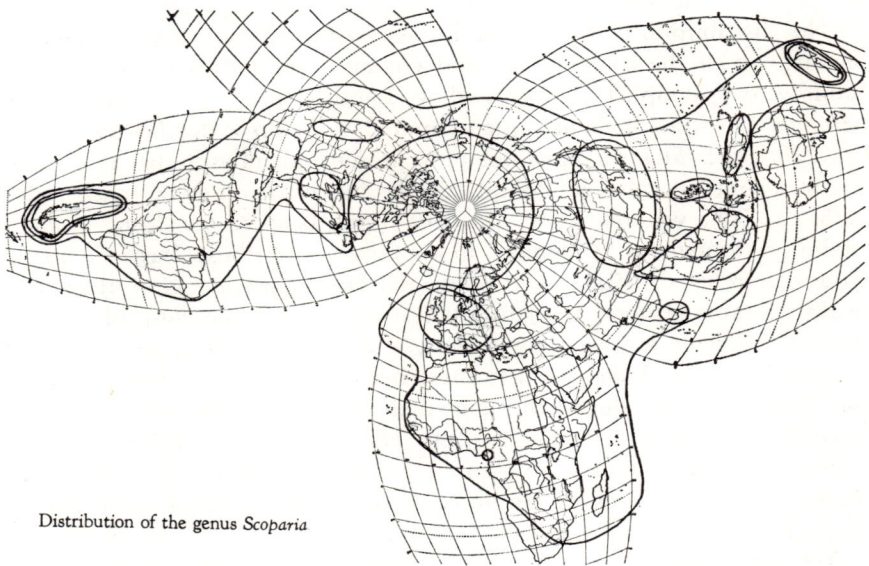

Distribution of the genus *Scoparia*

Figure 4-63 Generic area and "centers of endemism."
The nearly worldwide distribution of the large moth genus *Scoparia* [Pyralidae, Lepidopt.]. There are about three hundred known species in this genus. The total range is outlined; the circled areas are "centers of endemism"; those circled twice have particularly large number of species. From Munroe 1958. (*Courtesy of E. G. Munroe*)

the community, and this area is comparable to the areas of the species. There is, then, both *floristic* area study *and vegetational* phytogeography. The interaction of these two approaches becomes a fruitful, or at least fertile, attempt when the floristic areas of the community members are analyzed, compared, and evaluated. It would also be desirable to pursue these lines of research in the zoological-faunational aspects of biogeography.

However the "minimum area" concept of the plant synecologist should not be confused with our areographic endeavors, for this term has an entirely irrelevant meaning. "Minimum area" is the least extent of a biotope in which individuals of all characteristic species of the community occur, and thus it is a measure of the homogeneity of dispersion; it also indicates the minimum size of a proper sample area in field analysis of a stand of a community.

AREAS OF AQUATIC ANIMALS

Though we are discussing the zoogeography of land animals, it is useful to glance at the situation in area study of aquatic animals.

The distribution of freshwater forms is affected by ecological factors, among which the temperature and chemical properties of water (low salinity, for example) are outstanding. Land formations constitute much stronger barriers for many aquatic animals than do water bodies for the land forms; land formations not only present an inhospitable ecological environment but also hinder physically the locomotion of most water animals. Thus water dividers are very strong barriers—as strong as their stability throughout geological time. Because of these barriers, many young or relict distribution areas do not encompass a region of ecologically suitable waters. Dispersal upstream against the flow of a river may be difficult for weak swimmers, and waterfalls present additional physical barriers. Lake dwellers must resort to passive means of dispersal. Most lakes are short-lived, and their inhabitants are basically widespread, euryoecious animals. Ancient lakes are rare, but the few that have been studied abound in highly localized endemics—e.g., Lake Ohrid in the Balkans, Lake Baikal in Asia, the large lakes of East Africa, Lake Titicaca in the Andes (cf. Hubendick 1962).

Yet the dynamic history of the continents, which explains the present shape of all land areas, so profoundly influenced the dynamics of water systems that areas of water animals strongly resemble those of the land forms. These areas are generally drawn like the land areas previously discussed (Figure 4–44). However, in detail mapping, continuous occupancy of rivers is indicated by linear symbols (Figure 4–64) following their course, in the same manner as the coastal, linear areas discussed earlier. Figure 4-44 demonstrates the similarity between the distribution areas of a land animal, a river fish, and a land plant. All three are the result of geography, history, and—one might say—the technique of presentation. August Thienemann discusses these problems of freshwater zoogeography in an outstanding monograph (1950).

Two main area types can be distinguished among marine animals—they may be termed *epicontinental* and *oceanic*. Animals of the tidal zones and of the adjacent shallow-water zones have linear distributions that follow the configurations of the coast lines. But most of the presently existing coastal waters, the waters covering the *continental shelf* (the "epicontinental sea" of Ekman 1953), are also basically linear, though a closer scrutiny may reveal considerable width as well. The linear areas of littoral and shelf organisms are zonally limited; this fact is best expressed on the largely longitudinal coasts of the major continents which cut across the principal climatic belts. The various forms

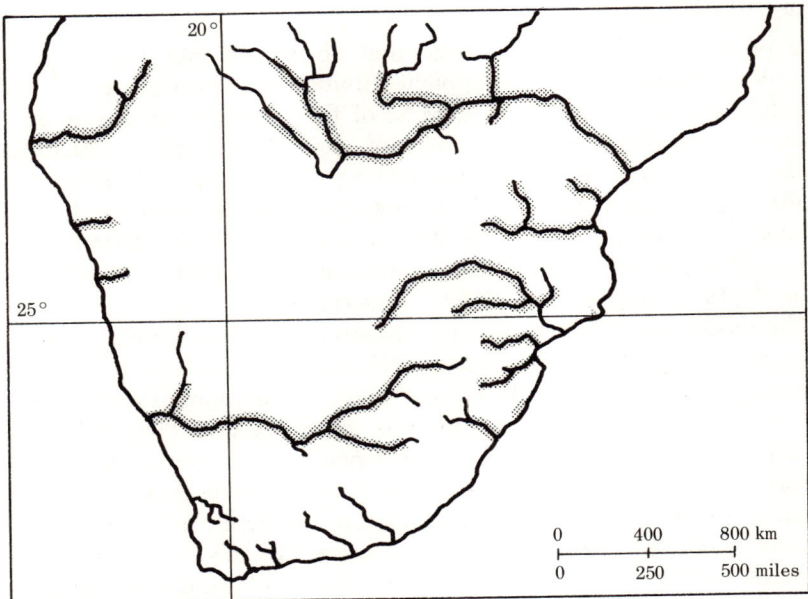

Figure 4-64 Mapping of fish distribution.
Disjunct distribution area of the fish *Barbus paludinosus* Peters [Cyprinidae, Cypriniformes] in South Africa. After Farquharson 1962. Sections of rivers containing this fish are indicated by dotting.

of discontinuities, the concept of relictness, endemicity, and so on—all apply to these areas as well. Coordination of their study with that of coastal land areas makes possible many valuable inferences and conclusions.

There appears to be very little knowledge of oceanic distribution areas, although the results of oceanographic surveys are rapidly accumulating and are being synthesized (e.g., Dunbar 1963). Salinity, temperature, mineral and organic content of the sea water, and currents are some of the known major limiting factors. It is, however, surmised that other factors are ultimately responsible for the shape of oceanic distribution areas (Bary 1963). In spite of the steady movement of water masses, marine organisms do have definite distribution areas in the ocean (Hedgepeth 1957, Glover 1961). Fluctuation of the ranges is not so much due to active movements of their inhabitants as to the fluctuation of water masses with distinct ecological character. Because of the existence of barriers and historic factors, disjunctions, relicts, and the like are discernible among oceanic area types as well.

PLANT AREAS

The point was made at the beginning of this chapter that plant geographers introduced the study of distributional areas much earlier than did their zoologist counterparts. Most of the principles discussed so far apply also to plants; indeed, many are derived from plant distribution studies. There is no particular need to repeat them here; modern treatments of phytogeography or plant geography may be consulted for more details, such as the works of Wulff (1933 in Russian; translated 1943), Cain (1944), Good (1964 edition), and Polunin (1960) in English; and Meusel (1943) and Walter (1954) in German.

The basic similarity between plant and animal distribution is striking (Figures 4–12, 4–13, 4–44). Wulff (*l.c.*) summarized the reasons for this similarity as follows. (a) Mutual dependence: the animal uses plants as its food, and therefore its distribution is correlated with that of the plant. On the other hand, many plants are dependent on animals for the dispersal of their pollen. (b) Plant and animal distribution is influenced by identical ecological factors, and as a result both are historically distributed from the same evolutionary centers (Wulff's "biocenters"). (c) Climatic and geological changes affect plants and animals similarly.

Without questioning the validity of any one of these statements, we may nevertheless restate the reasons for correlation between plant and animal distribution in a slightly different way, in accord with our present view of biological processes in both kingdoms.

(1) Despite profound structural and organizational differences, plants and animals retain identical cytogenetic mechanisms for achieving environmental adaptation.

(2) Therefore, based on largely identical mechanisms, evolutionary processes in plants and animals are parallel processes.

(3) Beyond these intrinsic attributes, parallel evolution is greatly enhanced (a) by identical extrinsic factors influencing plant and animal evolution at the same localities and in the same time interval; and (b) plant and animal evolutionary changes are mutually influenced by the plants and animals, and result in (c) the evolution of large, uniform ecosystem units.

(4) As pointed out in the earlier chapters, the means of *dispersal* and the ability to fill space (existence on an *area*) are specific biological characteristics subject to environmental selection.

(5) It is therefore a valid conclusion that *plant and animal areas are related systems which evolve under the influence of related and often identical mechanisms of selection and adaptation, and under mutual stimulation and dependence.*

A comparative and synthetical study of plant and animal areas and their mutual relationships would be a very rewarding branch of areography. The inferences and conclusions outlined above are based largely on an extensive knowledge of angiosperm areas and areas of land vertebrates, both of which have a recent and mutually influenced evolutionary history; invertebrates and non-angiosperm plants were not considered to the same extent. For these reasons, the inferences and conclusions given above are not demonstrably valid. However, any lack of validity is likely to be more apparent than real, for areography is based on *present* distributions, which are of course also the products of *recent* evolutionary history. Furthermore, the past distributionary history of invertebrates and non-angiosperms complicates the present picture much more than does the history of vertebrates and angiosperms. Insects and land plants had some 425 million years to distribute themselves; mammals and angiosperms, only about 180 million years.

CONCLUSIONS

Faunistics deals with the gathering of data about the occurrence of animals in space. The zoogeographer uses these data in determining the *distributional areas* of the species. The areas are the basic units in analytical biogeographical study. Research built on the analysis or synthesis of distributional areas is called *areography*.

The shape of the area and of its limits is determined by ecological and historical factors, and its study reveals the reactions of the species population toward these factors. Ecological valency (plasticity of tolerances and requirements, or width of adaptations) and mobility, together with the extent of available habitat, are important attributes that result in certain basic area types and structures. There is no directly apparent relationship between areas of species and of communities or lesser ecosystems; species areas may be smaller than community areas, or they may transgress community limits.

The different categories of discontinuities, further relictness, and endemism are important concepts of historically influenced area types. *The distributional area is a biologically determined attribute of the species, and is subject to evolutionary changes*. From this dynamic point of view may be distinguished constant, expanding, dislocated, contracting, potential, and extinct areas. Because of the temporal instability of the area, it is futile to search for centers of origin or dispersal within existing distribution areas.

Areas of higher taxa consist of the total of the areas of species they encompass. Such *composite areas* form abstractions which are extremely useful in revealing the evolutionary history and the past geography of the group.

The study of community areas is the geographic aspect of synecology (or syngeography), and comprises a useful and younger counterpart of biogeography.

In addition to the distributional areas of land animals, with which this chapter is principally concerned, there are freshwater, marine, and plant areas as well. It is entirely consistent with a dynamic and biological view of geographic distribution to find that the principles governing the demarcation of these area types are the same.

5 / Regional and Analytical Zoogeography

Ordo est anima rerum. (Latin proverb)

In Chapter 4 we became acquainted with the elementary principles governing distributional areas—the lowest units of zoogeographical research. Our task now is to discover what kind of pattern or order may be extracted from the great volume of information already available concerning these areas. We are looking for the composite units of comparative zoogeography, and it is highly important in selecting these units that we keep in mind the ultimate goal of comparative studies: causal analysis.

The casual observer as well as the field naturalist experiences the changes in the constitution of a local fauna as he progresses any distance in the field. Even though his general impression of the landscape and the appearance of the vegetation and faunation does not change throughout a large and uniform geographic area, the results of sampling show that certain species drop out while others appear on the habitat. When these observations are plotted, the distributional maps will show the overlap of species areas. The overlapping system of ranges is complicated, especially if the fauna is rich and the survey is complete, i.e., if animals of many phyla and classes have been observed.

Two attempts can be made after simple examination of the faunistic transect, but we cannot know in advance whether either will lead to a useful grouping, i.e., to finding composite units of ranges which are

242 Regional and Analytical Zoogeography

causally related. One basic observation will be that some ranges are small in diameter (and perhaps in extent as well) while others are wide and overlap with many ranges of the former group, and that still others are intermediate in extent. The second observation shows that area limits accumulate on certain parts of a transect and are scarce or even missing on other streches (Figure 5–1).

Actually, these simple observations did lead to past attempts at distributional grouping of biota. The small areas may represent endemics or relicts of local distribution; the large areas could indicate dominant, widespread species; the medium-sized areas may be good "ecological indicators," limited by some gross environmental feature. Overlapping of

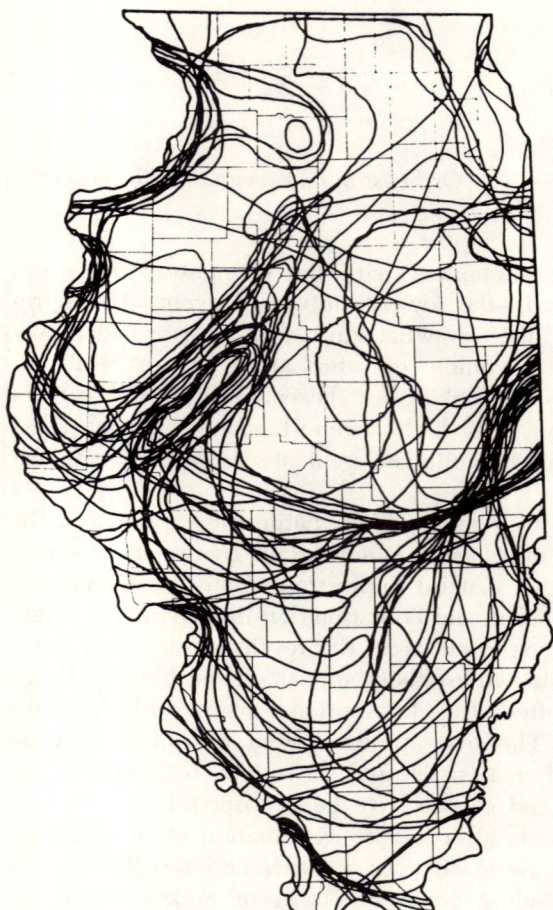

Figure 5-1 Clumping of area limits.

Outlines of the distribution areas of the amphibian and reptilian species in Illinois. From Huheey 1965. (*Courtesy of J. E. Huheey and The American Midland Naturalist*)—Where the boundary of this political district is not straight, it is formed by geographic features: Lake Michigan (northeast), Ohio River (southeast), Mississippi River (west). Note that distributional limits cluster at certain places, which represent natural barriers.

boundaries nearly always indicates a major break in the physical habitat, such as a mountain chain, the line where permafrost begins, or the margin of a desert basin. The first approach leads to consideration of the species, with its affinities, evolutionary adaptations, and historic background. The second approach leads to the assumption that distributional limits are contagiously distributed. The clumping of range limits has been proved statistically by Hagmeier and Stults (1964), and its causal analysis leads to ecological grouping of the distributions.

The multitude of early attempts toward grouping of animals on a geographical basis all had a rather negative approach in common. None was based on an assessment of the dynamic biology of animals in their environments, for such a consideration is the result of contemporary thinking, and has been built up bit by bit. The early attempts show that most zoologists had in their minds only systematics, phylogeny, geographic (regional) cataloging, or ecologic classification when dealing with animal distribution.

Before we proceed to a discussion of the biogeographical analysis of distribution types, it would be well to present a conspectus of these analyses.

1. From the *ecological point of view*, animals are grouped according to climatic valency and other environmental variables; this is also called the *eco-geographic approach*.

2. From the *geographic or chorological point of view*, faunas are grouped according to geographic regions. Inasmuch as the limits of large geographic units (oceans, continents) are also historic and/or contemporary barriers to animal distribution, these regions show some causally valid correlations of distribution; conversely, distinct faunas occur on areas which often coincide with geographic (physiographic) entities. Faunistic data serve as the basic for these analyses.

3. The basis of *areographic analysis* has been a combination of geographic, ecologic, and historic parameters of species areas.

4. In the *analysis of faunal diversity*, only qualitative diversity has yet been studied; consequently, our discussion will be restricted to species diversity of faunas.

5. A further and important analytic method—*historic analysis*—aims at understanding the causation of animal distributions, and will be examined in Chapter 6.

REGIONAL SYSTEMS BASED ON ECOLOGICAL DISTRIBUTION

The plant cover of the biosphere forms an almost continuous mat over the land. Early man lived primarily on this natural vegetation, and the vocabulary of every language reveals subtle distinctions between the types of vegetation cover according to the dominating life form, leaf form, and height of plants (e.g., scrub, shrubbery, thicket, copse, grove, wood, forest); even the developmental succession stages are named. The dependence of this plant cover on climatic and edaphic conditions is known to every farmer and, indeed, to everyone raised on the land. The experience of many generations shows which soil grows what, and under what climatic conditions a good harvest may be produced. The naturalists of past centuries, before our general urbanization, acquired most of this knowledge empirically and extended it to the wild vegetation as a matter of fact. Linnaeus, for example, commented on the dependence of vegetation upon habitat factors—climate, soil, and elevation—in his *Flora Lapponica* (1737). Alexander von Humboldt (1807) forcefully described the latitudinal climatic belts of the vegetation and the corresponding altitudinal belts of the South American Andes (see p. 113), and his writings laid the foundations of the science of plant geography.

The role of the environment in animal distribution did not make an impact on zoology comparable to Humboldt's on botany in the first half of the Nineteenth Century. Yet the example of the botanists, the accumulation of faunistic and taxonomic material, and thorough comparative studies resulted in the recognition of climate-related morphological differences (Bergman's, Gloger's and, later, Allen's rules). The impact of Humboldt's and de Candolle's plant ecological and geographical discoveries undoubtedly influenced several zoologists to try to determine simple continental patterns by using the gross climatic pattern as a major divider of animal distribution. The first such attempt was made by Pompper (1841). A. Wagner (1844–46) recognized the uniformity of the Eurasian and North American polar areas as well as the affinities of the north temperate regions of these continents, all of which he included in a Northern Zone. Agassiz (1850), Leunis (1860), and Günther (1870) continued to use the major climatic belts as dividers. Inspired by them, J. A. Allen made a practical attempt to divide the world into faunal zones based on the zones of climate (Figure 5-2).

Allen's Ecogeographic System

In 1871, J. A. Allen outlined a system in which temperature and humidity were considered the main differentiating factors of distributional entities. His system may be set forth substantially as follows.

Distributional entities follow the circumpolar climatic zones. Well-distributed species characterize the continent-wide belts; more restricted ones indicate subunits of lesser extent. From north to south, the forms

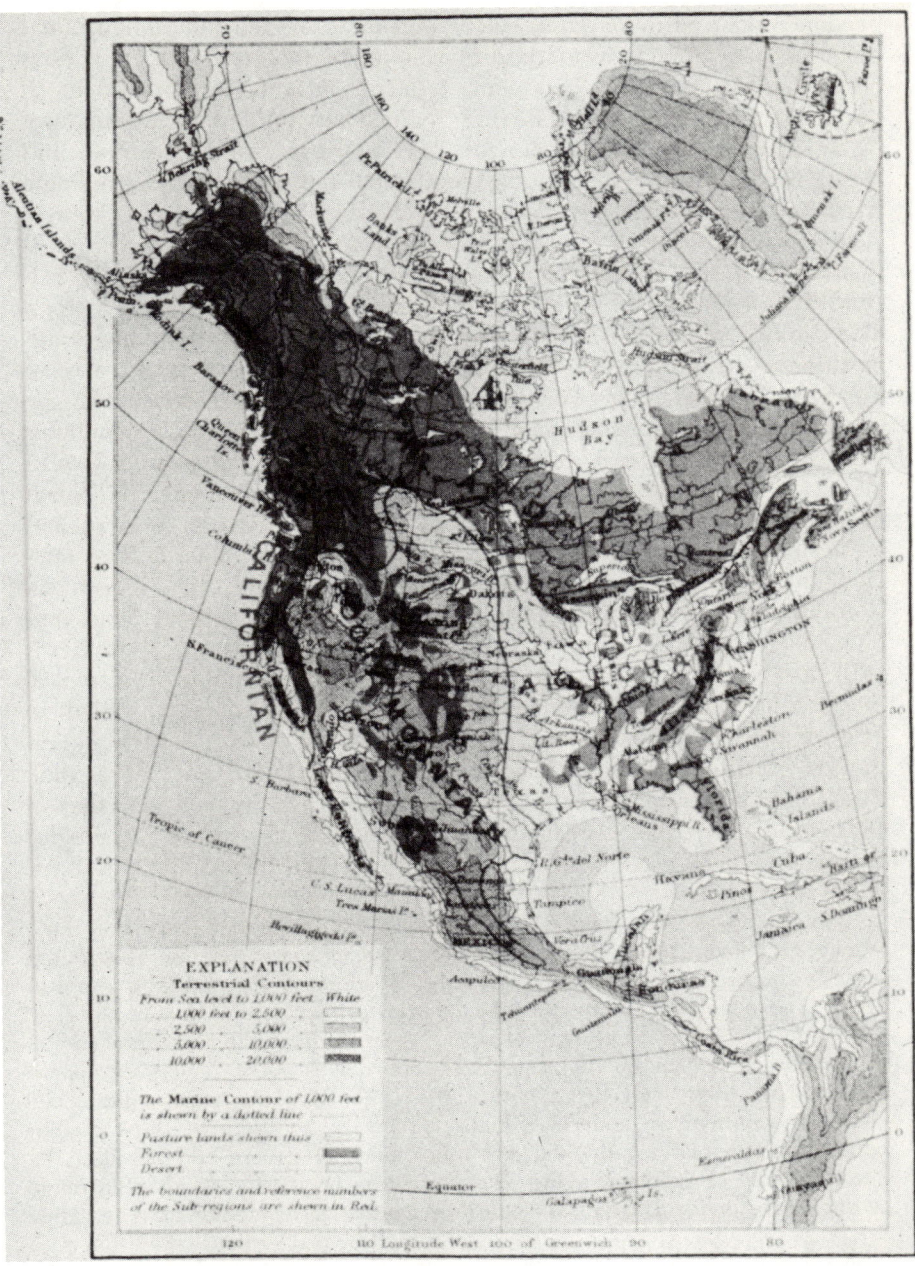

Figure 5-2 Allen's ecogeographic system: The North American Temperate Region. From Allen 1871.

become more and more divergent; within the same belt, the longitudinal divergence of animal forms is proportional to the size of the dividing ocean barrier; within a continent, it is proportional to differences in humidity. There are eight "Realms," which are limited by oceans and by belts of major climatic changes. The uniformity of arctic biota and the similarity of north temperate biota impel the recognition of a single Arctic (I) and North Temperate (II) Realm. The American Tropical Realm (III) is separated by the Atlantic from the Indo-African Tropical Realm (IV). The Arctic–North Temperate boundary is the 50° F (10° C) mean annual isotherm, or the northern limit of forest trees. The tropical realms are bounded by the 70° F (21° C) isotherm. South of this are a South American Temperate Realm (V) and an African Temperate Realm (VI). The Antarctic Realm (VII) is restricted to the islands surrounding Antarctica. The Australian Realm (VIII) includes New Zealand and New Guinea. The North Temperate, Indo-African and Australian Realms are each divided into two regions. The North American Temperate Region is further subdivided into an Eastern and a Western Province, each characterized by its own endemic bird species. In a final subdivision, the Eastern Province is divided into six "Faunas," each encompassing those species of birds, mammals, and reptiles limited by the isotherm chosen as the faunal boundary (Figure 5–2).

In this system, Allen recognized important ecogeographical principles, but he did not elaborate upon them. Although his system failed to achieve recognition outside contemporary North America, it is nevertheless worth our while to read some of his arguments in the original. They seem remarkably modern, especially so when we consider that in 1871—only two years after Haeckel had introduced the term—"Oekologie" had not yet been recognized as an important discipline.

Rarely is any species limited to a narrower area than that of two or three faunae or florae. Hence faunae and florae—which terms, in their restricted sense,* are properly applied only to the smallest of the onto-geographical divisions—are determined by the peculiar association of species; and not by the range of a single or a few "restricted" species; hence their general facies. Provinces and realms, on the other hand, may have species, and even genera and families, exclusively distinctive of them. As there are cosmopolitan, circumpolar, continental, and other kinds of species, so there must be cosmopolitan, circumpolar, continental and other kinds of genera and families; the latter, as well as species, having each a definite or specific geographic range as distinctive of them as any biological or anatomical character may be. They

---

*Allen proposes a hierarchial biogeographical classification into realms, regions, provinces, and—finally—floras and faunas.

are each circumscribed within definite areas, beyond which their special adaptation to their natural surroundings forbids their extension, unless aided by extraneous and unusual circumstances. (Allen 1871, p. 378.)

Life Zones

Although Allen's "Law of Circumpolar Distribution of Life in Zones" was never accepted as the basis of zoogeographical classification of the earth, it inspired an ecological classification system of biota which was modified and restated several times by its founder and finally almost completely refuted after half a century of use and controversy. This was Merriam's system of "Life Zones" (1892–'98).

Merriam continued Allen's work in discerning the coincidence of distribution limits with climatic and especially with temperature belts. He believed that temperature controls distribution, and he formulated the following two laws (1894).

Animals and plants are restricted in northward distribution by the total quantity of heat during the season of growth and reproduction (p. 233).
Animals and plants are restricted in southward distribution by the mean temperature of a brief period covering the hottest part of the year (p. 234).

Merriam believed that he could establish the zones of life according to certain isotherms which seemed to coincide with concentrations of plant and animal species limits and which also formed the boundaries of recognizable vegetation formations such as tundra, coniferous forest, etc. (Figure 5–3). This system of Life Zones worked fairly well in the uniform (and flat) central part of North America. They are especially well correlated to the altitudinal belts of the high Western mountains, some of which rise from the hot southern desert and reach beyond the belt of the alpine tundra. Differences in aridity and humidity were, however, not taken into account; neither was it ever proved that other minimum factors—e.g., winter cold (cf. Daubenmire 1938)—did not have a greater influence on distribution. Gross technical errors in Merriam's statistical calculation also became apparent. After stimulating geographically oriented research in ecology by a generation of North American biologists, the whole life zone system has been dropped as a geographically and ecologically unsound attempt at regional grouping. It was never applied to other geographic regions.*

---

*Merriam's Life Zone terminology is useful, and still used, in southwestern North America, where temperature-controlled zonation of the vegetation and faunation is a reality.

Ecogeographic Grouping of Plant Distribution

While these zoologists tried in vain to achieve a regional, geographical classification on an ecological basis, botanists such as Kerner (1863)

Figure 5-3   Merriam's life zones. (From Merriam 1894.)

Ecogeographic Grouping of Plant Distribution 249

and Warming (1909) were more fortunate. Plants are indeed more immediately controlled by climatic and edaphic, environmental factors in their distribution. Moreover, unlike in animals, members of one major taxon, the seed plants (angiosperms), dominate the land except on some

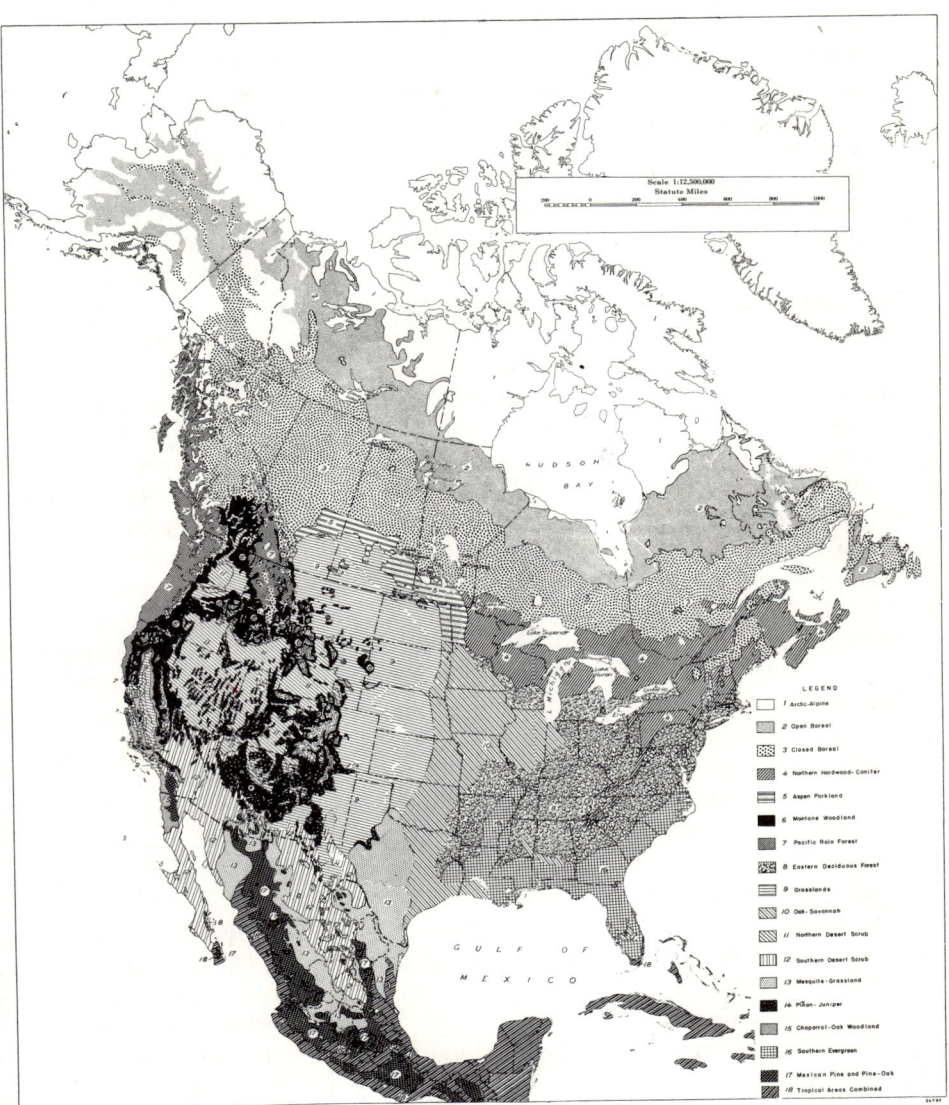

Figure 5-4 Biomes of North America. From Aldrich 1967. (*Courtesy of J. W. Aldrich and the Wildlife Society*)

poor habitats which are relatively insignificant in extent. These plants show certain major structural-biological adaptational types corresponding to the crucial environmental factors; these types are largely though not exclusively characteristic of phylogenetic branches of the angiosperms. Major climatic zones of the present are thus characteristically covered with certain dominating *plant formations* (Figures 5–4 and 5–5). Their geographical grouping is therefore much more natural than any similar grouping of animal distribution. Plant formations are stands which are dominated by certain *life forms* and which live under similar habitat conditions (Ellenberg 1956). It is not the distribution of taxa that delimits a formation, for unrelated or only remotely related species may belong to the same life form. Life form—this so-called physiognomical characteristic—is really a morphological manifestation of evolutionary adaptations to prevailing environmental conditions. Thus, for example, many unrelated plants have evergreen, needle-shaped leaves or succulent, xerophyllous assimilating organs. The basis of the widely accepted life-form classification of Raunkiaer is the structure and position of the organs through which the plant survives the most adverse conditions of its habitat. The system of physiognomical-ecological plant formations is practical, for it expresses a certain tangible uniformity which is evident and easy to describe. It has been adopted by most of the world's botanists, and even serves the field zoogeographer who is not versed in the intricacies of other, more complicated methods of vegetation analysis but who needs to place his study area in a vegetational system.

The distribution of the plant formations generally coincides with the climatic divisions of the earth. Since the macroclimate generally influences the existence of animals through its effect on the distribution of plant formations, the combination of climatic zones and vegetational formations would be adequate for the broad chorological grouping of animal habitats (Figure 5–5).

The distributional study of plant communities also provides a system of chorological division of the land which has been found suitable for the grouping of animal distributions. This approach of the botanist is based on the smallest tangible vegetational units which form stands in nature—namely, the communities or associations. A community of plants is a combination of individuals of different species occupying a common habitat—the "biotope"—and influenced by the same set of environmental factors. These plants influence one another as well as the environment through their effect on space, soil, and the microclimate and other properties of the biotope. Since animals interact with plants and chiefly depend upon them, the whole interacting system of plants—animals—physical habitat is considered a naturally associated unit—the ecosystem.

Ecogeographic Grouping of Plant Distribution 251

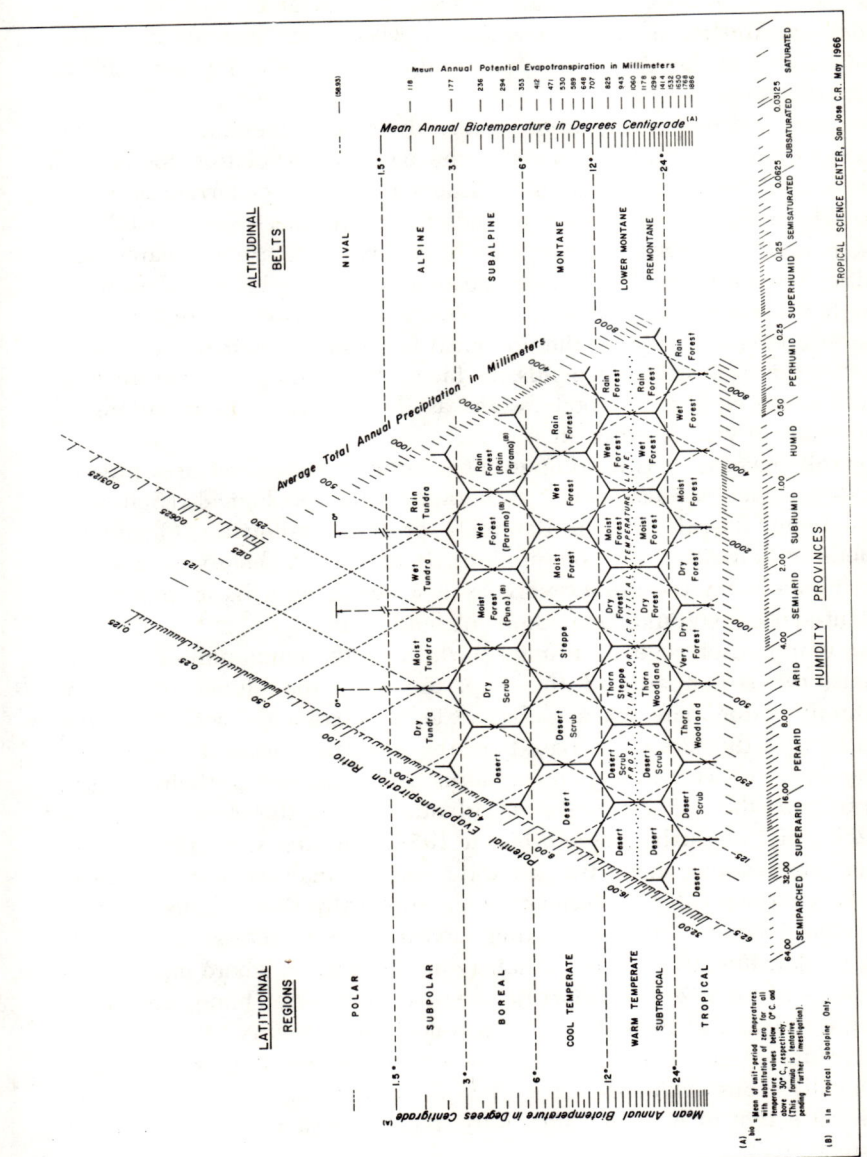

Figure 5-5 Diagram for the world classification of plant formations in correlation with climatic factors. From Holdridge 1967. (*Courtesy of L. R. Holdridge*)

Whether the communities are coadapted systems of an almost organismic nature, or are merely regularly repeated combinations resulting from the coincidences of similar biotope, species area, and history, is not decisive for our point of view (cf. p. 357). Such communities exist in nature, repeat themselves in space, and are classified in a hierarchical, empirical system.

The mapping of the vegetation of a study area is essential for regional zoogeographical studies. It serves as the basis of correlation for animal distributions; the plant communities indicate possible occurrences of animals where these have not been studied or have been overlooked. Finally, it must be borne in mind that in reality there are no plant communities but rather biotic communities consisting of plants *and* animals. Though the survey is usually done first by botanists, zoologists often take the liberty of adopting the botanical findings for their biota as well. Whether this is justified depends on facts of community structure that are often not well understood insofar as they concern interrelations of the animals in widely different phyla.

For this purpose, it is not important whether the botanist uses physiological-ecological (Sukatshev), areographic and ecological (Schmid), physiognomic-floristic (Braun-Blanquet), or dynamic-floristic (Clements) methods in delimiting the communities. If necessary, the zoogeographer himself should lay the groundwork by first acquiring a basic knowledge of plant synecology and its methods of field study.

The mapping of vegetation units starts from the communities or associations; the larger scale of the map, the more the higher "taxa"* of communities dominate; on a global scale, they coincide with the *plant formations* if the latter are based on the climatic zones of vegetation. If the basis of community recognition and analysis is their floristic composition, the highest units are identical with the *floristic regions* (p. 266) of the earth (Schmidthüsen 1959). In this way, plant geographers are working along two separate lines of analysis: the ecological, and the chorological. *Vegetational* areas, especially their higher entities, are determined by currently acting environmental factors; they show, in particular, the influence of climatic zones and areas where topography also plays a decisive role. *Floristic* areas, on the other hand, are based on higher taxa, and their distribution reflects past history. The finer the subdivisions of the two regional systems, the more they overlap geographically; thus when the size of a floral unit approaches the geographical extent of a plant community, the two systems merge (Braun-Blanquet 1932).

---

*These are: alliance, order, class.

Although the chorological analysis of vegetation forms the basis of field synecological and population ecological studies of the zoologist, the correlation of these studies with animal distribution has not been generally attempted, except for certain animal taxa on certain limited areas. The biogeographic study of biotic communities by zoologists has not advanced considerably in the past thirty years, especially in its terrestrial aspect. This is a promising field of future studies and syntheses.

When we evaluate the advantages and disadvantages for the zoogeographers of the chorological attempts of the plant geographer, our final judgment is that these are not particularly well suited for animal distributional grouping. Plant formations do not necessarily correspond to animal "formations" living in the same broad habitat entities, because of the differential spreading ability of plants and animals and because of differences in their evolutionary history. Looking for divisions of smaller geographic extent, we find that the chorology of plant associations has not yet been accomplished by the botanist. We cannot expect the zoologist to pioneer in the field of the plant geographer. Most plant community analyses are done on a scale far too detailed for the zoogeographer, while the system of plant formations is far too broad. This judgment applies only to the application of floristic plant geography to animal areography. The geographic grouping of faunations—paralleling vegetational plant geography—could be coordinated with, or based upon, the working concepts of the vegetation geographer.

Bioecology

On the basis of their empirical knowledge of nature and a priori reasoning, many biologists—beginning with Humboldt in the early Nineteenth Century—maintained that animals and plants both followed similar "rules" of ecological distribution. However, a convincing critical analysis of this matter has not even yet been undertaken. It is usual to point out the similarities (Palmgren 1930), to adapt the methods or the results of the plant geographer, and to call the result a kind of biogeography. The best of these efforts is the cooperation between one North American school of plant ecology and another of animal ecology, resulting in the terminology of *bioecology*. Plant ecologist F. A. Clements (1905, 1916, and so on) developed the thesis that the soil formation in a given climatic region tends to allow the development of a stable community, and that this climax community will be achieved through serial replacement of shorter-lived communities, no matter what the starting habitat.* This concept of successional and climax communities

---

*A further discussion of the climax concept is found, e.g., in Odum 1959, pp. 266–70.

earned wide recognition as well as criticism. It is now applied—with many modifications and exceptions—to regions that are largely homogeneous in climatic and geologic structure and history. Climax and successional communities actually exist, notwithstanding the fact that under other than purely climatic influences different kinds of "climax" will develop and persist. Clements postulated that animal members of plant communities must conform to the cyclic, successional development of their vegetation environment, and thus suggested that uniform governing rules could be found for the totality of the biotic community. He called the major unit in these associations of plants and animals *biotic formation* (1916). Two categories (or ranks, for the one includes the other) of biotic formations were recognized: the *biotic community* and the *biome* (Figure 5-4). The latter is formed by related communities within the same climatic belt, and is described and characterized by the climax rather than by the successional communities of its components. Kendeigh (1961) defines the biome as a biotic community characterized by distinctness in life forms of the important climax species. The biome is an abstraction (like orders, classes, and other taxa in systematics), a higher category which, however, contains communities *(biociations)* that may consist of taxonomically unrelated forms, since the life forms that distinguish it may result from the analogous (convergent) evolution of unrelated plants. For example, the deciduous forests of North and South America or the dominant plants of the South American, African, and Australian deserts are quite unrelated phylogenetically; yet each of these ecologically similar areas is classifiable under the same biome. The zoologist V. E. Shelford, cofounder with Clements of the biome concept, considers the biomes plant formations integrated with animal constituents. His criteria for considering animals among the dominant members of a biome are: they influence the composition of the vegetation; their behavior fits the environment; they are physiologically suited to the extremes of the physical habitat; and finally—and as the most practical criterion—they are distinctive and present throughout the biome (Shelford 1963).

The biome concept was introduced (Clements and Shelford, 1939) with a sound theoretical foundation. It is difficult to deny on ecological grounds that certain animals exert a dominating influence on other animal species of the biotic community if they take a crucial position in the food chain or in some other vital interrelation of the community members, whether plants or animals. It may also be argued that animals, as we have seen of the plants, can be classified into life forms. Such systems have recently been attempted (Remane 1943, Kühnelt 1965, Formosov 1966). Structural uniformity, means of locomotion, feeding habits, and feeding habitat are among the criteria that, in combination, provide an organism with the elementary potential for utilizing certain environmental niches (See Chapter 3). The representatives of life forms are often unrelated animals, which, as in plant life forms, have resulted from convergent evolution under similar environmental influences.

Nevertheless, in the decades since the introduction of the biome concept, only slow and late attempts have been made to determine the animal dominants of representative biomes and of their biotic communities. In the North American grassland biome or plant formation (which largely inspired the concept), the dominant role of the bison could not escape attention. However many other representative grassland biomes do not possess a similarly gregarious, abundant, and influential grazing mammal, even though the corresponding plant formations are basically similar in structure. Indeed, in other biomes it is often difficult to demonstrate the influence of a single large-bodied animal species comparable in its role to the bison of the North American grasslands. The coniferous forest formation, for example, has none. As an approach, the biome concept must be considered neither fully established nor in general use. North American ecologists often use the word "biome" in the sense of "plant formation."

The Concept of Biotic Provinces

Yet another attempt at biogeographical classification has been based on biotic communities. However, in the system of *biotic provinces,* the basis is not chorological but ecogeographical. After several earlier attempts—and especially inspired by Vestal (1914)—Dice (1943) developed a geographic division of the North American continent into 27 provinces (Figure 5–6). He said:

Each *biotic province* is characterized by the occurrence of one or more important ecologic associations that differ, at least in proportional area covered, from the associations of the adjacent provinces (p. 3). [Thus he tried to delimit considerable geographic areas] . . . over which the environmental complex produced by climate, topography and soil is sufficiently uniform to permit the development of characteristic types of ecological associations (p. 5).

Since some communities modify their biotope and, through succession, are able to spread, Dice also thought of the biotic province as a center of ecologic dispersal. The three essential features of this system are: it is geographic; it delimits areas of typical composition of biotic communities; and since such areas have biologically and topographically sharp boundaries only at large bodies of water, all other limits are arbitrary. The description and mapping of the North American biotic provinces reveal that in the northern and eastern parts of the continent they coincide more or less with the zonal plant formations and are of large extent. In the West, where topography, altitudinal, and climatic influences are extremely complicated, the biotic provinces diminish in size and increase in number, and their boundaries, following certain topographical features, become highly complex. The accompanying description in Dice's work is very broad, and includes almost casual references to certain typical plant formations and communities inhabiting the area. Dice admits that some of the boundaries could have been shifted fifty miles without offsetting the concept, and assumes—but does not elaborate—that animals would show similarly corresponding centers of ecological dispersal.

The concept of biotic province has seldom been used in its original sense. Goodman and Moore (1945) used it in Mexico; MacLachlan and Liversidge (1962), in South Africa (Figure 5-7). Most of the other

Figure 5-6 The biotic provinces of part of North America. Reprinted from *The Biotic Provinces of North America*, by L. R. Dice, 1943, by permission of University of Michigan Press. Copyright © 1943 by the University of Michigan.

works that mention biotic provinces use the term in a slightly different sense; the authors do not consider ecological similarities but only the faunistic uniformity of an area. Regarding this simplified usage, critics of Dice's concept maintain that range limits form clines—the faunal list changes gradually from one locality to another—and thus that districts cannot be delimited on such a vague basis. These critics maintain also that taxonomists might misuse the biotic province concept by assuming that species distributions conform to province boundaries. In recent applications of this concept, the ecological criterion has remained descriptive and highly arbitrary and the faunistic criteria have been emphasized.

A modern approach, reminiscent of the biotic provinces of Vestal and

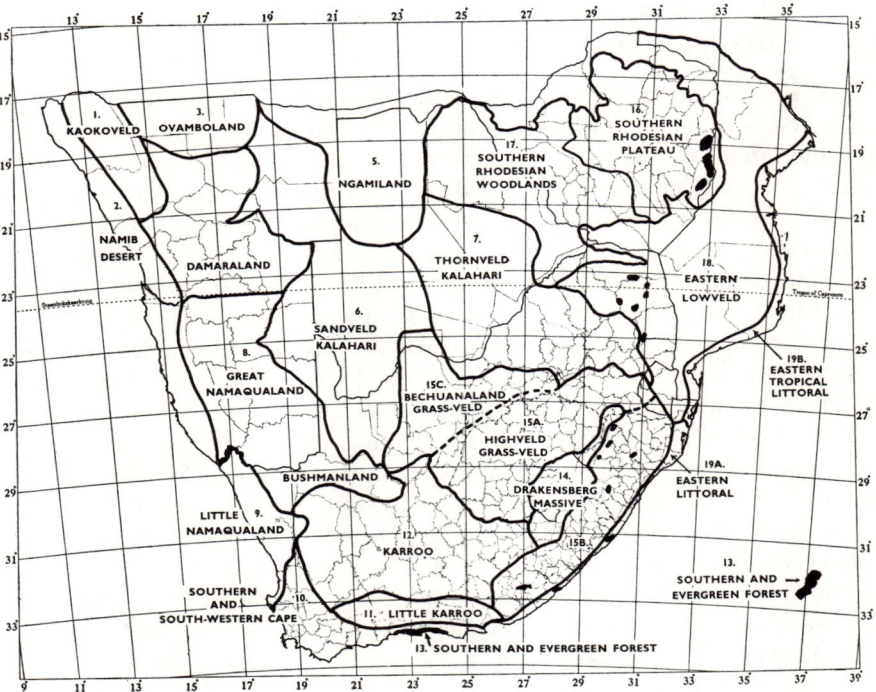

Figure 5-7 The biotic provinces of Southern Africa.
Prepared by G. R. McLachlan and R. Liversidge, from Liversidge 1962, Liversidge points out (1962 and *in litt.* 1968): that vegetation maps and the distribution areas of 423 bird species were considered in preparing the map and delimiting the provinces; that the boundaries cut across major climatic regions, that the names and areas applied to these provinces are in common use; and that the boundaries are in need of certain minor modifications. (*Courtesy of R. Liversidge*)

Dice, is followed by the Yugoslav zoogeographer Matvejev (1961), who presents a complete regional classification of his country in a hierarchial system. His basic unit is the area covered by an ecosystem, which he calls a province, although such "provinces" tend to overlap as the ecosystems mingle, especially in mountainous areas. Those entities which are higher than the province—the subkingdoms and kingdoms—are "predominantly characterized by historical moments of the more ancient past, and the lower ones—[he calls these subprovinces, regions and districts]—by more recent ecological moments" (Matvejev 1961, p. 186). The successively lower entities also include as characteristic features the unique or endemic taxa of lower categories: genera, species, subspecies. Basically, Matvejev's higher entities are geographic and faunistic, while his basic and lower entities are ecological.

REGIONAL SYSTEMS BASED ON GEOGRAPHIC DISTRIBUTION

The need for a geographical grouping of animal distribution arose in the first half of the nineteenth century. Sufficient geographical knowledge of the inhabited parts of the earth had already been accumulated, and a sufficient number of animals had been described and fitted into Linnaeus' system of classification, so that a few synthetizing biologists thought that the necessary basic knowledge was at last available. The earliest attempts were the catalogings of the faunas of the continents and seas by Illiger in 1815, by Swainson in 1835, and by Rütimeyer in 1867. Unfortunately, none of these works merits serious attention today; the faunistic knowledge on which they are based was highly inadequate.

Another approach was based on the distribution of the major taxonomic units. This method sought for regional entities with uniform faunal distribution, as characterized by the presence of unique families or even of orders or genera. Woodward (1851) used the phylum of Mollusca, and based 27 zoogeographical regions on them; Schmarda (1853) characterized his 21 land "realms" by various leading species or higher taxa of mammals, birds, and insects—largely corresponding to the subregions of Wallace (1876). The pulmonate snails of Keferstein (1866) yielded 34 "provinces" of the land area of the world. In the meantime, the idea of evolution became part of the general interest of zoologists, first through Robert Chambers' *Vestiges* (anomymously published in 1844) and then through the famous presentations of Darwin (1858, 1859) and Wallace (1858). Nevertheless, the first widely recognized regional zoogeographical system, based on the distribution of avian families by Philip Luthley Sclater (1858), was not influenced by the Darwin-Wallace theory, but rather stressed the separate creation of the faunas in the

Americas and on the continents of the Old World. The *Palaearctic, Aethiopian, Indian,* and *Australian* regions of his *Creatio Palaeogeana,* and the *Nearctic* (North America) and *Neotropical* (South America) regions in the *Creatio Neogeana* continue to be recognized as major regional units (Figures 5–8, 5–9).

What made Sclater's system almost perfect by the standards of his time was the fact that certain major barriers separate the avian and mammalian faunas of the five populated continents, and that these barriers—having been extant for a long geologic time—allowed substantial faunal separation. Tropical Asia and Africa are separated from the temperate Palaearctic by the ecological barrier of the transcontinental dry belt, which thus forms a major faunal divider. (The major dividers along the North Atlantic and North Pacific were genuine weaknesses in his regional system, but were soon corrected). Sclater based his regions on the *differences* in the distribution of the avian taxa considered—not on the affinities.

Sclater's regions were examined by A. Günther (1858) in a paper entitled "On the Zoogeographical Distribution of Reptiles." (In reality, the only reptiles dealt with were snakes; the work also discussed anuran, urodelan, and caecilian distributions). All these conform, according to Günther, to the regions of Sclater.

The important criticism started a decade later. In 1868, Thomas Henry Huxley wrote a paper on the taxonomy of an avian order, the henlike birds. In it he suggested major changes, using birds as the chief examples but invoking other evidence as well. He maintained that the Nearctic was

> ... really far more closely allied with the Palaearctic than with the Neotropical region, and that the inhabitants of the Indian and the Aethiopian regions are much more nearly connected with one another and with those of the Palaearctic region than they are with those of Australia (p. 314).

Thus, instead of Sclater's longitudinal, north-south major division, Huxley emphasized a latitudinal one separating the "north world" *Arctogaea* from the "south world" *Notogaea.* He, like others before him, recognized the *unity* of the northern circumpolar areas, and therefore suggested the addition of a *circumpolar* region to Sclater's Palaearctic, Nearctic, Aethiopian, and Indian regions within Arctogaea. Notogaea would thus include the "Australasian" region (Australia with New Guinea and the eastern half of the neighboring archipelago, where Wallace [1860] had already found a major faunal divider); New Zealand, and South America. These areas, as had already been surmised by Huxley and his contemporaries, had been biogeographically separated, during long geo-

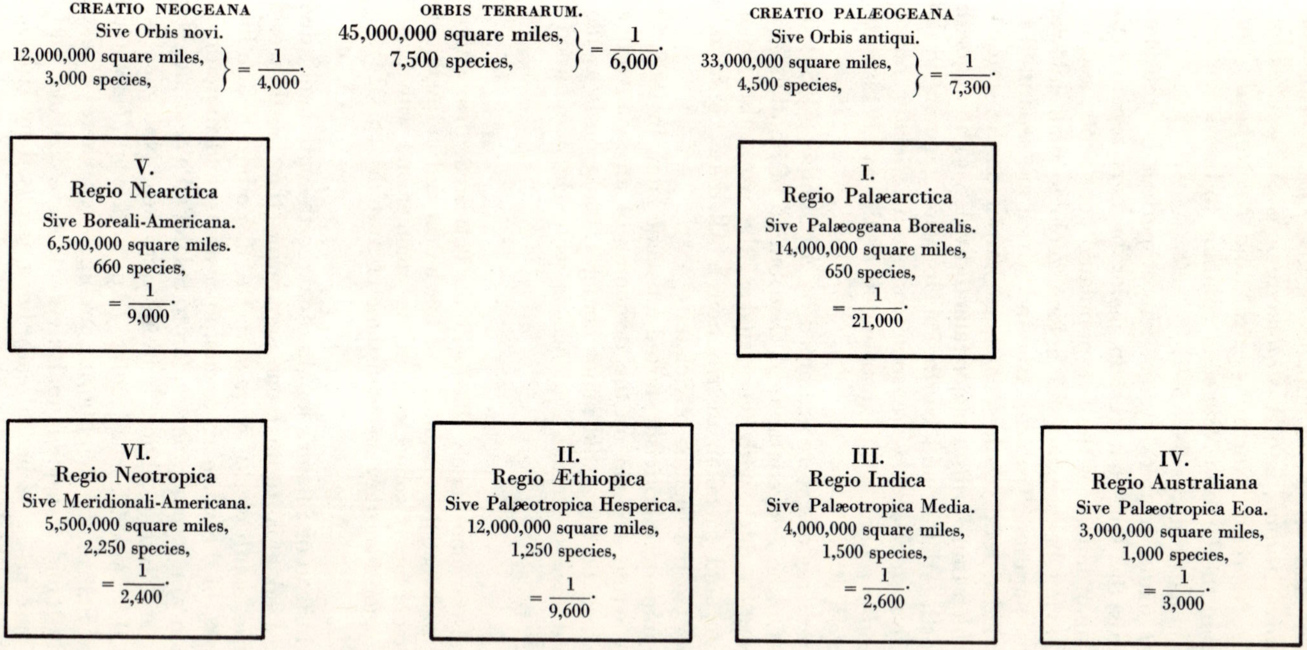

CREATIO PALÆOGEANA
Sive Orbis antiqui.
33,000,000 square miles, $\left.\begin{array}{c}\\\end{array}\right\} = \frac{1}{7,300}$.
4,500 species,

I.
Regio Palæarctica
Sive Palæogeana Borealis.
14,000,000 square miles,
650 species,
$= \frac{1}{21,000}$.

II.
Regio Æthiopica
Sive Palæotropica Hesperica.
12,000,000 square miles,
1,250 species,
$= \frac{1}{9,600}$.

III.
Regio Indica
Sive Palæotropica Media.
4,000,000 square miles,
1,500 species,
$= \frac{1}{2,600}$.

IV.
Regio Australiana
Sive Palæotropica Eoa.
3,000,000 square miles,
1,000 species,
$= \frac{1}{3,000}$.

ORBIS TERRARUM.
45,000,000 square miles, $\left.\begin{array}{c}\\\end{array}\right\} = \frac{1}{6,000}$.
7,500 species,

CREATIO NEOGEANA
Sive Orbis novi.
12,000,000 square miles, $\left.\begin{array}{c}\\\end{array}\right\} = \frac{1}{4,000}$.
3,000 species,

V.
Regio Nearctica
Sive Boreali-Americana.
6,500,000 square miles.
660 species,
$= \frac{1}{9,000}$.

VI.
Regio Neotropica
Sive Meridionali-Americana.
5,500,000 square miles,
2,250 species,
$= \frac{1}{2,400}$.

Figure 5-8 Sclater's 1858 schema of zoogeographic regions of the earth, based on bird distribution. The species figures given are now obsolete, and we show the whole schema for its historic value.

logical periods, from the circumpolar continental masses of the Arctogaea. Huxley's considerations were therefore considered sound by many of the subsequent regional analyses (Figure 5–9).

In J. A. Allen's (1871) attempt (See p. 246) to create a regional faunal system of the world on an ecological, climatic basis, he called his eight major units "realms," whereas these are called "regions" in Sclater's and Wallace's better-known systems. Allen's Arctic and Antarctic realms both include the circumpolar land areas. The northern, temperate, regions of North America and Eurasia are united into one realm, and the American and Old World tropical zones each form a realm. True to Allen's ecologic principle, temperate South America, South Africa, and the Australian-Oceanian archipelagoes comprise his Southern Temperate Realm—and remind us of rather similar divisions in contemporary floristic realms (p. 267).

Wallace's Work on Faunal Regions

Allen's ecological faunal system had little effect on contemporary thinking, which centered on Darwin and Wallace's evolutionary approach. In 1876, Alfred Russel Wallace published *The Geographical Distribution of Animals*, in which he supported Sclaters' regional system as preferable to that of Allen, a view with which most of his contemporaries agreed. In spite of the fact that Allen's work now appears in a better light, Wallace's deserves consideration because of its historical impact on eighty years of zoogeographical thought. This is indeed a monumental work, though Wallace modestly referred to it as a mere elaboration on the two chapters dealing with distribution in Darwin's *On the Origin of Species*.

Wallace emphasizes historic and evolutionary factors when he states in his preface (1876, p. v), "The amount of difference that exists between the animals of adjacent districts is closely related to preceding geological changes. . . ." and concludes (pp. 5–8) that "besides climate and vegetation, geological history and the resulting animal evolution influences distribution of animals."

The following statements reveal that Wallace had a good understanding of biotic communities (although he did not "discover" them for posterity; it is usual to attribute this concept to K. Möbius' 1877 work).

> Every change becomes the centre of an ever widening circle of effects. The different members of the organic world are so bound together by complex relations, that one change generally involves numerous other changes, often of the most unexpected kind. We know comparatively little of the way in which one animal or plant is bound up with others, but we know enough to assure us that groups the most apparently disconnected are often dependent on each

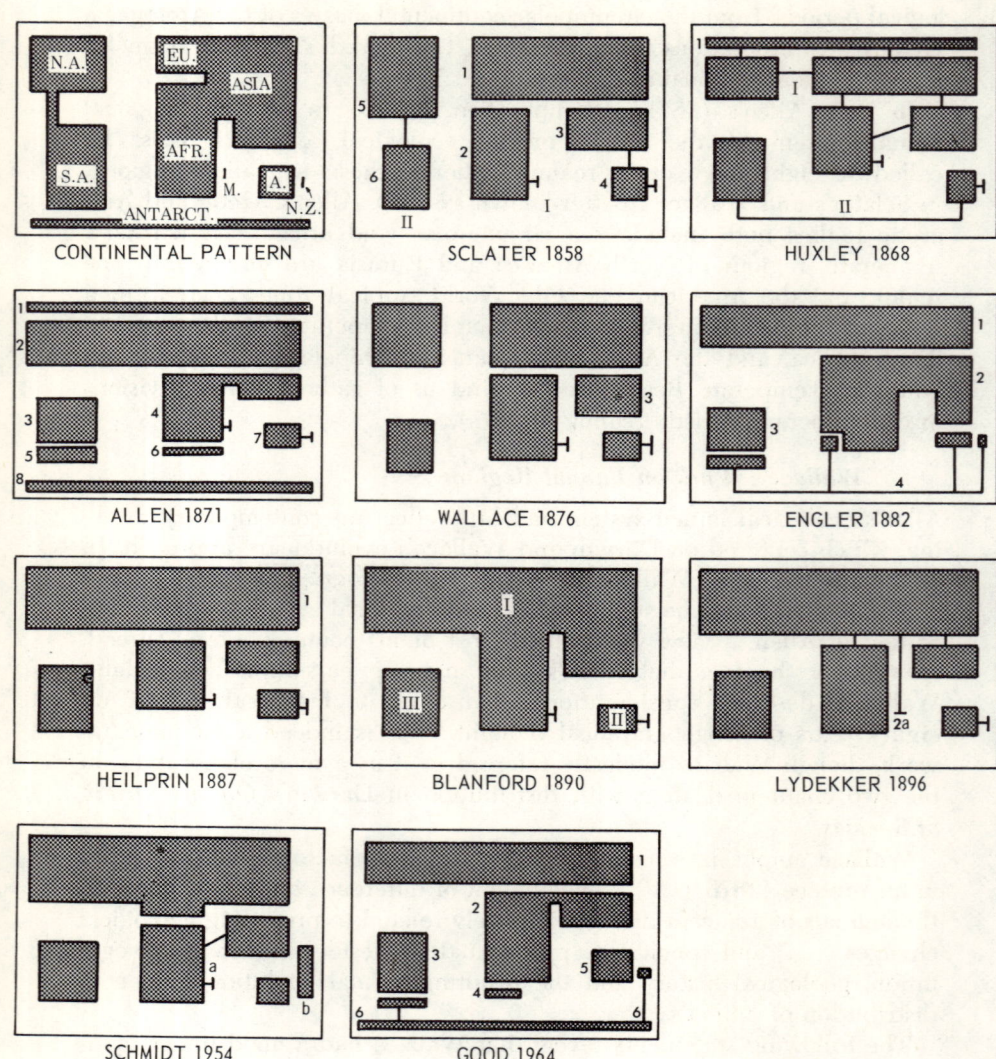

Figure 5-9 Faunal realms and regions; floral kingdoms.
Explanations:
Sclater: *I.* Palaeogean Realm. *1.* Palaearctic, *2.* Aethiopian, *3.* Indian, *4.* Australian Regions. *II.* Neogean Realm. *5.* Nearctic, *6.* Neotropical Regions.
Huxley: *I.* Arctogaea, *1.* Circumpolar Region. *II.* Notogaea.
Allen's Realms: *1.* Arctic *2.* Northern-Temperate *3.* American tropics *4.* Indo-African *5.* South American Temperate *6.* African Temperate *7.* Australian *8.* Antarctic.
Wallace's Regions, like Sclater's, except: *3.* Oriental.

other. . . . we shall be led to conclude that the several species, genera, families, and others, both of animals and vegetables which inhabit any extensive region, are bound together by a series of complex relations; so that the increase, diminution or extermination of any one, may set in motion a series of actions and reactions more or less affecting a large portion of the whole, and requiring perhaps centuries of fluctuations before the balance is restored. (pp. 44, 46.)

Wallace summarizes the difficulty of the ecological approach:

In many cases we should find that no satisfactory division of the earth could be made to correspond with the distribution even of an entire class; but we should have the coleopterologist and the lepidopterologist each with his own geography. (p. 58.)

He makes the point that even within the Coleoptera, the Longicornes depend on woody vegetation, the Carabid beetles on "barren regions," the Scarabaeidae on the occurrence of herbivorous mammals—thus giving examples of the vegetational, physical, and symbiotic factors that we now call the factors of ecological distribution.

Wallace maintains that since regional animal geography is a necessity, these regions should be based on the animal ranges themselves and not on climatic zones or geographic divisions of the earth. He was aware that such a regional system would have its shortcomings. In fact he discussed exactly those shortcomings which the subsequent critics and changing emphasis of our present zoogeographical thinking find questionable in Wallace's system. The more important of these are the following. (1) There is an inequality of the rank of several regions. (2) There are major differences in the distribution of different animal groups. (3) The use of positive characters (occurrence of characteristic families or genera) and negative ones (lack of characteristic higher taxa) is intermixed. (4) It is futile to draw sharp regional limits when many intermediate districts exist between the regions. (5) The regional system is not in accord with the past distributional conditions. (6) The size of an area has an influence on the number and importance of its fauna.

Engler's Kingdoms: *1.* Northern Extratropical *2.* Paleotropical *3.* South American *4.* Old Oceanic.
Heilprin: *1.* Holarctic Region.
Blanford's Realms: *I.* Arctogaea *II.* Notogaea *III.* Neogaea.
Lydekker, like Heilprin, except: *2a.* Malagasy Region.
Schmidt, like Heilprin, except *a.* Paleotropical, *b.* Oceanian.
Good's Kingdoms *1.–3.* like Engler's, *4.* South African *5.* Australian *6.* Antarctic.

Wallace's choice of the mammals in verifying the six regions of Sclater's avigeographical study was also deliberate. Günther (1858) showed that these regions fitted amphibians and reptiles as well. Wallace considered mammals a still better characterizing group because they are earthbound, well known, and conform to the main distributional barriers. Wallace admitted that North America, Eurasia, and Africa show greater affinity (through the Palaearctic) than do the two Americas for each other than does Australia with the rest of the Old World; his emphasis, however, is not on Sclater's or Huxley's realms but on the *regions*. All six regions have equal value in Wallace's system. He renamed the Indian region of Sclater the "Oriental region," as it has since been called (Plate I).

The Sclater-Wallace system of zoogeographical regions stirred up much controversy about where to draw the most appropriate regional boundaries. The only major "improvements" occurred in the years immediately following the publication of Wallace's book. Even these changes concern the larger entities rather than the 24 subregions Wallace assigned within his six regions; these subregions all have some important features, and are generally marked by certain major geographic and faunistic boundaries.

Heilprin (1887) suggested that the affinities of North America (the Nearctic) and temperate Eurasia (the Palaearctic) are greater than the differences Wallace emphasized; his term *Holarctic* for this united region has persisted, while Pal[a]earctic and Nearctic are still useful and in general usage when the corresponding areas are in any way treated separately (Figure 5–9). In spite of the general criticism of any regional system based on sharp faunal boundaries, the long-lasting continental connections during the major part of the Mesozoic and Cenozoic of the Palearctic and Nearctic regions is now proved beyond doubt. We are even more conscious than Heilprin of the close relations of the faunas and floras of these regions.

In using Wallace's principles when grouping the faunal regions, it must be realized that during the early Tertiary, when the northern temperate belt and tropical belts were much more northerly situated than today, the temperate and tropical faunas of the northern continents had quite free faunal interchange. During these same periods, South America and the Australian region largely followed their own evolutionary course in isolation. Therefore, three major divisions—realms—reminiscent of Huxley's 1868 suggestion, are justified: the *Arctogaean* (for the Holarctic, Aethiopian, and Oriental regions), the *Notogaean* (for Australia, New Zealand, and certain neighboring islands), and the *Neogaean* (South America) (Figure 5–9). This usage became common in the

1890's (Blanford 1890, Anon. 1893, W. L. Sclater 1894, Lydekker 1896, Jacobi 1900).

Madagascar, a large island with a unique biota that evolved in isolation, has also been a center of controversy. Blyth (1871), Allen (1878, 1892), Reichenow (1888), Möbius (1891), Lydekker (1896), and others considered it a separate region (sometimes called Lemuria or the Malagassy region, Figure 5–9). However, the peculiarities of its mammalian, avian, and insect fauna hardly justify its admittance to regional rank (cf. Millot 1952). If such an area is to be considered a region, then even smaller islands with peculiar faunas must be ranked similarly, as were the Hawaiian Islands by Holdhaus (1929).

In concluding our review of the history of faunal regions, let us keep in mind that we have mentioned or discussed 26 more or less different approaches, almost all of them formulated before the end of the past century. There were many more. F. A. Schilder (1954) presents, in a monographic statistical paper, 45 more or less original approaches and 17 duplications—and his list is by no means complete. Efforts in the present century, judging from Schilder's maps, have been largely duplications of, or variations on, what had already been accomplished by pioneering zoogeographers. Most of the older viewpoints that did not agree with the accomplishments of the Sclater-Huxley-Wallace-Heilprin-Blanford line stemmed from those zoologists who worked with details of fish, mollusk, or insect geography and found that their groups did not conform to the pattern derived from the evolutionary history of the land vertebrates. Yet they did not totally reject the dominant system; they either recognized only minor regional faunal divisions (e.g., the 34 molluskan provinces of Keferstein 1866 and the 13 oligochaetan regions of Michaelsen 1903) or emphasized the affinities of the southernmost temperate land areas—Patagonia, Tasmania, New Zealand (for example, Günther 1870, Gerhard 1883, Fischer 1887). The most particular geographic-evolutionary approach, that of Handlirsch (1913) on insects, corroborated rather than contradicted the existence and foundations of the Wallacean regions.

Essentially, the modified Wallacean regions reflect the contemporary faunal picture. The regions are based on: *the great oceanic barriers*, as longitudinal dividers; within the great northern land mass, on the *subtropical–warm temperate dry belt* separating the Palaearctic from the Ethiopian and Oriental regions; on the ocean barrier of Tertiary to recent duration having isolated the two southern—the Neotropical and the Australian—regions. Antarctic affinities of the southern temperate lands are not indicated. Wallace recognized that the evolutionary history of land vertebrates conforms to this pattern. The post-Wallacean concept of three major realms reflects, principally, a greater emphasis on the geologically long-standing isolation of South America and the general area of Australia and New Zealand (cf. Schmidt 1954 for a good summary and for his modification of the Sclater-Wallacean regions, as shown in Figure 5–9).

Floral Regions

It is useful for a complete understanding of the regional geographical systems to examine what botanists have accomplished in the way of floral geography. Fortunately for them, the majority of land plants belong to a single phylum, the angiosperms, which arose later than the land vertebrates but achieved early dominance over the gymnosperms, Bryophyta, and other lower phyla. Vertebrates share land habitats and resources with the generally much older insects. *Land plant geography is much more angiosperm geography than land animal geography is vertebrate geography* (regarding post-Cretaceous backgrounds of the recent and present distributions).

It is a facile excuse of zoologists that our colleagues devoting themselves to the plant world have much easier work when pioneering in the field. Once plants have been discovered, they are unable to escape the museum collector. Of the larger specimens, small fragments can be pressed to provide satisfactory samples for floristic documentation. Thus we should not hesitate to say that floristic knowledge has always proceeded faunistic exploration. The beginnings of regional floristic syntheses were largely contemporary with those of the first world faunists. Yet by the middle of the nineteenth century, floristic geography and paleobotany were much more advanced, and it is perhaps only an accident of human history that Adolf Engler, the Wallace of regional plant geography, happened to be younger than Alfred R. Wallace and that Engler's *Versuch einer Entwicklungsgeschichte der Pflanzenwelt* appeared from three to six years after Wallace's *Geographical Distribution of Animals*. For practical purposes, these works were contemporaneous, and their authors were inspired by the same *esprit du temps*.

Engler's treatise is a model of a well-developed, well-written work. Volume One (1879)—*An Attempt at an Evolutionary History of the Extratropical Floral Areas of the Northern Hemisphere*—deals with the historic plant geography of what we now call the Holarctic. The origins of Engler's floral areas are the old Tertiary and Cretaceous flora as known at that time. Engler discusses the historical unity of the north temperate circumpolar region, the isolating effects of the Ice Age, and the historic reason for a division into floral regions, provinces, and districts of this vast territory. While Wallace was truly a pioneer, Engler availed himself of the pioneering work of Hooker, Asa Gray, and other excellent contemporary workers. His Volume Two (1882) treats the tropical and south temperate areas, and concludes with the plant geographical groupings of the earth (Figure 5–9). Engler distinguishes four climatically determined groups of plants that had already differentiated in the early Tertiary. He termed these four groups *floral elements*. (1) The *Arcto-tertiary* elements dominate among the present arboreal and shrub flora of North America and temperate Eurasia. (2) The *Palaeotropical* and (3) *Neotropical* elements include such plants which have a long-standing fossil history in the tropical areas of the Old and New Worlds, respectively. (4) The *"Old-Oceanic"* elements consist of those plants which mark the floristic affinities of southern South America, southern South Africa, southern Australia, New Zealand, and the islands of the southern oceans (Turrill 1959 calls them *Southern* elements). Engler's four floral *kingdoms,* or realms, correspond to the four element groups (Figure 5–9). The *Northern Extratropical* kingdom has circumpolar arctic and subarctic, five Eurasian, and two North American regions. The *Paleotropical* kingdom includes all the tropics of the Old World, even those of tropical Australia and Oceania (ten regions). The *South American* kingdom (five regions) excludes the temperate extreme of this continent, which, dominated by the common southern floral element, is classed with the above-enumerated areas of south temperate affinities as *Old Oceanic* Kingdom with eight regions.

Engler's regional phytogeography has survived to the present day as the generally accepted classification with only few modifications of its major categories (Walter 1954, Good 1964). These modifications concern the southern kingdom, and divide it into three parts: the *Australian* kingdom now encompasses the whole of Australia and Tasmania; the Cape (of Good Hope) Peninsula (with a rich and unique flora in a Mediterranean climate) becomes a small, independent *South African* kingdom; and the Chilean-Patagonian temperate area, New Zealand, and Antarctica with its islands comprise the circumpolar *Antarctic* kingdom (Figure 5–9).

The South African kingdom is a unique feature of the floral distributional picture. The uniting of the other southern lands is another major deviation from the Sclater-Wallace zoogeographical presentation—but accepted by zoogeographers as well—based on the widely discontinuous distribution of certain animal taxa (for example, Huxley 1868). This treatment is indeed based on a still unsolved major problem of historical biogeography—the question of the origin and importance of a southern circumpolar fauna (*cf.* Darlington 1965, Brundin 1967).

Floral geographers also use further subdivisions of Engler's regional system (Figure 5–10); there seems to be as much disagreement among them about regional boundaries as there is among zoogeographers among their systems. Nor is the botanist more advanced than the zoologist in the use of statistical methods for finding lines or areas of substantial floral change which would be suitable for delimiting entities of a regional system (see p. 278).

Subdivisions of the Zoogeographic Regions

Wallace and the followers of his zoogeographical system were not content with six broad regions, but made many attempts to subdivide these regions into smaller entities on a hierarchical basis. The criteria used were often purely geographic, because faunal knowledge was insufficient or only locally available. Thus, e.g., Wallace's separation of the European and the Siberian subregions of the Palearctic, along the Ural mountains and the Caspian, was based on insufficient faunistic evidence; this fact was soon pointed out by the Russian faunists Severtsov (1877) and Menzbier (1882).

Where the then extant faunistic evidence was used, it consisted mainly of the presence or absence of certain widespread and numerous taxa and of certain outstanding endemisms. In using the original descriptions of subregions and provinces from the past literature, it is often impossible to find detailed documentation in the form of faunal lists or distribution maps. On the other hand, such works as Van Dyke's (1919, 1942) on the regional distribution of the North American beetle fauna produced workable maps with the boundaries of subdivisions well marked. Along major topographic or ecologic barriers, there are major faunal breaks, which can be recognized without particularly accurate analytic methods. A system of zoogeographical subregions developed—along with the floristic regional mapping activities—in every country where zoogeographers were active. A complete discussion of all such subregions would require a descriptive zoogeography of the entire earth, and therefore we can refer to only a few examples here (Figures 5–11, 5–12).

With the increased faunistical knowledge we now have of several groups of animals in different areas of the earth, it is possible to base the geographical *grouping* of animals on an accurate knowledge of the actual distribution of all or most species and higher taxa within a well-known group. In treating a large number of data, when we want to know the average, or to compare and group sets of data for correlations and underlying, complex causes—even when we want to predict probabilities—we often use statistics.

Without endeavoring to present a thorough survey of the methods and achievements of statistical zoogeography, we shall discuss some of

Subdivisions of the Zoogeographic Regions 269

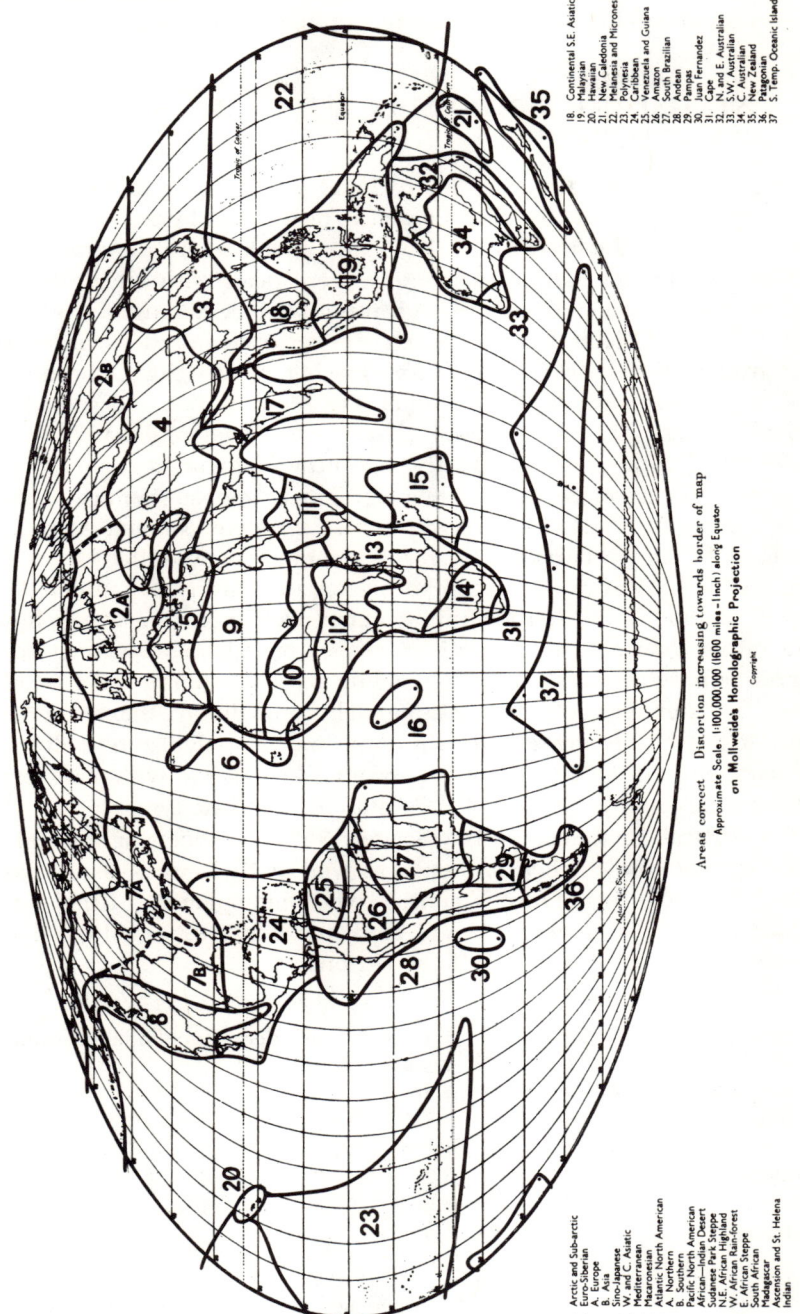

Figure 5-10 Map of the floral regions of the world. From Good 1964. (*Courtesy of Longmans, Green and Co. Ltd., London*)

the pertinent efforts. Reference will be made elsewhere to recent work on the predictability of certain zoogeographical phenomena (p. 374).

STATISTICAL METHODS IN REGIONAL ZOOGEOGRAPHY

Neither the ecologically founded regional systems nor those based on evolutionary relationships—the uniqueness of regional faunas—made much use of statistics, chiefly because they came about before the era of exact biology and before natural phenomena were dealt with on the basis of populations. Taxonomy was typological until the beginning of the Twentieth Century (Kleinschmidt, Rensch); genetics acquired its population background in the 1920's and 1930's (Fischer, S. Wright, Dobzhansky), as did plant ecology (Braun-Blanquet, Du Rietz, Rübel, and so on) and animal ecology (Lotka, Volterra, Gause). Thus we see that the disciplines allied with biogeography also had little mathematical background.

As mentioned before, the entities of regional zoogeographical systems were set up on the basis of positive and negative characters, i.e., the presence of some characteristic forms (endemics) and the absence of forms which occur in the neighboring areas. This is a genuine typological method, indicated also by a preference for finding a few higher taxa rather than a large number of species or genera to characterize a region or province. Instead of depending on a repeatable mathematical basis, the soundness of such an approach depends on: the background knowledge of the investigator—in that era of polyhistors (all-round scholars), a man able to grasp all essential knowledge within his broad field—and an intuitive choice of the right characteristic types.*

A statistical treatment of the Sclater-Wallacean zoogeographical regions largely verified their strengths as well as their weaknesses. Anton Handlirsch, the ingenious Viennese paleoentomologist (1913), catalogued over 16,000 insect genera of the world and established the degree of insect endemicity for each of the regions.

Region	Number of genera	Endemic genera	Endemic/nonendemic ratio
Neotropical	5,617	3,437	1 : 0.63
Nearctic	3,467	797	1 : 3.35
Palaearctic	4,956	1,859	1 : 1.67
Aethiopian	3,968	2,249	1 : 0.76
Oriental	4,137	1,641	1 : 1.52
Australian	3,101	1,400	1 : 1.21

*"Intuitive" implies here long previous experience as well as subconscious choice.

PLATE I The zoogeographical regions of the world. From Wallace 1876.

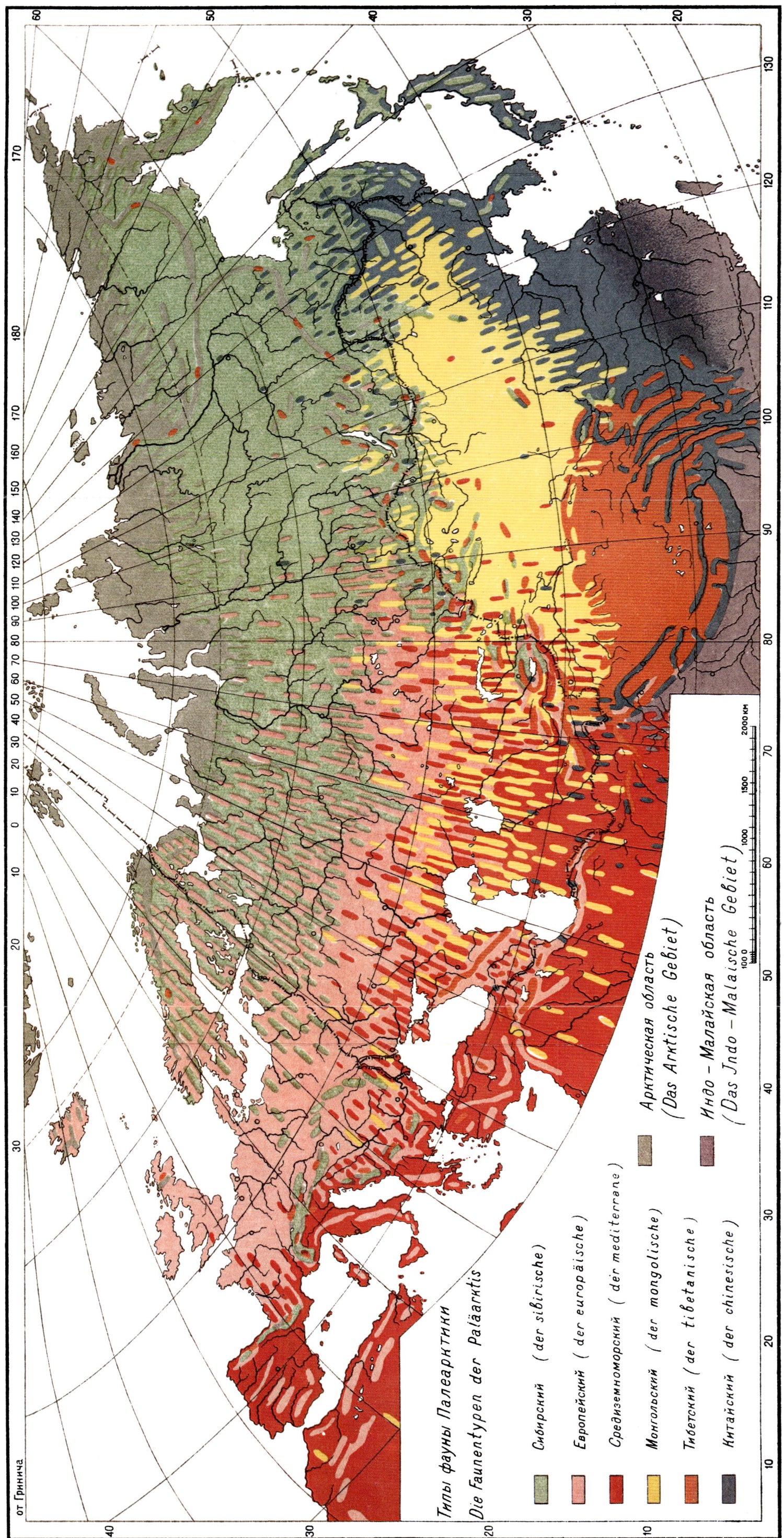

PLATE II The fauna types of the Palearctic Region. From Stegmann 1938. Gray: Arctic Faunal Region; Violet: Indo-Malayan Faunal Region. Within the Palearctic Faunal Region, Pink: European F.T.; Red: Mediterranean F.T.; Orange: Tibetan F.T.; Yellow: Mongolian F.T.; Green: Siberian Fauna Type; Blue: Chinese F.T. (Courtesy of B. Stegmann)

The degree of endemicity expresses the distinctiveness of the region; the Neotropical, Aethiopian, and Australian regions lead in this respect. From the treatment of genera common to several regions, Handlirsch concluded that the strongest ties are between the Palaearctic and Nearctic, the Nearctic and Neotropical, and less strongly between the Palaearctic and Oriental and between the Oriental and Australian regions. The Aethiopian is very distinct from both the Palaearctic and the Orien-

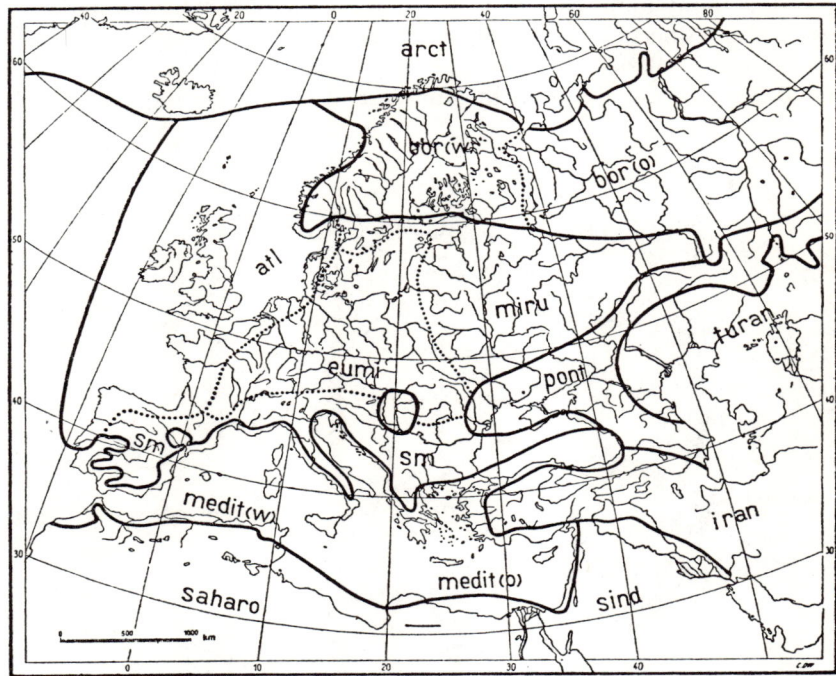

Figure 5-11 Biogeographical provinces of Europe. (From Freitag 1962, following Walter 1954). Walter mapped the geographic elements of the flora—species groups with similar or identical distribution areas were established. Freitag modified Walter's elements on the basis of the distribution of the land vertebrates. These districts are not called provinces by their authors but "geoelements"—geographic area groups. (*Courtesy of Gustav Fischer, Verlag, Stuttgart*)

atl: Atlantic
arct: Arctic
bor: Boreal
eumi: Central European
miru: Middle Russian
saharo: Saharan
w: West
oval area without sign in
   Hungary: Semiarid Pontian
and Pannonian endemic species
sm: Submediterranean
pont: Pontian
turan: Turanian
medit: Mediterranean
iran: Iranian
saharo-sind: Saharo-sindian
o: East

tal regions. These ratios serve to confirm the validity of Wallace's regions, as qualified by later amendments; an exception is the strong affinity between the Nearctic and the Neotropical regions shown by Handlirsch's data. This can be explained by the fact that insects are older, and radiated earlier, than mammals (Halftter's 1964 analysis of Mexican insect faunas implicitly corroborates this explanation and Handlirsch's finding).

Handlirsch's method offers the basic entities of zoogeographical statistics that render clearer the delimitation of faunal regions. These entities are: the unique features of the fauna (Handlirsch's endemics); the ties (common occupancy) with other, mainly neighboring faunal areas; and the forms missing here but present in neighboring faunal areas. The faunal list provides the taxa; the simplest way is to work on the species level.

Figure 5-12 Mammal provinces of North America. (From Hagmeier 1966.) Broken lines indicate subdivisions of provinces; dot-dash-dot lines, political boundaries. (*Courtesy of E. M. Hagmeier and Systematic Zoology*)

The basic formula gives the percentage of common species among the compared faunas. If the number of species in the larger fauna is a, that in the smaller fauna is b, and the number of commonly occurring species is c, then

$$\frac{100 \times c}{a + b - c} = R \qquad (1)*$$

where R stands for *faunal resemblance*. This percentage index alone is not very revealing; to appreciate its value, we must be able to compare it with indices of other, neighboring faunas and must use the series of indices as the measure of faunal similarity within a larger geographic area. Even in these cases, formula (1) usually gives a biased index, for the two or more faunas compared rarely contain a similar number of taxa. If b is much smaller than a, the role of the common taxa—that is, of c—is much greater in b than in a; but this index does not indicate that fact. It also often happens that the smaller fauna is represented by a smaller, incomplete species list, and thus the bias becomes still greater. It is likely that the more widespread species (which are thus in c) are also the more abundant ones and, are found more easily by the faunists; it is therefore wiser to express resemblance in the percentage index R_b of the smaller fauna alone:

$$\frac{100 \times c}{b} = R_b \qquad (2)**$$

For further statistical work, and especially to eliminate the bias of the difference of magnitude among the faunas, we may express the size (S) of the smaller fauna (b) as the difference in number of taxa, when

$$\frac{100 \times (a - b)}{a} = S \qquad (3)***$$

Regional zoogeography in the past has been so remote from statistical thinking that few workers went beyond the use of the basic procedures outlined above. Perhaps the major psychological barrier was that, as Simpson (1960) pointed out, faunal lists are in reality only incomplete samples of the actually occurring taxa of the district. Furthermore, the results of indices (1), (2), and (3) have meaning—as we have already suggested—only in a broad, synthetic approach. Then, on a larger area, not only does the sampling of the fauna become uneven but error creeps

---

*Jaccard's (1902) coefficient of community.
**Simpson's (1943) formula.
***Schilder's (1955) formula.

in through the uneven taxonomic status of the basic entities. What one faunist describes as a species another may degrade to the subspecies level, or vice versa. Moreover, large areas have large faunal lists, and there the mere numerical species listing—faunal magnitude, as in (3) above—loses its importance as the occurrence of phylogenetic entities (that is, major taxa) also becomes statistically comparable. But the incorporation of higher taxa into the statistical comparison again increases the bias of incorrectly or unevenly assigning the phylogenetic rank. European mammalogists, for example, often work with a different species concept from that used by scientists studying the North American mammalian fauna, where the worst results of the "splitting" era have not yet been fully eliminated. Burt (1958) gives good examples of this discrepancy.

Most of these critical aspects have been considered by the Swede Sven Ekman (1940), who elaborated upon the basic resemblance index (1). He introduced the principle that each species and each higher taxon characterize to a certain degree the fauna in which it occurs. The degree of importance will depend on the circumstances of its distribution within as well as outside the area under scrutiny. Therefore, each taxon of the faunal list must be given a *basic zoogeographical value*, and this value must be modified according to the status of the species in the area considered. The basic value depends in part on the taxonomic rank; a genus has twice the value of its species; a family has six times the value of the number of species it has on the faunal list; and so on. But these full values are assigned only to the unique elements in either fauna.* Commonly occurring taxa have diminishing values—for example, a commonly occurring genus counts only one-half, a family only one-sixth, of the value of their commonly occurring species. Another component of the assigned value reflects the endemicity of the taxon (high value) versus a broad distribution range outside the compared areas (low value). Besides these measures of specificity of the compared faunas, conspicuous lack of forms that occur in the other surrounding areas also negatively influence the total zoogeographical value of an area. Ekman also gave attention to a point that many zoogeographers missed—the limits of the areas under comparison. A boundary can be drawn between two zoogeographical areas where the greatest *faunal change* (F) occurs; this can be expressed, following Ekman's index, by

$$\frac{A + B}{C} = F \qquad (4)$$

*I.e., to those taxa which do not occur in the other fauna under comparison.

where A represents the total of the positive zoogeographical values of fauna a, B is the total of those of fauna b, and C is the total of those of the taxa in common between a and b. We arbitrarily divide the total major geographical area into subareas and then determine the index (4) for two neighboring areas; a major boundary then lies between those two areas in which F has the highest numerical value (which indicates the highest specificity of the two neighboring areas and/or the lowest number of identical taxa).

Ekman's unique but cumbersome method is based on a large number of elements in which subjective judgment is needed, and therefore its accuracy may often be challenged. Nevertheless, it does show the way toward desired, statistical methods based on *correlation analysis*.

Schilder (1955, 1956) compromises between the accurate but complicated analytical procedure of Ekman and simpler but exact formulas. The greatest value of Schilder's method is not in its theoretical approach but in the effectiveness of its graphic presentation (Figure 5–13). He recommends the simultaneous use of formula (2), which expresses the qualitative affinity of the smaller fauna to its larger neighbor, and formula (3) for the quantitative differences of the faunas, based on the larger one. Then he combines the two indices in the form of a symbol connecting the core of the two areas on a map. Completing this procedure for all neighboring part areas of a major geographical unit, we have a visual picture of the numerical relationships and specific affinities of the whole region.

Schilder also gives a method of determining faunal boundaries based on the percentage occurrence in the connecting area of the forms characteristic of the two regions. If we compare area A with area Z, and the intervening subareas are B, C, D, and so on, Schilder's index could be expressed:

$$P_A = \frac{100 \, (x - x_{min})}{(x_{max} - x_{min})} \qquad (5)$$

where P stands for *Percentage Occurrence Index* and where the x values are calculated as follows. The series of subareas are tabulated from A to Z, and the number of A species in the subareas B, C, D etc. to Z are tabulated; similarly, the number of Z species *missing* in areas A, B, C, etc. are tabulated. Then x is the sum of these two numbers. For example, x_A equals the numbers of A species in area A plus the number of species characteristic in Z but missing in A area. Then x_{min} is the smallest of all the x numbers, and x_{max} is the largest. This calculation expresses P on the basis of percentage of fauna A; we can also calculate

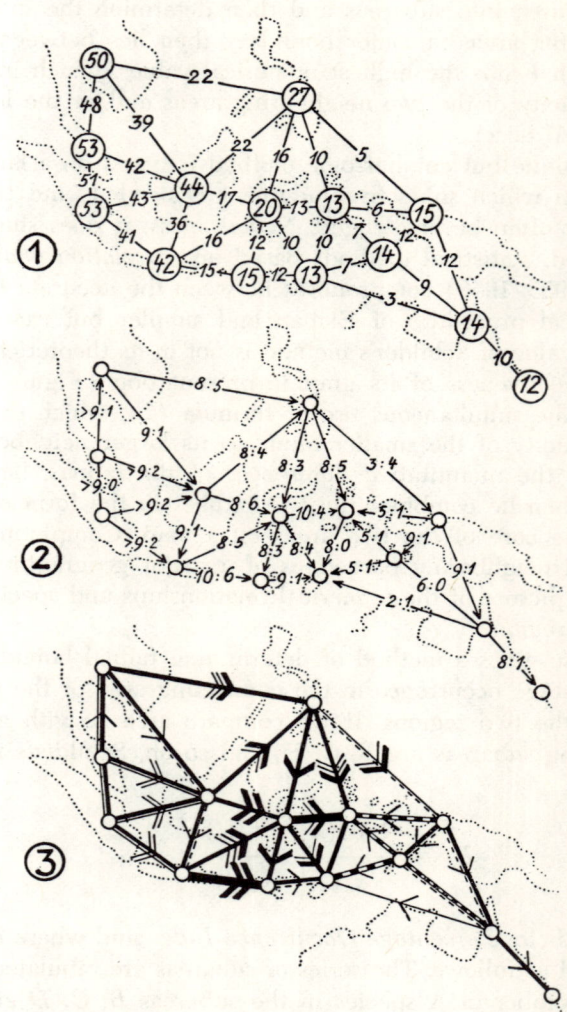

Figure 5-13 Schilder's (1955, 1956) graphic representation of faunal resemblances, using the mammalian fauna of Australasia. (*Courtesy of F. A. Schilder and G. Fischer Verlag, Jena*)

1. Numbers in circle indicate size of faunas on islands or mainland geographic entities (or checkpoints). Numbers on the lines connecting geographic entities signify the mammal units in common between the two areas. Each mammal unit means the occurrence of a species or a group of related species.

the species resemblance in terms of fauna Z; the two resemblance values complement one another to 100. The index can be tabulated:

	A	B	C	D G	F	H	EI	J	Z
P_A	100	80		60	40		20		0
P_Z	0	20		40	60		80		100

which illustrates of the relations between the two end faunas (A and Z). The sequence on the graph, as in the theoretical example above, is not necessarily the geographic sequence of the subareas A to Z—just as area A may not always contain the maximum number of A species, and the total number of A (or Z) species may be larger than the number of A or Z species in any one subarea. The result certainly indicates the subareas of least resemblance to the two distant source faunas (A and Z), provided the whole district is well known faunistically and if the samples taken are true representatives of each local fauna (as, for example, in an archipelago connecting two continents). For more details of this interesting method, see Schilder's works (1955, 1956); he calculates, as an example, the faunal resemblance of the theriofauna of the Indonesian Archipelago.

Schilder's method excludes the common species, and is based only on the specific elements of the two faunas under comparison. It presupposes that there is a gradual decreasing of the end elements from the one end of the area chain to the other. It also indicates when such is not the case, thereby calling to our attention extinction or regression from, or local evolution upon, any of the local districts. It does not reveal whether the basic difference between the two end faunas is caused by altered ecological conditions or by differential immigration. It merely shows the percentage differences, and enables the pinpointing of a boundary line. Simpson (1961) applies formula (2) to each two of the subareas, and calls the boundary line the "Faunal Balance Line."

The foregoing discussion basically summarizes or refers to the independent work of four zoologists: Ekman (1940), Simpson (1947, 1960, 1961), Schilder (1943, 1947–48, 1955, 1956), and Burt (1958), within

---

2. The two resemblance indices 2 and 3 (see p. 34) are on the line connecting two faunas; an arrow points from the larger toward the smaller of the two compared faunas.

3. The two indices are now replaced by symbols in the following way. Index 2 is expressed in the kind and thickness of the connecting lines; index 3 is expressed by the number and thickness of the arrowheads. Comparison of diagrams (2) and (3) makes further explanation unnecessary.

On the basis of this presentation, the strength of the 'lines' dividing the Oriental and Australian mammalian faunas can be discussed and compared with other studies. (see Mayr 1944, Rensch 1950.)

two decades of the mid-twentieth century.* Among botanists there is a desire to base regional floristic entities upon the *uniqueness* of the flora. Walter (1954) suggests that one compare the number of distinctive taxa of two neighboring areas. If one measures this "floral contrast" at, for example, 100 km intervals, a curve can be drawn; this curve will be rather level and low-lying where the floristic difference is slight between neighboring check points. A steep slope points toward a major break in floral continuity, caused by a present or historic barrier of some kind, which is the place to separate the neighboring floras. Walter recommends the use of species lists for the lowest regional entites and lists of increasingly higher taxa for the higher ranks in the regional system.

Recently, interest has arisen in the use of statistical methods in delimiting *faunistic provinces*. Webb (1950), Barrera (1962), Ryan (1963), and Huheey (1965) all based their methods on one or another variation of the faunal resemblance formula (1) (as presented on page 273). Several mammalian zoogeographers made use of the distribution maps which Hall and Kelson (1959) published in *The Mammals of North America*.

Hagmeier and Stults (1964) and Hagmeier (1966) also utilized these range maps, but used different statistics. Their survey of the previous attempts to designate regional units of North American animals points to the criticism and ambiguities caused by naming both the ecological (Dice 1943) and faunistic (Webb 1950) entities as *provinces*. If one ranks all mammalian distribution areas according to their size, the frequency curve thus obtained shows a nearly Poisson distribution, with a slight skew toward smaller size. The statistical details indicate that the distribution of almost all members of the fauna is limited to North America; the most common distribution area size is about 500 km in diameter (the range is from 100 to 4,200 km), and *the distribution areas do not occur at random but are clumped* according to certain variables. Thus, an attempt can be made to find the lines of greatest faunal changes which delimit areas of greatest faunal homogeneity. The lines of greatest

---

*It is interesting and informative to note how much repetition occurs here. Ekman's book is the earliest and most extensive, but Schilder is the only author who cites it. Schilder apparently did not know of Simpson's earlier paper, but developed his formula (2). Burt does not quote either Ekman or Schilder, but his reasoning is very suggestive of both. Simpson's 1960 paper discusses Burt's objections (besides giving many new and interesting points that are beyond the scope of the present work) and carries them further, almost to the point that Ekman reached in 1940. Undoubtedly, a "dynamic" barrier (war), language barrier (two published in German, two in English), and geographic barrier (Sweden, East Germany, United States) were—and are still—hampering the unity of our "international" science. All the above four authors worked on mammalian faunas; we can imagine that the communication hiatus is likely much greater among specialists of different animal groups.

changes were located with the aid of a 50-mile square (80 by 80 km) grid system. The species (n) were listed in each block, the number of range limits per block (L) were counted, and an "Index of Faunistic Change" was computed. Not only were the areas of greatest faunal homogeneity found and named as zoogeographical provinces, but even the degree of faunistic affinity was computed, using indices (1) and (2), between provinces. The faunal affinities were subjected to a cluster analysis (cluster analytical methods are summarized in Sokal and Sneath 1963). This method results in mean percent similarity values for each pair of units—in this case, for each pair of the faunistically closest related provinces. Hagmeier and Stults found that the mean percent similarities of their mammalian provinces (our formula [1]) are in close agreement with Preston's (1962, and earlier) theoretically derived "Resemblance Equation." As a result, Hagmeier (1966) concluded that the faunas of two geographic areas may be considered homogeneous if the mean percent similarities are larger than 65 percent (Figure 5–14). If the calculated similarities are less, the compared areas rank as separate faunal provinces. Mean percent similarities of unit pairs result in a dichotomous, hierarchical system from which one may select higher-ranking groups—superprovinces, subregions, and regions—for the continent which has been analyzed. It is proof of the importance of ecological affinities over historic and taxonomic relations that these higher entities of Hagmeier and Stults' North American theriofaunal provinces are strongly related to ecogeographical entities (Figures 5–12, 5–14) as found by Dice (1943) (Figure 5–6) and revised by Kendeigh (1961). Usefulness of the method is further proved by the results of its application to marine molluscan distributions on the continental shelf (Valentine 1966); the distribution areas fall into patterns reminiscent of the hydrographic patterns governing temperature regimes on the shelf.

The statistical method explicitly or implicitly seeks to verify the existence of faunal regions or smaller geographic entities by comparing lists of species or higher taxa on any two or more geographic areas. Some of the proponents of these methods have tried to establish an undisputable boundary to these regional units on the basis of faunal statistics. We should, however, keep in mind that the regional approach is not the only one to which statistical methods can give understanding of the geographical (chorological) attributes of mixed animal populations throughout the biosphere. Statistical methods can be useful in any form of exact zoogeography, whether one uses abundance of individuals, ecofaunas, biotic communities, area correlations, or any other basic unit.* Statistical zoogeography is the methodology of the future.

*The general theory of island biogeography, as approached by mathematical methods in MacArthur and Wilson's recent book (1967), is a good example.

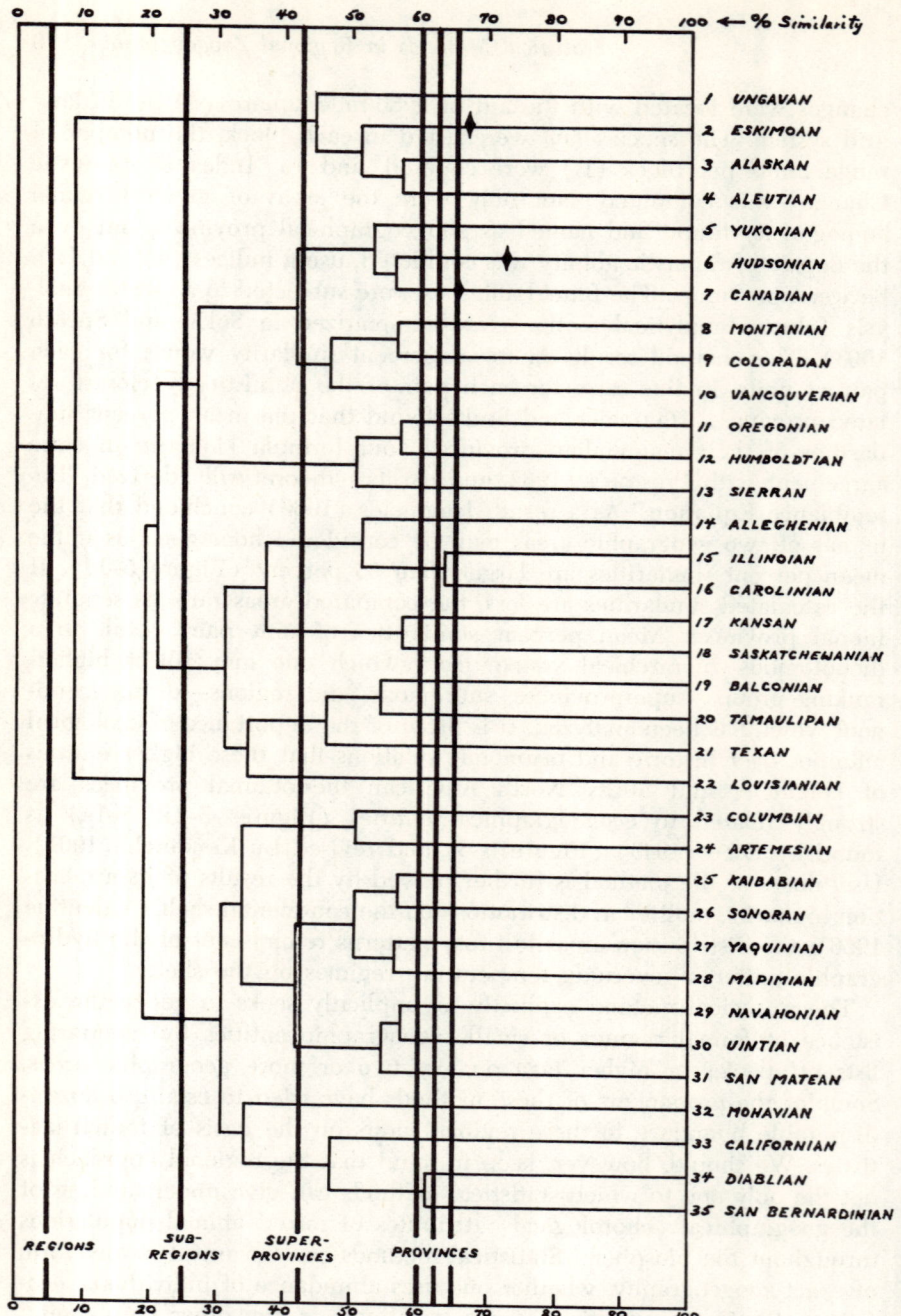

Figure 5-14 Percent similarity values of theriofaunal provinces of North America. (From Hagmeier, 1966.) The paired grouping of the provinces gives decreasing

AREOGEOGRAPHIC ANALYSIS

The ecogeographical and purely geographical grouping of regional faunas is a classificatory device for the convenience of the analyst who wants to explain the causation of faunal patterns. Once such patterns are recognized and the regional entities are delimited, there opens a second avenue of analysis—that is, to discover the different *common denominators* of the members of a regional faunal list. The simplest common denominator is the *phylogenetic relationship* (the common ancestry), because it denotes a related distributional history (Figure 6–33). Historic groups of nonrelated species may be tied to one another by common *geographic origin* or by common routes of *immigration* or *retreat*. Biogeographers often use the concept of historic "elements" within a biota.

Such an approach may be a powerful tool in the hands of a taxonomist in historical zoogeography. However, the basis of such a study, unless it uses one of the above-listed techniques, remains the ill-delimited arbitrary *geographic region* and the number of taxa that live upon it. The only geographical analysis to consider the total world distribution of the species of a region is Ekman's comprehensive method. Space on land is occupied by the vegetation and the faunation, and the regional check list of the faunist gives only *point checks* of its qualitative diversity. The continuum of the often intergrading and interdigitating graduated faunations can be resolved by an analytical method based on the distribution areas of the species. The use of the species areas as a basis for analysis enables us to scrutinize geographic regions; the total areas of each occupant will of course usually extend well beyond the limits of the regions in an orderly way, indicating affinities between those species groups which have similar or identical areas. Once we recognize the importance of the species area as an analytical tool in one kind of regional entity (the faunal region or "province"), we may go a step further and utilize this method in the causal analysis of another spatial entity—the ecosystem. Zoogeographers do not often take this initial step, although phytogeographers regularly avail themselves of it in analyzing the *floral elements* of plant associations.* As a matter of fact, analysis of

---

*The word "element" is applied here with particular emphasis to those divisions or constituents of the species list which reveal common geographic origin or ecologic adaptations.

---

degrees of faunal resemblance between higher ranking pairs of province groups (see p. 279), enabling thereby the use of terms like "superprovince," "subregion," and "region" for these ranks, and thus giving numerical value and numerically derived boundaries to these old terms. (*Courtesy of E. M. Hagmeier and Systematic Zoology*)

the distributional area is a well-developed branch of plant geography, especially in the Old World; its methods and practical results may be profitably studied in, for example, the texts and monographs of Meusel (1943), Cain (1944), Walter (1954), and Polunin (1960). Some zoogeographers recognize the importance of faunal analysis, but do not make use of the comparative area study as a tool of this method. For instance, Uvarov (1938) considers *ecofaunas* as the lowest units of which a geographic fauna consists. He stresses that, locally, each species is the member of only one ecofauna, and that members of ecofaunas may belong, from the historical point of view, to various historic faunas, such as Ethiopian, Lemurian, and the like. Nikolsky (1947) calls the components of faunal lists *faunal complexes*, consisting of species which are ecologically similar and have a common geographic origin.

Analysis of a local (regional, faunational) fauna list will show that the constituent species fall into groups with respect to the shape of their geographic areas. Although some authors call these groups "faunas," this usage is not recommended because the word "fauna" already means too many things. *Fauna* (or *faunal*) *element* is a meaningful expression; its counterpart, the *floral element*, is the accepted term in phytogeography. "Faunal group" is often applied to these elements. "Faunal analysis" of this kind is so seldom used among English-language zoogeographers that it is worthwhile to propose a uniform nomenclature, before the above-cited terms gain wider usage and lead to further interpretations.

Fauna may be understood to have three distinct meanings: (1) the species list of the total faunation of any geographic entity, delimited naturally (topographically) or arbitrarily, e.g., by political boundaries; (2) the species list of a single major taxon (order, class, etc.) on a geographic entity (also called avifauna, malacofauna, entomofauna, and so on); (3) the species list of an ecological unit (a community, biome, and the like).

All these "faunas" may be analyzed for the distributional area types of the species, and each species (with its area) is an *element* of a fauna. A simple example would be a small area *A* on the west coast of a continent. A number of species of the faunal list terminate just south of *A*, but are distributed much farther northward; we call these the *northern elements*. Other species live in the interior, but reach the coast at region *A*; these are our *eastern elements*. Still other species are distributed more extensively to the south than to the north or east of area *A;* we call these the *southern elements*. A fourth group lives only in district *A*, which has a peculiar environment, but some of its constituent species are found overseas on another continent. These latter species form the *transoceanic element* of the fauna of district *A*, and those which are restricted to district A (endemic to A) will be customarily named after the locality; if A stands for, say, Arcadia, these will comprise the *Arcadian endemic element*. The faunal elements are often named after the locality of the focal point (center of gravity, major extent of area) of their distribution. This practice appears sound, but there are instances in which it is difficult to assign a species to any geographical faunal group or in which a species may be distributed in several geographic entities. The least grebe *Podiceps ruficollis* Pontopp. [Podicipitidae, Aves] is widespread in Eurasia; hence, in Australia this grebe is called an *Old World,* or Palearctic-Oriental, element. However, it recently spread back to Indonesia, where it now lives side by side with the ancestral form, *P. r. vulcanorum,* behaving like a separate species (*P. novaehollandiae*), and where it should be considered an *Australian* faunal element (Voous 1959).

This simple example outlines the beginning of the procedures used in analyzing a fauna for its elements. Most instances, however, are not so simple. The following routine is generally recommended. Map the distribution areas of the species. Superimpose all species areas, or, group by group, superimpose those species areas which show a superficial relationship or which overlap. Separate the different "geographic" elements resulting from the overlaps. Compare the members of such a "geographic element" according to their historic and ecologic relations, in order to draw conclusions or inferences about the causation of the existence of each element in the geographic region under study, in order to analyze the whole regional fauna by a numerical comparison of its elements.

The following discussion of several particular studies will clarify the use of faunal elements. One of the chief advantages of the area-analytical method is that *geographic regional boundaries become unnecessary;* the ratio of occurrence of the faunal elements characterizes each geographic district, with these ratios forming graded transitions toward neighboring districts.

Hultén (1937) introduced the theory of *equiformal progressive areas.* If a number of species start to spread from a common center of dispersal under ideally uniform conditions, each species will spread radially, but their areas will be of different sizes, owing to their specific differences in spreading; however, all will continue to occupy the common center from which they spread. Such an ideal example would hardly occur in nature, but the common center and the progressively widening and gradually differing distribution of the species are empirical facts demonstrable by analysis of the biota of a large tract and by the completion of the mapping of the world area—that is, the total distribution area for

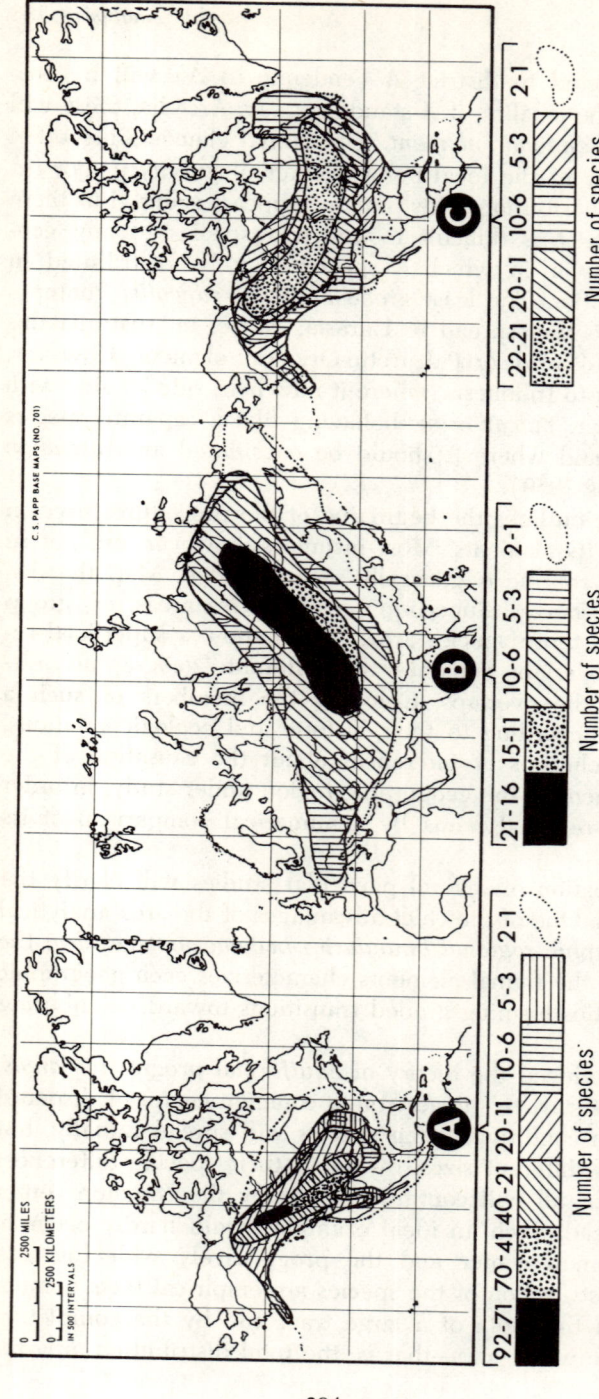

Figure 5-15 Examples of Hultén's grouping of the arctic and boreal biota into elements according to their "equiformal, progressive areas." (After Hultén 1937.)

A. Equiformal progressive areas of ninety-two Western American plants of continental character.
B. Equiformal progressive areas of twenty-two Boreal American plants of continental character.
C. Equiformal progressive areas of seventy-eight Boreal Eurasiatic plants of continental character.

each member of the biota. Hultén calls geographically similar but not identical areas "equiformal progressive areas"; the center of their distribution is where most or all of these species occur, and they spread from this general area. Neither Hultén nor his followers believe that this central area is a "center of origin," but rather that it is the *center of radiation;* in our interpretation, this is the area covered with the ecosystem which provides the prime habitat of these species and which *at present* occupies the area central to the studied distributions. The biota whose areas constitute such an orderly geographic pattern may belong to widely different phylogenetic groups, thus proving that the orderliness and reality of pattern is *geologic* as well as *historic.*

Hultén analyzed nearly 2,500 northern Holarctic plant areas, and found that the centers of radiation are all in regions which were not glaciated during the Pleistocene; therefore, he considers these central areas as more or less corresponding to the glacial refugia of the presently arctic and boreal biota (Figure 5-15).

Stegmann (1938) mapped about 900 bird species of the Palearctic region. His six centers of the distribution areas correspond to six major climatic-ecological units; geographic analysis shows that each part of the Palearctic region harbors a mixture of some of these six faunal types. Stegmann draws the important conclusion that any one geographic district is characterized not by a unique or uniform fauna but by the characteristic mixture of the six basic faunal types. The result of such analysis—the mosaiclike map (Plate II) derived from it—shows the ecological and historic relations of the local fauna; moreover, it eliminates disputable, rigid boundaries or "transitional" zones of biogeographic districts, for boundaries are artifacts and do not conform to the real situation of the mosaiclike blending of animals of different origins and ecologic affiliations.

Reinig (1950) finds that Stegmann's bird areas correlate well with those of insect examples. He finds over 40 area nuclei ("Arealkerne") in the Palearctic, many of which show mutual relations and indicate preglacial or pre-Pleistocene affinities, as is true when disjunct areas of certain species encompass two or more of the centers of postglacial dispersal. De Lattin (1957) uses the Holarctic Lepidoptera, but also considers certain bird and mammal distribution areas. Following Reinig (1950, and also 1937), de Lattin divides the Holarctic fauna into three broad ecological groups. Species of the *Arboreal Fauna* live in forest, woodland, and scrub. Animals of the open grassland and desert comprise the *Eremial Fauna.* Tundra and other arctic habitats harbor the *Boreal Fauna.* De Lattin verifies Reinig's area nuclei (which remind us of the botanist Hultén's centers of radiation), and calls these *dispersal centers*

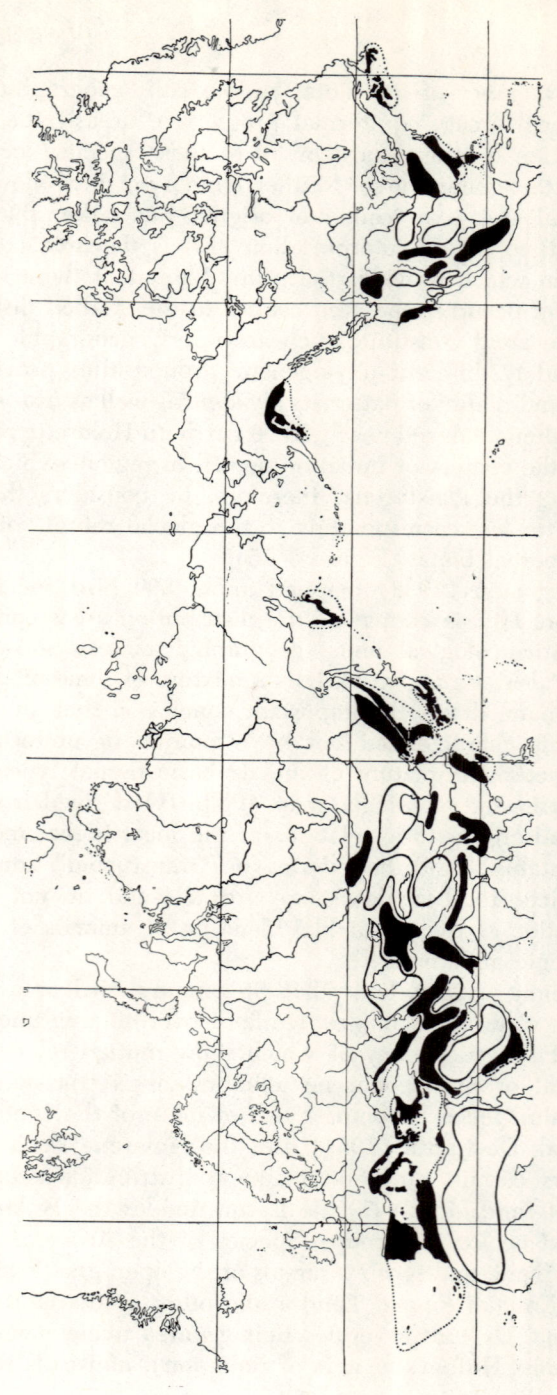

Figure 5-16 Area nuclei (dispersal centers) of the arboreal and eremial faunas of the Holarctic. (From de Lattin 1957.) Eremial centers outlined; arboreal, in solid black. Courtesy of G. de Lattin and *Akademische Verlaggesellschaft Geest & Portig, Leipzig*)

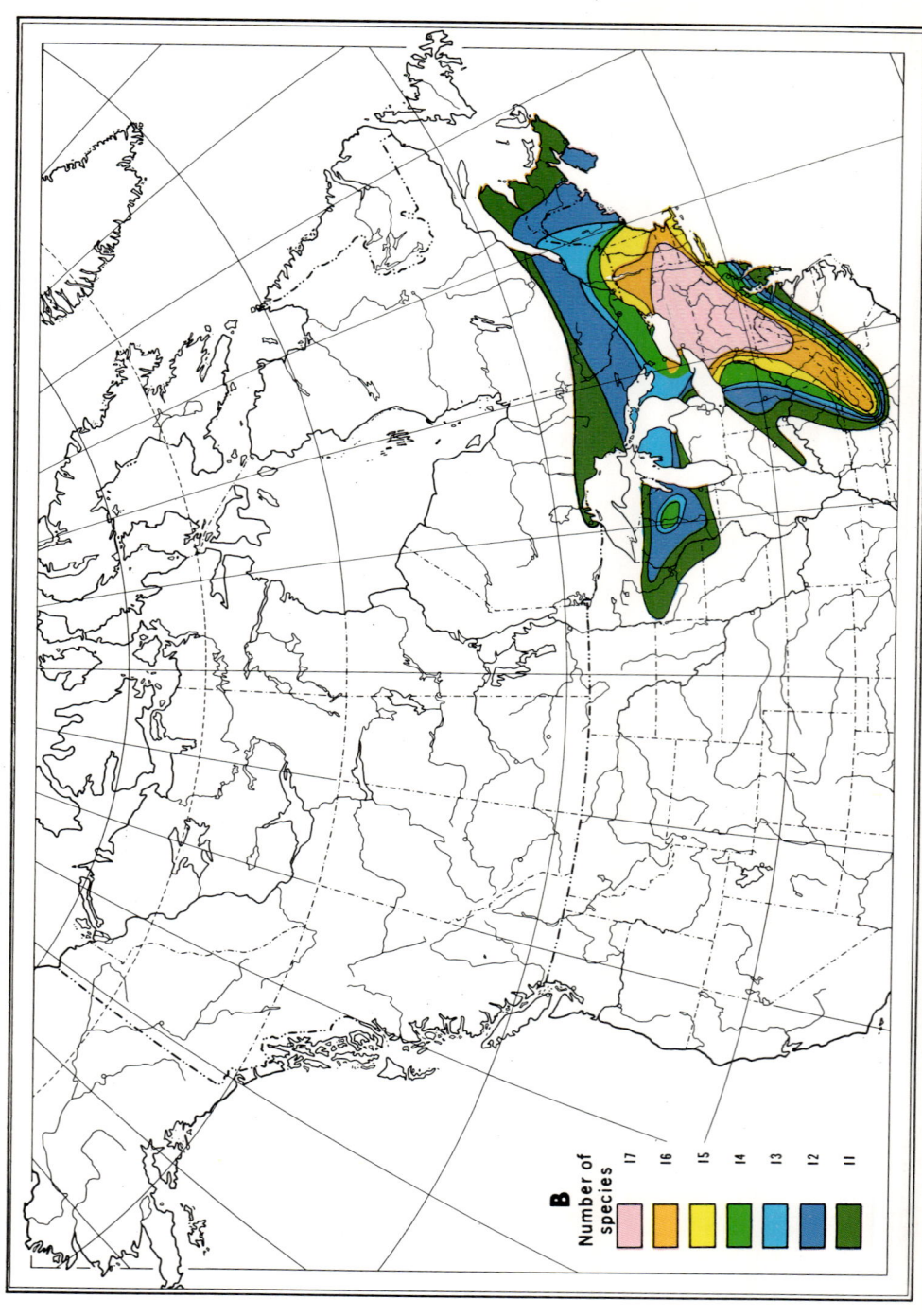

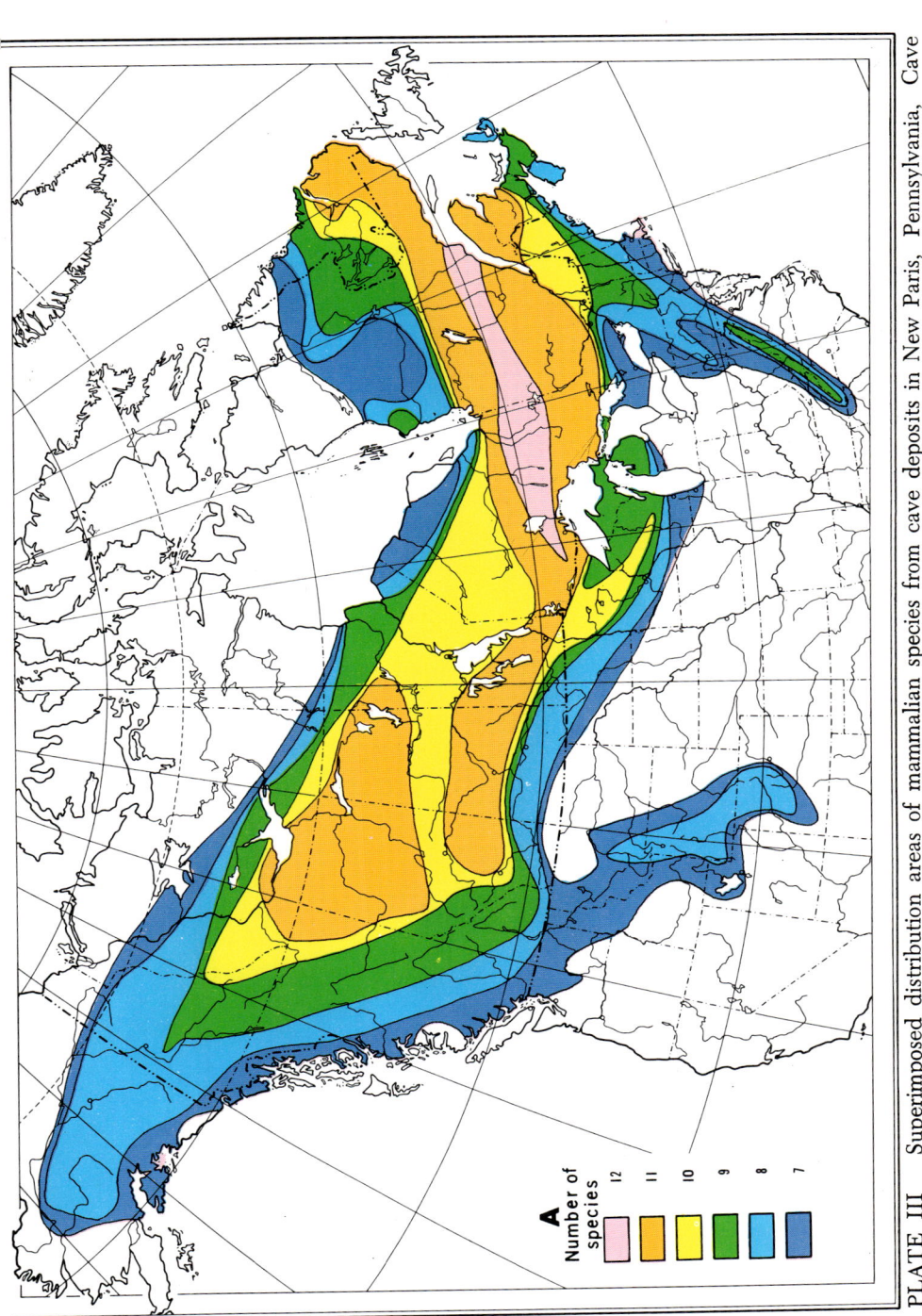

PLATE III Superimposed distribution areas of mammalian species from cave deposits in New Paris, Pennsylvania, Cave No. IV. A: Fauna of cave fill from 6 to 9 m. B: Fauna of 0 to 6 m. From Guilday, Martin, and McCrady 1964. (Courtesy of J. E. Guilday and the National Speleological Society of the United States.)

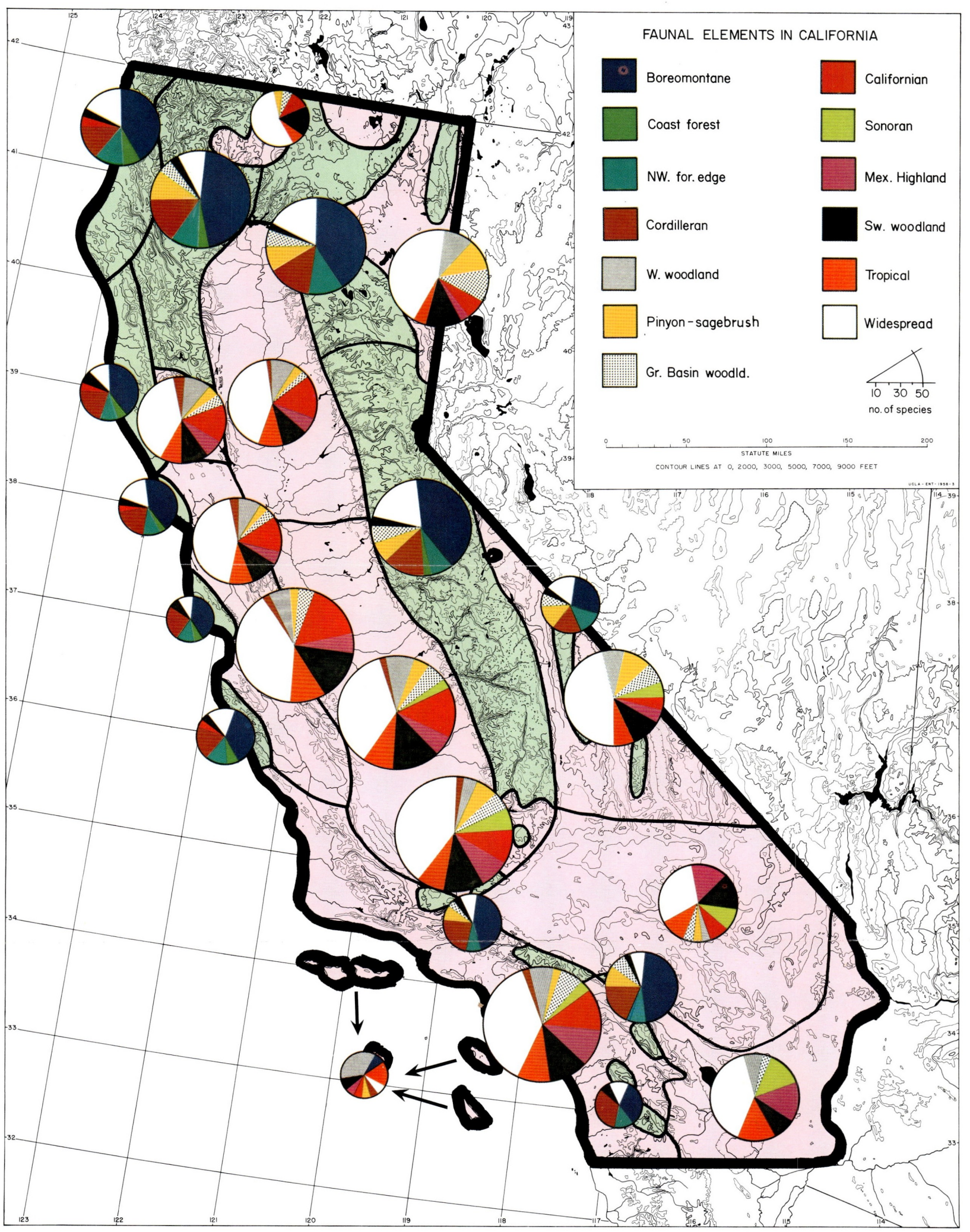

PLATE IV Ecofaunal analysis of the birds of California. This plate continues the sequence of Figures 17 through 19. The ecofaunal groups, or elements, have been based on the passerine avifauna (Udvardy 1963b). The faunal provinces of California (Miller 1951) are analyzed, and for each province a pie diagram shows the percentage of the participating elements.

(Figure 5-16). He distinguishes 24 dispersal centers of the Holarctic "arboreal" fauna, fourteen "eremial" centers, and five "boreal" centers. A species whose area includes only one of these centers is *monocentric;* such are most Holarctic species with small or moderately large distribution areas. Widely distributed species are *polycentric;* de Lattin suggests that, during alternating glacial and interglacial periods, species belonging to the arboreal and eremial area centers, on the one hand, and the boreal centers, on the other, alternately expanded their areas or contracted them again. He believes that, at the present time, the arboreal and eremial species areas are in an expansive phase, while the species of boreal affinities have regressed into boreal refugia. De Lattin's scheme of alternating expansion and regression of certain ecological faunas and floras following gross climatic alternations is well in accord with similar theories of other biogeographers (p. 318), but he arrived at it by areographic analysis rather than by study of fossil biota.

De Lattin believes that the arboreal and eremial centers—the core areas of the forest and open-country faunas—did not move at all during the Pleistocene, while, on the contrary, the boreal centers show decisive evidence of movement. All or most of the present area of the boreal species is often in regions which were wholly glaciated during the Pleistocene glacials.

In this latter respect, however, there are many who believe that nearly *all* "dispersal centers"—to use de Lattin's expression—moved geographically following the climatic fluctuations of the Pleistocene. We shall refer to the paleoecological teamwork of Guilday, Martin, and McCrady (1964) because, in addition to the fact that it provides the best kind of evidence—namely, a comparison of fossil record and present distribution—it gives a good example of the use of the detecting of core areas of faunal elements by superimposing distribution maps (Figures 4-61, 4-62 and Plate III). These authors analyzed the fossil content of a cave fill nine meters deep in a sinkhole at New Paris, Pennsylvania. The pollen and vertebrate remains that have accumulated there from about 11,300 ± 1000 years B.P.* are indicators of woodland and forest environments. The bottom third of the fill is dominated by fossils of woodland vertebrates whose present center of distribution lies southeast and west of Hudson Bay, corresponding to the "Boreal Forest Biome" (Figure 5-4 and Plate IIIA). The upper layers of the fill contain a forest fauna of postglacial times; the core of the modern distributions of its members (Plate IIIB) lies well to the south of the distribution area of those species which were dominant during the previous millennia. The sequence

---

*Radiocarbon dating. B.P. is the palaeontologists' abbreviation for "before present."

of the two faunas thus witnesses the substantial displacement of the pine-spruce community in the course of 11,000 years.

Voous (1955, 1960, 1963) advocates the use of faunal analysis and of the concept of elements (his "faunal types"). He distinguishes, provisionally, 19 distributional area types within the Holarctic (1963). Voous' main argument for the use of this analytic approach is that while *there are distinct faunas* based on, and demonstrable by, the related shape of distribution areas, *there are no distinct zoogeographical regions.* Voous, the taxonomist-zoogeographer, emphasizes the historical and distributional-historical clues given by this method, though he does not deny the importance of the method in corroborating ecological affinities. An ecologically inclined researcher will soon realize that this method in reality points out "ecofaunas"—that is, mostly ecologically related species tied to the same biome or ecosystem, thence having similar area shape. When I mapped the passerine bird distributions in North America (Udvardy 1963a) and started to compare the the distribution maps, the groups which could most easily be discerned were those of species strongly tied to single biomes; the less easily analyzable distributions were of ecotone species which follow widespread and more general ecological situations. Finally, coast-to-coast distributions of widespread and ecologically versatile species did not fit any subcontinental pattern whatsoever. Miller (1951) applied the "faunal group" concept to an analysis of the Californian avifauna. He defined four such groups consisting of species which have "strong or repeated associations" and "similar centers of distribution and probably often similar areas or origin" (p. 582). One of the groups contains the endemics of the area, and three groups have affinities north, east, and south, respectively, of California; a final, residue group includes very widely distributed forms. These groups stand up well when compared with others based on the life zones, biotic provinces, and biome categories also applied to California. Application of my "ecogeographic faunal groups" to the biotic provinces of California (using these as conveniently small districts with available faunal lists) indicates that ecogeographic grouping and analysis reveal much more of the ecological and dispersional affinities of local faunas than any of the other methods Miller attempted (Figures 5-17, 5-18, 5-19, Plate IV).

The most fruitful analytic method seems to be the work with faunal elements based on species distribution, as this provides clues to the dynamism of ecological as well as historical distribution. It is curious that its use has been restricted largely to biogeographic analysis of the Palearctic region (but see Winterbottom 1965). Palearctic biogeographers used the concept of geographic provinces much earlier and to a greater extent than anywhere else, and realized that these provinces, divided

Areogeographic Analysis 289

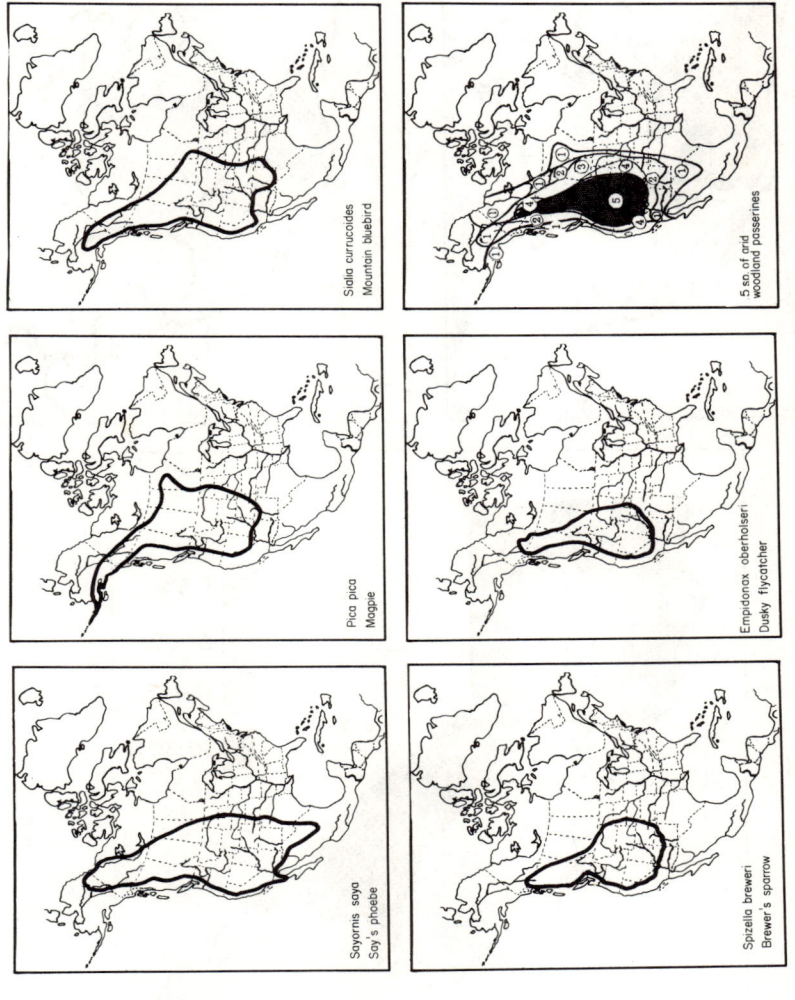

Figure 5-17 Distribution area of five North American passerine bird species. Superimposed, as on lower right, they form the Northwestern Arid Woodland unit of the Great Basin Ecofaunal Group.

290 *Regional and Analytical Zoogeography*

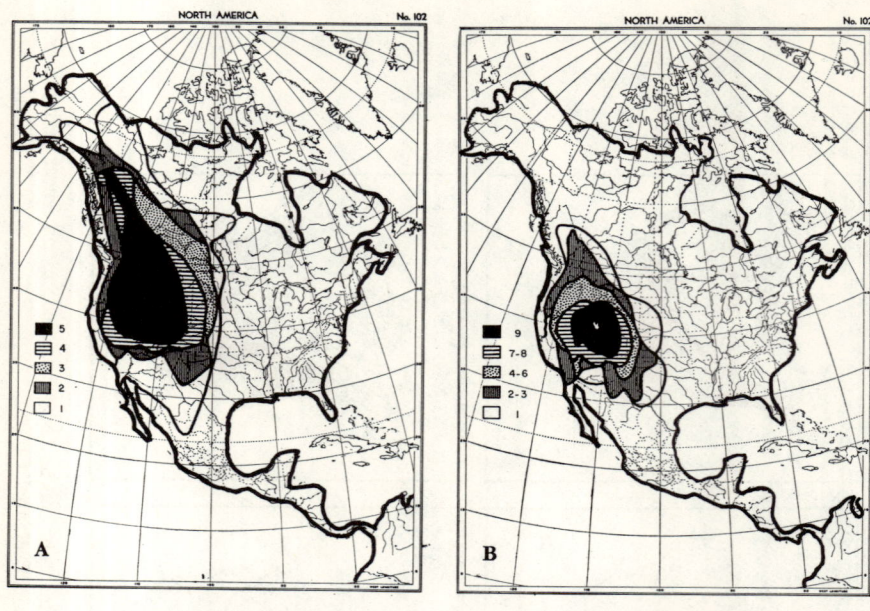

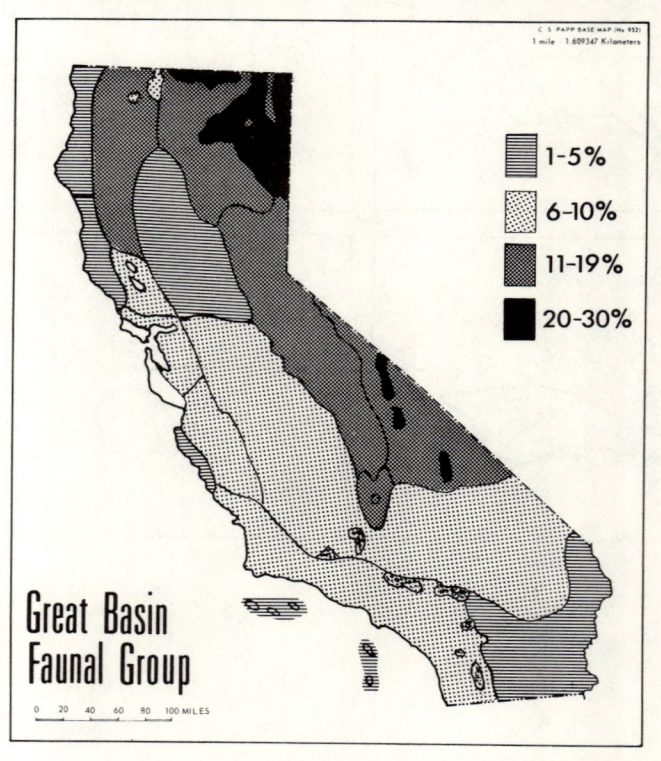

into too small districts, coincide with the entities of ecological formations and become per se relatively useless. The example of the concept of floral elements which Old World botanists frequently use also acted as inspiration (see e.g., p. 267, our reference to Engler's historic floral elements of 1879–82). Finally, the original argument for an analytic approach

Figure 5-18   Ecofaunal groups in California.
Some of the Ecogeographic Faunal Groups of North America and their percentual participation in the regional units (following Miller 1951) of the Californian avifauna. A. Superimposed distribution of five northwestern arid-woodland species of the Great Basin Ecogeographic Faunal Group. (See also Figure 5-17). B. Superimposed distribution of nine sagebrush-arid-woodland species of the same faunal group. (A and B Basin Faunal Group (viz., A and B combined) in the regional from Udvardy 1963b). C. Percentual participation of the Great avifaunas of California. D. Percentual participation of the Californian endemic passerine bird species (seven species) in the same regional avifaunas.

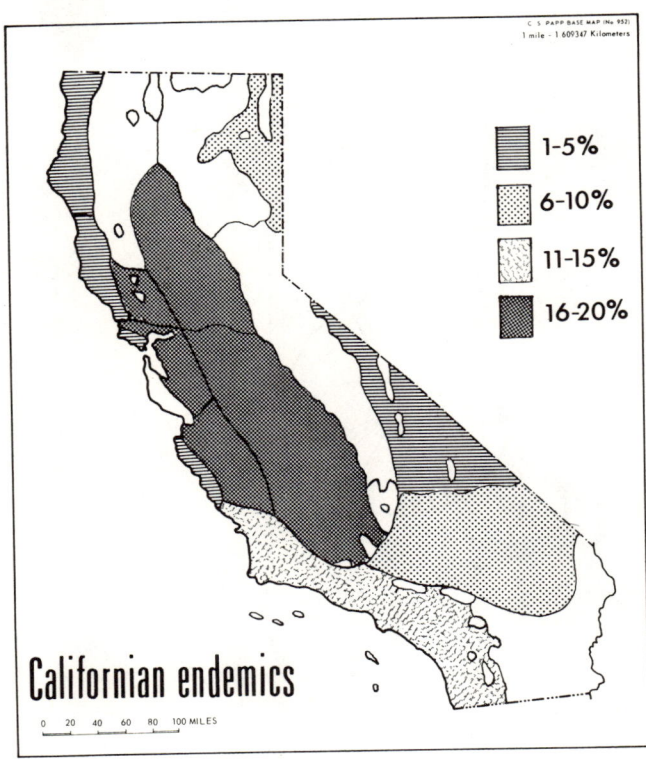

D

292 Regional and Analytical Zoogeography

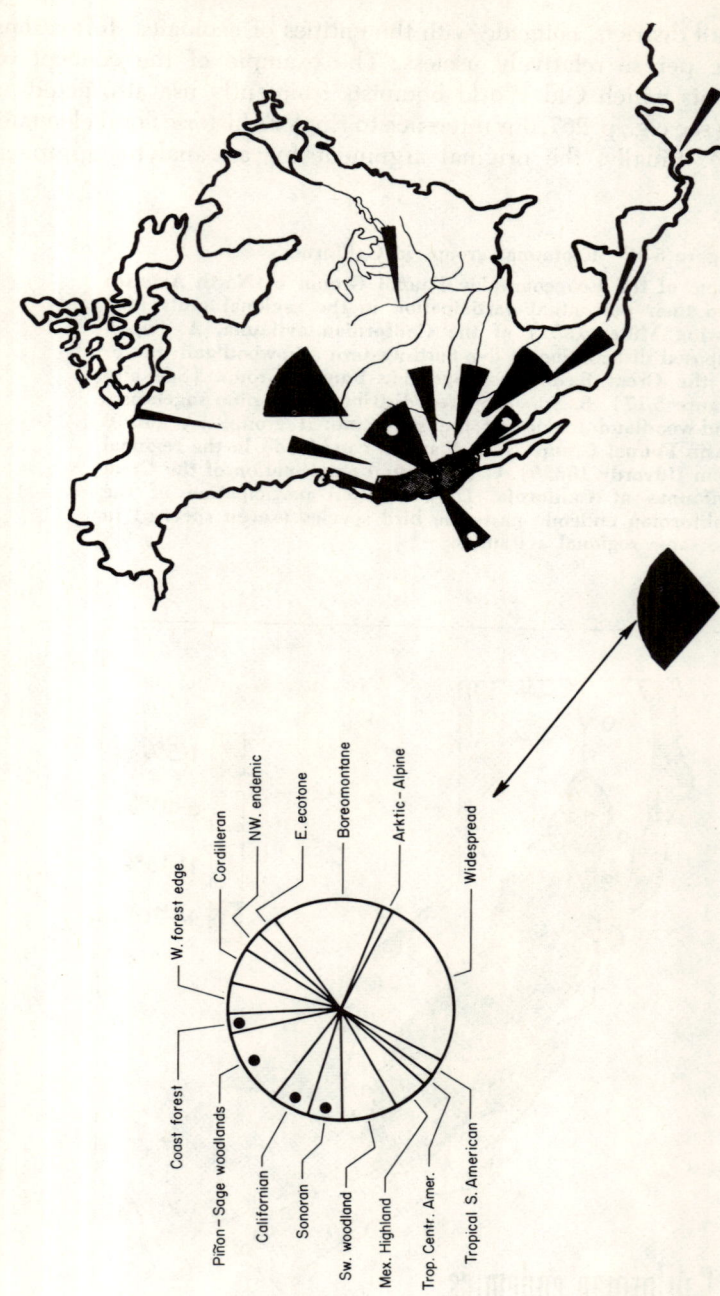

Figure 5-19 The composition of the Californian passerine avifauna. The pie diagram shows the percentual participation of the different ecogeographic faunal groups. The map indicates the approximate geographic position of the center (largest species density) of each of these faunal groups.

based on species areas stems from the early Swiss zoologist Rütimeyer, who in 1868 suggested the "drawing of circles" around the domiciles of animals, for every species. According to Rütimeyer (1867, pp. 147–48, original in German),

From the complicated meshwork of such lines the so-called natural distribution areas will emerge. One may surmise that these districts do not only serve as present domiciles of the group but that they also served the developing of its present peculiarities and modifications.

He hoped by this method to find the distributional and evolutionary centers of groups, especially in areas of varied terrain and sharp geographic limits. He also argued for the simultaneous use of the province concept for extensive areas of rather uniform topography where "the lines [area limits] become so numerous that their meaning becomes obscured" *(l.c.)*, though he realized that in such circumstances the delimiting of provinces becomes vague and arbitrary.

All three of the analytical methods—the geographic, the historical, and the areographic—are based on qualitative characteristics of the list of occurrence. The frequency of occurrence—in other words, the absolute or relative density of the species populations—has not yet been utilized in zoogeographical studies. Its importance in population dynamics is recognized; ecologists of ecosystems measure, compare, and analyze density relations, and plant ecologists have built their concept of ecological dominance on relative frequency. Zoogeographers should be aware that data on population density are accumulating for many animal species and for many ecosystems. Commonness or rarity are important "life historical" characteristics of animals which contribute greatly, as we have already seen, to the success or failure of dispersal and expansion. Biostatistical methods, such as cluster analysis and the like, could easily be used in regional or group analysis based on the relation of area to *density*.

FAUNAL DIVERSITY

When we analyze faunas, it becomes apparent that the number of species is different at every locality. As each species represents a *different* form occupying a *different* niche, the numerical (faunal) list also becomes an indicator of *faunal diversity*. Diversity is the condition of having differences—variety. The term is less fortunate when applied to a taxon to mean the size of the taxon—the differences in numbers of units within that taxonomic rank. *Fauna size*, when applied to the fauna of a geographic area, may mean several things. It may mean diversity at any one locality. It also may involve many regional faunas combined into a single faunal list. In this latter instance, closely related, sibling, and/or vicarious forms may fill the same niche in the various reginal faunas, and consequently the diversity would not increase greatly though the total *fauna size* would be large. Therefore, one could not use geographic differences in the size of a taxon for proof of faunal, taxon, or geographic diversity.* The size of a taxon and the size of a regional (larger geographic area) fauna should be kept distinct from *faunal diversity*. Trends in *species diversity of ecosystems* also exist. The varying numbers of species therefore interest zoogeographers and ecologists alike.

Ecologists were the first to offer an explanation of the phenomenon of varying species diversity of ecosystems. They were impressed by the strong limiting effect of extreme physical factors in the northern temperate and subarctic habitats and also in the corresponding extreme habitats of high mountains. An empirical rule has been suggested by Thienemann (1918): that the more extreme the living conditions of the habitat, the fewer the species which can stand these conditions, but that there will be more individuals (of these species) per area unit supported by the habitat. In such circumstances, as Thienemann postulated, only a small number of species become adapted to the harshness of the physical habitats, and it follows that these few species will not create unduly large numbers of opportunities for other species and that therefore fewer niches will be available. The producer component of the ecosystem is also simpler; the plants possess structurally less varied organs, not suitable for providing many feeding strata and opportunities, as plants of lower latitudes do. For a long time, this simple ecological reasoning satisfactorily explained the *proximate* reason for poor diversity of these communities and, implicitly, the meager faunal lists in extreme alpine and subpolar areas and habitats.

Since the time of Wallace and the other tropical naturalists, ecologists have been fascinated by the high diversity of plants and animals at low latitudes. If one associates the tropics with the many-storied tropical rain forest, the increased number of ecological niches, the year-round growing season, the moderate temperature regime, and the long, undisturbed geological time, these factors all provide a ready explanation for the high faunal and floral diversity compared even with that of tem-

---

*For example, Darlington (1965) tabulates the decreasing number of species in some vertebrate and invertebrate taxa of southern South America; but, instead of localities, he gives a north-south extent of 2° to 6° latitude. One does not know how much of the decrease is due to decreased diversity of any one local fauna, and how much to fewer kinds of faunas and to the size of the area from which the species list has been assembled.

perate forest habitats. Notwithstanding, we know that the tropics harbor many simpler communities and sharp seasonal contrasts, and also that the dynamic changes of the past have affected them as they have affected the higher latitudes, and that therefore these explanations are not always valid. Recently, interest has arisen in the intricacies of tropical species diversity, owing to such pilot theories as Dobzhansky's (1950) and Hutchinson's (1959) and also owing to the availability of census data from tropical areas which quantified, at least for some organisms and habitats, tropical diversity (e.g., Udvardy 1957). Several recent ecological studies have focused on species diversity (Klopfer 1959, 1962; MacArthur and MacArthur 1961; MacArthur 1964; Connel and Orias 1964; Paine 1966). New ecological theories have emerged from these studies, and many or all of these are plausible explanations for some aspect of the *proximate reasons* (*sensu* Baker 1938; see p. 117) for tropical ecological diversity; aside from having more ecological niches, the niches may be better exploited, or divided, by several organisms. Competitive processes, rather than physical hardships, limit animal populations, and thereby subtle evolutionary differences appear; predation is also more diversified, and thus potential competitors, if in the prey category, are better able to coexist. These are some of the theories that Fisher (1960), Pianka (1966), and others have discussed.

The ultimate reasons for the ecological diversity of faunas—either plant diversity, as supporting the animal world, or direct reasons—have often been discussed, but were interspersed with the above, proximately acting factors, with no distinction being made between the two. Ultimate factors may be found among the historic and dynamic biogeographical parameters of faunas and floras. Simpson (1964) demonstrated diversification trends among mammals (Figure 5-20), and also summarized many of the arguments of Fisher (1960); but while Fisher claimed that the temperate faunas have been impoverished, Simpson thinks that warm-temperate habitats have, since the early Tertiary, enjoyed as much uninterrupted time for evolution as have the tropical ones.

We now present arguments which point to increased opportunity for tropical and subtropical regions to increase the diversity of their faunas and, thereby, the diversity of their communities. Most of these arguments are old; others seem self-evident when applied to this problem.

The tropical belt receives more and retains more of the sun's energy than subpolar areas, and is therefore able to sustain and generate more organic material. Larger plant biomass and more stratified plant communities are produced, and these support more animals, quantitatively as well as qualitatively.

Rate and end results of evolutionary processes depend on time and

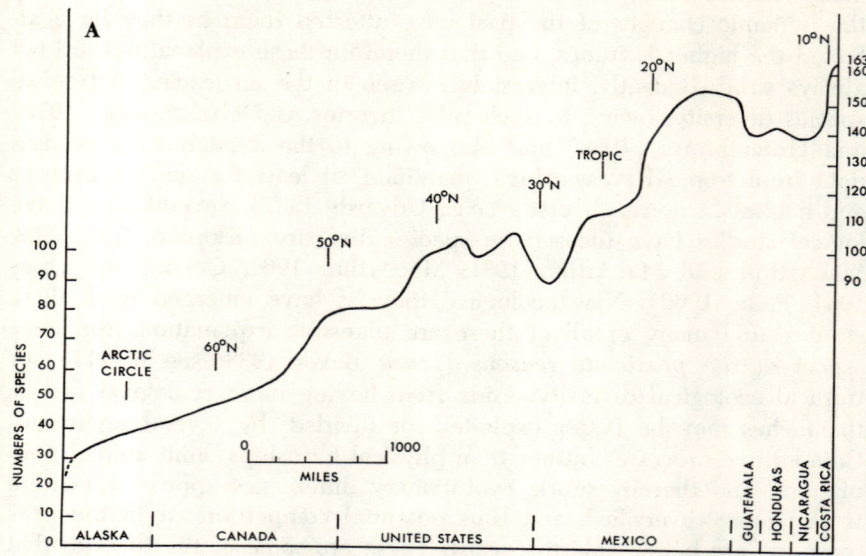

Figure 5-20 Faunal diversity pertaining to mammals. From Simpson 1964.
A. Latitudinal transect of North and Central America along the main backbone of the continent, with high mountain ranges and arid belt within the transect.
B. A transect of North America along the 100th meridian, where no major mountain range complicates the trend and magnitude of diversity. (*Courtesy of G. G. Simpson and Systematic Zoology*)

geographic area. During the past ten million years, the tropical-subtropical belts have had a longer and less interrupted evolutionary past than has the temperate and cool habitats, and most of the existing species are of Pleistocene age. During this epoch, the tropics were less drastically affected by climatic changes than were the subpolar areas. The geographic extent of the tropical-subtropical belt was once much larger than it is now. At present, the ratio of the total extent of cold, temperate, subtropical, and tropical vegetation belts is, in rough proportion, 1:2:3:2½ (Good 1964). Even now, with one-tenth of the land area still in a glacial stage, nearly one-third of the land is in the tropical belt and a full one-third is subtropical. It has often been claimed that area size and evolution have intimate relations (for example, by Darlington [1959] in speaking for the tropical origin of land vertebrates). Even if this is debatable, larger area enables more speciation by isolation and more diversification of the isolates, owing to the variety of climatic factors and

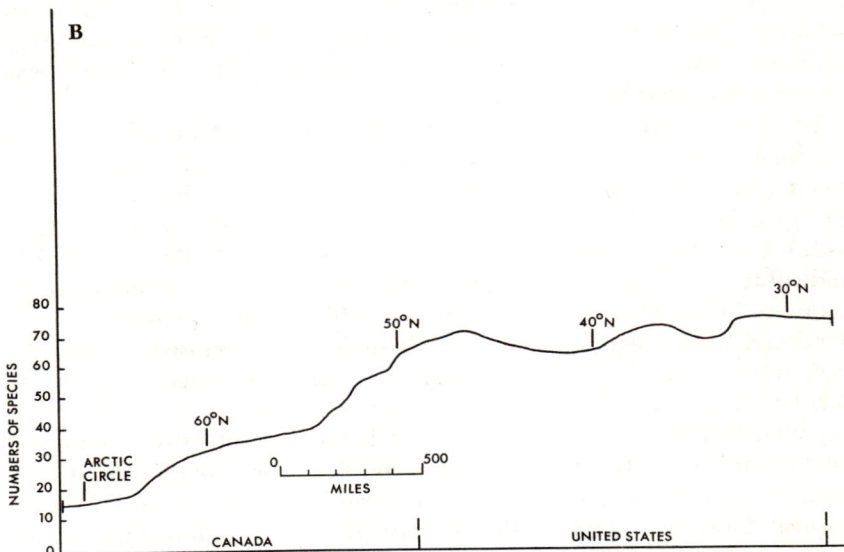

Figure 5-20 Continued.

also often of the topographic factors on the large area. Sympatry and niche sharing, with all the postulated or observed ecological adjustments, may then set in.

Not only are the tropical vegetation belts the most extensive in size, but they also possess outstanding subdivisions—splittings by barriers longitudinally as well as latitudinally. Ocean barriers cause the main longitudinal divisions. However, ocean barriers are very slight in the northern circumpolar land areas which presently harbor a very uniform subarctic and boreal biota. There is ample paleontological evidence that these same northern circumpolar land areas had intercommunicating warm-temperate and tropical biotas before the late Tertiary cooling trend decimated their members. The many circumtropical organisms (which have not changed) indicate that other likewise circumpolarly spreading, tropical organisms have changed and enriched regions outside their origin. The same reasoning applies to tropical biotas of the Southern Hemisphere. It is not important in this context whether their connections were across the sea or by land routes; intermittent connections and disjunctions could have caused alternation of isolation and connection, favoring exchange of evolutionary duplications. Furthermore, latitudinal barriers, such as the former Tethys Sea in the Old World, the former Central American sea barrier between the Americas, and the present Australasian sea barrier, have caused evolutionary duplications. Many of these achieved sympatry (see, for examples, Simpson 1965 for mammals, Mayr 1946 and 1964 for birds, Halftter 1964 for insects of the Americas), even though extinctions have also occurred.

To this "spatial heterogeneity" (Pianka 1966) must be added the altitudinal diversity of the tropics, as opposed to the temperate and cool zones (Simpson 1964). The presence of this zonation (Figures 3–14, 3–15) greatly increases evolutionary opportunities in the tropics (Figure 5–21). One-fourth of the present land area in tropical (20° N to 20° S) and subtropical (20° to 40°) latitudes is covered with temperate, arctic-antarctic, and alpine ecosystems which add to, and stimulate, tropical floral and faunal diversity. Only a little more than three-sixteenths of the temperate vegetation belt is covered with alpine and arctic or subarctic vegetation and flora (Good 1964), but even these are rather sterile, impoverished biota. Therefore, vertical heterogeneity provides—in both quality and quantity—less favorable conditions for evolutionary diversification in the nontropical belts. Furthermore, the reciprocal effect of tropical biota moving into the altitudinal zones is altogether missing here. (cf. Figures 6–23, 6–27).

It follows that Pleistocene speciation phenomena caused by the shifting of altitudinal belts were felt much more in the tropics than in the subpolar belts.

The catastrophic faunal disturbances and eradications of the Pleistocene glaciations mostly affected the subpolar belts of life. It seems that historical biogeographers largely agree—at least as far as the Northern Hemisphere is concerned (Deevey 1949, Wright 1961)—that no climatic-vegetational belt was wiped out or totally mixed with others, and that the vegetation of the present subtropical and tropical belts shrank but did not die out and did not move at the rate at which the subpolar belts were transposed when ice covered a large extent of their present areas. Nevertheless, displacements, shrinkings of tropical ecosystems and subsequent speciation occurred at a high rate. There are little data to suggest that this postglacial and interglacial evolution had a rate in the tropics different from that (p. 386) in temperate and semi-tropical areas. Some biogeographers (for example, Dunbar 1960, 1963), however, point out that evolutionary and adaptive processes in recent arctic habitats have been extremely slow; this view—perhaps a reformulation of Thienemann's previously cited principle—again adds to the balance of tropical versus subpolar diversity.

In areas where the Pleistocene glaciations decimated the biota and in other areas where the biota is very recent and incomplete (such as recently deglaciated areas), the diversity of the fauna is very low. It fol-

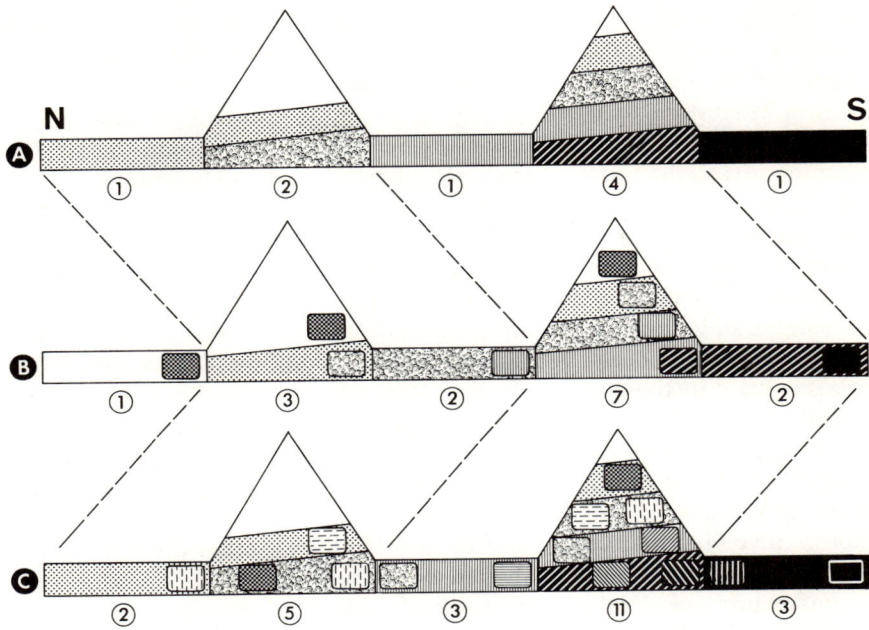

Figure 5-21 Hypothetical schema of climatic changes and altitudinal zonation contributing to faunal diversity through increased rate of speciation in southern areas.

Numbers in circles indicate number of species in the corresponding climatic zone. White area is the nival belt, devoid of life.

A. This is a north-to-south transect through five latitudinal climatic belts. Each belt harbors one species (1 through 5) and each altitudinal belt also harbors one species (also numbered from 1 up). The northern mountain has only one altitudinal zone of life, the southern one has three (in addition to the zone at sea level surrounding the mountain).

B. Between periods A and B a glacial shift of climate occurred: each climatic zone shifted southward by a total zone width. In a locally favorable habitat a relict population of the previous inhabitant survives. Altitudinal zones also move downhill, and on southern slopes, relicts remain. Faunal diversity increased as a result of the climatic change and geographic movement of the zones, for most of the isolated relicts speciated (square inserts).

C. In an interglacial period C, the climatic belts shifted a full width northward relative to their position in B, restoring their position during period A. The zonally distributed species shifted their areas in a corresponding manner, but again left isolates behind. For the sake of simplicity we did not move the preglacial relicts, did not consider extinction either; the square symbol of relict occurrence is retained for the new glacial relicts. Faunal diversity has greatly increased again.

Two additional suppositions can be made: 1. If we suppose that half of the isolates get extinct or re-united with their parent stock without speciation the diversity numbers of the five habitats would run at the end, i.e., at present, like (1)-(3)-(2)-(7)-(2), in other words, very high still in southern, montaineous areas. 2. If we would allow time for mountain zone species to spread to, and adapt in, lowland conditions, the diversity of the southern lowlands would also greatly increase.

300 *Regional and Analytical Zoogeography*

lows from the previous points that such conditions of low faunal diversity are exceptional and recent. The "rich" faunas of tropical areas represent the more "normal" diversity situation.

It is interesting to find that not all groups or taxa of animals follow the latitudinal trend of increased diversity. One such group among the limnic organisms is the basommatophoran snails (Hubendick 1962). These live in relatively ephemeric habitats in which radiating evolution and multiplication of the species are hindered by the short age of the lakes and water systems, in which the diversity trend does not increase and sometimes even decreases in a north-south direction. Among seabirds belonging to various avian orders, diversity in the North Pacific increases and culminates at boreal latitudes; there is a much smaller corresponding phenomenon in the North Atlantic (Udvardy 1969*b*), although the food supply is equally adequate there. Opportunities for multiple speciation and for adaptive radiation are better in the North Pacific than in the North Atlantic or on the more southern coasts of these oceans. Thus, we have given possible explanations for the two diversity trends, shown above, involving snails and seabirds; it is likely that other diversity trends could also be analyzed and their reasons explained.

Past causes of diversity grew out of historic-evolutionary conditions, and the presently acting reasons for the maintenance of diversity are ecological. In this latter respect, *faunal diversity* (faunal size) and the total population *of a faunation* have strong correlations; therefore, one may expect more clues to the causes of faunal diversity to emerge as the number of animals living in a place are better studied and understood.

These considerations are quite hypothetical; there is little concrete proof to support them, because the pertinent synthetic, faunistic, synecological, and other data have not yet been assembled. They serve here to indicate that several phenomena may be adduced as independent causes of diversity, and that their effect might prove to be influential, especially if they act concurrently. They may also serve as stimulants either to prove or disprove these surmised causes, and thereby the better to clarify the ultimate reasons for diversities, which are themselves so fascinating to observe and to study.

CONCLUSIONS

Since distribution areas do not occur at random but are clumped, a regional grouping of distribution areas is justified; limits have presumably been affected by the same factors acting on different animal species.

J. A. Allen's regional system of the world is based on the climatic zones; Merriam's Life Zones single out certain climatic correlations as delimiting groups of animals of equal tolerance. These systems have not won recognition among zoogeographers of other continents, and not even in North America are they now in any except limited use.

Major plant formations are likewise the results of zonally acting climates, and these formations have been recognized as convenient regional entities of the vegetation. Postulating that animals are dependent on the distribution of the plant formations for their food, shelter, and so on, Clements and Shelford presented the biome concept. While botanists everywhere have adopted and amalgamated Clements' system, zoogeographers and ecologists are slower in thinking in terms of biomes of the world.

Early in the Twentieth Century, the analysis of contagious or clumped distributions led to the establishment in Europe of a system of floral as well as vegetational provinces within major floral regions, on the one hand, and formational entities, on the other. In North America, Dice's system of biotic provinces was based on the distribution of biomes and of ecological and topographical barriers of minor rank. Hagmeier's more objective method analyzes the statistical clumping of limits.

Sclater's regional system is based on differences in major taxa of birds between continents and, in Eurasia, also between tropical and extra-tropical parts of that continent. Though T. H. Huxley emphasized the unity of the Northern and Southern worlds, the accepted regional system became Sclater's system, amended by Wallace. The Sclater-Wallacean zoogeographical regions are longitudinally separated by oceans, transversally by other major barriers. Wallace, like his botanist contemporary Engler, emphasizes the common history of the faunas inhabiting his regions. Seventeen other historic-geographic approaches are mentioned or discussed; they differ mainly in details from the ones mentioned above.

The use of statistical methods in zoogeography is discussed briefly. Such methods were employed for separating areas of faunal uniformity and for finding the natural limits of these.

Areographic analysis aims at defining groups of species with geographically (and therefore ecologically and/or historically) related distribution areas. It is believed that ecologically related animals followed climatic fluctuations more or less as a group, that therefore they had much in common in the past as well (the further back in time, the less they may be ecologically related). Moreover, analysis and grouping of ecofaunal elements do away with rigid and disputable regional boundaries. Hultén's, Stegmann's, and de Lattin's approaches are discussed together with those of the author.

Geographically different trends in faunal diversity are the results of geographic and historic factors which are summarized.

6 / Dynamic Zoogeography

> *Behold, there come seven years of great plenty throughout all the land of Egypt:*
> *And there shall arise after them seven years of famine; and all the plenty shall be forgotten in the land of Egypt; and the famine shall consume the land . . .*
> Genesis 41:29,30.

In the earliest writings of thoughtful men, we find considerable awareness of the vicissitudes of nature in any given time span. Perception of shorter time periods is largely subjective, based to a great extent on the regularity of body function, our "biological clocks." But, as we grow older, we perceive time more objectively, and perception is further expanded by the mental integration of the experiences of our elders. To assist perception, we resort to numerical comparisons, graphic means (Figure 6–1), and the like. Thus, the understanding of longer time periods is no longer subjective, but depends on such things as training in numbers, reading graphs, and making comparative abstractions.

A dichotomy undoubtedly existed in the past between biologists who use the *functional* approach and those who use the *evolutionary* or historical approach. The latter group were seeking, among other things, a time perspective on the origin of functional systems (Table 11). Since Darwin, scientists of this type have been called "evolutionary scientists", and by now, evolutionary ideas have penetrated even such functional fields as physiology, ethology, developmental mechanics, and the like.

In environmental biology (zoogeography is, ab ovo, an environmentally oriented science), the time or *temporal aspect* of the inquiry is an entirely natural orientation. This fact has not long been recognized. Wallace, the celebrated "father" of the subject, was aware of the importance of

Dynamic Zoogeography   303

the historical aspect of zoogeography; yet his most lasting contribution remains the division of the globe into zoogeographic regions. These regions represented the status quo in distribution, and the boundaries he drew were, for the most part, considered static, absolute lines; refining

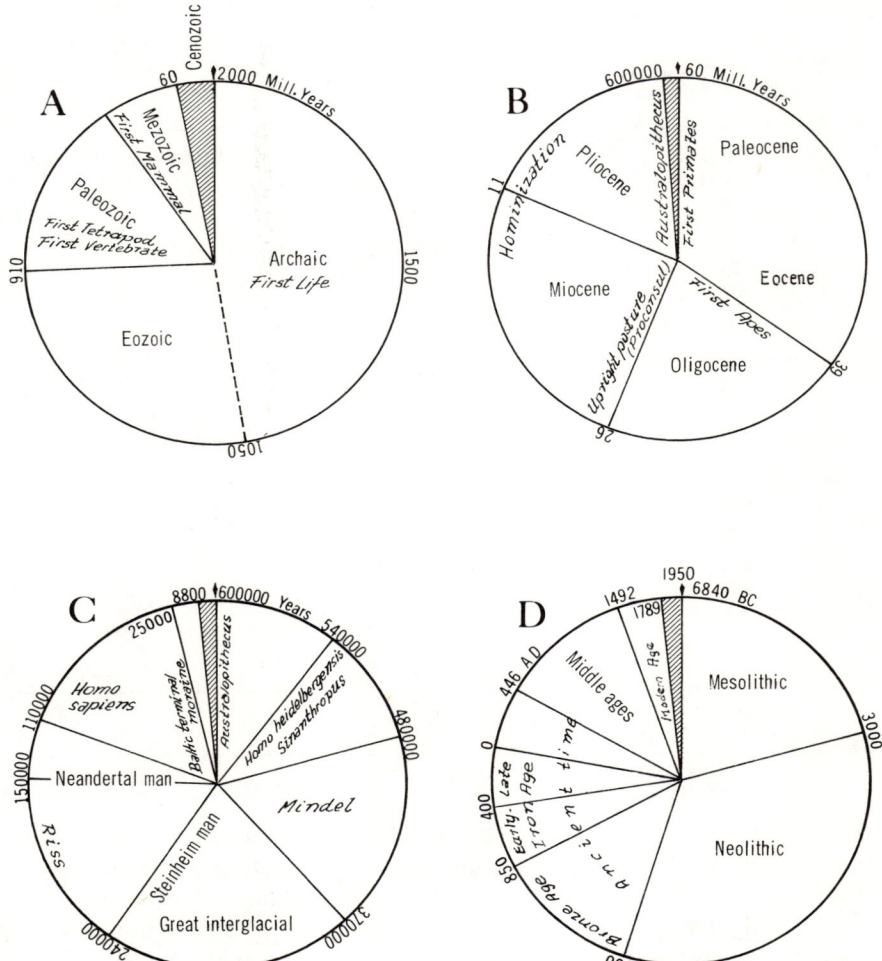

Figure 6-1 Approximate relation of the duration of the different geological and archeological time periods. After Wagner 1950. The arc of the shaded segment in each circle is magnified to form the total circumference of the next circle.
A. The history of earth and the Cenozoic.
B. The Cenozoic and the Pleistocene.
C. The Pleistocene and the Recent (Holocene).
D. The Holocene and the history of the United States.

304 *Dynamic Zoogeography*

Table 11   The geological time scale.
(From Simpson and Beck 1965.) Pleistocene modified.

READ TIME SEQUENCE FROM BOTTOM TO TOP: OLDER TO YOUNGER.

Approximate time since the beginning of the period, in millions of years	Era	Periods (for Mezozoic and Paleozoic) Epochs (for Cenozoic)
0.01		Recent (=Holocene, Alluvial)
±1.0–2		Pleistocene
10		Pliocene
25	Cenozoic	Miocene
35		Oligocene
55		Eocene
70		Paleocene
135		Cretaceous
180	Mesozoic	Jurassic
230		Triassic
280		Permian
345		Carboniferous
405	Paleozoic	Devonian
425		Silurian
500		Ordovician
600		Cambrian
>3000	Pre-Cambrian	

their details became, for at least forty years, the chief occupation of many scientists. It took two paleontologists, representing a truly historical discipline—Anton Handlirsch, of Austria, and William D. Matthews, of the United States, to redirect the preoccupation of zoogeographers toward the role of past climatic and evolutionary fluctuations. In spite of the influence of these men, the majority of animal geographers still worked in two separate chambers, the geographers of recent fauna concentrated upon the status quo, while the paleontologists were interested only in faunas of the remote past.

The intention of this chapter is not to present "historical zoogeography" or "animal geography of the past" as a discussion of facts pertaining to past distributions. As far as vertebrates are concerned, Darlington's 1957 book, *Zoogeography* offers all facts, interpretations, and broad conclusions necessary for those who wish to learn in detail how a historical zoogeographer utilizes both fossil and recent evidence in forming hypotheses. There is, unfortunately, no comparable work on any invertebrate group, though such a work is badly needed to counterbalance the often justified criticism that vertebrate zoogeography only tells one side of the story.

The main object here is to present a synthesis of the dynamic, ongoing processes which form and modify animal distribution throughout all time. Certain portions of this approach will be a distillation of what has been stated in earlier chapters; other portions, pertaining to evolutionary and geographic syntheses related to the time factor, will be considered here for the first time.

INTERACTING ENTITIES THAT INFLUENCE ZOOGEOGRAPHICAL PROCESSES

Geographic Entities

Changes in animal distribution are caused by the dynamic existence of physiographic entities composing the surface layers of the earth upon which or in which animals live. Our primary interest—the land surface—consists of the *continents* and those islands which are situated near them. Most of these *continental islands* are on the *continental shelf*.

The permanency of the continents and their location is a subject of violent controversies among geologists and paleographers. Because the level of the oceans and the height of the land fluctuate independently, the inundation of low-lying land areas by the oceans often results in the isolation of continental islands, whose life is, geologically speaking, ephemeral.

Oceanic islands, on the other hand, rise up from the ocean floor and are of volcanic origin. Most of these are far from the continents. Their land biota presumably arrived by overseas dispersal, unlike the animals and plants on most continental islands, whose ancestors had a land route available (see also p. 370).

The surface of the continents is not at rest, but is in process of either elevation or of being leveled. Tectonic movements and plutonic activities result in *mountain systems* and *plateaus. Plains* may originate as elevated sea bottom or be formed by wind action or by wind- and water-borne deposits.

For the zoogeographer, the entities of the hydrographic series are

306 Dynamic Zoogeography

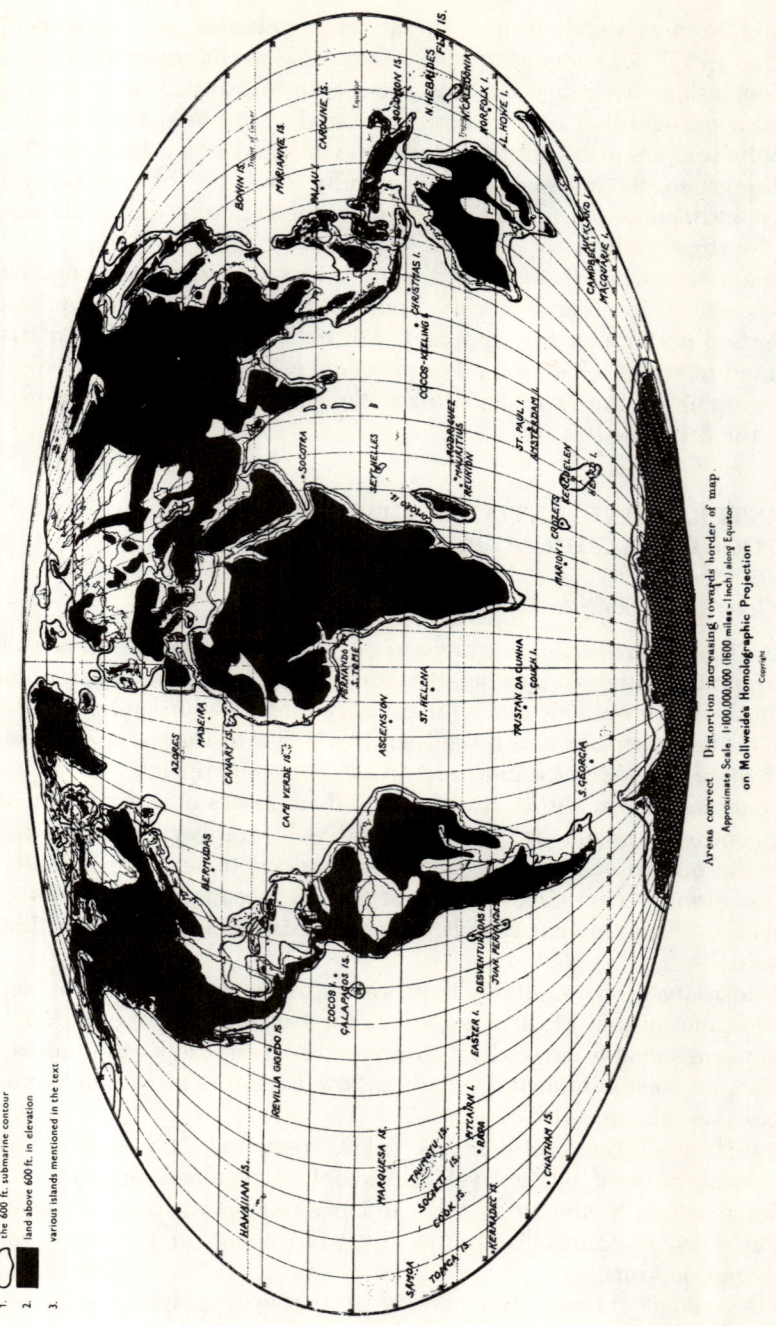

Figure 6-2 Map of the world showing the 600-foot (about 200 m) submarine contour, and land above 600 feet in elevation. From Good 1964, slightly modified. (*Courtesy of Longmans, Green and Co. Ltd., London*)

equally important (we have already emphasized their role in barring land animals from expanding their areas). The *oceans* themselves are the strongest barriers; even though the continents they separate may once have had—or may now have—land connections, these always impose ecological limitations on spreading. Geologically speaking, oceans at times penetrate land masses, especially at those places which we call intercontinental bridge areas. These more ephemeral *epicontinental seas* induce important isolation of land forms, usually followed by speciation and/or diverging faunal evolution (Figure 6–2).

Inland water bodies are the *lakes* and the *river systems*—drainages—which were assessed for the land zoogeographer in Chapters 2 and 4. Ice fields, because they are solid, can also be considered land formations. However, because of their role as barriers for most land animals, and because they themselves do not support terrestrial life, we may consider them part of the hydroseries.

Temporal Entities

If we think of biological phenomena as adaptations to changes wrought by the passage of time, it is easier to discuss the causal relationships of such phenomena to those short-term cyclic changes in the environment which are *reversible*. Longer cycles are more difficult to study, because we must rely on historical accounts and on circumstantial rather than direct, empirical evidence. When the span of the cycle is longer than mankind's accumulated experience, the conjectural nature of the evidence makes the dating and even the nature of the cycle uncertain. Techniques for precise dating are improving rapidly and, within the limitations of these techniques, environmental events may be delimited in considerable detail. More ancient events can be delimited with only broad approximations.

Circadian changes affect most animals and plants, at least when their life span consists of several days. However, the study of these changes is not within the direct scope of biogeography, except in the context of their ecological implications. *Seasonal periodicity* affects distribution relative to length of daylight and darkness, and these are different at different latitudes. Seasonal periodicity affects distribution but these relationships, seasonal migrations, seasonal areas, and the like have already been discussed. The change of seasons especially affects the distribution of the short-lived and the seasonally adapted forms and may, in addition, cause irreversible environmental changes. The *annual cycles* are attributed to seasonal change, but there are variations from year to year (summers may be warmer or colder; snow cover or annual flooding may be greater in one year, less in another). These variations may be part of a cycle, but their causes become increasingly difficult to ascertain.

Multiannual cycles, according to some scientists, include: the three-and-a-half- and eleven-year ones which manifest themselves in the peak populations and in the spatial expansion of certain arctic and boreal animals; the cone or seed-bearing cycles of many temperate zone trees; and the two-, three-, four-, seven-, thirteen- and seventeen-year cycles of abundance in salmonid fish, melolonthid beetles, and cicadas, caused by the life span—the reproductive age—of the organism. The periodically occurring eruption periods of volcanoes are examples of multiannual, irregular environmental changes. Other examples are: the series of dry years, causing increasing drought in the arid belts of the continents, followed by a number of rainy years; and, unusually warm winters followed by drastically cold ones, such as those in Europe in 1928–29, 1939–40, 1941–42, 1946–47, 1955–56, etc. in which no regular cycles were apparent. We do not know the reason for the regularity or irregularity of such "acyclically" recurring events; it is possible that they are the result of the superimposition of several cycles of shorter oscillation length (Palmgren 1949).

The most immediately evident multiannual cycles are the climatic ones; rhythmicity in the dynamism of the earth's crust (volcanism, tectonic movements, earthquakes) is much less understood. In the climatic picture of any one area, there are *secular* trends—long-enduring processes in which changes are measured in centuries. There are the fossil indicators (pollen; river, glacial, or wind-borne sedimentation) of past climates, which are reasonably accurate for the past few ten thousands of years, and within which one may discern millennial climatic trends.

Phenomena which endure for secular or longer periods cause *irreversible* biological change. For example, continental islands have been alternately attached to and detached from continents. In the long period of attachment many changes in biota have resulted, and these plants and animals have had effects on even such basic conditions as that of the soil. Because of these effects, the now detached islands cannot revert to the conditions of an earlier detachment.

Thienemann (1950), following Gajl (1924), admits the concepts of *eustasy* and *astasy* into the vocabulary of the zoogeographer. Environmental phenomena (such as the physical characteristics of deep lakes or the level of the world ocean) are *eustatic* if they change only slowly and if they gradually but imperceptibly influence the life of any one generation of animals or any successional status of a habitat or community. Phenomena that influence the constancy of environmental conditions are termed *astatic*. Considering even this dichotomous division—which of course represents two extremes of a graduated scale—we are able to categorize environmental changes according to their effects on the existence and distribution of animals throughout time (Table 12).

Table 12 Temporal units of environmental changes as they affect animal distribution

			Nature of Changes	
Examples	Duration of Changes	Name of Change	Eustatic[1] (slow gradual)	Astatic (sudden drastic)
Photoperiod, Tide	Day to day	Dial		
Frozen Ground, Flooding	Up to several months	Seasonal		
Precipitation, Nut Crop	Year to year	Annual		May be cyclic (reversible)
Drought, Sunspot Activity	Several years	Multiannual		
Glacial Advances & Retreats, Plant Succession	In terms of 10 years	Decennial	Eustatic	
Lake Filling	Hundreds of years	Secular		
Sea-Level Changes	Thousands of years	Millenial		Irreversible May be cyclic[2]
Glacials & Interglacials, Mountain Building	From many thousands to millions of years	"Geologic"		

NOTE:. [1]Eustatic changes may be reversible, but this is not important because of the slowness of changes. [2]Astatic changes of long duration may be cyclic, but this is not important because dynamic evolution during the period involved makes their effect irreversible.

Organismic Entities

The lowest-ranking unit of interest to the biogeographer is the *individual*, for a single dispersing individual may become the founder of a population.

The basic organismic entity in evolutionary biology is the *species*. Our earlier discussion of distribution areas was based on the area of the species. However, if our interest is focused on the geographic aspect of dynamic history, the criteria of the species will not be entirely satisfactory as the basic unit of biogeography. We have seen that geographically widely disjunct part areas are considered as being inhabited by populations of the same species, if they are capable of interbreeding and are reproductively isolated from all other animal populations—these are the two crucial points of our definition of species. However, such geographic *isolates* which are too far from one another or from the main distribution area of the species to exchange genes may genetically diverge from one another under different selection pressure and may reach subspecific or even specific rank. (In the latter case, we no longer speak in terms of disjunct areas.) Yet during the time they undergo genetic changes—i.e., while they speciate—their isolated area is, biogeographically speaking, the same entity, whether it is a disjunct part of the total area of a species or a subspecies or whether the inhabiting, geographically isolated population is considered a full species by the taxonomist. Furthermore, it will have the same position and influence in the faunal and faunational assemblance where it lives. Therefore, the useful entity for the biogeographer analyzing an existing distributional picture is not the species area but the area of a reproductively isolated *population,* whether it be the total area or the part area of a species or subspecies.

The situation is different in historical studies when we study past geographic movement and the origin of distributions. A main point of conjectural evidence in the historical analysis of a distributional area is the phylogenetic relationship of the inhabiting population to other contemporaneous populations. Important in this context are the *total areas of all taxa,* such as subspecific and specific areas, areas of sibling and vicarious species (species pairs, triads, swarms, etc.) or superspecies, genera, and families, insofar as these taxa truly represent evolutionary entities (see p. 231).

A good example of the monographic treatment of the zoogeography of a single species is found in the work of Thienemann (1950) on *Mysis o. relicta* Lovén [Mysidae, Malacostraca]. This short monograph explains the *present* distribution of this crustacean in Europe on the basis of structural and physiological, ecological, geographic, and geologic-historic evidence. Monographic treatment of a higher taxon may often be even more rewarding, because many comparisons can be made and patterns can be detected on the basis of the analysis of group and single areas (*cf.* the review of the insect order of stoneflies [Plecoptera] by Illies 1965).

Biogeographical Entities

In Chapter 5, I suggested that the biogeographical character of a region can be best clarified through a consideration of the *combined* ecological, evolutionary, and historical attributes of its faunation and vegetation. In the modern "historical" approach, increased emphasis is placed on the dynamism of such entities as plant formations, biomes, floral or faunal elements, or total biotas; the dynamism of faunas is the inquiry most often pursued. Yet no one is actually able to treat *all* plants and animals or even all plant and animal species of a large geographic area; the study must usually be based on only a few groups of organisms, and necessarily misses the broadest comparisons from which a synthetic study would benefit most. Keast (1962), in his study of geography and speciation in Australia, encompassed all land vertebrates of that continent; but as a vertebrate zoologist, he was not able to consider insects. The well-coordinated teamwork which produced the *Symposium on the Biogeography of Baja California and Adjacent Seas (Systematic Zoology* 9 (1960), 47–232) closely approaches the ideal. The synthetic ecological method, infrequently used but very promising, is the treatment of the dynamism of *ecofaunas,* i.e., of a group of species which live under similar climatic influences and which are usually members of the same biome, with at least part of their geographic areas overlapping and therefore with at least part of their history in common. One of the earliest works to follow this method is the summary treatment of the Palaearctic avifauna by Stegmann (1938). Another example of this method is Smith's (1957) discussion of the historical aspect of an ecofauna: on the "Prairie Peninsula" of the northeastern United States (see also p. 218).

VECTORS AND MODES OF DYNAMISM IN ZOOGEOGRAPHICAL PROCESSES

The discussion of the dynamism of a distributional area (p. 225) emphasized the importance of those cosmic, crustal, climatic, biotic, and symbiotic environmental elements which affect the size and limits of the area. These same factors of course underlie the temporal sequence of distribution changes. Although they have acted on past distributions simultaneously and in a complex way, they will be dealt with here under separate headings.

Short-Term Changes of Distribution

It should be kept in mind that since many species exhibit vagility in their individuals, and that since in most species the young do not settle exactly in the home range of the adult, the distributional picture changes in its minute details year by year and season by season. Every detailed study of the marginal populations of land animals reveals this phenomenon. It had earlier been concluded that relatively stable boundaries occur only at such area limits where the limiting barrier is a completely foreign ecological formation and where no intermediary habitat types or zones occur—such as between land and water, at the borders of icefields, or in rocky or precipitous mountain chains. Even these conditions change

—waters rise or recede, icefields melt or advance—but here the pace of change is *eustatic,* and is hence too slow to be observable on a short-term scale. A fluctuating limit is established where habitats blend or mix, and this is the place to look for trends in the change of distribution. The adaptability of marginal individuals and their microevolutionary changes (see p. 142) is an important intrinsic attribute that influences both expansion and regression. A further, important dynamic factor pointed out by many zoogeographers is the effect of *population pressure* —the pioneering of surplus individuals because they have no room to settle or because crowded populations produce vagile, pioneering genotypic or phenotypic combinations in many individuals. However, high population density is most often the result of supernormal, optimal habitat conditions within the marginal population; the same environmental factors condition these habitats and likewise those which were beyond the limits of habitability in more unfavorable years. In the final analysis, therefore, *trends in environmental conditions seem to be decisive for area fluctuations*. Environmental trends may be followed by slow expansion; generally speaking, long-range pioneering is the exceptional mode of dispersal. Animals which live on particular habitats of discontinuous occurrence are especially likely to exhibit this kind of pioneering. Within a local geographic area, they may show large limit fluctuations in a short sequence of years. In these instances, however, the total distribution area is geographically large; because *vagility, area size, and distributional changes are often correlated,* long-distance pioneering and fluctuations of the limit must be considered on different scales.

In addition to fluctuation in the capacity of the habitat to support a denser or sparser population, there are two external conditions that may force pioneering and thereby a possible extension or regression of the range.

One is the *chance-governed destruction of a habitat* which has harbored a reproducing population. The population is either extinguished (*cf.* MacLulich 1957) or compelled to pioneer elsewhere and even to accept an inferior habitat. Volcanism, avalanches, climatic catastrophes, landslides, floods, fires, and man's activities may destroy any kind of habitat. Fires and insect gradations are especially important factors in forests; droughts, sandstorms, and peat fires wreak devastation on arid open habitats and woodlands; exceptional water levels and drought affect wetlands. Habitat is thus created on the one hand and destroyed on the other. For any one species, catastrophic events may affect marginal as well as central areas; though their effects are generally localized, they may give rise to local trends of advancement or retreat of the distribution. In insular distribution, we know of several recent catastrophes that extinguished certain insular species (p. 316). Certain populations are so delicately small and have such a limited distribution that they are constantly in peril of extinction from even minor catastrophes. The change in the environment brought about by a catastrophe and by the extinction of a population may, however, enhance the spread of other populations into the area.

Plant ecologists have intensively studied the effect of forest fires, and have found that "fire successions" (fire subclimax, Shelford 1963) are regular features in affected areas. It has been observed that many plants show special adaptations toward fire-resistance; but the corresponding literature on animals is scarce, and the effect of recurring fires on their distribution has not yet been assessed.

The effects of vegetational succession are slower. The turnover of the animal inhabitants of successional community stands is well known, and examples can be found in every ecology textbook. The points to emphasize here are: that regular successional changes occur in all but the longer-lasting, stable habitats in terms of year-to-year trends; that these affect the local animal populations; and that these influence the dynamics of the distribution limit. However, the more lasting effect of these changes is also eustatic, for it manifests itself in longer periods of time, measured in decenniums or centuries.

An interesting aspect of these phenomena was brought forward by Schmidt (1945), who hypothesized upon the evolutionary importance of two kinds of animal dispersals. Dispersal into the "familiar" habitats is an observable process in "historic" (relatively short) periods of time, where each species—owing to successional trends—tends to keep its proper domicile within the geographically changing stands of these habitats. This geographic dynamism (which will ultimately result in the gradual shift of the distribution area) results, from the viewpoint of the evolutionist, in homeostasis of habitat occupation and therefore in conditions conducive to specialization, i.e., to improvements useful in the exploitation of a stable habitat. This progressive specialization trend is, in Schmidt's concept, not observable, because it requires a "geologic time scale" for its accomplishment; but during such long time spans, an opposite evolutionary trend also arises—the radiating tendency of adaptive evolution. Schmidt reconciles these two geographically based evolutionary trends in the following way.

Certain organisms adapt more and more to "an ecological formation that may be distinctively subject to succession" (p. 789), while others "adjust themselves to life in new habitats quite without reference to the direction of succession" (p. 790). For example, the originally riparian turtles radiated into freshwater and marine habitats and also into grass-

314 *Dynamic Zoogeography*

lands and deserts. Thus Schmidt discerned (Figure 6–3) two contradictory tendencies: that dispersal into a proper, specific habitat enhances specialization; and that dispersal into related, inferior, but temporarily advantageous habitat—for example, a habitat offering ample food supply

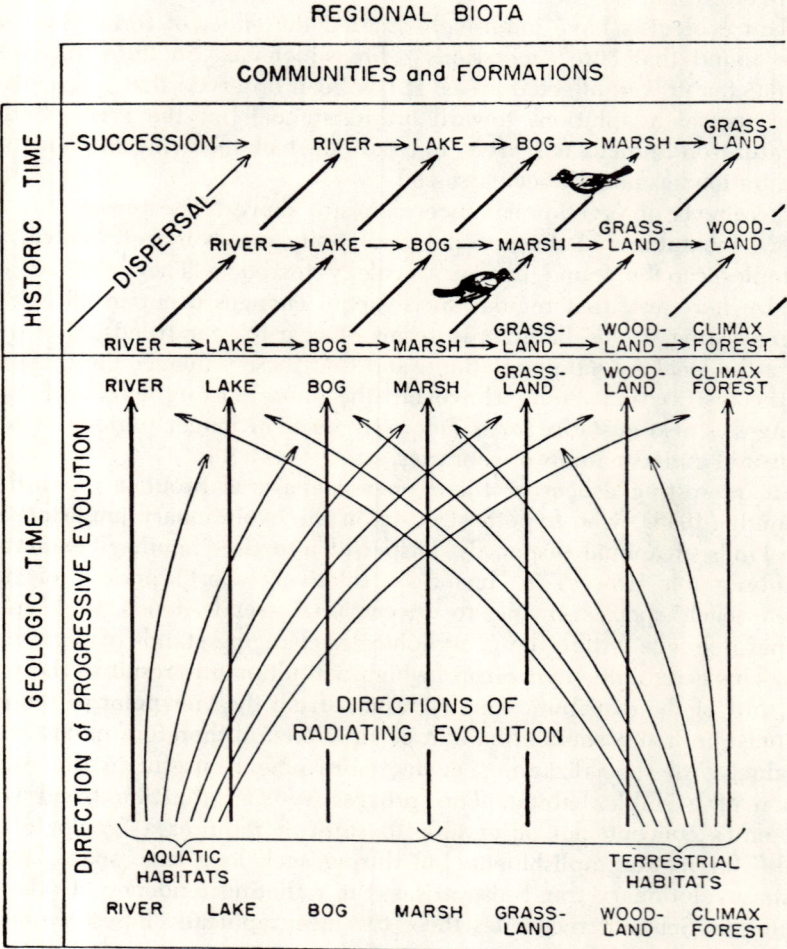

Figure 6-3 "Evolution, succession, and dispersal."
The upper half of the diagram shows that dispersal into the specific, optimal habitat retains the species within its adaptive niche and enhances specialisation. The lower half shows that dispersal into related habitats (if followed by isolation) may lead to adaptive radiation. From K. P. Schmidt 1945. (*Courtesy of American Midland Naturalist*)

—extends the range and may, by isolation, eventually lead to divergence.

It is challenging to discover an ecogeographic basis for the two types or degrees of evolution. Examples of the second type clearly exist: e.g., when members of arboreally nesting bird groups (buzzards, storks, herons) abandon their ancestrally acquired instinct for high perches and nesting platforms (on treeless expanses where food abounds) and begin to nest on the ground. Tenuovo (1963) discusses this habit of the crow *Corvus corone cornix* L. [Corvidae, Aves], and Drent *et al.* (1964), of *C. caurinus* Baird—both on treeless coastal isles. Similarly, the marine-adapted cormorants [Phalacrocoracidae, Aves] as well as certain versatile tree nesters of that group accept the ground as nesting substratum when settling on coastal rocks (van Tets 1965). The endurance in, and adaptation to, brackish or even salt-water habitats by amphibians and other freshwater animals has been discussed elsewhere (Neill 1958). Adaptation of soil organisms to cave environment is also a case in point (*cf.* Vandel 1964). Evolutionists emphasize both of these processes when claiming that both necessitate the additional mechanism of reproductive isolation (Mayr 1963*a*).

The ideas of Schmidt were put forward at a time when our present interpretation of the evolutionary theory was still in the making, and are in agreement with botanist Good's contemporaneous ideas about the relations between the factors of evolution, environment, and dispersal. Good (1964) maintains that evolutionary changes, changes in environmental conditions, and the speed of dispersal are interrelated temporal phenomena, each with a different rate. Good's concept postulates that the current rate of evolution is much slower than that of the other two factors. The discussion of these three rates will be confined mainly to decennial and secular manifestations of distributional change. At these levels, we can base our knowledge on tangible facts rather than on scarce fossil evidence and hypotheses.

Of the factors of evolution, environment and dispersal, the second is the most important. Changes in the environment usually precede and cause changes in the rates of evolution and dispersal. The rates of expansion and retreat of species are also related to environmental change.

Changes in Environmental Conditions

The most violent changes in the physiography of the earth are associated with volcanism. The only effect of volcanic outbreaks on large, continental areas is through their long-range and cumulative influences on the gross climate. Auer (1958) finds that periods of volcanism in Patagonia were followed by periods of humid climate and of forest advancement toward the pampas. Volcanism of oceanic islands, especially on archipelagoes, greatly concerns us, as whole islands may be quickly altered by volcanic activities which wipe out old habitats and create new ones. Although the geological age of many volcanoes and lava fields is negligible, it is nevertheless long enough for vegetation and faunation to have become established on them. These faunations are usually peculiar in their faunistic composition because they are composed of pioneers and are ideally subject to local evolution; furthermore, the localized distributions which volcanism creates may be wiped out again by renewed volcanic activity. Brattstrom (1963), for example, accounts for the destruction of a young local endemic, the wren *Salpinctes obsoletus exsul* Ridgeway [Troglodytidae, Aves] by an eruption in 1952 on San Benedicto Island, off the west coast of Mexico.

Ice sheets once covered large areas of North America, northwestern and northern Eurasia, southern South America, and New Zealand; they of course still persist in Greenland and Antarctica. It would perhaps be well to review here some of the geological effects of their retreat.

The tremendous weight of an ice sheet several kilometers deep causes the crust of the earth to warp. The crust of the surrounding ice-free areas responds by a slight rise in altitude. When the ice finally melts, seawater first invades the meltwater lakes in the deepest troughs and basins. However, as these have now become rid of their extra weight, the pressure from the surrounding areas starts a leveling process. The deglaciated areas rise and the surrounding elevated country sinks until an equilibrium is reached. Water is drained off from all but the deepest basins in the course of the uplift. A good example of rising land is the Laurentian Shield in North America, with the Hudson Bay region in its basin (Figure 6–4). In Europe, the Baltic lies in such a glacial depression. Evidence of the rising of its coasts is abundant, as is seen in the ancient former harbors and fishing villages now far inland from the coast and in the high-water marks on sea walls now several meters above sea level. In fact, the average land rise in southern Finland is easily observed within a single generation, and now amounts to about half a meter per century. (This rate of uplift was, according to Flint [1957], much greater immediately after the deglaciation of some ten or twelve thousand years ago.)

When I visited Finland in the 1940's, the older inhabitants pointed out to me the changes which had occurred in the geography of the coast within their lifetime. What had been reefs—according to names on the map, and in their own memories—had become groups of rocks; the former rocks appeared as dry *skerries*, except during the most violent winter storms. Formerly low, barren skerries were covered with meadows, and the grassy islets of the immediate past had become sheltered thickets or coppices of pines. As the land rises, the mainland increases in area, a process which is augmented as some of the coastal islets become continuous with it. In the last stages of transformation the expanded mainland and the coastal islets, though still rising, exhibit a diminishing pace of environmental changes. This pace will now depend largely on regular plant succession, which will be governed by the rate of soil accumulation. The floristic and faunistic effects of the uplift process have been well studied in the Åland Islands in the Baltic Sea (cf. Palmgren 1927); these islands are a prime example of the changes wrought by that process.

Climatic Changes and Their Rates

The study of present and past climates is called "climatology," of which there are excellent discussions by Brooks (1949), Dorf (1959), and Zeuner (1958). The history of climatic changes interests the paleoclimatologist, the paleogeographer, the geologist, and even the paleontologist.

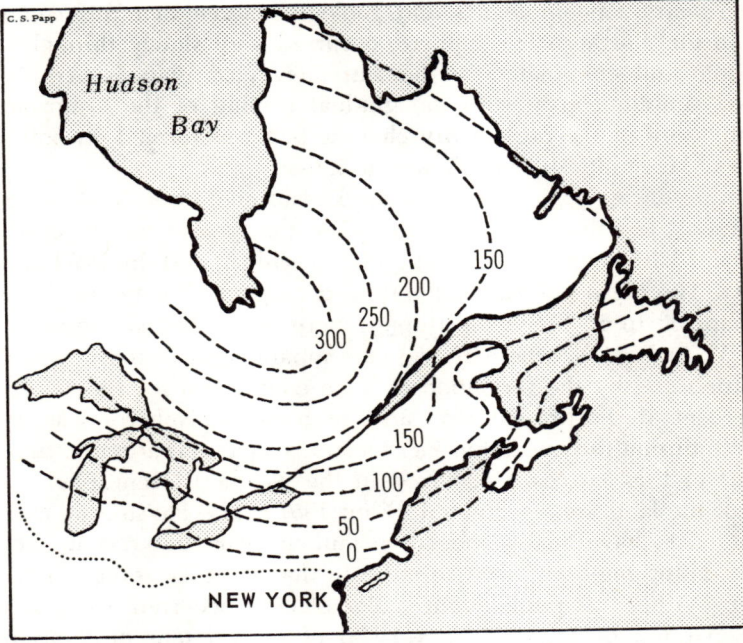

Figure 6-4 Rise of the Laurentian Shield after it had lost its Wisconsin ice sheet. The depression was deepest at the southern end of the present Hudson Bay where a rise of about 300 m occurred during the past (roughly) 10,000 years. After Antevs 1931.

Documentation of climates in past geological periods often stems from analysis of the known fossil biota of that period. The farther back in time we delve, the more uncertain and generalized become the findings about past climates, especially about their dating. The closer to the present we proceed, the more evident will become the rate of climatic fluctuations. We present a survey into the vicissitudes of climate, beginning with an area which is well known—Central Europe (Deevey 1949, Wright 1961, Freitag 1962).

During the first, and probably the shorter, part of the Cenozoic Era, the climate of Central Europe was tropical, with a well-defined alternation between dry and rainy seasons and with predominant biota such as today dwell only in tropical climates. A mean annual temperature of 22° C, such as prevails today in the tropics, is assumed for that era. There was, however, a gradual cooling in the Upper Tertiary, with the mean annual temperature dropping in the Miocene to about 18° C and, during the approximately nine million years of the Pliocene, down to about 10° C (Freitag 1962). The vagueness of these estimates becomes evident when we consider that the Miocene lasted approximately twenty million years. Paleontologists and geologists agree that this cooling, at least in the Northern Hemisphere, occurred very slowly throughout the Cenozoic and culminated in the extremely cold glacials. Aridity progressed simultaneously with the gradual cooling of the Cenozoic, and the precursor of the present dry climatic belt was formed on both hemispheres (Macginitie 1958) (Figures 6–5, 6–6).

The present geological epoch is the Pleistocene; often—and inaccurately—the past few thousand years are called the Recent, or Holocene. During the Pleistocene, the cooling process culminated in glaciations, or *diluvia*. The mean annual temperature decreased all over the earth; less (perhaps 3° to 5° C) in the tropics, more in the higher latitudes (perhaps 5° to 8° C), where it was accompanied by extensive mountain glaciation and by the formation of huge continental ice sheets. Roughly 32 percent of the present land area of both hemispheres was covered with ice during the maximum Pleistocene glaciation; naturally, the greater part of this ice cover occurred in the Northern Hemisphere, where most of the land area is situated (Flint 1957). The Pleistocene is usually estimated to have lasted about one million years but recent, more accurate dating methods seem to indicate that the glaciation began much earlier. Within this period, four major glacials occurred, each of which was separated by a warm interglacial of several tens of thousands of years' duration. Each glacial consisted of alternating advances and retreats; this large-scale cyclicity of climate characterizes the Pleistocene. We should add that as the climatic belts shifted southward during gla-

Climatic Changes and Their Rates 319

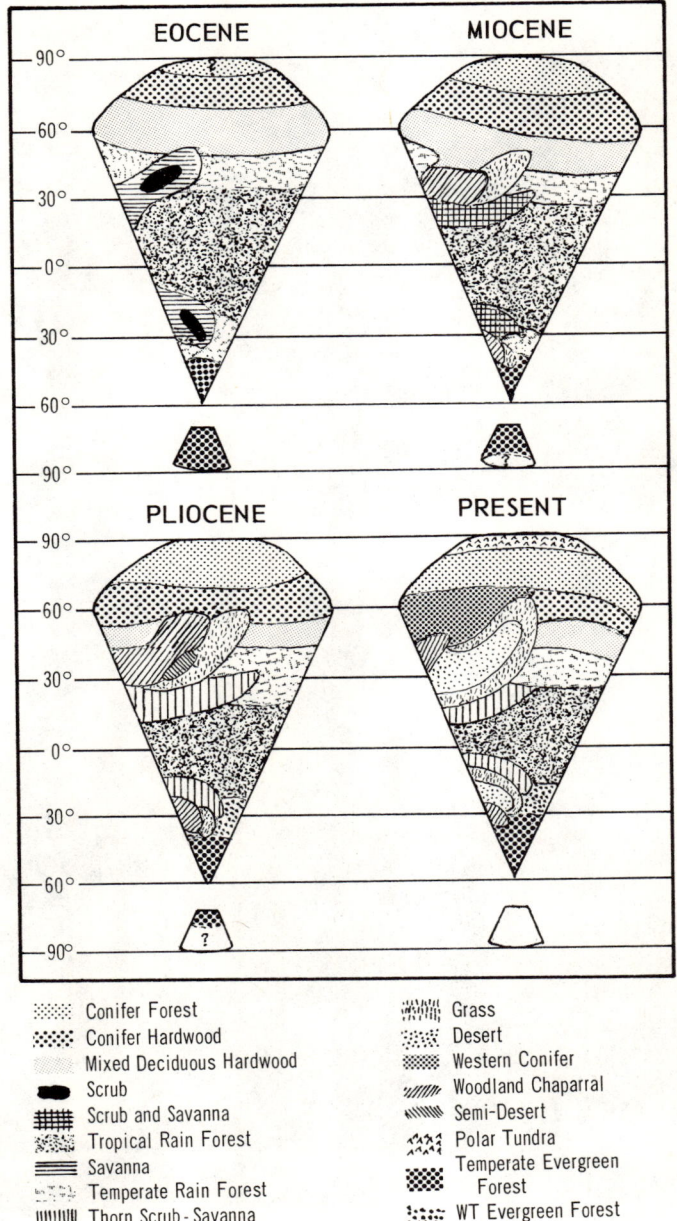

Figure 6-5 Changing patterns of some generalized Tertiary environments. After Axelrod 1952, modified by D. Axelrod (*in litt.*) 1967.

Figure 6-6 Changes of a particular environment during the late Tertiary.
(A) View of the present vegetation (desert scrub) at Haiwee Dam, Owens Valley, on the east slope of the Sierra Nevada, at an elevation of 1,340 m. An arrow marks the site where Axelrod and Ting (1960) uncovered a fossil flora from the late Pliocene. These paleobotanists conclude that at that time the elevation of the area was about 760 m, with valleys running westward and moist coastal air and ample precipitation creating a vegetation reminiscent of the Sierran pine-fir forest. Such forest (B) grows at present on the west flank or the range, much to the northward. (*Courtesy of Daniel Axelrod*)

ciations, more or less simultaneous *pluvials*—rainy, wet climatic belts—have prevailed on many of those areas of the middle latitudes which now lie in the warm-temperate zone but now have very arid and extreme climates. These pluvials and the resulting large lakes and forested vegetation areas in southwestern North America are especially well documented (Morrison 1965), as is the pluvial history of the Sahara (Moreau 1963, 1966) and Australia (Gentilli 1949). Indeed, it is a safe generalization that the whole land and water surface of the earth was affected by the climatic changes of the Pleistocene (Figures 6–7, 6–8, 6–9, 6–10, 6–11).

The most recent glaciation (named *Wisconsin* in America and *Würm* in Europe) occurred more than forty thousand years ago, and began its rather sudden retreat some 10,000 years ago. Considering that about 10 percent of the land area is still covered with ice (Antarctica; Greenland; parts of the Alaskan, British Columbian, and New Zealand mountains; and many other, smaller mountain-glacier areas throughout the world), and that about 25 percent is underlain by permafrost (Péwé et al. 1965), we are presently living in an interglacial of uncertain duration. Seen in this perspective, the climatic history of the immediate past is a history of steady improvement, but with advances and recessions. Actually, the greatest millennial oscillation of climate occurred between 6000 and 8000 B.P.,* when it was as warm as, or even warmer than, the warm, dry years of the 1930's. This was the time of "postglacial climatic optimum." These millennia, or parts of them, are also described as "hypisthermal" or "xerotherm" periods.

While the glacials and interglacials of the Pleistocene are documented by various and numerous geological data, the climate of the past few millennia may be deduced from the present extensive knowledge of the alternations of plant formations and also from the presence of animal fossils, the approximate age of most of which can be determined by chemical dating techniques using radioisotopes. Archeology also contributes to our knowledge of climatic fluctuations within recent millennia.

Further to narrow the time scale and also to emphasize the secular fluctuations (Figure 6–12A), may be recalled that several anthropologists have attributed the decay of the Roman empire to a wave of warming which made life in the Central European forest easier and hence the German expansion more possible. However, a more acceptable correlation would seem to be between a warm climatic period and the 900 B.P. spreading of the Vikings. They colonized Iceland and even Greenland, and their sagas described the southwestern Greenland coast as truly

---

*B.P. stands for "before present" and is a term used by geologists and biogeographers.

322   Dynamic Zoogeography

Figure 6-7   An approximate representation of two climatic extremes in North America during the Pleistocene. From Mengel 1964, after Dorf 1959, modified. *(Courtesy of R. M. Mengel and the Cornell Laboratory of Ornithology.)* Climatic zones: *1.* arctic;

Figure 6-7 (Continued)
2. subarctic; 3. temperate; 4. subtropical; 5. tropical. A: Climatic conditions in a period of maximum glaciation; B: during the climatic optimum of an interglacial.

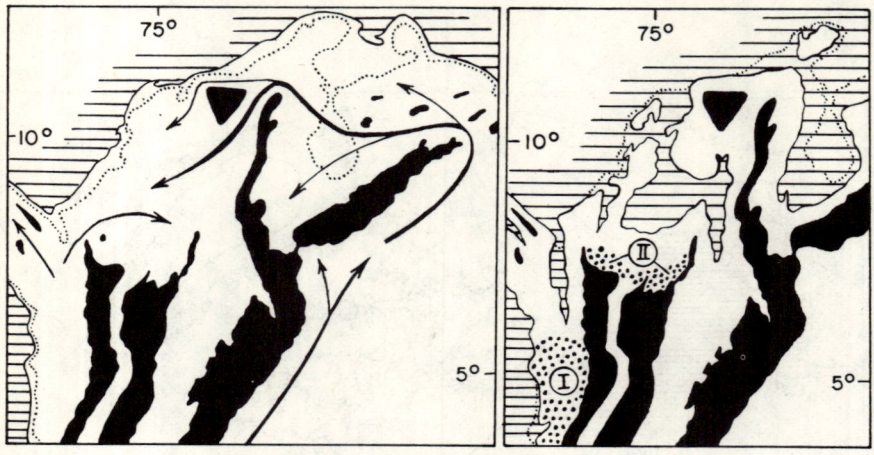

Figure 6-8 Northern Colombia during the Pleistocene.
The present coast line is dotted; mountain areas presently over 1000 m high are in black. *Left:* During glacials. The sea level was lowered by about 100 m. Arrows indicate immigration routes of cis-Andean and trans-Andean forest avifaunas which advanced during this humid climate period. *Right:* During interglacials. The sea level was about 30 to 50 m higher than during the glacials, and the climate was dry in the northern half of the area. Birds of the humid forest found refuges in the Chocó area (I) and Nechí area (II), as witnessed by the present occurrence there of numerous endemic birds, according to Haffer (1967). (*Courtesy of J. Haffer and The American Museum of Natural History*)

green with vegetation. A worsening of the climate in the Fourteenth Century followed, and the Norsemen of Greenland fared as many isolated populations do in a rapidly changing environment (Koch 1945). They could not adapt to arctic life, as did their foes, the slowly advancing Eskimos. Ice conditions prevented supplies and reinforcements from reaching the Vikings from Scandinavia, and so they died out. Permafrost preserved the last unburied bodies, as it had those of the mammoths, thousands of years earlier. During this same period, alpine glaciers overran whole settlements in Central Europe.

The next cold period lasted over 200 years—(from about 1600 to about 1850)—is sometimes called the "little ice age," referring to the increase of polar and montane icecaps and the long, cold, icy European winters. The tales of the Alaskan coastal Indians tell of times in this period when glaciers covered their villages. Napoleon's retreat from Moscow in 1812 made one of those extraordinarily cold winters memorable. But less extreme years followed, and the general rate of improvement of climate recorded both in Europe and North America from approximately 1870 to 1960 was at first slow, then rapid, and then again slow (Figure 6–

Climatic Changes and Their Rates 325

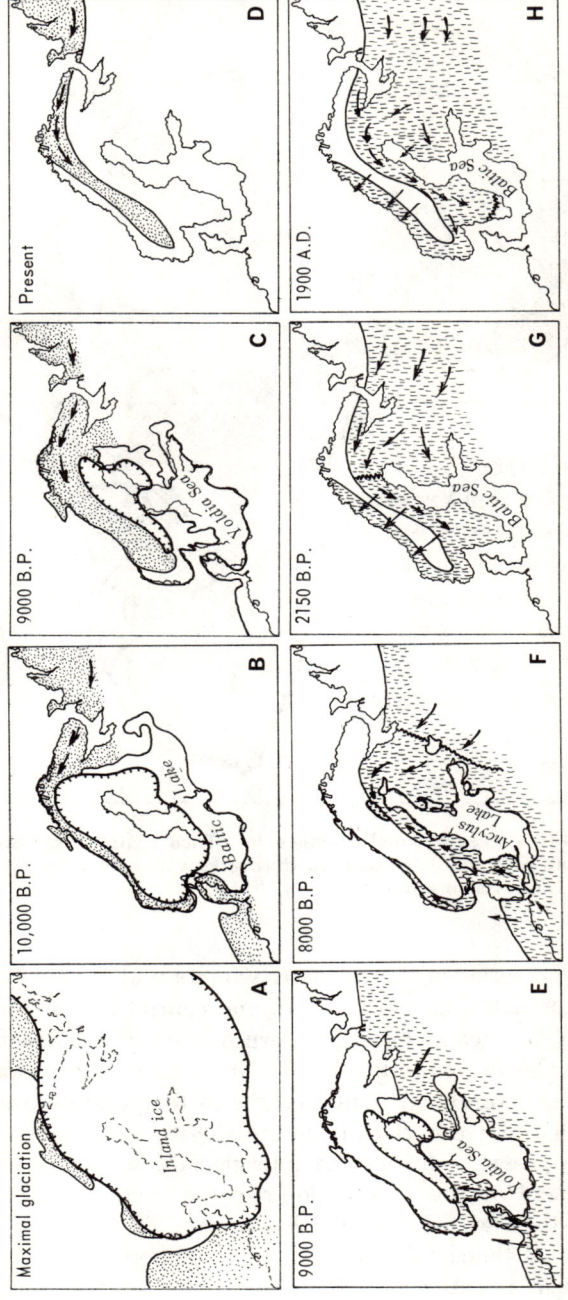

Figure 6-9 Millennial dynamism in northern Europe. (After Örn 1961.) *Above:* Presumed immigrational history of the tundra avifauna; tundra belt shaded. *Below:* Presumed immigrational history of the coniferous forest avifauna; taiga belt shaded. The western limit of spruce indicated by zigzag line on maps *F*, *G*, and *H*.

326 Dynamic Zoogeography

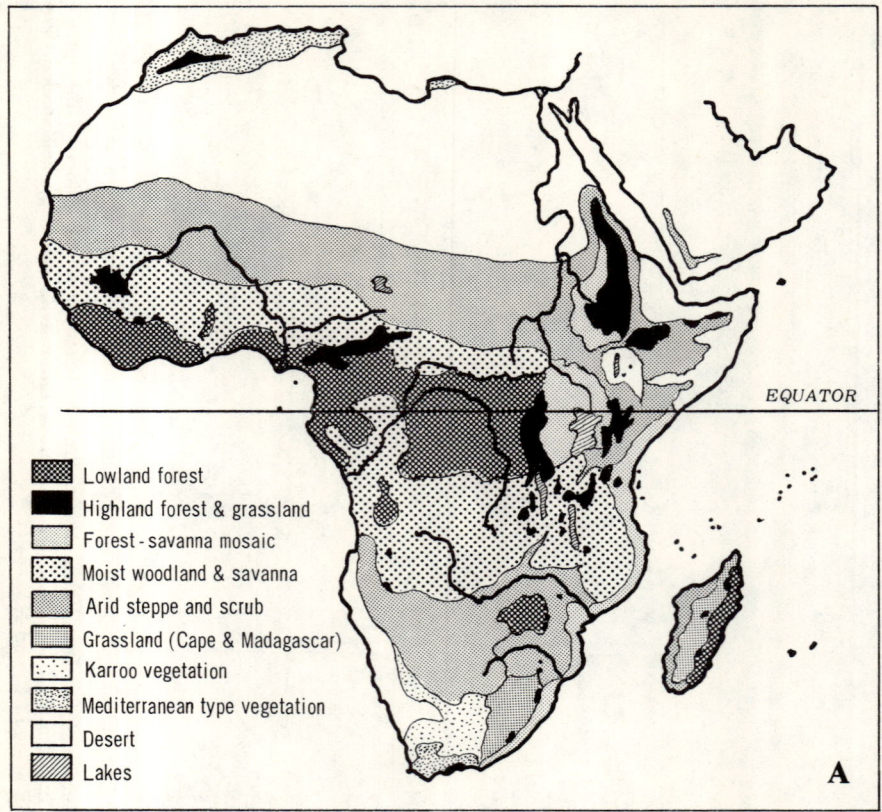

Figure 6-10   Vegetational changes in Africa. After Carcasson 1964. A: Present vegetation; B: probable vegetation during Pluvial maximum; C: probable vegetation during interpluvial maximum.

12B). The series of warm, short winters and relatively cool, rainy summers culminated in northwestern and central Europe in the 1930's, while in North America and southwestern Asia there was drought.

Atmospheric humidity also has its secular fluctuations, but these are not exactly correlated with changes in temperature and are less exactly known. During the glacials of the Pleistocene, the subtropical dry belt was compressed; during the interglacials, it expanded again. It seems that present conditions are closer to the average of the past three interglacials than to the glacials themselves. The increase in aridity in the past twelve thousand years has been studied, especially in North America, Africa, and Australia. Less bioclimatological research has been done

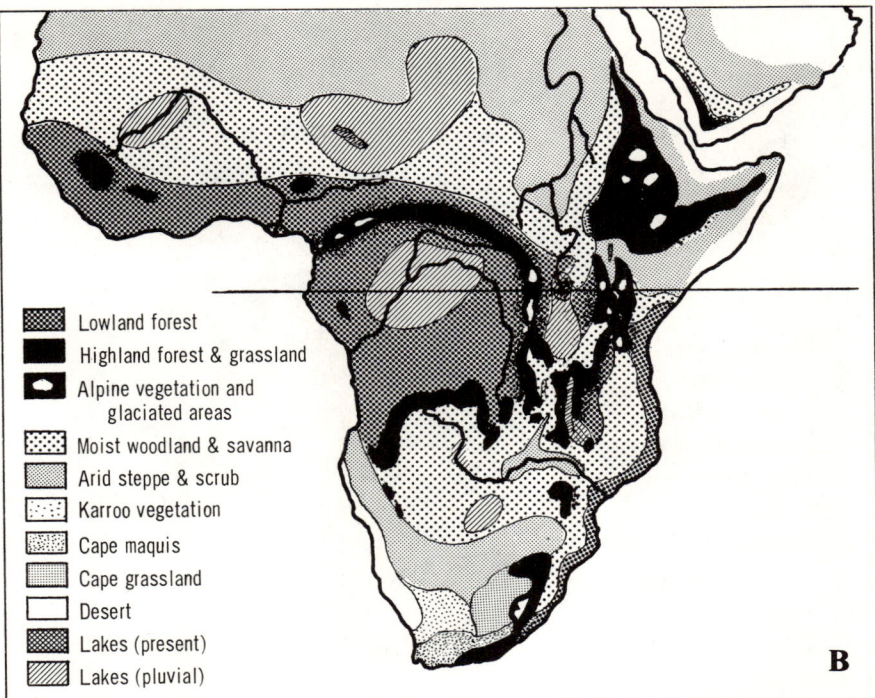

Figure 6–10 continued

in the Asiatic dry belt—especially with respect to faunal implications—but historical documentation there substitutes for biogeographical evidence. The evolution of dry-adapted vegetation (Axelrod 1950) and its expansion in southwestern North America (Martin and Mehringer 1965), together with other directly climatological evidence, lead many to believe that the present long-range trend is toward increasing aridity in the transcontinental dry belts (Hubbs 1960).

Dynamic Changes of the Vegetation

Past vegetational and floral changes are at present better understood than are faunational and faunal changes on land. This is because plants, especially vascular ones, have structural elements—such as pollen, fruit, and seed—that easily fossilize in quantities, and thus the vegetational, or at least the floral, composition of past environments can be reconstructed from their study. An additional advantage of paleobotanists and plant geographers over zoologists is that the present vegetation, at least in the temperate latitudes, is more studied and better understood than are the corresponding faunations. For the more recent past, however, the student

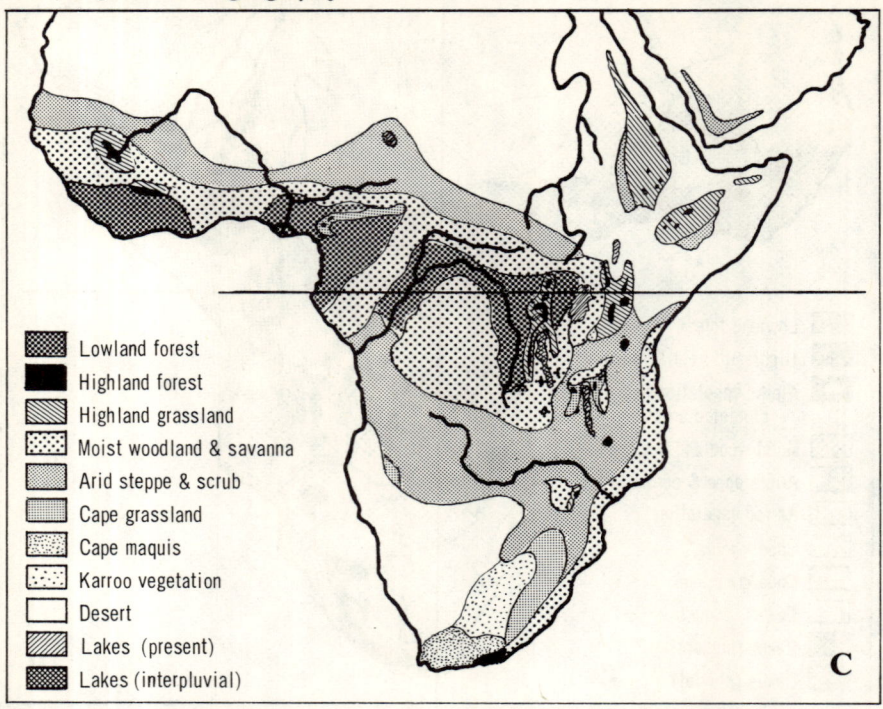

Figure 6–10 continued

of phytogeographical changes is also handicapped. One handicapping circumstance is the influence of mankind upon the natural vegetation. Burning, cultivation, reforestation, and other alternations are as old as civilization. Cultivated plants and their accompanying weeds are carried about by man to an extent comparable with, if not surpassing, the rate of accidental animal introductions. The earlier tribal wanderings and wars, as well as the later building of railroads and highways, promoted the anthropochore expansion of plant areas (*cf.* p. 49).

Plants are dependent for their existence on the soil. Soil conditions therefore influence vegetational changes just as much as do geological, climatic, and anthropogene changes. Soil, in turn, is derived from, or sheltered by, the vegetation. Each plant formation and each community occupies, builds, or maintains its own soil zone corresponding to the prevailing climatic conditions. Plant successional changes—especially on climatically marginal habitats—are affected or even guided by changing soil conditions. Successional alternations of biotic communities are slow, secular processes. For example, a lake fills in and becomes meadow, woodland, then forest. Successional changes, however, take place simultaneously with other environmental oscillations. The climatological trend of the broader area may be from forest to grassland. These coinciding cycles add to the predicament of the plant geographer who wishes to distinguish between these two different types of dynamic change.

Each plant community contains a number of species which occur together because of their history, tolerances, and requirements. Each species, however, is distributed in its own way outside the stands of the particular community; each one reacts in a different way to climatic changes. If these changes mean a deterioration of climatic conditions, the species will retreat from the area below its limits of tolerances. If climatic conditions improve, the species may advance, if other conditions are satisfactory—if suitable ecosystems on suitable soils are able to incorporate the species' pioneering individuals. Analysis of fossil floras (Chaney 1940, Axelrod 1952, Figure 6–5) indicate that major vegetational formations are very old. They respond to long-range climatic changes by spreading or receding due to coadaptation and similar, rather broad, tolerances and requirements of their member plants (see p. 357).

Vegetational movements often have a relatively slow, secular pace, because many plants survive under very inferior conditions, even after they have ceased to reproduce sexually. The dominant plants of many major communities are woody perennials with a long individual life span; in marginal conditions, they reproduce slowly. Climate-induced distributional changes in plants are evident in several recently deglaciated areas of arctic and montane habitat zones. In the Alps, the number of pioneering plant species has increased on certain peaks (Gams 1949, Braun-Blanquet 1955). The recent regression of the earth's valley glaciers has been followed by pioneering forest stands (see e.g., Lawrence [1958] for this phenomenon observed in Alaska). Another example is to be found in Scandinavia, where the tree line is slowly advancing northward. Sirén (1961) summarizes this latter phenomenon, and shows that seed years of the northernmost pine woods occurred eight times between 1910 and 1960, in contrast to the long-term average of only three or four times in a century. In these fifty years, not only has the number of seed-producing years increased but there has also been an increase in seedling survival. As one noticeable effect of the climatic improvements in this century, Erkamo (1952) mentions that since 1910 in southern Finland, the maple *Acer platanoides* L. [Aceraceae] has begun to produce seedlings, the wind-dispersed rush *Typha latifolia* L. [Typhaceae] has spread widely, and other, xerophilous, plants have established themselves in more localized areas. All this results in a gain of only a few meters in a century, but, more significantly, the annual growing season has expanded from an average of 90 days before 1910 to 120 days within the next fifty

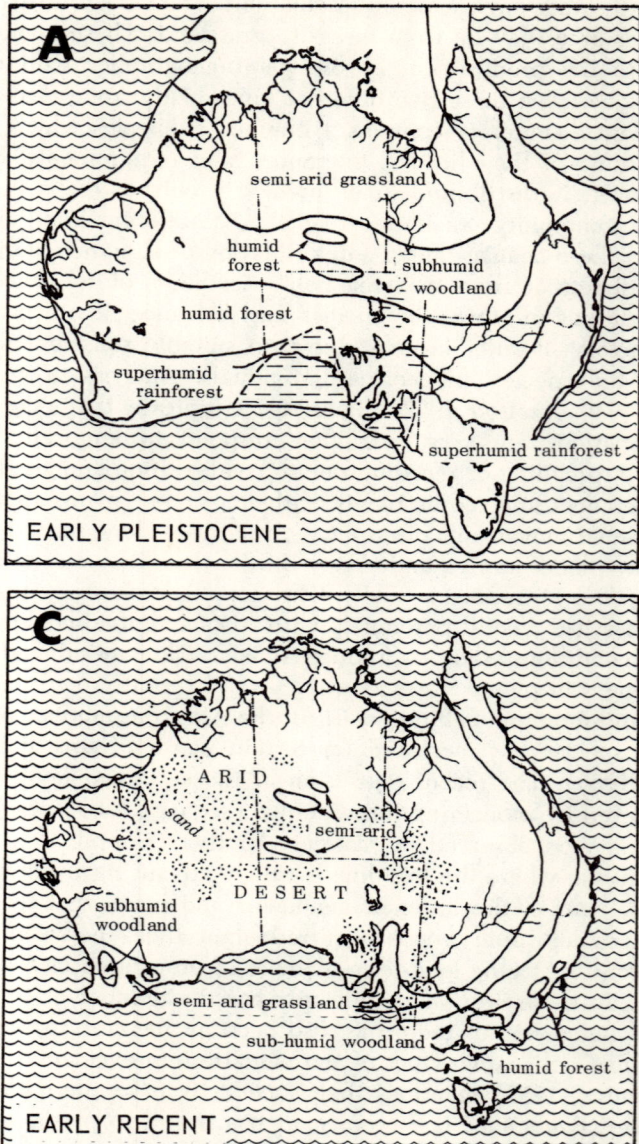

Figure 6-11 Extreme phases of the Australian environment during the past million years. Expansion, shrinking, and splitting of the area and extinction or renewed expansion were the fate of the biota (cf. Figure 6-32). When xerophilic species expanded, the humidiphilic ones were on the retreat, and vice versa. A   During Mindel

Dynamic Changes of the Vegetation 331

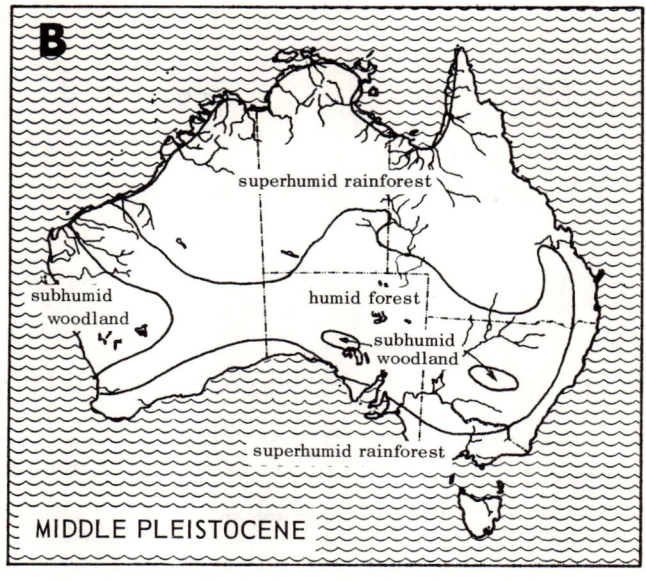

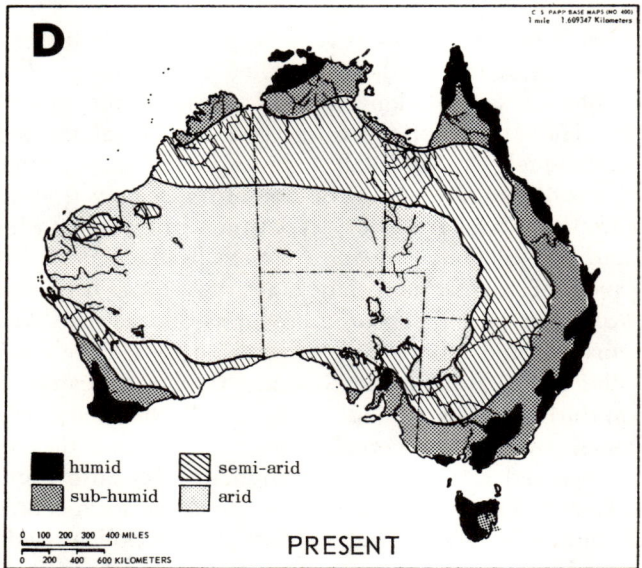

Figure 6-11 (Continued)
glacial (continental shelf part of the land); B hot, humid interglacial (Mindel-Riss); C thermal maximum and aridity of early Recent times; D present. After Gentilli 1949.

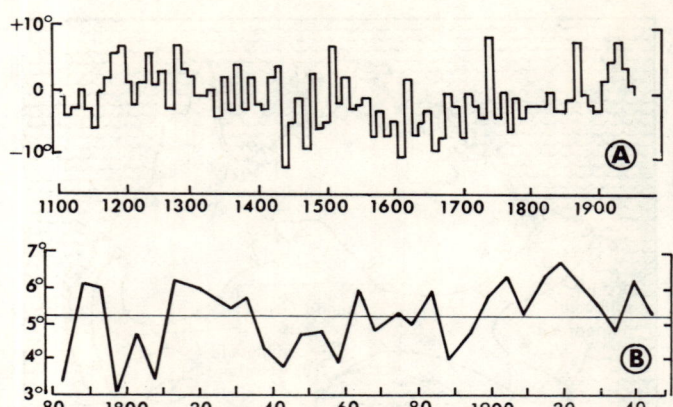

Figure 6-12 Climatic changes during the past millennium.
A The relative severity of winter in N.W. Europe since the twelfth century. After Kalela 1952.
B Monthly mean temperature curve for March in Budapest 1780–1945; the horizontal line is the mean of the whole period. From Keve and Udvardy 1951.

years. The resulting habitat improvement changed the undervegetation of the pine stands at the northern tree line to a more productive type by an extension of several dominant species from the more southerly communities. The canopy of the forest still consists of the same, long-lived pine individuals as it did half a century ago. Antevs (1947) calculated that the southern borders of the boreal forest moved about 400 km in 3,000 years to the vicinity of 61° N latitude in Sweden, which would average about one kilometer advance in each five-year period. However, the actual pace of vegetational movement may have been more rapid, because during this period repeated advances and retreats occurred, as the pollen analysis and other fossil evidence suggest. Berg's (1950) estimate that the permafrost of the Siberian tundra retreated northward about 40 km during the past hundred years indicates a somewhat slower process than advances of the boreal forest but still faster than the rate of glacier advances and retreats in high mountains (*cf.* also Figure 2–22).

The best-known vegetational changes are those of north-central Europe and the northeastern United States. Most of the pollen evidence preserved in post-Wisconsinian lakes and peat bogs has been well studied. At the present time, southwestern North America is also under intense scrutiny. In South America, vegetational changes due to climatic influences have been extensively studied by several Finnish expeditions. According to Auer (1960, 1963), the Andean forest started an advance from

Dynamic Changes of the Vegetation 333

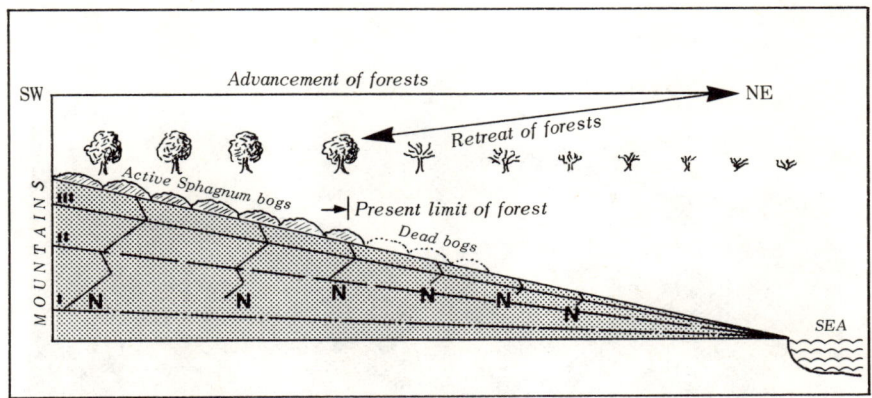

Figure 6-13 Evidences of changing environments in southern South America. Schematic profile of the sphagnum bogs in Tierra del Fuego, to show the vegetational and climatic changes of the postglacial millennia which are based on fossil evidence. (From Auer 1965.) A transect is shown from the Cordilleras to the Atlantic coast on the right; the triangle-shaped transect indicates the thinning of sphagnum peat with three volcanic ash layers (I to III) from the mountains to the sea. The corresponding eruptions occurred about 9000, 4480, and 2240 B.P.; by their aid, the deposits have been timed and then the timings have been confirmed by radiocarbon analysis. N indicates the limit of *Notophagus* (southern beech) pollen, taken as the limit of forests toward the pampas grassland. The present limit of forests and active bogs is shown on the surface of the transect; subfossil remnants of the forest trees can be found in the pampas. The arrows show the postglacial advancement and, thereafter, the retreat of forests which continues at the present time. (See Fig. 6-14)

Pacific coastal refugia towards the plains of eastern Patagonia about nine thousand years ago. When the climate was most humid and rainy, about 2300 B.P., the forest zones almost reached the Atlantic coast. Since that time, drought has set in and the forest has retreated westward (Figures 6–13 and 6–14). During the period of advance, zonal forest communities (each zone requiring progressively less and less moisture) migrated eastward *in an orderly way*. The retreat eliminated these zones, and closed forests were presently replaced directly by a belt of steppes 50 to 100 km wide. E. K. Kalela (1941) attributes the present extinction of these relatively drought-tolerant communities to the prevailing wind, which opposes the spread of their seeds but at the same time kills the old stands of mesic forest, which are then replaced by steppe. Similarly, the postglacial dry, warm period exterminated many relict stands of the spruce-fir forest community in the southern Appalachian Mountains of North America by elevating their lower limits above the height of the mountain peaks. Although the climate there is at present suitable for the growth of

334 Dynamic Zoogeography

Figure 6-14 Parkland in Tierra del Fuego on morain soil. The grassland advances from the left foreground, and surrounds some old stands of the forest which are slowly dying out. (*Courtesy of V. Auer.*)

this type of forest, its place has been taken by other plant communities (Whittaker 1956 and Figure 6–15).

Apart from the rate of change of the vegetational entities, one finds that single plant species are able to expand or contract as rapidly as do most conspicuously mobile mammals, birds, and butterflies. Though most examples pertain to a rapid invasion of newly accessible, virgin areas (Burges [1960] gives a good summary of these), the spectacular spread of adventives (Meusel 1943) and natural immigrations to marginal areas (Good 1936, 1964, Erkamo 1956) are also documented (Figure 6–16).

Dynamic Changes of Animal Distributions

In our discussion of dynamic area limits, we have employed a deductive documentation of distributional changes of animals throughout time. We will now present an inductive discussion of the same phenomenon, dealing first with single animal species, then with faunas, and finally with faunations, i.e., the assemblances of animal constituents of ecosystems. We will start from the present and work our way back along the increasing scale of the system of temporal units.

Single Animal Species The decennial fluctuations of many insect spe-

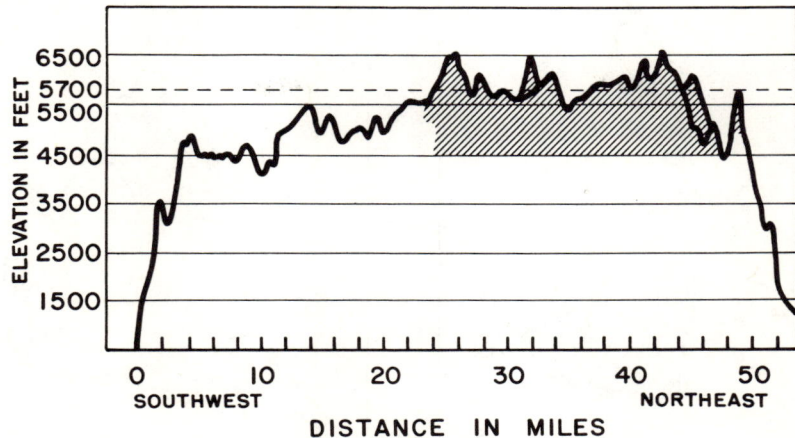

Figure 6-15 Profile of the Great Smoky Mountains of the southeastern United States to illustrate Whittaker's (1956) hypothesis on the southern limit of the Appalachian spruce-fir forests. Diagonal shading indicates the distribution of this community on higher peaks of the northeast half of the mountain range. Climatic warming during the xerothermic period, sufficient to displace the lower limits of spruce-fir forests upward from 4,500 ft. to approximately 5,700 ft. (1,370 m to 1,430 m), would account for the present distribution of this community. According to this hypothesis, the southwestern part of the community became extinct; whatever plants or animals were restricted to this area only (because of their ecologic requirements) must have faced extinction or, perhaps in rare instances, relict occurrence on restricted refugia or amalgamation into another community if they were able to adapt to it. From Livingstone et al. 1966. (*Courtesy of R. H. Whittaker and the Ecological Society of America*)

cies have already been documented. During the severe drought of the years 1931 to 1936, the economically important North American grasshopper *Aulocara elliotti* Thomas [Acrididae, Orthopt.] advanced its distribution area northward from central North Dakota to southern Manitoba (Bird 1937). Between 1947 and 1957, the Macrolepidoptera fauna of Sweden increased by 36 new species. Although a majority of these could have been in the country previously and not been discovered, and although some may have been only casual visitors, Svensson (1957) has presented evidence to indicate that eighteen of these species were newly establishing themselves; five had moved in from the southeast, and eleven from the south. One could cite many more such examples from Central Europe—especially from the more southern latitudes—but Warnecke (1961) rightly points out that since most butterfly species show minor fluctuations of their borders, one should not consider these as true range extensions unless they remain in a new area for at least sixty years. Good

336 Dynamic Zoogeography

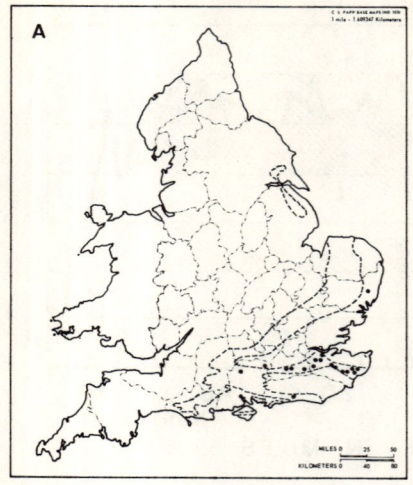

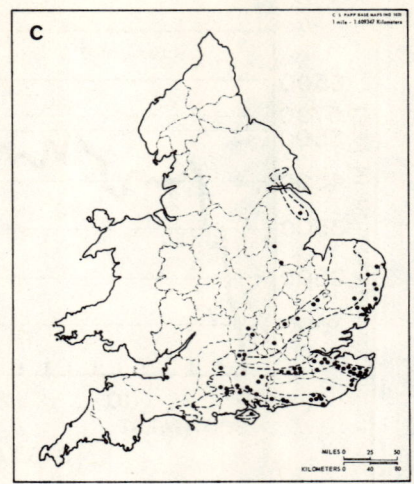

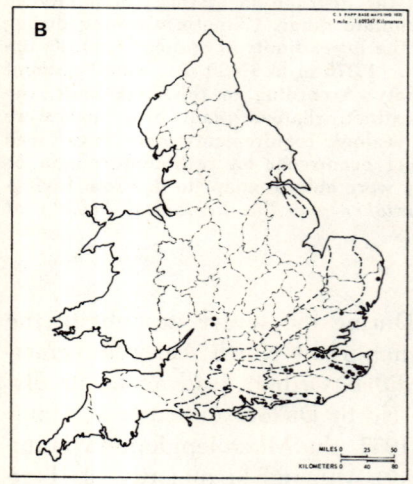

Figure 6-16 Distributional changes of plants and animals. Changes in the distribution of the lizard orchid *Himantoglossum hircinum* (Koch) [Orchidaceae] in England. After Good 1936. A Prior to 1899; B 1900–1919; C 1920–1933.

Changes in the distribution of the Syrian woodpecker *Dendrocopos syriacus* Hempr. & Ehrenb. [Picidae, Aves] in Hungary. After Keve 1960. D First observations of pioneering in this area, 1937–40. E Distributional data 1940–1954. F 1954–1960.

examples of such fluctuations can be observed in Ford's (1957) work on British butterflies. *Aporia crataegi* (L.) [Pieridae, Lepidopt.], which was widely scattered throughout southern England during the Nineteenth Century, has not been found there since the 1920's; but the white admiral *Limenitis camilla* Schiff [Nymphalidae, Lepidopt.] has extended its range throughout the same area in the past few decades. In the scarabaeid beetle fauna of England, according to Johnson (1962), the extinc-

Dynamic Changes of Animal Distributions 337

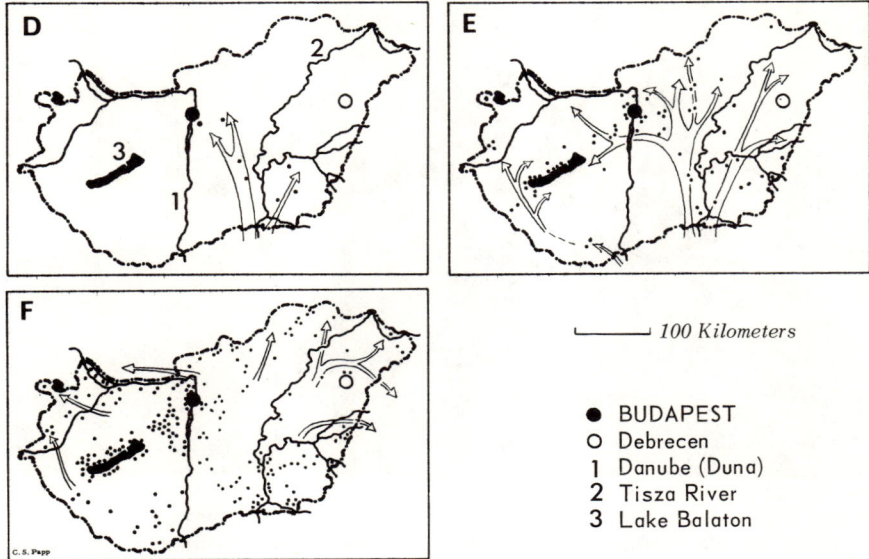

Figure 6–16 continued

tions of the past hundred years (20 species) predominate over the newly observed two species. Many of these beetles were restricted to localities which are now highly disturbed by industrialization or which have been transformed by urban residential developments. Lindroth's (1949) monograph of the carabid beetles of Fennoscandia discusses a dozen species of ground beetles that may temporarily appear in this area but do not establish themselves. Eight or nine species are late arrivals there, true immigrants of the past 30 or 40 years. Sixteen species have considerably increased their distribution area in northern Europe by advancing northward. Nine species greatly decreased and withdrew from this area or died out. Kaisila (1962), in analyzing the expansion of Lepidoptera in Finland during the past hundred years, found that the climax of the range extensions occurred in the 1930's and was apparently correlated with the climatic optimum. In addition to the many southern visitors—which, like the ground beetles in Lindroth's study, appeared oftener during this warm period—Kaisila carefully studied the dynamics of 58 species of butterflies and moths which expanded their regular reproductive populations to Finland. Among many of these butterfly species, increased population on their marginal areas preceded mass exodus and pioneering beyond the previous northern limit. Since not all common species expand, and since in some expansive species (noctuids) poly-

338   Dynamic Zoogeography

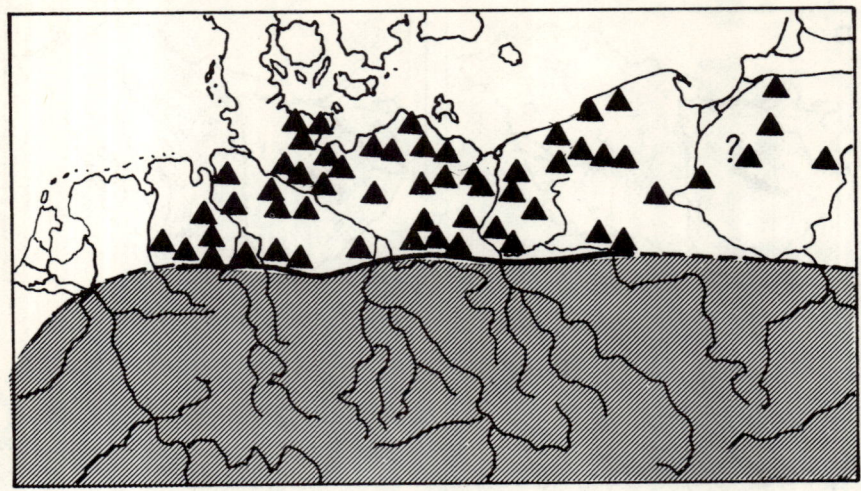

Figure 6-17
Advance of the northern limit of the marbled white butterfly *Melanargia galathea* L. [Satyridae, Lepidopt.] in Central Europe. (After Warnecke 1961.) Hatched area was occupied before 1870; triangles show localities where pioneering colonies were found after 1870.

morphy can be discerned, Kaisila concluded that population genetical characteristics play an important role in expansion. The total Macrolepidoptera fauna of Finland in 1960 consisted of 902 species; about one-fourth of these (approximately 230 species) appeared in Finland between 1860 and 1960, while two species are definitely known to have disappeared from the country during this time (Kaisila 1967, *in litt.*).

Similarly conspicuous expansions and regressions in the same area are known to have been associated with short-term climatic changes. Other expansions (e.g., in northeastern North America) of mammals which live in successional communities and ecotones are clearly coupled with the alteration of habitats caused by civilization. Perhaps best known are the secular and decennial distribution changes of the avifauna, especially when the number of professional and amateur observers is great, as it is in western and northern Europe and in eastern North America. These areas have all been affected by climatic changes as well as by cultivation and urbanization; therefore, the causation of the avifaunal changes was not generally recognized until the 1930's and 1940's.

Outstanding examples are shown in Figures 2–6, 2–20, 4–58, 6–16, 6–17, 6–18, 6–19. Reference should also be made to a few summarizing publications: for European regions, Kalela 1944, 1949, Salomonsen 1948, Keve

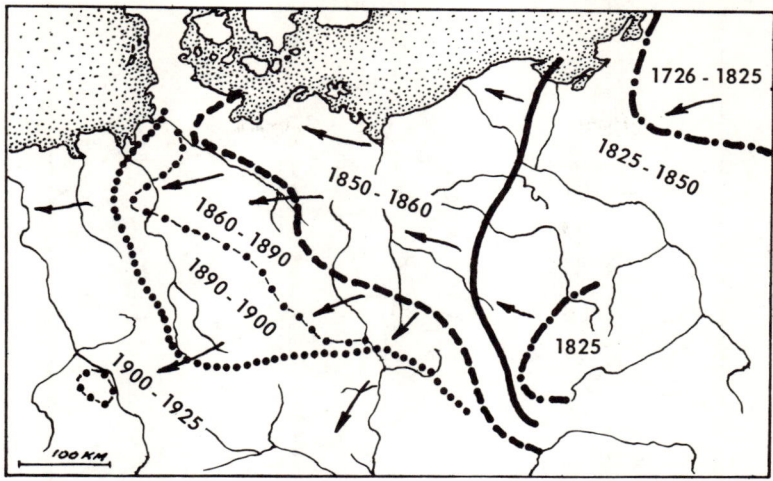

Figure 6-18 Advance in Central Europe of the East European *Senecio vernalis* W. & K. [Compositae] between 1726 and 1926. In addition to its anemochore dispersal this plant is also spread by anthropochore means—it follows railroad lines and is also carried by man in cereal shipments. This type of expansion area is reminiscent of that of the butterfly in Figure 6-17. In such instances, it is difficult to separate the ultimate (habitat improvement due to climatic dynamism?) and proximate (anthropochore carrier?) factors enabling the area expansion. After Meusel 1943.

and Udvardy 1951, Niethammer 1951, Curry-Lindahl 1961; for North American regions, Boyd 1962 and de Vos 1964a.

These examples, and most of the data on range fluctuations come from the temperate and cold-temperate regions of the Northern Hemisphere. Details of animal distributions in the tropical areas are less exactly known, or the knowledge is too recent to be compared with the distribution of several decades ago. Scarcity of observers is another handicapping circumstance. I have searched, with little success, the pertinent literature for South American, African, and Southeast Asian distributions. Faunistic knowledge must become more accurate; only then will it be possible to compare data on advances and retreats. According to R. H. Carcasson (1965, *in litt.*), there are two examples in East Africa which could constitute a genuine extension of range. Both species involved are forest insects; thus human influence is ruled out. *Precis sophia infracta* Rog. [Nymphalidae, Lepidopt.] was not known on the eastern side of the Rift Valley, although it was common in Uganda and western Kenya. Recently it was found to be common east of the Rift, in most of the forest around Nairobi. *Mylothris yulei ertli* Suff. [Pieridae, Lepidopt.] advanced westward about 240 km from its nearest known localities on the eastern scarp of Kenya to Nyasaland and has recently reached Nairobi.

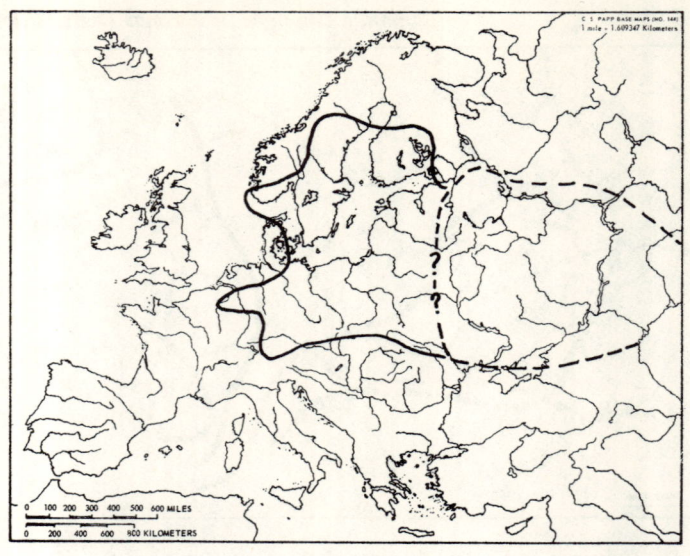

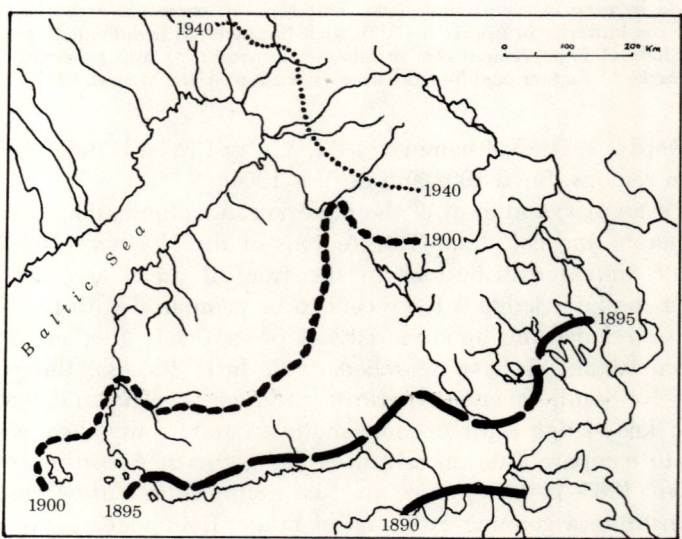

Figure 6-19
A Expansion of the butterfly *Eupithecia sinuosaria* Ev. [*Hydriomenidae, Lepidopt.*], an Eurosibirian faunal element, to north central Europe. From Warnecke 1961.
— — — Approximative distributional limit about 1890.
——— About 1960.
Near its western limit—for example, in Denmark, Bavaria, and Austria—only scattered, single occurrences are known.

B Details of the expansion of the same species from Finland. (From Kaisila 1962.) It colonized the cultivated, agricultural areas of Finland with a remarkable speed, and became common at many places.

Dynamism of Faunas The ecofaunal affinities of Holarctic birds are well known, and this group serves as our best example of the concept of faunal dynamism. Every local fauna at any one time has a *dynamic potential*—that is, a certain proportion of its members are actually or potentially expansive or regressive. The magnitude of the dynamic potential of a fauna depends on the geographic and ecologic position of its members.

We shall consider the avifauna of a well-delimited geographic entity in a continental area—that of the Carpathian Basin in Central Europe. Data on the distribution and abundance of the 218 species existing there are, for the most part, limited to the period between 1860 and 1960. About 38 percent of this avifauna—83 species—have their distributional limit across, or on the margin of, the Carpathian Basin. It is upon this *marginal element* of the avifauna that we will focus our attention. Several of these species have been inconspicuous, neither well known nor studied, especially not during the past century. Yet at least 37 species—43 percent of the total of marginal species—have been observed to exhibit dynamism of their limits, i.e., they were either advancing or retreating during this hundred-year period. If we scrutinize the list of these species, we find some examples of species of sporadic occurrence which would breed at a few localities for a few years, only to disappear again for decades. Such instances remind us of the irregular visitors among the Finnish butterflies discussed above. Some of the larger water birds and birds of prey succumbed to intensified habitat alteration or direct persecution by man. In the majority of cases, however, correlation with climatic changes can be assumed (Keve and Udvardy 1951). Such is true, for example, of the five species which advanced their northern limit in this area and for which the increasing spring temperatures may have meant prolonged reproductive seasons. In the opposite direction, cooler and wetter midsummer months seem to be correlated with the southward advance of seven northern species.

As the data in Table 13 show, the mobile element in the avifauna of the Carpathian Basin consists mainly of those species which have their northern, southern—even western—but not eastern limits within this area. The southward advance of northern elements is simultaneous with their expansion toward the Atlantic coastal areas of Europe, and is correlated with the cooler and wetter summers of Central Europe (the so-called maritimization of the summer climate). The northward advance of a more numerous species group of southern affinities clearly correlated the longer "growing" season with milder and shorter winters.

Scandinavian—especially Swedish and Finnish—scientists pioneered in correlating faunistic changes with climatic fluctuations. For example, according to Linnaeus' *Fauna Suecica* (1745), the nightingale *Luscinia*

Table 13 Analysis of the intrinsic dynamic potential of the avifauna in the Carpathian Basin 1860–1960.

Species with distribution limit on or near this region compose the *Intrinsic Dynamic Potential* of the fauna. Species which moved their limit during the time period considered represent the *Realized Dynamic Potential;* species which did not move their area limit to an observable degree form the *Unrealized Dynamic Potential.* N, W, S, and E mean that the species has its northern, etc. limit in this region; consequently its distribution is southern, etc. in relation to our region. (From Udvardy 1969a.)

Distribution limit in the region	With-drew	Receding	Sporadic breeder	Newly arrived	Total realized dynamic potential	Total unrealized dynamic potential	Total intrinsic dynamic potential
N	3	2	1	4	10	13	23
W	3	7	2	4	11	22	33
S	1	1	3	5	10	10	20
E	–	1	–	–	1	1	2
	7	11	6	13	37	46	83

luscinia (L.) [Muscicapidae, Aves] nested around Stockholm and Uppsala, in central Sweden, in the middle of the eighteenth century. Jägerskiöld (1919) found that this bird had retreated southward during the century after Linnaeus, and that by the 1850's its Scandinavian distribution had become restricted to southern Sweden, as had that of the roe deer discussed earlier (Figure 2–20). During the past hundred years, however, this nightingale has expanded its area again to the latitudes of Stockholm and Uppsala. Jägerskiöld correlates these changes with fluctuations of the thermoclimate; Lönnberg (1924) also considers the effects of drought on the displacement of waterfowl from the arid interior of Eurasia.

In the Finnish faunistic literature, Siivonen (1943) found mention of eight new bird species up to 1880 and 44 new species appearing between then and 1940. Among the first eight species, four were of northern, arctic derivation, while the other four had advanced from the southeast. During the latter period—that of climatic amelioration—only six arctic and Siberian species pioneered in Finland; the majority of the newcomers were southeastern and southern faunal elements. Kalela's (1952) synthesis points out that, in addition to basically new pioneering attempts, there were several other evidences of changes in the nature of the bird and mammal fauna. These may be summarized as follows. Species of mammals and nonmigratory birds spread northward and northeastward, apparently benefiting from the mild winters. Migratory birds were able to spread because of the higher spring and early summer temperatures (cf. p. 84). Aquatic birds of the southeast European steppe pioneered into the area, presumably because of drought in their former distribution area. Many northern species declined at their southern limits. Many previously migratory birds began to winter in Finland or at least to arrive there substantially earlier than before. There was an increase in the number of reported accidental species (potential long-range pioneers).* Kalela's most significant conclusion is: "In some cases it can be directly proved that the recent changes in the fauna are a link in an alternating advance and withdrawal of the limits" (p. 51). As an example, the presence of the polecat *Mustela putorius* L. [Mustelidae, Mammalia] was documented in Finland between 1539 and 1579, although it is known to have disappeared sometime between the end of the sixteenth and the latter half of the eighteenth century; since then, it has again immigrated, to approximately the same areas it had occupied four hundred years earlier.

We have expanded on certain avian and mammalian examples because these are the best documented. The general conclusion from floral and faunal analyses is that decennial fluctuations of area limits occur regularly. These fluctuations are most marked and observable in species with good dispersal capacities (flying beetles, butterflies, birds, plant species with good seed dispersal, and larger mammals), especially those which are generally studied because of their economic importance. These species comprise a significant part of the total fauna in the areas studied. Although changes in composition of the fauna during an eighty-year period are considerable, their impact on the life of the biotic community has not been studied or assessed.** Many of these changes are fluctuations with decennial or secular rhythmicity.

*At this point, one must be cautious. The number of observers, being on the lookout for accidental "rarities," also increased greatly in Scandinavia and perhaps account for the increased number of observations (von Haartman 1967, *in litt.*).

**The long-term study of Wytham Hill, near Oxford, England, is expected to yield additional pertinent data to those already published by Elton (1966).

Although the best examples of faunal and floral changes are found in areas deglaciated as recently as nine thousand years ago, the corresponding fluctuations noted in Central Europe and scattered data from elsewhere permit at least the hypothesis that all temperate faunas are dynamic. There is room for much future research in this particular field.

There is no evidence of secular and millennial oscillations based on direct observations except for such historic documentation as the representation of animals in the decorations of ancient buildings and in the cave paintings left by prehistoric man. Undocumented interpretation of such evidence often diminishes its credibility. Subfossil and fossil remnants of certain elements of the fauna are studied by the paleontologist; their geographic and faunistic interpretation is the task of the paleobiogeographer. Deevey (1949) scrutinized all available geological, climatological, and biogeographical evidence pertaining to Europe and North America in the Pleistocene. The emphasis is on vegetational changes, as might be expected in view of the state of knowledge two decades ago. Little was then known of the millennial and secular faunal dynamism of the late Pleistocene. Of the rich literature that followed Deevey's review, the work of Guilday and his associates (1964) deserves special mention (Plate 3). This study demonstrates that a single rich fossil source may throw light upon, and provide evidence for, the movement in time of a whole ecofauna. A recent symposium on the Pleistocene of North America (Wright and Frey 1965) will bring the reader farther up to date on the faunal history of this continent.

Area Analysis.—Today's zoogeographer rounds out evidence of the paleoecologist and the evolutionary paleontologist with inferences drawn from area study, taxonomy, and the ecological relations of the species or faunas as they now exist. This section will elaborate upon the working methods of the vertebrate zoogeographer, because he is in a much more favorable position than the student of land invertebrates with respect to the data of present distributions.

The climate of a particular region depends primarily upon its geographic position, but it is also affected by other factors, including relationship to surrounding land masses and water bodies and relative elevation (relief). Therefore, secular and millennial climatic changes do not appear to be correlated in all regions of the earth. This conclusion is reached through inductive reasoning and is based on contemporary observation. Some parts of the earth are rising, while others are sinking (Figure 6–4). The mean temperature has recently been rising in many areas, is unaffected in others, and has been falling in still others. The Hawaiian Islands are an example of a volcanic island chain, in which some islands are aged and rapidly eroding away, while others, containing active volcanoes, are building up. This diversity, at any one time, of dynamic factors in the physical environment, influences changes in the biotic environment as well. It is a chain reaction—not all plants will react to changes identically, but their reactions will influence the changes in animals through the chain reactions of an ecosystem. The net result will be a confusing array of distribution areas in which, as we have seen earlier (Chapter 5), an order can be found only with painstaking analytic methods. From the analysis of the present distributional and ecological patterns, we may infer that several local discrepancies, inconsistent with past climatic oscillations, may be explained by a differential local manifestation of a global phenomenon. One finds, for example, that while the drought years of the 1930's critically affected the central and southwestern portions of the United States and the dry belt of western Asia, the summer precipitation increased during the same period in western and northwestern Europe. The paleoecologist, tracing similar oscillations on the millennial level, finds that the postglacial climatic "optimum" of eastern North America and Western Europe did not seem to have a clear counterpart in the arid southwestern United States (Martin and Mehringer 1965).

Consistently emphasized throughout this book have been the speed with which certain animals are able to spread and the speed with which they have actually advanced or retreated during recent times. Although we have no reason to doubt that such distributional changes have occurred in the immediate past, few zoogeographers realized the great mobility of faunal elements until only a few years ago.* The explanation is that zoogeography was (and still is) pursued by specialists rather than by synthesists. When a scientist of the 1930's or 1940's spoke of glacial or even preglacial relicts, he often did not consciously correlate the relict occurrence with any one of the four major glaciations of the Pleistocene. Neither did he consider that his relicts may have occurred because during interglacials, or even during the post-Wisconsin thermal optimum, thermophilous and southern faunations might have spread to, and remained on, his study area. The taxonomist-zoogeographer was handicapped by a lack of knowledge of the overall importance of geographic speciation and evolutionary rates.

The significance of the geographical or chorological aspects of taxonomic species analysis emerged only in the late 1940's. Even today, many zoogeographical works disregard the geography of microevolution. On the other hand, many philogenetic approaches lack clear geographic as-

---

*The important exception of the Scandinavian biogeographers has been discussed earlier (p. 341).

pects. In these circumstances, L. Croizat (1966) is justified to advance, expand, and reiterate the thesis that evolution, time, and geographic dispersal are most closely correlated phenomena—that they are, indeed, different aspects of the same phenomenon.

According to the evolutionary thesis, animal species arise through the mechanism of genetic variation and environmental selection. If the environment does change beyond the given fitness of the species (as most environments do in the course of time), any species population may follow one of three alternatives. It may not be able to adapt, and so becomes extinct. It may, if the geography of the environment allows, move gradually, following the environmental changes, and survive elsewhere. It may react by adaptational changes, thereby evolving into a different form. In exceptional circumstances, the environment may not change and the species may survive unaltered through geological time periods. If environment changes on only a part of the geographic area of the species, it will respond, if it is flexible enough, with corresponding adaptations; these differences will occur in parts of the species area, and will usually be gradual and of a clinal nature. If the affected part of the species population occupies a part area which is discontinuous with the rest of the distribution area, the adaptational changes exhibited by the isolated population will be sufficiently abrupt to call them taxonomically subspecific or even specific differences. The emphasis in this geographic version of the evolutionary theory is on the *extrinsic* environmental influence, although such *intrinsic* factors as mutation rate, recombination potential, available gene pool, population size, and dispersability are equally important.

Let us now pursue the situation which the analytic zoogeographer faces in scrutinizing distribution areas, using the results of this scheme and relying on the taxonomist's treatment. The basis of geographic analysis will be the distribution area of the species. It has been established earlier (Chapter 4) that the extent and geographic configuration of an area depend, *ecologically*, on the breadth of the species' ecological requirements or adaptations, on the extent of the available habitat, and, further, on the mobility of the species. Our immediate task is to discover the importance of these dynamic factors.

The history of the species will be reflected in its taxonomic structure. Monotypic species are either old forms on the retreat—often with a relict distribution as judged from evidence of former, wider occurrence—or new, recently isolated, and speciated forms, of which evidence may be provided by the presence on or near the distribution area of the parental or sibling stock from which they became isolated. Polytypic species might reflect either widespread but philopatric forms or diversity of the habitat with recent or immediately past isolation. In either instance, geographic discontinuity of the present distribution area means isolation by a barrier or a fairly recent crossing of the barrier. In any event, the *isolates* on discontinuous part-areas are prospective new species, since each isolate is under different environmental influence and since each population isolate is also likely to be different in its genetic structure. It is true, as asserted (Sibley 1957, Mayr 1963), that most of such isolates (the *peripheral isolates* of the evolutionary taxonomist) may die out or, through the dynamism of the barriers and available habitats, may find itself reunited and reattached to the main distribution area of the species. However, microevolutionary divergence sets in, as it were, immediately after completed isolation; therefore, we may consider all isolates potential subjects of perfected speciation.

The North American avifauna is a suitable example. It is taxonomically well explored and distributionally well known, except at its southern limit in Mexico. Among 521 land bird species of North America (north of the Mexican border), approximately one-half—269—are *monotypic* species, without an existing, closely related "vicarious" species to match it—that is, they do not belong to a "superspecies." Of the remaining half of the avifauna, 125 species are subspeciated, but on a continuous distribution area. This quarter of the total bird species, though it has the indicia of incipient speciation, is not likely to become considerably differentiated without further geographic isolation. The fourth quarter of the avifauna either has a geographically discontinuous area or already exhibits different stages of recently completed speciation. This last area type—offering the most tangible evidence of dynamism in the immediate past—includes 56 species—nearly 10 percent of the total avifauna considered. This last quarter of the avifauna may be termed the *dynamic element*. While it is true that these 56 species are truly speciated, it is likely that many of them have passed into a static stage, and that just as many other species will be affected by ecologic changes and will respond dynamically with area movements.*

---

*Because I am not a taxonomist, I have been conservative in assigning species to recently separated pairs or triads—the number given is the minimum number. It may therefore be significant that the taxonomist Keast's broader analysis of the Australian avifauna (1961) presents comparable and still more pronounced results. Of 425 species of Australian birds, only 45 percent are monotypic. Of the polytypic majority, less than half (23 percent of the total) show clinal variation only, and 136 species—32 percent of the total—consist of geographically discontinuous and thus geographically isolated subspecies, which are subject to speciation or to the other alternatives of dynamic changes. K. H. Voous likewise analyzed an avifauna. According to his data (as related by Sylvester-Bradley 1963), "some indication of post-Tertiary species-formation can be detected in 43 percent of the birds at present breeding in Europe."

Table 14 Geographic and taxonomic indications of present or recent speciation in the western North American herpetofauna (north of Mexico).

	Amphibians		Reptiles	
	No.	%	No.	%
Total No. of Species	61	100	142	100
Geographic Disjunction	30	49	54	38
Subspeciation	17	28	67	47
Disjunction and Subspeciation	12	20	35	26

It is often claimed that amphibians, insects, and other, older phylogenetic groups differ in dynamism from the more recent—geologically—birds, especially the passerine birds, which are only now at the height of their adaptive radiation. As far as geographic analysis can demonstrate, no major difference is manifest in the distributional area types and species structure of the samples thus far examined (Table 14).

Thus, in three continental-subcontinental avifaunas and a subcontinental herpetofauna, from 25 to 43 percent of the species displayed geographic and taxonomic indications of recent or present dynamic changes culminating in speciation, areal movement, or extinction. This analysis could lead to still more valuable and thought-provoking results if other, *invertebrate* taxa, with different evolutionary, adaptational, ecological, and historical backgrounds would also be scrutinized and the results of such future surveys coordinated. The morphological difference, combined with the discontinuity of geographic area displayed by a considerable proportion of the present vertebrate fauna and also, presumably, by the insect faunas, indicates the dynamism of the fauna in general. Discontinuities in which the part areas are inhabited by morphologically identical populations would—other things being equal—be very recent. Those discontinuities which separate subspecifically different populations are older; again, the barriers separating members of a superspecies are older. The timing of these separations is most often possible only by inference, using collateral evidence of a taxonomic or faunistic nature.

There are an increasing number of taxonomical studies, mainly of vertebrates, in which sound zoogeographical inferences are drawn from the analysis of species areas and populations. A good example is P. and H. Smith's 1962 paper on the lined snake *Tropidoclonion lineatum* (Hallowell) and the western hognosed snake *Heterodon nasicus* Baird and Girard, both of the family Colubridae. The lined snake occurs in the United States on the low plains from South Dakota to south-central Texas; the other species is more widespread in the same general area. Both snakes also have disjunct populations in Illinois and eastern Missouri. The main population of these snakes exhibits north-south clinal differences in several morphological characteristics; hence these different populations were gives subspecific names and status. P. and H. Smith studied these disjunct populations and found that in both instances they agree with the Oklahoma populations of snakes rather than, as one would expect, with their nearest neighbors at the same latitude in Nebraska. The inference is thus made that these disjunct areas are inhabited by relict populations dating from the time of the postglacial thermal optimum, when arid conditions were much more widespread, when the prairie extended eastward, and when the total area of both species was probably displaced northward and more widespread. This distributional evidence is well in line with the earlier assumption that isotherms and animal populations were transposed from 160 to 480 km north of their present positions during the postglacial thermal optimum period.

Often geographic discontinuities are not extant, but steep clines and morphologic discontinuities reveal that such discontinuities had occurred in the past, and that the once separated populations had reinvaded the intervening area of the former barrier. Stresemann's studies of European bird subspeciation (1919a and b) were the first to note that these forms recently spread and met from Pleistocene refuges. Among others (Rand 1948), there is a very well-studied example from North America, the flickers (*Colaptes*, Picidae, Aves) which Dillon (1956) outlined and Short (1965) analyzed in great detail with respect to the genetic consequences of their contact following Pleistocene isolations (Figure 6–20).

Extinction and Zoogeography.—Having considered certain zoogeographic aspects of evolution, we must now examine a negative result—extinction—of the same dynamic phenomenon. No definition is necessary; the paleontological record unequivocally shows that extinction has parallelled evolution throughout the entire history of the animal kingdom. It is likewise a postulate of the evolutionary theory—since this is a gradual and orderly phenomenon, and since evolutionary discontinuities as well as geographic discontinuities exist between ancestors and progenies (the

350 Dynamic Zoogeography

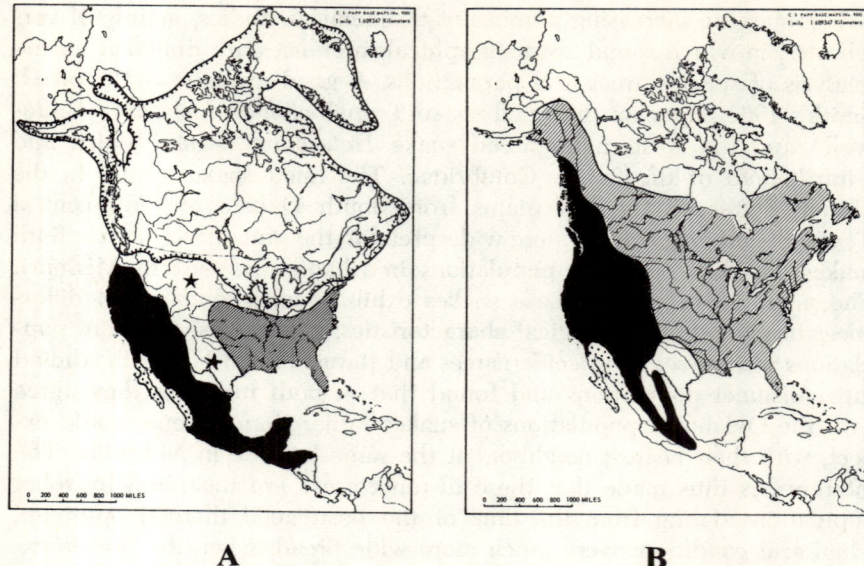

Figure 6-20 Former discontinuities revealed by taxonomic study.
Distribution of flickers *Colaptes auratus* (L.) [Picidae, Aves] in North America. Hatched area: distribution of the *auratus* group; black area: *cafer* group; dotted area: *chrysoïdes* group.

A Hypothetical distribution at the height of the Wisconsin glaciation. White area marked with asterisks denotes the mountain barrier isolating the *cafer* and *auratus* groups from one another. After Dillon 1956.

B Present distribution, After Short 1965. *Cafer* and *auratus* now have no isolating barriers, and hybridize widely in a zone 350 to 800 km wide along the dividing line shown on this map; heavy introgression of characters occurs in a still much wider belt of the areas of both subspecies. *Chrysoïdes* is better isolated by deserts and other, unsuitable habitats, and shows less hybridization.

historical argument again) as well as between members of any taxon—that in the past these were connected by now extinct forms and distributions. Extinction has a large, though scattered, literature in evolutionary, zoogeographical, and paleontological studies. To clarify the relevant concepts, some initial decisions must be made.

Extinction can be viewed with two objectives in mind. Extinction of many animal *species* has occurred in the past; such processes may be going on among many species at the present time; and much extinction will be accomplished in the future. This is the concept of *singular extinction*. One may treat singular extinctions in a synthetic way if there is reason to surmise that these have some underlying factor in common—

that is, if they are due to the same extrinsic or intrinsic causes. In discussions of past extinctions, the method of tackling the problem *eo ipso* becomes synthetic, because this is the chief working method of historical surveys in biology. The concept of *group extinction* concerns the biotic communities or the extinction of taxa—that is, of evolutionary entities—admitting that a common cause (or set of causes) initiated the group extinctions throughout a considerable span of geological time.* For many readers the two phenomena (singular and group extinctions) may merge and do not become easily separable because the terminology of expression is identical. Yet a fundamental difference arises: when the paleontologist speaks of extinction of the dinosaurs—meaning a process involving several millions of years in time and all members of several reptilian orders; when the zoogeographer of the immediate past speaks of extinction of the camels in North America, which took place within the past 20,000 years; and when the faunist speaks of the extinction of Steller's sea cow in 1768.

(A) *Singular extinctions.*—We can treat here only the contemporaneous or nearly contemporaneous examples of extinction in our endeavor to arrive at some general principles, because in any discussion of the historic (geologically historic) past we immediately turn to the paleontologist's comparative and generalizing working method. We have already given a zoogeographical interpretation of the phenomenon of extinction when pointing out that distribution areas may shrink or expand and that if they shrink to nothing, the extinction of occupants is a natural corollary (Figure 4–31). This *geographic aspect* of extinction can be analyzed, following the method of the environmental, organismic biologist, along two lines. The extinction of a species population throughout its entire range will occur when either the rate of its adaptation to changing environments (evolutionary adaptations) or the rate of its dispersal (whereby the species could transpose its populations to other localities still suitable for its existence) is slower than the rate of environmental changes. These are the *ultimate causes* of extinction. If the rate of evolutionary adaptation is greater than, or equal to, the rate of environmental change, the species has a chance for survival. This, however, is only a *chance*, for constellations of the environmental conditions may be such as accidentally to block the "escape route" of the species, which may thus *become extinct by chance alone*.

Among the proximate causes of extinction may be included any or all

---

*A third kind of extinction occurs when a form disappears from the fossil record because of its evolutionary transformation into a different, later progeny (Rensch 1959). We may call this kind *pseudoextinction*.

of those environmental factors which influence continued existence. Some of the proximate reasons for extinction invoked in actual—or sometimes merely postulated—cases are: the collapse of a biotic community when a key link of the food chain is removed; the arrival by barrier crossing of a new member of the faunation which is directly detrimental to the species in question (disease vectors, predators, parasites, competitors, and other interfering organisms); the climatic change from which no escape is possible because of insurmountable barriers (Figure 6–15).

Among the intrinsic, ultimate factors, such general evolutionary changes have been proposed which are likely to have occurred, especially in their environmental interrelationships, and having caused, in the long run, extinction. Thus Reinig (1938) has suggested "decrease in ecological valency"; Rensch (1959), "overspecialisation." It is also easy to speculate on the effects of population dynamics, such as too low a population density, preventing the meeting of the sexes necessary for reproduction (Kurtén 1958) or preventing the forming of viable social units, where such is a prerequisite of existence. Reduced mutation rate has been suggested, at least for species confined to near-doomed refugia (Hultén 1937). This argument is none other than the above-mentioned specialization principle. A species which has given up the trial-and-error method of evolutionary experimentation may have done so by restricting its variability, owing to, for example, constant counterselection during unique and extreme conditions which permitted only one method of survival; or it could have done so by evolving some major adaptive changes which fitted it for that particular situation in which it proved successful—that is, by specialization—which made versatile mutations prohibitive. Among vertebrates, such a specialization could be, for example, large size. Large size has the adaptational consequence of reduced tolerance to hot conditions (Cowles 1939). Large size also results in smaller population size, reduced reproduction, and increased longevity. It would follow that a small population size and a large interval between generations of a large animal would decrease the effective rate of mutations. The evolutionary trend called gigantism alone, without environmental changes, does not necessarily lead to speedy extinction, but it may lead to easier fossilization and recovery; the small-sized fossils, though more numerous, may be difficult to retrieve.* Gill (1955) extensively discusses the extinction which is said to have occurred among the giant forms of Australian marsupials. He points out that many smaller forms have likewise become ex-

---

*Hutchinson (1965) states that a planktonic organism as rare as one individual per m^3 of water may never be found by the sampling limnologist, even if it actually numbers in the millions in any one lake. This may apply, *mutatis mutandis*, to small-sized and uncommon animals of the past.

tinct. Climatic changes, man's arrival, and predation by the dingo (a wild dog) have all been suggested as causes of these exterminations, but each of these arguments is refutable. Gill proposes, in addition to these factors, the possible effect of behavioral specialization—namely, the loss of behavioral versatility in times of stress on the population.

(B) *Synthetic extinction.*—Many historical biogeographers recognize so-called *faunal extinction*, i.e., group extinctions which affect faunas of distinct geographic areas. Simpson (1965) discusses the high extinction rate after faunal mixing after an ecogeographic barrier has been removed or crossed, resulting in changes in faunal composition, *inter alia*, by extinction of formerly common or even dominant groups of vertebrates. Mayr (1963, 1965) cites the vulnerability of insular faunas to drastic geographic changes on islands. This insular extinction parallels the ease with which insular evolution occurs during a relatively short time because of geographic isolation. Ross (1962) discusses wholesale regional extinction—for example, the extinction of a great number of Arcto-tertiary forest biota in Europe and western North America and the relatively lesser decimation of the same biota in the eastern Palearctic and eastern Nearctic. We believe that this differential extinction was based on the differential vicissitudes to which these biota were exposed, owing to local modification of the impact of the Pleistocene climatic shifts.

A mass extinction process of singularly large dimensions is that caused by the ever increasing impact of the currently dominant species—(*Homo sapiens* L.). Although never calculated, the rate of extinctions caused by man is perhaps equivalent to the geometric rate of increase of human populations.

The most common textbook examples of group extinctions are the *phyletic extinctions*, because evolutionists, taxonomists, and zoogeographers alike are interested in this phenomenon. The life of a major taxon, such as an order or class, seems to be composed, like that of a species, of: a small-scale beginning, which is often obscure because of its localized nature; a massive geographic spread coupled with adaptive radiation; and finally, regression. Thereafter, the expansive phase may be observed again, or the decline and extinction—or possibly the relict distribution of a few survivors—can be expected.

The explanations that are offered often lack clarity with respect to the ultimate or proximate reasons involved. When it is claimed that dinosaurs died out because small, nocturnal, and cunning mammals devoured their eggs, evidently a proximate factor is hypothesized. When climatic changes are mentioned as leading to the extinction of dinosaurs, evidently the lack of climatic adaptability—an ultimate factor—is suggested. For instance, the Cenozoic climate might have caused their extinction since, lacking endothermy, they could not keep warm enough. Or conversely, during the warming trend of the late Mesozoic, lack of heat dissipation facilities in their large bodies might have been the crucial factor. High ambient temperatures might have affected their spermatogenesis, reducing their reproductive rate. It is not the task of the general zoogeographer to indicate his choice among these explanations. The same ultimate factors may be postulated for the extinction of a whole major taxon as for the extinction of single species: major environmental changes; specific environmental changes affecting the particular taxon; specialization with inherent lack of adaptability or versatility; and general decrease of ecological valency, perhaps independent of the rate of structural or habitat specialization.

A highly favored principle—and one which is generally invoked in discussions of phyletic extinction—is *competitive exclusion,* or the *ecological replacement principle.* Often one group disappears and another seems to take its place in a corresponding or identical major habitat or in a niche. Nicol (1961), for example, invokes a striking correlation in the evolutionary history of cephalopods. Several times throughout the geological ages, expansion of the ammonoid cephalopods was simultaneous with regression of the nautiloid group, and vice versa. Mayr (1963, 1965), Rensch (1959), and Simpson (1965) speak of competitive replacements having taken place when two faunas mingled. Darlington (1948, 1957) went so far as to postulate, following Darwin: that new forms must always replace older ones; that for every species that evolves, another becomes extinct; and that a new group cannot evolve unless an old one dies out. Darwin reasoned correctly that rarity precedes extinction, that rarity is a common phenomenon, and that consequently extinction should be common as well. He was not correct when he postulated that in speciation there is a struggle between the most closely related forms, and that the strongest or fittest wins and the other loses and becomes extinct. Darwin said: "Each new variety or species, during the progress of its formation, will generally press hardest on its nearest kindred, and tend to exterminate them" (1859, p. 110). Thus, he believed in competitive extinction, phylogenetic extinction, and the notion that old species must give way to new ones; but these concepts do not hold true in the light of the new knowledge of the past hundred years.

The theory of competitive exclusion states that when a requisite is used simultaneously by two (or more) kinds of organisms and it fails to satisfy their combined need, the better-adapted one will eliminate the other. Ecologists, looking for manifestations of assumed competition, make the following assertions.

In many instances, competition is not present. Replacement occurs as one environment slowly changes into another through time; while the one species decreases, the other appears and increases in numbers, i.e., their existence did not depend on the commonly utilized requisite.

When real competitive situations occur, the outcome is not as clear as it would be in experimental populations or in theoretical equations, owing to many intervening, modificatory environmental factors.

Often the competitive situation seems to resolve itself in diverging adaptations. In this way, competition may become an important factor in diversification.

Many interspecific relations, even though harmful for one or both partners, do not qualify as competition (Udvardy 1951; Birch 1957; Hutchinson 1957, 1965; Hardin 1960).

In what way can these principles and their ramifications be applied to faunal and phyletic competition? When two faunas meet—i.e., when a barrier is removed between two geographic areas—the members of each will begin an expansion into the area of the other. The exchange of species will naturally be greatest between two related communities. Proof of this historically common situation is found in the diversity of biotic communities often containing many closely related forms that must have evolved in geographic isolation. We may assume that many of the newly arrived species use requisites which are already used by the old members of these communities. The relations of the old and new members of these communities will be resolved and readjusted according to the principles outlined above. Since the whole community of the area will take part in the readjustments, some members will alter the course of their adaptive evolution; others will disappear, but in only a few instances will this occur as the result of competition. Many other kinds of interaction may directly or indirectly affect the members of a community which is exposed to the intrusion of new species. Savile (1959), in a relevant discussion of the penetration of barriers, stresses the potential role of introduced diseases in faunal replacement. There are, indeed, examples of this kind (see p. 361 on the introduction of avian diseases to isolated islands). In addition to its direct effect on the host organism, the most plausible effect would be the one which is indirectly harmful—for example, when a plant which has served as food for one organism becomes exploited by another. The chestnut blight *Endothia parasitica* (Murr.) P. J. & H. W. Anderson [Diaporthaceae, Ascomycetes], a disease of the Oriental chestnut tree, was accidentally introduced into North America, where it wiped out the native chestnut tree, which had no immunity. Squirrels which had fed on the fruit of the North American tree were now deprived of a portion of their diet. If the number of squirrels did not decrease in response to this forced change in their diet, they most likely took more food from another plant, which thus also became affected by the change. But, what about competition among members of a single major phyletic line—e.g., marsupial and placental mammals? Knowing that in the absence of placentals the Australian marsupials successfully radiated in the main mammalian habitat (and food) opportunities but did not do so elsewhere, one might consider competition as *one* of the possible avenues of replacement. Yet other possibilities might be considered even more important—for example, predation by behaviorally superior placental predators or more effective parasitism by their common parasites and diseases. It would be more cautious for the ecologically inclined zoogeographer not to use the term "competition" in the phyletic, paleontological sense at all; instead, he might objectively and more "descriptively" speak only of *replacement*.

Diversity of ecological communities, both in the same and in adjacent climatic zones, is often due to faunal mixing. Although competitive exclusion probably occurred in many instances as a proximate factor of extinction, readjustment (diversification) was probably the result in others. Thus, competition is only one of several major factors active in faunal mixing.

The principle of community evolution and diversity can be usefully invoked in the discussion of phyletic extinction and competition. The energy resources of a biotic community are provided by the plants, and these resources are finite in any one community. One may reason that the exploiters of the resources—the consumer animals—should consequently also be finite in number. However, cycling of the resources is variable, and differs from community to community. It is also asserted that the stability (or we may say, evolutionary success) of communities increases with the diversification of the routes of cycling: the more diversified and the more numerous pathways that exist to counter a disturbance, the more stable a community becomes. Therefore, one can no longer say, as Darwin did, that there is no room for new forms unless the old ones disappear. On the contrary, the history of biotic communities is that of diversification and enrichment with new species and new pathways to cycle energy. Though direct proof in the form of historical evidence is not extant (paleoecology is a very young science and has not yet assembled the proof of the paleontological record) we do find, for example, that the present north temperate habitats are in the process of becoming enriched by the postglacially expanding species (following a period of severe exterminations); undoubtedly, evolutionary adaptations of species are continually occurring and, consequently, are occurring in communities as well. Analysis of the rate of speciation in tropical faunas is needed to show that these too are not yet saturated.

This trend of thought leads us to the assumption that the evolution of the species and the diversification of form are the consequences of evolution and diversification of communities.

The phyletic evolution and extinction which occurred in the animal world paralleled that which took place among the plants. Throughout the late Paleozoic, Mesozoic, and Cenozoic, plant evolution progressed side by side with animal evolution. The producer members of the community were governed by the basic influence of climatic and geographic changes and by their own competitive and other relations. Thus, a likely cause of the phyletic extinction of animals could have been the change of community patterns.

Dynamism of Communities

There are several schools of thought about biotic communities. One of these thinks of biotic associations as coadapted, fairly stable, "superorganismic" abstract systems manifested in communities (individual stands of the specific plant-animal association). Climate and a few leading organisms should have dominant roles in its composition. This concept leads us to assume that the most important dynamism (apart form oscillatory and homeostatic processes) is the expansion or retreat of whole communities, chiefly of their dominants. Other ecologists nearly reject the community concept, and point out that where external conditions do not change abruptly, vegetation and faunation form continuums. Species limits may be found almost everywhere, and thus the composition of biota changes, as it were, clinally. A brief examination of the dynamic aspects of the community concept may reconcile some of these opinions into a compromise.

Since species, even less widespread ones, usually occupy more than one habitat, their abundance, and thereby their impact on the cohabitants, will be different in each habitat they frequent. Moreover, species that now live together often show widely different distributional, faunistic, and evolutionary ties. Notwithstanding these differences, a multitude of mutual adaptations has been demonstrated among the members of a single community (Elton 1966), suggesting that they lived together for a long time. Dynamism of individual species, now as in the past, is demonstrably different, and varies with the passage of time and with changes in ecological conditions. One school of ecologists attributes great importance to ecological dominants—the numerous, the large-sized, or the synecologically influential and vital members of main food chains in communities. Others, however, point out that these dominants are important mainly in temperate communities; many communities, particularly in the tropics, have too large a number of such dominant members.

Apart from a few exceptional situations, plants as producers dominate in terrestrial communities. Plants—especially those which form the most stable communities—those which are adapted to live in the habitats which change most slowly under the influence of climatic, tectonic, and edaphic events—are much longer-lived than the average land animal. This is commonplace knowledge regarding trees, but even shrubs and grasses may live for centuries.* It follows that dynamism—evolutionary, coadaptational, and geographic—must be much slower in these plants than it is in animals. Then, communities could be, on the average, much older than many of their animal members. These animal species, in turn, could better adapt to the plants than vice versa because of their shorter individual and evolutionary life span. There is almost no pertinent literature, and therefore it is not possible to test these postulates. One future avenue of research could be surveying the differences in rate and mode of evolutionary and distributionary adaptation to successional (therefore, in time and space, ephemeric) and "climax" (more mature and stable, therefore probably older) communities.

It would be worthwhile to explore the ages of these community types in tropical (therefore much older) and temperate (phylogenetically recent and, moreover, decimated by recurrent warm interglacials and cold glacials during the Pleistocene) environments, for there has been little research in this direction. The literature that does exist on the subject is solely botanical and often controversial (Axelrod 1950, 1958; Braun 1950; Chaney 1940; Firbas 1949, 1952; Whittaker 1959).

It is well established by fossil evidence that many plant species which are now dominant members of associations of a geographic region lived during the Tertiary, or even during the Pleistocene, in other geographic regions where they now are extinct. Other species which are now subordinate or only locally dominant seem to have had a dominant role in the geologic past. If large size in animals is a sufficient criterion of ecological dominance (which is not proved but is often tacitly assumed by paleontologists), then the widespread Pleistocene and recent extinction of gigantic mammals would serve as another proof that dominance relations in communities—and thereby fundamental trophic and structural relations as well—change as rapidly as any other aspect of the geography of biota. As noted in Chapter 3, no information source focuses on the age of communities. Therefore, much of what has been said of their past *en bloc* movements is hypothetical. Nevertheless, their movement on the

---

*Because the concept of individuals and individual life differs in botany from that in zoology, a tussock of bunchgrass or a scrub in the desert may be considered as one individual. These constantly replace lost parts with new shoots, and thus their life is indeed very long.

millennial and smaller chronological scales is well documented. The zoological evidence is more circumstantial, and is based on faunal assemblages (e.g., Guilday *et al.* 1964; see Plate 3). Plant associations are known to follow the retreat of glaciers (Heusser 1960). Those forming altitudinal zones descend and ascend the mountain chains following climatic trends (Martin and Mehringer 1965, Whittaker 1956). Forest belts give place to grasslands in xeric periods, and advance again when mesic conditions prevail (Auer 1951, 1960).

Summarizing these observations, postulates, and hypotheses, we find that communities are dynamic systems on a high level of organismic complexity. Though they are by no means superorganisms, the mode of their dynamism is closely comparable to that of their members, the coexisting plant and animal species. Communities in time exhibit geographic movements, advances, and retreats; while doing so, they undergo dynamic, evolutionary changes in their structure and function by acquiring new elements and dropping others. Thus, in the course of their geographic history, they change location, become extinct, or transform (evolve) into new communities.

Man as a Dynamic Agent in Zoogeography

Primeval man had not much more influence on the ecosystems in which he lived than did other large primates. His tool-using behavior, the social systems of his populations, and his interspecific relations (feeding, escape and defense behavior, and others)—all had their roots in similar but simpler primitive behavior patterns and environmental influences. All this changed as a result of his use of fire, increase in numbers, spread, barrier crossings, agricultural and pastoral practices, and, later, as a result of all aspects of his civilization, urbanization, and total harnessing or altering of the earth's resources. In sum, mankind's role supersedes *all* natural agents which influence animal dispersal, distribution, and faunal dynamism.

We have previously considered man as a carrier of dispersing animals, and observed that there now exist almost unlimited possibilities for faunal exchange due to *anthropochore dispersal* (Chapter 2). Domestic and game animals, pets, and many others have been intentionally introduced into remote areas. The weeds of cultivated crops, the fauna of the ruderal and agricultural areas, and the parasites of the introduced animals spread together with these and with the cultivated plants (Figure 6–18). The work of the zoogeographer often becomes increasingly difficult in the face of the manifold opportunities for introduction and extension through human carriers. An example is provided by the lumbricid earthworms. They spread with earth attached to rooted plants that man transports across natural barriers. Gates (1966) asserts, on the basis of his faunistic work in and around greenhouses in northwestern North America, that European lumbricids could have been brought across the Atlantic for over 350 years, and that for 250 years many of the twenty kinds now known in North America could have been unintentionally propagated in greenhouses. He believes that a considerable portion of the present distribution area of European lumbricids in North America is due to such introductions and not to a transatlantic land bridge, which other investigators maintain (cf. Omodeo 1963).

We again refer to Elton's (1958) treatise on the introduction of animals and its consequences. But from the point of view of faunal dynamics, we must emphasize that while many of the introduced animals are known to have spread rapidly and to have increased from a handful of pioneers to very conspicuous and dense populations, *an even larger number of introductions failed*. Again, in other instances, the initial increase was followed by a period of scarcity, and the new species seemingly settled into the ecosystems of its new habitats in a more balanced condition. Elton's recent long-term study (1966) of an oak woodland community in England provides the best examples.

Most introduced animals can gain a foothold only in habitats which have been distributed by civilization or which were the results of man's activities. Lindroth's study in Newfoundland, discussed at length in Chapter 2, again bears this out. Before the 1940's, when the relationships among habitat, population dynamics, and ecosystem metabolism were much less studied than they are today, the mutual influence of introduced or adventive animals and the native faunation was often misunderstood. Exaggerated reports without any ecological backing spread unfounded opinions about the introduced or adventive species "displacing" or "outcompeting" members of the native fauna. Naturally, there are such instances, but more often the native faunation and vegetation—that is, the ecosystem—repel the intruders.

This process is also often called "competition." Emerson (1955) treats the geographical dispersions of termite genera. He emphasizes that some twenty species of termites have been inadvertently introduced to climatically suitable regions, where they inflict great economic damage on human constructions; but, "typically the introduced species are unable to invade the native environment, but remain confined to the constructions of man. It seems probable that competition with native species is a biotic barrier to the spread of introduced species." (p. 474.) However, Emerson hastens to add, "predators, parasites, and other phylogenetically unrelated organisms, besides competition with related species, may be involved in the production of biotic barriers."

It has been said that the native birds of New Zealand suffered a great deal because of introductions. Hesse (1924) and the translators and reworkers of his book (Hesse, Allee, and Schmidt 1937, 1951) say:

When by accident or intent, modern animals are introduced in the isolated areas, they promptly demonstrate their competitive superiority over primitive animals. Thus the marsupials of Australia are disappearing before the introduced cattle, sheep, rabbits, housecats, and foxes of the settlers. The endemic birds of New Zealand give way before the buntings, starlings, and goldfinches. (1951 p. 113.)

The introduced land vertebrates of New Zealand spread mainly into the man-made and man-altered habitat, from whence the native bird fauna disappeared *before* the arrival of alien species (Thomson 1922, Wodzicky 1965). In fact, cattle and sheep alone did not crowd out the marsupial grazers of Australia; it was the kangaroo hunts (massacres) of the settlers that cleared the ground for the domestic stock.

It has also been said that introduced birds displaced the now extinct species of drepaniids and other indigenous Hawaiian birds. The introduced and widely dispersed passerine bird, the mejiro, or Japanese white-eye *Zosterops japonica* Temminck and Schlegel [Zosteropidae], lives in many kinds of native wooded habitats in the Hawaiian Islands. Experimental studies by Warner (1968) proved that this introduced bird is a host to the avian malaria and bird pox, that the accidentally introduced tropical mosquito *Culex pipiens quinquefasciatus* Say [Culicidae, Dipt.] is the vector, and that these native birds have no immunity against such diseases. Thus, lack of immunity against an introduced disease, rather than competitive replacement, exterminated these native birds. In all parts of the Southern Hemisphere, the native earthworms disappear with the introduction of Lumbricidae. It may be remarked at this point that Murchie (1956) has written about an analogous situation in North America—that is, of Michigan earthworms:

No evidence was obtained which would support the notion that the peregrine species of earthworms, presumably introduced from Europe, have replaced endemic American forms. It appears to me more likely that changing farm practice and land use may be of greater significance in controlling the nature of the earthworm fauna within a given area. . . . Therefore, instances of apparent earthworm succession, although possibly quite real, should, I believe, be attributed to changes in habitat availability and the introduction of new forms by man, rather than direct competition between earthworm species. (p. 69.)

Although these examples are cited to show man's activities as one cause of dynamic faunal changes, they also reveal the importance of the habitat and of the interspecific relationships—such as the parasite-vector-host relationship—in being instrumental in singular as well as faunal replacement. It may be concluded, that where ecologic replacement occurs between native animals and introduced forms, it is preceded and aided by habitat alteration by man. Statements citing competitive replacement by calling either the adventives or the native animals "successful competitors" should be avoided if they are not backed by factual observations.

Islands are especially vulnerable to man-fostered dynamism. Insular faunas do not have the proportion of participating taxonomic groups that exist on mainland areas and habitats of comparable size (see p. 372). Islands have *unbalanced faunas* with *short* faunal lists; and if a species is introduced or a habitat-altering activity of man takes place, it usually affects several or most members of the small indigenous fauna (Figure 6-21). Many modern examples have been reported in a recent symposium (Fosberg 1963). On the main island of the Three Kings group, not far from the northern tip of New Zealand, the Maoris cultivated forest clearings until about 1840, when they moved away. Forest growth then reclaimed these clearings, and forest succession proceeded toward the climax forest type. In the 1880's, goats were liberated on the island. The goats devoured and exterminated most of the endemic plants; a scrub forest of but a single dominant tree grew up, because this tree was unpalatable to the goats. In 1947, the goats were eradicated. For a second time, natural succession set in, and already there is a mixed forest with endemics reinvading from the outlying islands and with a new faunation exploiting the secondary forest (Turbott 1963).

All temperate continental regions are so intensely settled and cultivated that no faunal change can occur without human interference. For example, in Scandinavia, where spectacular spreads were described as a result of climatic improvements (p. 341), the same natural cause brought about a great advancement of cultivated areas toward the North (Hustich 1952). De Vos (1964*b*) describes range extensions of eight conspicuous mammal species in the Great Lakes region of North America. Perhaps all but the spread of the raccoon are ascribable to habitat alterations. In such regions, where a sparse, pastoral or gatherer-hunter native population has recently been replaced by agricultural and industrialized-urbanized human societies, the student of dynamic zoogeography finds outstanding opportunities to study the transformation of the local habitats and faunations. Tischler (1965) briefly mentions several of the avenues an animal may use when changing from natural to man-made habitats. Turček (1966) stresses the role of "borderland," including ecotone and edge habitats, as a mediator of the spread of faunal elements from adjacent communities. Agricultural borderlands now penetrate natural communities almost everywhere; these borderlands are used to exchange in-

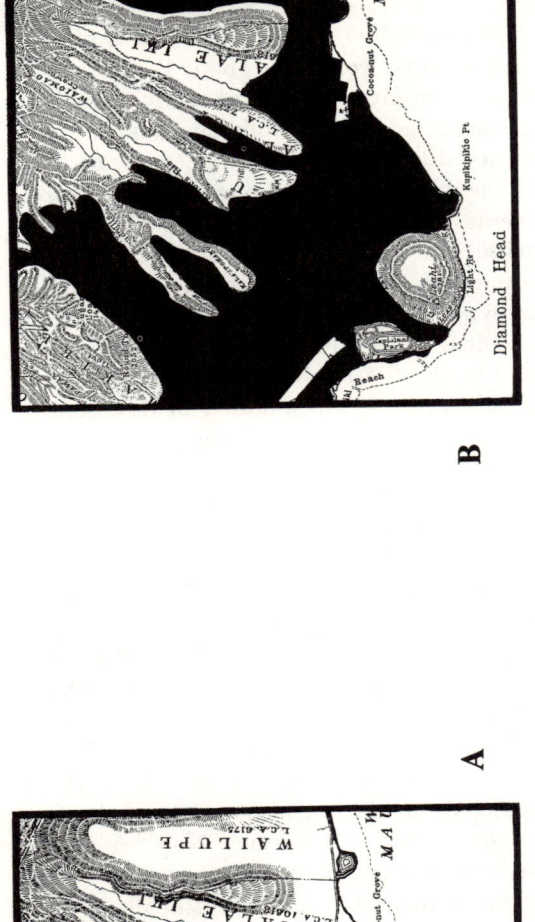

Figure 6-21 Man's impact through time.

Map of a portion of southeastern Oahu, Hawaiian chain. The 1966 map is based on *Bryan's Sectional Maps of Honolulu*. (Courtesy of E. H. Bryan, Jr.)

A. *In 1902.*—At the left margin, above, Oahu College (the present University of Hawaii Campus) represents the easternmost end of suburban Honolulu. Between that city and Waikiki Beach (lower left) with Kapiolani Park, there was countryside. The village of Kapahulu is shown. The upper right corner of the map contains a short section of the Koolau range, the backbone of eastern Oahu, reaching here a height of 799 m (2621 ft). These mountains were covered with *ohia* forest, the western slope of the ridges with *koa* forest, the eastern slopes with *koa* woodland, *staghorn fern*, grassy, baldy spots or barren desert. The valleys (Manoa, Palolo, Waialae) were also forested, with *kukui* growing along the streams. Much of the coastal alluvium was by 1902 turned to canefields.

B *In 1966.* The blackened area—all the flat land, valley floor and not-too-steep hillside has been turned to suburbs or to intensely cultivated smaller meadows and parks, the former awaiting subdivision that has happened since 1966. In the upper left corner, Mt. Tantalus has several villas along the highway. The forest of the ridges is fenced around and preserved as water conservation area; Diamond Head (an extinct crater) is a military reserve, and grown with grass and the adventive *pukiave* scrub.

digenous and introduced faunal elements, usually at the expense of the indigenous fauna and faunation.

There seems to be a lack of pertinent literature about North America and the other newly civilized continents, especially about their tropical zones. I shall mention an example from my own field of study. Swallows and swifts in the Mediterranean area had several millennia—and, in central and northern Europe, a great many centuries—to adapt and move to human dwellings for nesting. The black redstart *Phoenicurus ochruros* (Gmelin) [Muscicapidae], a Mediterranean rock-breeding bird, invaded central and western Europe within the past 120 years (Mayr 1926). It reached the borders of the Hungarian plains along the foothills of the Carpathians, where it was known as a rock dweller until the 1930's. When the first pioneers of these birds arrived in Debrecen, on the plains, they nested initially on the higher levels of tall downtown buildings, which provided them with a nesting habitat that corresponded in its essential elements to their earlier environment, rocky cliffs. Because of a population increase, the black redstarts shortly began to nest in suitable crevices of one-story suburban brick buildings. Next, they chose even wooden structures. Thus, the species adapted to life in a garden-woodland habitat which was not available to it before, through the gradual change of the nesting substratum—from tall stony structures to one-story brick buildings—to low wooden sheds, verandas, and the like.

In eastern North America, the chimney swift *Chaetura pelagica* (L.) [Apodidae] and the purple martin *Progne subis* (L.) [Hirundinidae], both hole-nesting birds, became adapted to nesting, respectively, in chimneys and in breeding boxes. Distributional changes presumably followed this adaptation, as among the European swift *Apus apus* (L.), which nests not only on houses in forest clearings but also in cities where it would not occur naturally (Koskimies 1956). In western North America, the purple martin and the Vaux's swift *Chaeture vauxi* (Townsend) still nest in natural woodlands, but the martin also nests occasionally on tall buildings in large cities. The urbanization process thus has already started, and will likely continue in the coming decades, again with accompanying distributional changes. In this same area, half a dozen bird species from the open woodlands have extended their distribution area northward along the Pacific coast during the past few decades. This expansion occurred simultaneously with, or shortly after, the great deforestation of the early twentieth century, when the dense forest was replaced by the small, "forest clearance"-like settlements and meadows. These meadows and villages now form a more or less continuous habitat bridge across the rain-forest belt from the open natural areas of the Pacific coast. But an opposite, man-induced trend of faunal change is also noticable.

A few bird species of the coniferous forest are spreading south of their previous limits in California, owing to the planting of coniferous trees in suburban gardens. The presence of such trees is apparently a necessary prerequisite of their habitat selection.

As a global forecast of mankind's continuing role in faunal dynamism, we may predict that most natural areas which still remain unsettled will be—wisely or unwisely—manipulated. Large-scale transplantings will continue, and accidental introductions will reach enormous proportions. *Antropogene zoogeography* is the future, already timely, applied branch of our science.

HISTORIC ASPECTS OF DYNAMIC ZOOGEOGRAPHY

In the first half of this chapter, we have discussed certain phenomena and principles pertaining to the single vectors of dynamism in zoogeographical processes. Now we will consider the *combined* effect of several or all of these vectors on animal *assemblages*. The treatment of groups serves two ends. First, the group can be studied for itself, analytically, although entities such as faunas and taxa are abstractions. In nature, only species populations occur. The second point in studying the dynamism and history of faunal or taxonomic groups of distribution areas (and the pertinent animal populations) is that the past distribution of single species ought to be based on fossil evidence, of which there is never enough to outline the past distributional area of a single animal for any one geological period. As a substitute, group treatment will be used. We will shortly expand on the principles underlying this method.

The questions asked in historical enquiries are: Where was it? When was it? Why? The spatial and temporal distribution of an animal species or of a group of species is the subject of *historical zoogeography*. In this field, there is an overlap with the working field of the paleontologist. *Paleontology* embraces the total knowledge of past biota, and has two main objectives besides the providing of a descriptive basis for knowledge about fossils. One of these objectives is the establishment of the history of life in the form of phylogenetical relationships; the other is the spatial aspect of the same. Historical zoogeography thus becomes synonymous with the zoological branch of paleontology, at least with the geographic aspect of paleozoology. If we refer once more to our introductory chapter and try to fit these concepts into the system of biology (Figure 1–1), we find that, among the dynamic objectives, speciation and phylogeny can be treated historically; and, among the spatial relations, the same treatment can be applied to environments, distribution areas, ecosystems, and other supraspecific levels. Moreover, one may, with great success, complete the foundations of phylogeny through the study of fossil structures. The study of the functional aspects of these objects is less rewarding; therefore, paleontology remains chiefly the study of "life's unbroken *continuity* through *time*, over *space*," as Leon Croizat (1958) has said of biogeography.

The Application of the Principle of Uniformitarianism in Historical Zoogeography

The dual task of paleontology can be resolved on the basis of studying fossil evidence and fossilized environment if one keeps in mind the maxim "The present is the key to the past," first set forth by James Hutton about two hundred years ago. To geologists this maxim aptly summarizes their principle of *uniformitarianism*—namely, that all past geological processes can be explained on the basis of presently manifested processes, their qualities, and their rates. It follows from the nature of the evolution of life that Hutton's statement is a truism, even for the zoogeographer. But, by the same reasoning, the opposite is also true— that is, the past is the key to the present. Present distribution is as much influenced by the past as it is by contemporary factors, and it is at this point that paleozoology and zoogeography overlap without a clear demarcation line.

Historical biogeographers and paleoecologists often apply this principle —consciously or unconsciously—when fossil facts are to be interpreted or when, for want of sufficient facts, working hypotheses are formed. Only when even such hypotheses fail are bolder explanations offered, as, for example, regarding the origin of life, which occurred in such a remote, geologic past that the basic conditions of life, if not the basic processes of the environment, might actually have been different. Uniformitarianism is the unwritten and unspoken principle by which, for example, the major role of "macromutations" and "hologenetic evolution" has been refuted, showing that the demonstrable, existing processes underlying microevolution are sufficient to account for major evolutionary changes.

Much of the earlier discussion in this chapter, relating to arguments about the dynamic nature of environmental changes affecting distributions, was preparatory to the discussion of uniformitarianism. The main purpose of this book is to elucidate the dynamic processes which underlie the instability of the present faunal and distributional picture. Our *reversed* uniformitarian principle states that *all of those dynamic processes which have been shown to have been active in the geological past are also acting in the present time.* The demonstration of these processes is sometimes difficult, because of the slow rate of changes; in other instances, it is easy; but there has not been enough attention paid to them by zoogeographers.

There are, however, two principles related to animals to which the uniformitarian principle should be applied with caution. These are the *constancy of ecological valency* and the *constancy of means of dispersal*. We may arrive at misleading conclusions in both instances when the principle is applied not to universally valid *processes* but, instead, to adaptational *characteristics* that are bound to change in time, as environmental conditions change.

Constancy of Ecological Valency

At the dawn of the ecological approach, Nehring (1890) concluded that the distribution of tundras and steppes in the Palearctic was completely different during the Pleistocene glaciations. He evoked the proposition that animals which are now characteristic of certain climatic regions were also characteristic of these regions in the past, if they were structurally completely identical to their recent relatives. But he hastened to add that this argument was useful only if simultaneously applied to a group of species which have identical climatic requirements at present. Warnecke (1936) draws attention to Nehring's proposition, saying that this seems to be a *principle*, not recognized but universally used by historical zoogeographers. Thienemann (1950) emphasizes at length the stability of ecological amplitude in time as well as throughout the distribution area of the species (see p. 112). He mentions that many relict species *adapted* to the changing environment as opposed to others, which are conservative, like our "refugional" relicts (p. 209). Paleontologists like Martin (1958) and Romer (1961) also see clearly the pitfalls of the strict application of this principle to explain paleoecological conditions.

In our first example, the reasoning is based on fossil evidence. Halftter (1964) discusses the origin of the different elements of the North American fauna. He mentions that the few insect fossils known from the late Pleistocene, Californian, Rancho La Brea asphalt pits all belong to now extinct species, and that thus they stand witness for profound faunal changes. In Florida, to the contrary, the known fossil Pleistocene entomofauna contains generally the same species that still exist there, suggesting the similarity of the present environment and that of their pleistocene ancestors.

Darlington (1961) discusses the carabid beetles of northeastern Australia and New Guinea. Both of these areas are now covered with rain forests, but their carabid fauna is different. In other, less humid habitats, there are carabids common to both geographic areas. Darlington first rules out differential dispersal across the Torres Strait. He then reasons that, since we know, on a geological basis, that the two areas once had a land corridor between them, beetles of both the rain forest and other habitats could have crossed through this corridor by slow penetration. Since the rain-forest carabids failed to do so, this bridge must not have contained continuous rain-forest habitat. Consequently, the differential distribution of the carabids—provided that the forms currently adapted to the rain forest were so adapted even in the past—offers a clue to paleoclimatological conditions, which must have been drier than they are now. In this example, the constancy is a cardinal supposition. Carabid beetles provide another example also, in Lindroth's (1963) treatise of the faunal history of Newfoundland, which is based almost entirely on an analysis of the present distribution of this insect family on that rugged, northern island. Both the "constancy of valency" and the "constancy of means of dispersal" principles are used implicitly. Lindroth asserts that the presence of nonflying or hardly flying species in coastal areas speaks for their survival there during the latest glaciation of the area rather than on nunataks surrounded by ice. Their now restricted distribution and brachypterous or dimorphic locomotor equipment seem to indicate—*if we assume that these structural-behavioral characters did not change*—that they have probably been present in this area since pre-Wisconsinian times. Lindroth infers that, judged from their present distribution, they are not able to endure a high-arctic climate. Therefore, it is not likely that they could have survived on nunataks which have—and had—an extremely high-arctic climate.

On the other hand, *examples of changes in "ecological valency" are well known*. Adaptational changes commonly accompany all degrees of speciation processes. Montane faunas provide good examples. Gressitt (1961) says that the high-montane insect fauna of New Guinea is derived from the surrounding tropical lowlands. Evidently, changes of ecological adaptation took place, on the genotypic level, for these insects now live in a cool-temperate environment. The same is said by Mayr and Phelps (1963) of most of the endemic birds in the montane avifauna of the isolated plateaus in northwestern Brazil and adjacent Venezuela. Their nearest relatives live in the tropical zone. According to ample fossil evidence, the two European marmots were vicarious, cold-steppe and tundra animals during the Pleistocene. The alpine marmot *Marmota marmota* (L.) [Sciuridae, Mamm.] now lives in the alpine tundra above timberline, while the bobak marmot *M. bobak* (Mueller) is distributed along the continental steppe belt (Figure 6–22). Mani (1962) discusses the changing adaptations that accompanied speciation in the Himalayas (Figure 6–23). Closely related forms can have as different ecological requirements yet as similar structural (at least osteological) characteristics

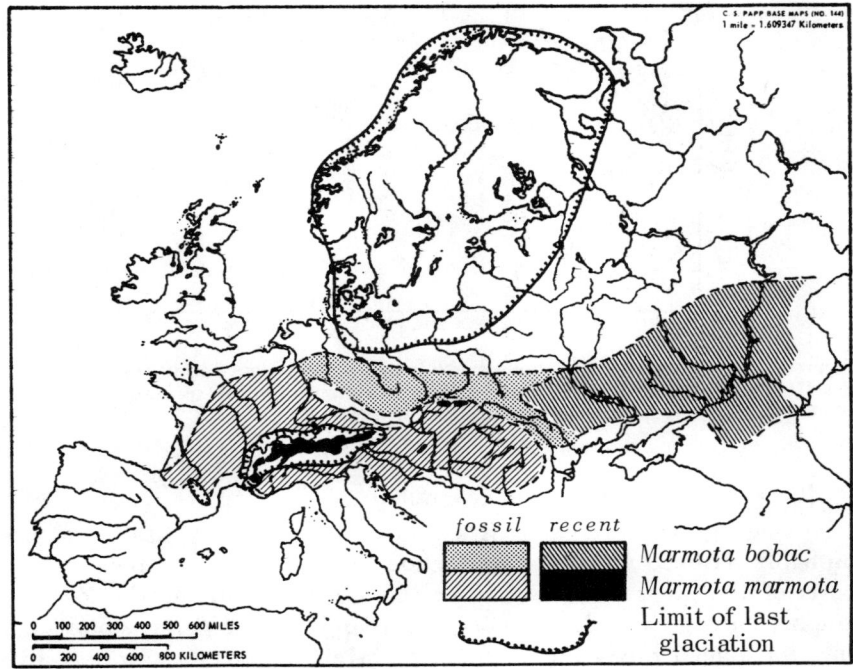

Figure 6-22 Change of ecological valency through time. Distribution of the Alpine marmot *Marmota marmota* (L.), [Sciuridae, Mamm.] and the bobak marmot *M. bobak* (Mueller) in Europe at the present time and during the Pleistocene, south of the glaciated part of northern Europe. After Freitag 1962.

as did the mammoth of the taiga woodland and its congener the elephant in the hot tropics.*

Constancy of dispersal capacities is such a closely related concept that a separate discussion is hardly necessary. We remember the curious change of dispersal mechanism in the collared dove (pp. 28, 203). This bird was known to live on one side of the Danube in central Europe for decades without showing any signs of being able to disperse across the mighty river. Since the 1930's, it has become the most expansive and explosive avian colonizer ever observed (apart from those transplanted by man).

*Botanists are also very much aware of the possibility of changes in ecological valency which may go undetected. "Change in tolerance may or may not be accompanied by morphological change, and morphological change may or may not be accompanied by change in tolerance. Morphologically similar species may show wide difference in tolerance, and species with similar tolerance may show very little morphological similarity." (Good 1931, p. 155; Good 1964, p. 417).

370 *Dynamic Zoogeography*

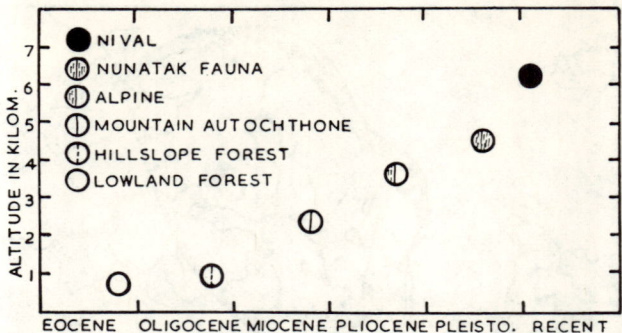

Figure 6-23 Changing adaptations accompanying speciation during the Tertiary in the Himalayan insect fauna leading to the evolution of nival species. From Mani 1962. (*Courtesy of Methuen and Co. Ltd., London*)

Dynamics of Island Zoogeography

The zoogeography of islands is as old as zoogeography itself, and has probably received much more attention than its share would be on the basis of the areas or the biota of the islands of the world. Zoogeographers are great travelers. Because islands are stepping stones, not only for animal travel but for human travel as well, they present two fascinating aspects to every traveling naturalist: the simplicity of island life, on the one hand, and—paradoxically—the diversity of strange life on some of the larger islands and archipelagoes, on the other. Ever since the memorable voyage of the *Beagle* (Darwin 1839), evolutionary theory and island zoogeography have been inseparably intertwined. In spite of the large amount of literature that followed, Wallace's *Island Life* (1880) is still worth reading. A modern synthesis of island biogeography, coupled with a descriptive account of the islands of the world, appeared recently bearing the same title (Carlquist 1965).

Interest in island phenomena increased greatly when the small volcanic island of Krakatau, in the straits between Sumatra and Java, exploded in a spectacular series of eruptions in 1883. All life was buried under glowing ashes. The return, by over-water dispersal, of the plants and animals has been periodically studied (Dammerman 1948*), and the process of settling virgin areas has thereby become much clarified.

We are already acquainted with the division of the world's islands into continental and oceanic ones (p. 305). This division was based on the assumption that islands near the present continents are ephemerically

---

*The data tabulated by him are discussed by Hesse, Allee, and Schmidt (1951) and further analyzed by MacArthur & Wilson (1967).

separated by epicontinental seas but that islands far away from the continents have different origins (Wallace 1880). Darlington (1957) objects to this distinction, and points out, rightly, that exceptions and transitional cases are numerous. From a strictly nonbiological point of view, it is indeed difficult to say what a continent is. For the geologist, a land mass formed of *sial* rock would be the correct answer. The noncontinental areas, of *sima* rock, whether completely covered by deep seas or containing ridges, seamounts, and other features that almost reach or even break through the surface of the water, would be considered oceanic areas. Yet geologists are not in agreement about the permanency of the present continents, and many say that some of the present islands or groups of islands may have been parts of continents in the past. Geographically, if we were to take size as a criterion, even Madagascar, New Guinea, and Greenland would possibly qualify as continents. If we consider the biogeographical peculiarities of islands, we may conclude that islands are water-surrounded pieces of land where the fauna is simplified, as compared to that of a continent. An island shows peculiarity due to local opportunities and evolution to a far greater degree than does an average part of a continental land mass. Munroe (1957) pointed out that such criteria, if applied to Cuba (geographically an island at present but part of a past continental mass, according to many geologists), would make it an island in the eyes of the ornithologist; he would base his views on the poor avifaunal list and on the discrepancies of the occurrence of avian life forms there as compared to those on the neighboring Central American mainland. However, the malacologist, forming his opinion from the age, variety, evolutionary diversity, and geographic subdivision of the land snails, would be justified in considering Cuba a continent.

Munroe's keen skepticism agrees with my conviction that the biogeography of both past and present is far too compartmentized to be able to reach generally valid conclusion.* *Faunal analysis,* albeit of a single or of a few groups of plants or animals, is the basis for our present concepts of insular zoogeography and its dynamics. On the basis of faunal and floral analysis, the biogeographer still has a use for the distinction between "continental" and "oceanic"; a *continental fauna* on islands is the remnant of a formerly continuous land fauna. An *oceanic fauna* arose in oceanic isolation, based on immigrants which reached there by long-distance dispersal.

Continental *island faunas* are simpler than faunas on comparable mainland areas. This simplicity pertains to number of species—that is, the faunal list—but can be analyzed for its causation in two ways. One

*Except by group analysis and the statistical approach.

may compare the two faunal lists species by species or group by group, and thus establish which species or higher taxa are missing on the island. One may also study the ecological faunal elements and find discrepancies in this way. An island that was formerly part of a continent had, at the time of their separation, a fauna which was closely related to the faunas of the mainland of comparable habitats. However, we realize that the two were never completely identical, since ecofaunas are delimited on the basis of only a partial faunal list. Many species limits fluctuate; the moment a barrier became established, the retreating members of the now "island fauna" or "island ecofauna" may disappear from the island, and the advancing species will have difficulty reaching it because of the barrier. Margalef (1961) thinks that, in addition to being limited by the absolute size of area necessary to maintain an ecosystem, island ecosystems are limited chiefly by this one-sided weeding effect of climatic fluctuations. Habitat modifications also set in, partly because the physical and biological habitat conditions are altered by exposure to onshore breezes, humidity, currents, tides, and other coastal influences which now may have their impact on the whole island. Such influences on a continental and ecologically nonisolated area work in two ways, but on an island their negative effect predominates. Again, more of the former inhabitants die out, and fewer new ones get across the barrier. In addition, species which are rare—and these form the majority of almost all faunas—run the risk of extinction due to population fluctuation and to lack of replacement or recruitment; new immigrants, even those possessing the potentiality of becoming common, initially face this same risk. As these circumstances are especially favorable for the random preservation of relicts (irrespective of their regular or relict occurrence on the neighboring mainland or their extinction there), Darlington (1957) asserts that one is able to distinguish continental island faunas by the evidence of irregular relict occurrences.

The causes underlying *faunal depauperation* of islands, as enumerated in the foregoing, are those causes which may be ascribed to size and isolation. The correlation between size and faunal richness becomes a statistically measurable phenomenon only if the statistical sample is large enough and if it follows those regressional equations botanists, zoologists, and biostatisticians have established regarding both mainland and island biota (see Munroe 1953, Niering 1956, Preston 1962, Hamilton, Barth and Rubinoff 1964, Mayr 1965). As an example, Figure 6-24 shows the simple correlation of island size and herpetological and mammalian faunal size (combined to give statistically valid results) on three continental islands in the Gulf of California (Lowe 1955).

Faunal disharmony is another negative aspect of the depauperate con-

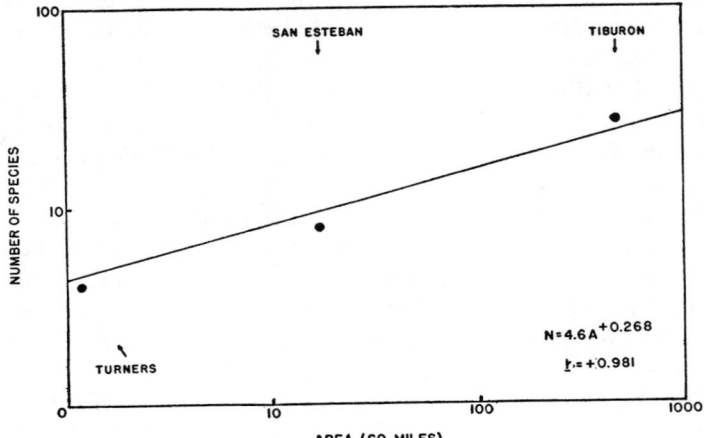

Figure 6-24 Correlation of island size and faunal size. Regression on double logarithmic scale of numbers of species (of land reptiles and mammals) on total island area for each of three adjacent islands in the Gulf of California. From Lowe 1955. (*Courtesy of C. H. Lowe and the Society for the Study of Evolution*)

tinental island faunas. Is is related to the status quo of taxonomic or ecological gaps in the faunal composition as compared to mainland-source faunas. Faunal disharmony is also a characteristic of continental faunas that, owing to a filter bridge, have been only partially and disharmonically replenished, from another source across a barrier. Disharmony is also characteristic of oceanic island faunas; but, as Carlquist (1966a) points out, a positive aspect of those faunas is that all their members possessed adequate means of overseas dispersal, at least until they succeeded in reaching their islands.

Depauperation manifests itself in disharmony on islands. Certain habitat types are either poorly developed or are missing altogether. Certain plants and animals never reach the islands across the water barrier. These initial handicaps will further screen the roster of possible immigrants and will also influence the evolution of those which succeed in obtaining a foothold.

The dynamism of the faunas of *oceanic islands* (or faunas of continental islands completely denuded of their former biota) depends on two circumstances. First, there is long-distance *dispersal* across water. This, as we have seen in Chapter 2, greatly restricts the number of possible colonizers from among members of source faunas. By inductive analysis of bug species lists (Hemiptera, Heteroptera) from several island groups, Leston (1957) calculates the "spreading potential," a relative, numerical index, for each family. His index, however, shows more than the mere spreading potential. Whether certain families reached more or fewer islands actually depends on the combined relative success of take-off, barrier crossing, and colonization. Another, and nonmathematical, method of ascertaining the possible overseas origin—and thereby the "oceanity" of an island fauna—would involve the same criteria, but would be judged according to the distributional characteristics of the taxa of the source fauna elsewhere within their total distribution area. A short survey of this nature by Caughley (1964) shows that the New Zealandic vertebrate fauna, except for some geologically old endemics and relicts, could very well have arrived there by means of overseas dispersal.

Mention has been made of the statistical probabilities of dispersal, and the importance of distance and time was emphasized (p. 53). Darlington (1938b) made the point that chances of successful crossing diminish at a geometrical rate with the distance. We shall use Inger's (1954) data in the following example. If a water barrier is M meters wide and the probability of crossing (denoted as P) has a value of X, then the probability of crossing a gap M meters wide will be determined by the equation

$$P_M = X \qquad (1)$$

and the probability of crossing a gap N meters wide will be

$$P_N = X^{N/M} \qquad (2)$$

Between Mindanao, the largest southern island in the Philippine Archipelago, and Basilan, a small neighboring island, is a water gap of 18 km in width. By arbitrarily assigning the probability of 1/1000 to the success of crossing by an amphibian (i.e., by a frog which is floating on a raft, or lifted by a typhoon) we get, by equation (1).

$$P_{18} = 1/1000 \qquad (3)$$

Negros, another small island, is 45 km from Mindanao. The probability of dispersal there, substituting in equation (2), is

$$P_{45} = (1/1000)^{45/18} = 1/31000000 \qquad (4)$$

Thus an increase of two and a half times in distance diminished the chances of the frog (originally, 1/1000) to reach Basilan by 31,000 times. Thus, great distance, like a huge sieve, filters out the prospective crossers (cf. the filter routes of faunal dispersal, Chapter 2).

A more sophisticated theory to explain this phenomenon has been presented by proponents of a branch of environmental biology who utilize biostatistical models to find possible rules governing numerical phenomena in the biological world. This theory is presented below, with certain modifications and reservations, as an example also worth following in continental zoogeography.

This theory (MacArthur and Wilson, 1963 and 1967) views island fauna as being in a dynamic equilibrium between newly arriving and already-established species, on the one hand, and species that become extinct, on the other. Actually, the larger the fauna of an island, the more species that belong to the rare category and the larger the rate of extinction among them. If the source fauna is the same, a nearby island will continuously receive a large number of the easily dispersing species; a more distant island will have a slower rate of saturation, as we have seen in the example of the frogs of the Philippines. The rate of extinction is dependent on the number of species on the island (see Figure 6-25), the result is a balanced immigration/extinction rate at a species density lower on the more distant than on the nearby island.

The rate of immigration depends on the rate of dispersing individuals

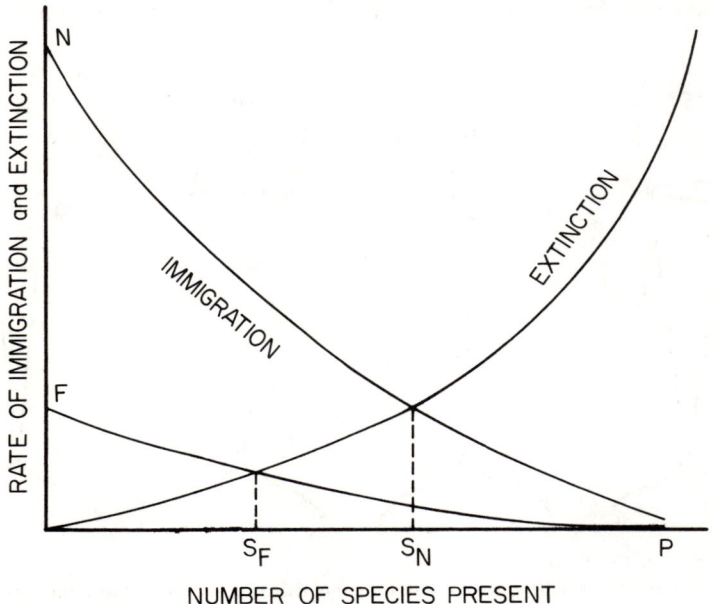

Figure 6-25 Equilibrium model of two island faunas, both islands of equal size. From island; F = Immigration rate of far island. P = Total number of possible immigrant species: S_F and S_N = stabilization points for the two islands, where rate of immigration balances rate of extinction.

reaching the island and on the number of species already present on the island. Additional individuals of species already present on the island would not count as new immigrants. The rate of dispersing individuals for a particular island depends on such important factors at the take-off as the size of the distribution area (source area), its direction from the island, and the size of the species population on the source area. During the passage, distance (p. 374) and wind direction or other weather conditions should be taken into consideration.

The rate of landings will depend on the size of the island, on its shape (the "catching angle," as Lindroth [1960] termed it, i.e., the broadest dimension to the direction of arrival of the immigrants) and also on its height (Figure 6–26). For small insects, the air current on the leeward side of mountains will facilitate landing. The catching angle for birds is larger than the actual size of the island; birds, being active flyers, often change direction and steer toward the island when they sight the cloud cap formed above it during the middle of a warm day. Such clouds are more visible from great distances than the island itself. The total breadth of an archipelago serves as a catching area for actively orienting immigrants, such as birds. Little is known about the flight behavior of wind-dispersed insects. Earlier, we discussed several examples in which active

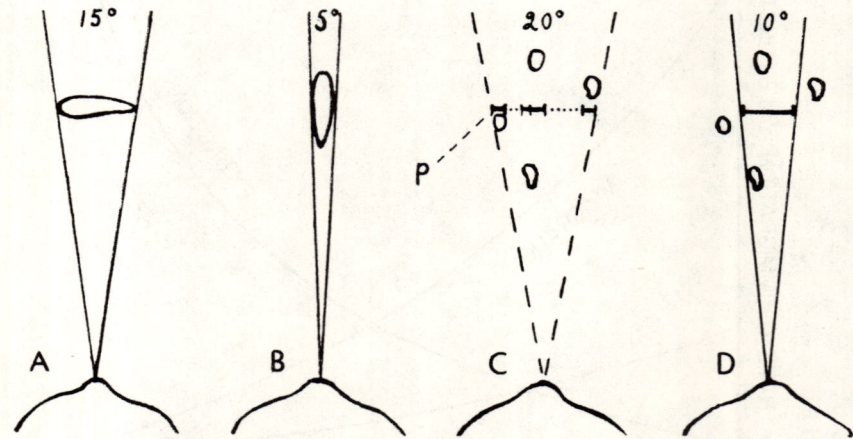

Figure 6-26 Catching angle of islands.

The rate of landings will depend not only on the size of the island but also on its catching angle relative to the direction of the source fauna, or the prevailing wind. From Lindroth 1960. (*Courtesy of C. H. Lindroth and the Municipal Museum of Funchal, Madeira*) A and D, single island in different exposures; C and D, Archipelago; C, seeming angle, D, angle regarded as real and calculated on line of projection, "p."

habitat selection influenced the termination of flight. It is very likely that long-distance dispersal does not deprive most insects of their navigational means, and that upon sighting land they would actively seek to alight. I myself have witnessed the landing of a flying winged ant on board a steamer in the open Pacific; upon landing, this insect shed its wings and was in the act of looking for proper shelter when I collected it.

Apart from the global statistical probabilities outlined above, the *success of immigrants* depends on their finding the proper habitat and other conditions of existence. The more varied the island habitats, the more opportunities for colonization; elevation plays an important role (Paulian 1961, Usinger 1963), not only by establishing belts of different humidity and temperature conditions but also by breaking the wind and dividing these belts into leeward and windward subhabitats. Island faunas are enriched by the evolution that often follows colonization. This phenomenon—*insular evolution*—does not necessarily depend on the age of the island or archipelago but, rather, on the diversity and uniqueness of its habitats and, in island groups, on the isolating factors and processes which enhance radiating evolution (Lack 1947, Hamilton and Rubinoff 1963, 1964). Most important among the special conditions of island environments which stimulate evolution is the simplicity of the environment as compared with that of similar mainland habitats: food chains are shorter, and resources are of lesser variety. Among the intrinsic factors, it is usual to emphasize the small size of the starter (colonizer) group, which increases chances of random fixation (genetic drift) and unusual mixing to occur within the depauperate gene pool of the colonizers. As much as these prerequisites likewise increase opportunities for adaptive evolution into the ecological niches of the island, they may also result in new gene pools which, in the long run, prove nonadaptive and die out (Mayr 1942, 1959), thus adding to the severity of the extinction rate, which is high enough by chance alone.

Crowell (1963) and MacArthur and Wilson (1967) focused attention on colonizing attempts which suffer periodic extinction. This phenomenon belongs to the class of periodic range fluctuations discussed earlier in this chapter; we have seen that advances are due to be followed, as a response to given environmental changes, by retreat—that is, by local extinction—of the former pioneer population. We have also pointed out that area extension, as a response to climatic or other environmental changes, is often preceded or initiated by long-distance pioneers, e.g., among birds and butterflies; we also observed that such long-distance pioneering—the incidence of "accidental visitors"—fluctuates in intensity (Kalela 1952, Salomonsen 1948, Kaisila 1962). This indicates correlation with other population phenomena on the source area as well as beyond its limits. We must transfer this dynamic way of thinking about accidentals in general to island accidentals—that is, immigrants.

Discussions of island evolution often point to the probability that a single landing may result in the formation of a population nucleus. Mayr (1943) showed that the endemic Hawaiian avifauna stems from fourteen ancestral species. Thus, successful landings by at least one pair of birds* of each of the fourteen species would have been sufficient to start an island fauna which has now increased to seventy species by evolutionary processes. Zimmerman (1948) estimates that about twenty-four ancestral colonizations gave rise to over a thousand native land mollusks; in insects, where one fertilized female would have been sufficient to start a colony, he surmises that 3,722 known endemic species were derived from some 254 successful immigrations. The successful establishment of single immigrants or of small groups is known among the adventives and introduced species, but equally well known are the instances of failure. One should not exclude the possibility that successful overseas colonizations (as with the anemo-hydrochore insect pioneers of Scandinavia; see Chapter 2) are perhaps the results of mass arrivals or repeated landings. There is need for future studies in which the frequency of dispersal and the frequency of landing could be correlated on the basis of empirical studies.

It is important to remember that colonization (long-distance dispersal) is a *dynamic phenomenon* and not a one-time occurrence. There is an impressive number of accidental visitors to those islands which have been intensively studied. Therefore, one may assume that each island receives prospective immigrants of *all* species, even those which had already immigrated before, at a regular and even predictable rate. An unfit immigrant population may become extinct immediately. Extinction could of course also happen by chance—that is, one may consider a "landed immigrant" as the prospective starter of an immigrant population. If the landing is unsuccessful, or if the immigrant dies before finding the proper habitat or other conditions for reproduction, the "prospective population nucleus" suffers extinction. New individuals—or, rather, groups or pairs—will, however, arrive, carrying different genes. There is a good chance that some of them will carry the right gene combination for the island—a combination which on the mainland would prove nonadaptive, inferior to the prevailing suc-

---

*In some birds, such as duck and grouse, a single, gravid female would perhaps suffice; in others, pair formation, nesting, and alternating incubation by both sexes are essential for raising young. For instance, pigeons and doves could not be raised without the male.

cessful combinations, or too infrequent to have an influence on the whole population. This dynamism of the colonizing approach is an additional factor which may in the long run promote the success of island speciation.

There are many aspects of *evolutionary adaptations* which are peculiar to, or common upon, islands. Such adaptations which alter the mobility and, thereby, the *dispersibility* (Carlquist 1966a) of animals are worth calling to the attention of zoogeographers. These mechanisms have already been discussed with regard to dispersal and to pioneering adverse habitats (p. 60 and 124). The loss (at least temporarily) of bisexuality is one of these mechanisms. Occurrence of monoecy in island plants, according to Baker (1955), is the rule rather than the exception. The loss of special organs which facilitate aerial locomotion has been stressed by Carlquist (1966a and b). First comes, as is usual in evolutionary processes, the adaptive change in behavior. An insular endemic bird, the Laysan finch *Psittirostra cantans* (Wilson) [Drepanididae], for example, flies reluctantly above cover, although its wings are perfectly developed. In some other island birds and insects, the reduction of flight behavior was followed by atrophy or shortening of the wings. The zoogeographic importance of this phenomenon is that the population would not suffer the loss of individuals which might fly or drift away. The production of such individuals would be a total loss for the island population; it is therefore likely that selection is speeded by such removal of the best (and voluntary) flyers. On the other hand, localized, flightless animals will not have much role in populating other islands or mainland areas, because of their reduced dispersibility. This circumstance may play a role, hitherto unrecognized, in preventing autochthonous island animals from invading mainlands, especially since behavioral restrictions of mobility and agility may go unnoticed.

Historical and Analytical Dynamics of Islands

Recent paleogeographic and oceanographic studies have added much to our undertaking of island biota and their historic origin. The main ocean basins have been in existence, basically unaltered, since the rise of modern continental faunas and floras, from which the islands have mainly been populated. The majority of oceanic islands arose as volcanoes along fracture lines of the sea bottom, eventually reaching the surface; after their extinction, erosion flattened them. When sunk beneath the surface, they are recognized as the guyots—flat-topped seamounts—so abundant, for example, in the Pacific (Menard and Hamilton, 1963). High and low volcanic islands, coral islands, and atolls are members of the dynamic sequence which are presently exposed. The differing past configurations of the island groups of the world's oceans, and not the postulated land bridges and sunken continents, provide the clues to many problems of historical island biogeography. It is just as possible to visualize survival of relicts by island hopping as to demonstrate insular evolution by this kind of dynamics and isolation. Many puzzles of island distributions—especially relict distributions (cf. Croizat 1958)—await further study by new methods of the paleobiogeographer. Our changing view of the rate of evolution, in accord with the geologist's concept of the youthfulness of the volcanic islands and archipelagoes of the ocean, strengthens the view that most of these islands have received their biota very recently, geologically speaking, by overwater dispersal (Zimmerman 1963).

The analytic study of the affinities of oceanic island faunas is a rewarding tool in the hands of taxonomists and zoogeographers. The faunal affinities of each island group will depend, first, on its geographic position and, second, on the dispersal ecology and history of each particular taxon. Because of the varying and sometimes considerable distances between source areas and faunas, the drawing of limits and naming of faunal regions become futile, especially for oceanic areas. Although they stir interest and stimulate discussion, the limits and names of divisions steadily change with increased knowledge (compare, for example, Gressitt 1956 and 1961 and Usinger 1963 on the changing regional divisions of Oceania).

The Insularity of Ecological Archipelagoes on Mainlands

In general discussions of the geography of evolution, it is customary to refer to archipelagoes as ideal for mass speciation and adaptive radiation, and then to point out that the same kind of geographical isolation induces similar speciation on mainlands. Although ecological archipelagoes—scattered and isolated stands of an ecosystem—are very common, the corresponding evolutionary phenomena are surprisingly scarce and occur with some frequency only on a semicontinental scale. An archipelago of montane evergreen forest occurs in tropical eastern Africa, and Moreau (1954) describes the isolated distribution there of ten different forms of the white-eye *Zosterops* [Zosteropidae, Aves] on separate mountain masses within a distance of about a thousand kilometers (Figures 6–27 and 6–28; other examples are shown in Figures 4–25 and 4–37). Because it is difficult to find examples of adaptive radiation in this frequently encountered field situation, we consider several factors that hinder the rate of speciation. Loss or reduction of mobility, which is important on islands, does not occur at the

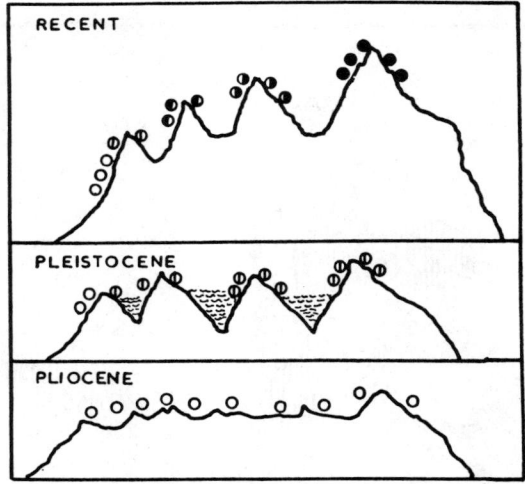

Figure 6-27 Speciation in an alpine ecological archipelago.

Isolating effect of Pleistocene glaciations and of the post-Pleistocene rise (which followed the unloading after the retreat of the glaciers) on speciation and subspeciation among nival insects in the Himalayas. The Pleistocene glaciers isolated and broke up populations of cold-adapted species into isolated units, which then speciated. From Mani 1962. (*Courtesy of Methuen and Co. Ltd., London*)

same rate in mainland habitat isolates, and therefore isolation may not be effective (except in certain insects on older mountain tops; see Figure 6–27). Most such habitat isolates originate by splitting or by the reduction of formerly continuous habitats; therefore, as on continental islands, the fauna is largely if not entirely residual. Consequently, such a fauna did not have the advantage of a small starting population, which condition often leads to speciation. On the other hand, this fauna is subject throughout time to the immigration of many animals from the neighboring habitats, which aids in reducing the uniqueness of the isolated community. Its age is also likely to be less than that of oceanic archipelagoes, as zonal and other environmental fluctuations occur at a relatively high rate. Reinforcements may reach the ecological islands more often than they do the true islands, since dispersal does not take place across a water barrier. This last reason would, however, apply only to flying or anemochore dispersal. Earthbound animals have far greater difficulty in penetrating ecological barriers on land than in dispersing accidentally across water.

A final point worth mentioning about the apparently inferior role of ecological archipelagoes as opposed to real archipelagoes is the human factor in study. Much more is known of island and archipelago evolution than of the mainland counterpart. Island faunas are limited; they are easy to study; there is the incentive that one may do a "complete" work amid those finite conditions; and scientists, as we admitted at the beginning of this discussion, like to visit islands and to have excuses to do so. One example is the outstanding speciation of reptiles on

382 Dynamic Zoogeography

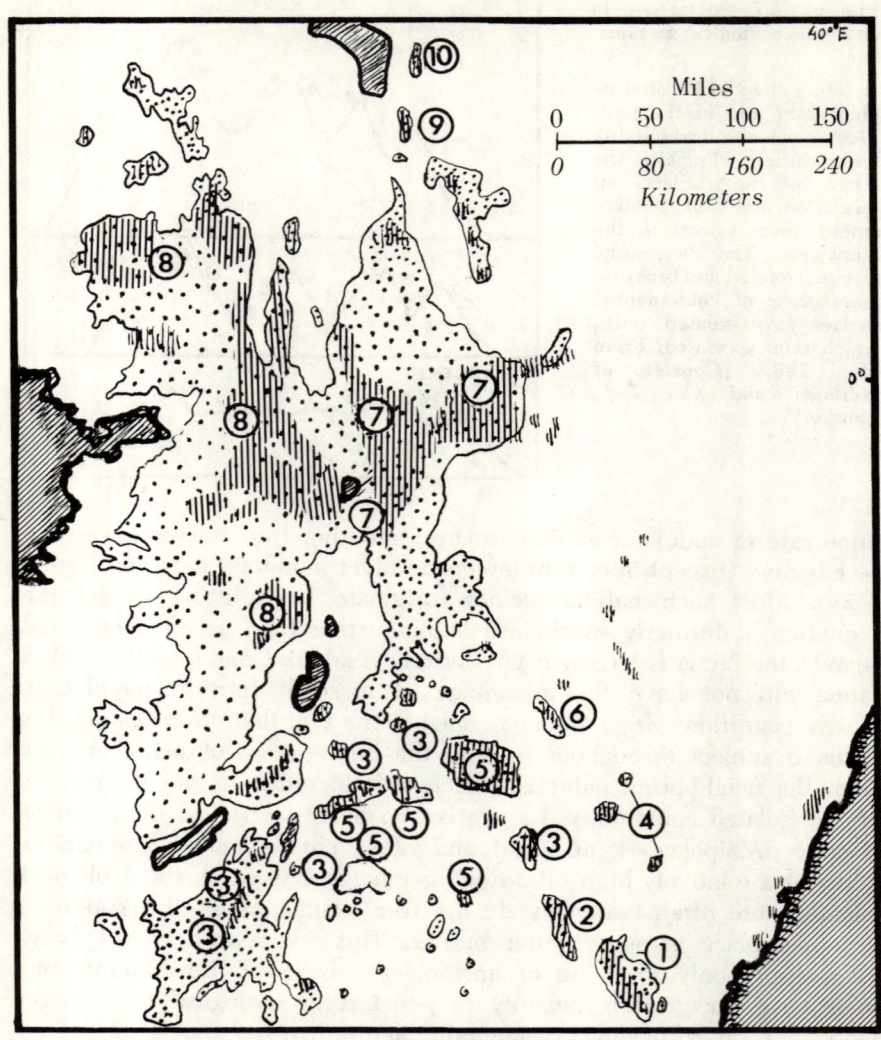

Figure 6-28 Insularity in an ecological archipelago in Africa.
Distribution of ten allopatric forms (subspecies and species) of the white-eye *Zosterops* (Zosteropidae, Aves) in the montane forest of eastern Africa. The montane evergreen forest is represented by dots; 5,000-ft contour, by a solid line. After Moreau 1954.

the islands of the Mediterranean. Two species are especially well known in this respect: *Lacerta pityusensis* Boscá [Lacertidae, Rept.] has 32 subspecies, and *L. sicula* Rafinesque has 39. Mertens and Wermuth (1961) remark that while no one would seriously attempt to name all the different populations of the common European land lizard *Lacerta muralis* (L.), this naming has, in fact, been done for every small island that has a lizard population somewhat isolated and therefore different from the others.

There is good reason to surmise that such explosive speciation as that on archipelagoes and islands may occur also on mainlands more often than is known today; in fact, it might be a regular phenomenon. The ecological island of the isolated, forest-clad volcano Mount Elgon, in Central Africa, is surrounded by desert. There Jeannel (1961) collected no less than 26 species of the carabid beetle genus *Trechus* Clairv., a basically Palaearctic genus. Jeannel thinks these species evolved there very recently from only few ancestors. Reading about this interesting example stimulates one to anticipate more such discoveries of species swarms in ecological archipelagoes.

Dynamic Zoogeography of the Continents

Analytic zoogeography, as discussed in Chapter 5, revealed certain repeated distributional patterns on a continental scale. The combination of geographic (chorologic) distributional patterns and the main barriers result in land vertebrate patterns known as Sclater-Wallace zoogeographical regions. Ecofaunal patterns and geographic entities are combined in the biotic provinces. Geographic and evolutionary affinities of several plant and animal species suggest common immigrational and expansional (or even regressional) backgrounds combined with related ecological modes of living. Historical background varies in almost every species of a locality, and only if *all* taxonomic groups could be considered together would an analysis show the true relations of the heterogeneous assemblage which we simply call a local fauna (Figure 6–29).

Our detailed discussion of the causes of dynamic changes in animal distribution shows clearly that the faunal picture of today is only a momentary "snapshot" of the changing, evolving system of animal assemblages of the land. As we acquainted ourselves with the various and different rates of dynamic changes, we began to realize that the basic faunal pattern of continental land areas is largely the result of the events of the past few million years—that is, of the changes wrought by the alternating glaciations and interglacials of the Pleistocene. We may be biased in our outlook, for we know the most about the faunal dynamism of the immediate past. However, there are some even more weighty arguments.

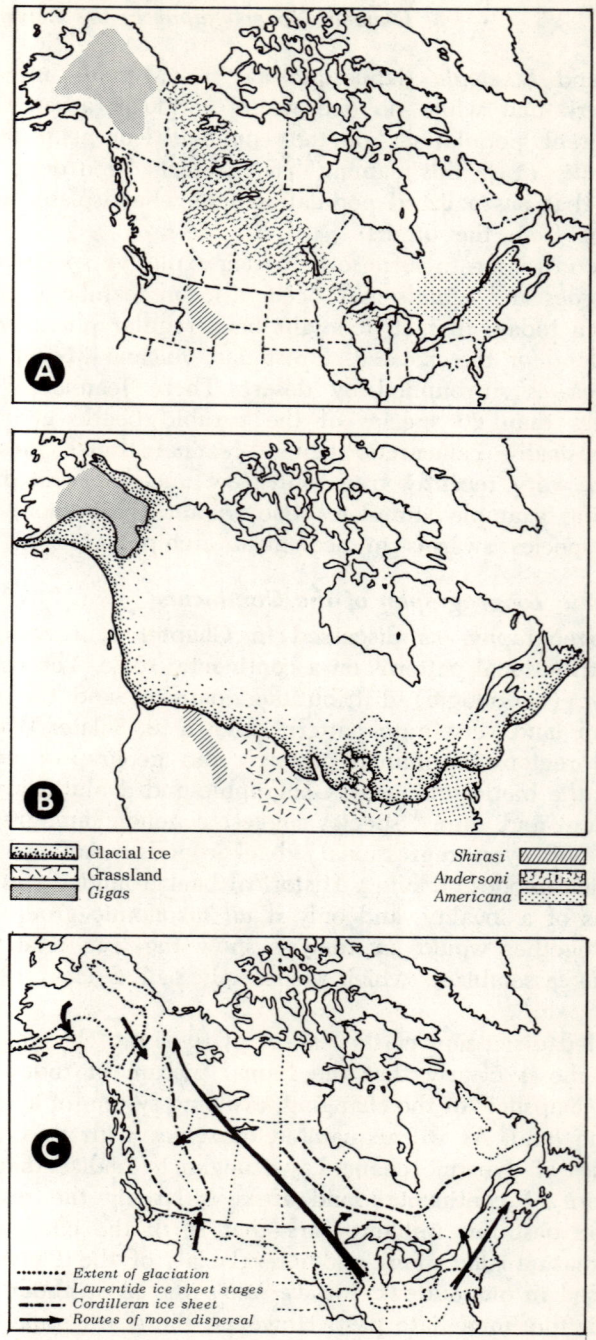

Figure 6-29

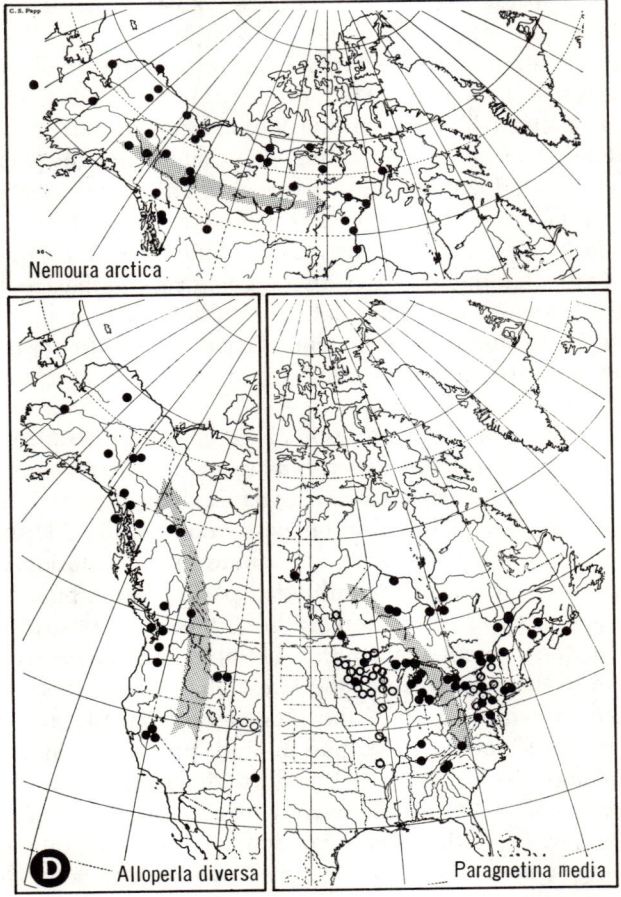

Figure 6-29 Coordination of vertebrate and invertebrate distributions toward a common causal basis. The present distribution and taxonomy of a mammal and some insects seem to point toward the same four Wisconsin refugia of these animals.

A. Distribution of the moose *Alces alces* (L.) [Cervidae, Mamm.] in North America about 1875, with four subspecies indicated.

B. Postulated distribution of the moose during the greatest extent of Wisconsin glaciation.

C. Postulated post-Wisconsin dispersal pattern of the moose.

D. Distribution data of several North American stonefly (Plecoptera) species, with postulated postglacial remigration routes. *Above*: *Nemoura arctica* Esbin-Peterson. *Left*: *Alloperla diversa* Frisom. *Right*: *Paragnetina media* (Walker). (A, B, and C after Peterson 1955; D after Ricker 1964, as modified by Illyes 1965.)

(1) The complete zoogeographical knowledge of the earth's surface at any given time is based principally on the distribution of *species* populations. Owing to the scarcity of fossils, the reconstruction of the past distribution of these entities will always be incomplete—the further back in time, the more incomplete it will be.

(2) The life span of the majority of terrestrial and freshwater species is less than one million years. Therefore, within this time span the species distribution pattern—the content of faunas—has radically changed.

(3) Rates of environmental changes are such that the changes cause alternately radical decrease or great expansion of the species areas within this time span. Even the accumulative effect of eustatic changes will become highly influential during such a long period of time.

(4) Notwithstanding the fact that geologists emphasize the rarity—almost the uniqueness—of such relatively sudden revolutionary changes as those occurring during the Pleistocene glaciations, *we do in fact live in the Pleistocene* and we are witnessing rapid and radical changes of environments as well as of faunas.

Therefore *the immediate causation of the present distributional picture occurred at the time of the Pleistocene glaciations.* It seems convenient to separate on the time-scale causal: dynamic biogeography, as restricted to the study of present distribution with roots in the Pleistocene past; and paleobiogeography, focusing on whatever distributional facts and causations can be found for the unlimited past of living beings. This is by no means a dogmatic separation; rather, it is a practicality. A person interested in the present distribution of biota seldom needs to go back further than to the dynamism of the latest phase of the Tertiary. He who is interested in evolution or in the biogeographical picture of the past, although he uses species distribution and present faunas as starting points for comparison, will work chiefly with distributions of the higher taxa of the remote past. Another word of caution regarding the above-enumerated points is that the evolutionary rate, the rate of environmental changes, and other related rates have seldom been studied, and we use them only in relation to the Pleistocene. It is beyond the scope of this discussion to attempt to determine whether these rates were different and if they were, to what degree during other and earlier geological epochs.

Continental zoogeography is based primarily on the geography of the continents. The recent zoogeography of Antarctica is negligible (for our present purpose), owing to the small marginal geographic extent of habitable area there and to the paucity of the depauperate Antarctic land biota. The geography of each of the three other southern continents is unique. The Cordilleras mountain chain is a uniting feature of the

Americas; it obviously served as a dispersal highway, but the extent to which the accompanying highland faunas are related is not known. Its history of vertebrates (Simpson 1950, Darlington 1957, Mayr 1963a) differs from that of scarabeid beetles (Halffter 1964). The geographic position and geologic background of the two northernmost continents are such that they strongly resemble each other. This—and the fact that they were periodically united in the past—accounts for the unity of a substantial portion of their biota.

Continental ice sheets diminished the habitable portions of North America, Eurasia, and South America during the Pleistocene. The treeless, cold tundra biome evolved and expanded, and temperate forest belts were displaced and shrank. In south-central Eurasia, central America, antitropical Africa, and Australia, pluvials were experienced which resulted in shifts, in the shrinking of the dry belts, and in the expansion of forested, wooded, and savanna habitats. Tropical montane forests descended and expanded; lowland forests shrank and became disjunct. Continental shelf areas became exposed during and following the glacials and became partially inundated again during the interglacials. Important land connections were thus established between Eurasia and North America (Beringia; see Hopkins 1959); between Australia and New Guinea in the north, and between Australia and Tasmania in the south (see Keast 1962); and in the Indomalayan Archipelago (see Mayr 1944). During the interglacials, aridity progressed to varying extents in the antitropical dry-belt areas of today; tropical lowland forests expanded, as did the temperate forests; montane forests diminished and moved altitudinally during the warm periods. Examples of the presently acting secular dynamics, and of the dynamics of the fauna during the past few millennia complete the grand kaleidoscope which the zoogeographer of a continental area is bound to use as the background for his survey. The accompanying faunal movements are well known for mammals and birds and for a few insect families. There is, however, no synchronized view of these movements; some even assert that all of our present zoogeography is vertebrate zoogeography, and that land invertebrates might present a different picture. During several years of source searching and extensive correspondence with systematic biologists and zoogeographers, I was not able to bring together enough published material emphatically to deny such assertion (but see Figures 6–30, 6–31, 6–32), especially concerning the mainly tropical continents. Most of the synthetic literature consists of evolutionary-taxonomic facts and interpretations and of paleogeographical inferences or speculations. Croizat's (1958, 1966) monumental volumes comprise a library of facts, but his emphasis is historical and so highly controversial that it has not yet been

388 *Dynamic Zoogeography*

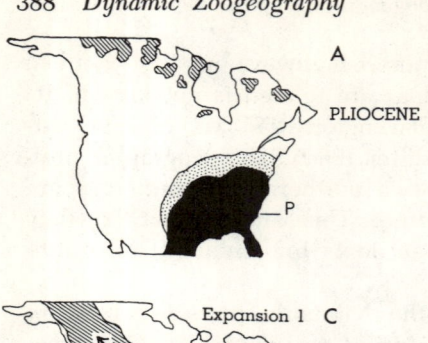
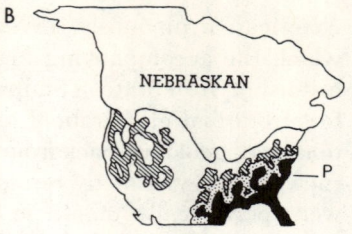
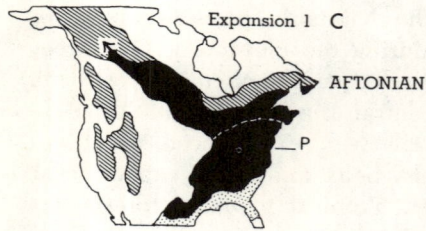
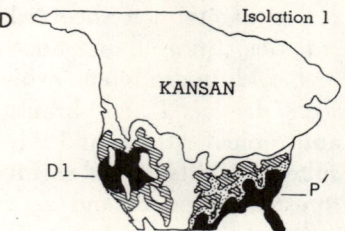
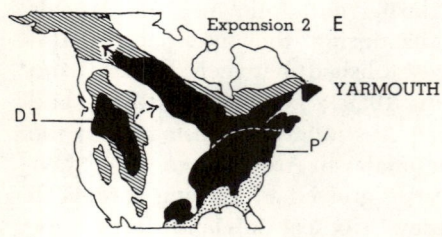
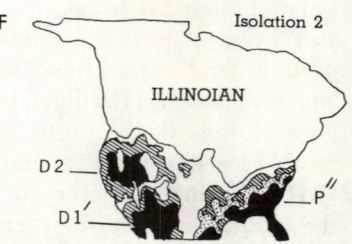
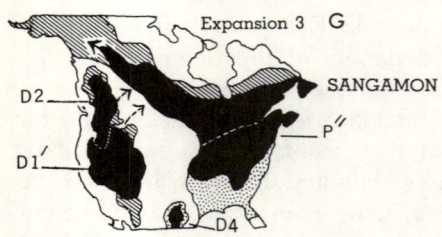
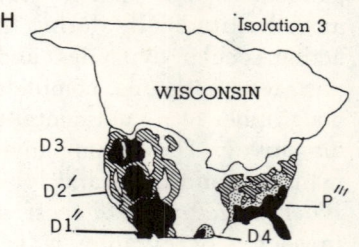
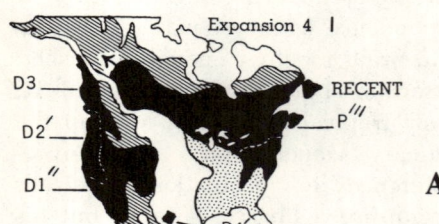

P = parental species
D = derivative species

Prime marks = number of glacial cycles
removed from origin

Hatched areas = boreal coniferous forest
Stippled areas = deciduous forest
Black areas = ranges of birds

Figure 6-30

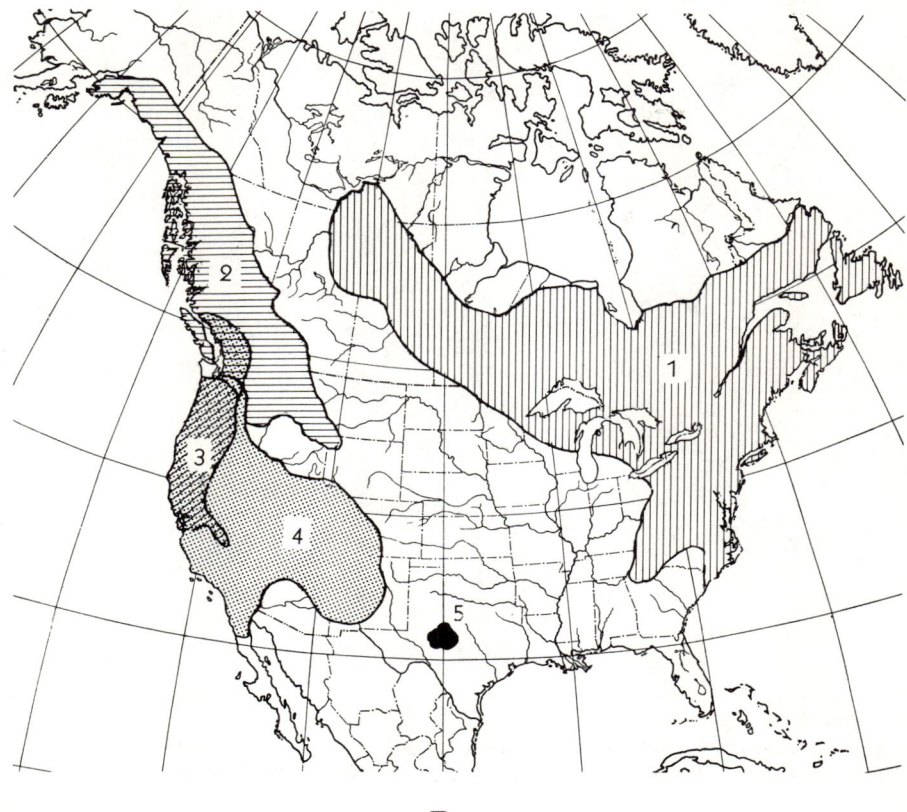

B

Figure 6-30 Millennial dynamics of habitats and accompanying distribution areas and evolution of the species. From Mengel 1964. (*Courtesy of R. M. Mengel and the Cornell Laboratory of Ornithology*)

A Speciation in wood warblers (Parulidae, Aves). A model sequence (A to H) showing the effects of "glacial flow and ebb" on adaptation and evolution of a hypothetical ancestral wood warbler and its descendant populations. The ancestor lived in the deciduous Arcto-tertiary forests of southeastern North America during the Pliocene. At the onset of glacial cooling, this forest became in contact, and mixed with, northern coniferous forest elements, and the warbler began to adapt to these. Upon retreat of the ice, this species followed the boreal forest northward and transcontinentally. At the next glaciation, its area split and speciation set in, enhanced by the shrinking, disjunct western part area during the subsequent interglacial. The whole process then repeated itself, resulting in the (hypothetical) pattern of one original and four derived species. B and C show the actual distribution of two superspecies of warblers.

B The distribution of the *Dendroica virens* group: *Dendroica* (1) *virens* (Gm.), (2) *townsendi* (Townsend), (3) *occidentalis* (Townsend), (4) *nigrescens* (Townsend) and (5) *chrysoparis* Sclater and Salvin.

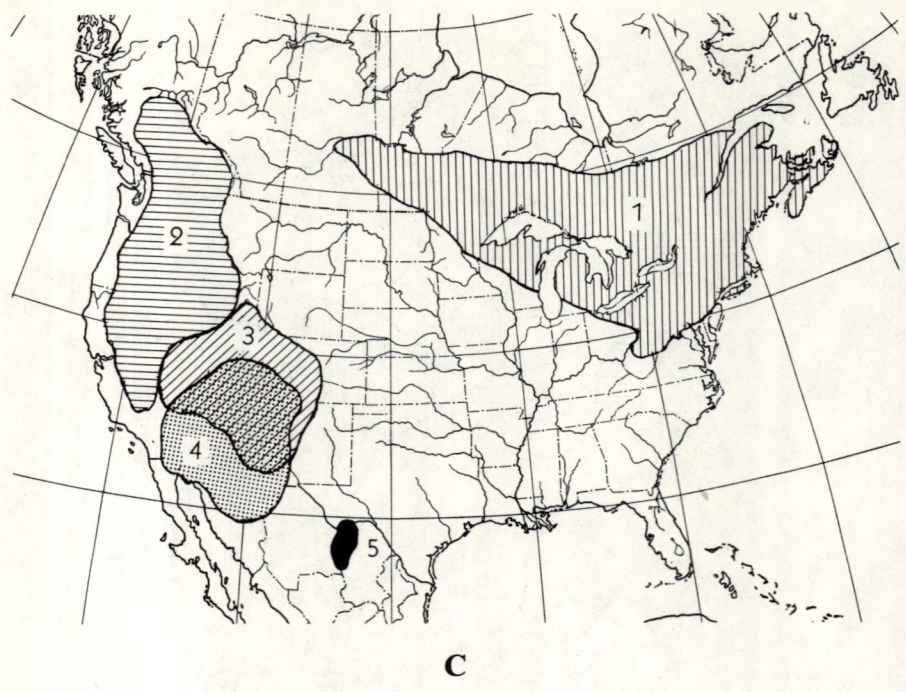

C

Figure 6-30 (Continued)
C *Vermivora*. (1) and (2) disjunct subspecies of V. *ruficapilla* (Wilson), (3) *virginiae* (Baird), (4) *luciae* (Cooper) and (5) *crissalis* (Salvin and Godman).

fully examined. But the facts he adduces would be enough for a score of young zoogeographers to build into orderly, analytic monographs.

Recent treatments of the causal zoogeography of the immediate past are based largely upon the tangible evidence of vertebrates and, above all, on the better-known Pleistocene history of vegetation. The emerging picture is that of periodic alternations of ecologically antagonistic major zonal formations which simultaneously expand and shrink. The accompanying fauna follows the vicissitudes of the vegetation. In this process, the "migrating" distribution area of its members becomes discontinuous and again reunited; relicts or isolates are left behind; and many of its members become extinct. In the western Palearctic and in the Nearctic, there is enough fossil evidence to attempt to document at least the latest expansion of the boreal biota of some mesic ecosystems (Wright *et al.* 1965, Hubbs *et al.* 1958, Freitag 1962, de Lattin 1957 and 1967, Moreau 1955 and 1966). Lönnberg (1929) and Moreau (1952) laid the foundations of a Pleistocene biogeography of Africa, but recent biogeographic

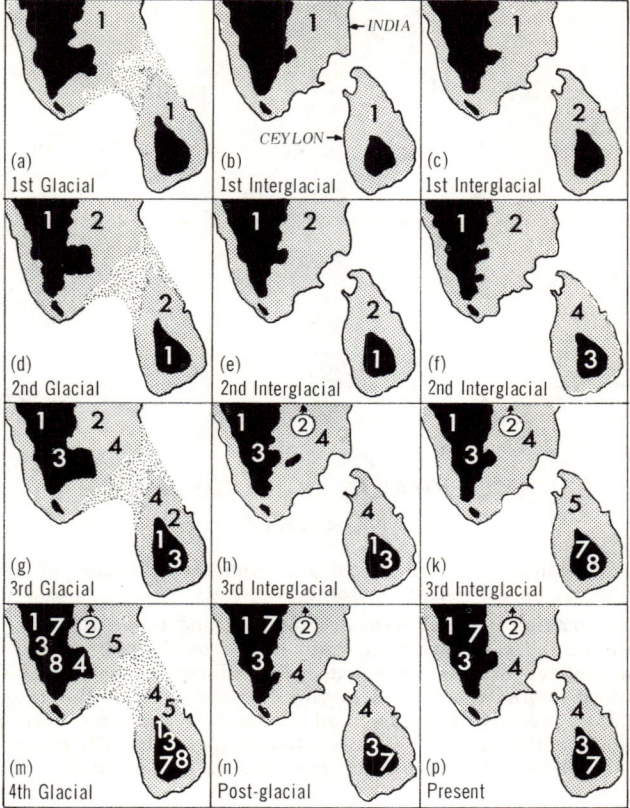

Figure 6-31 Hypothetical model for origin of five species of *Funambulus* [Sciuridae, Rodentia, Mamm.] squirrels. (After Moore 1960.)

From top to bottom: geographic isolation (sea and land habitat barriers shown have been active) in the sequence of the four Pleistocene glacials, and interglacials (each glacial and the following interglacial, with speciation shown left to right) resulted in five species being derived from one hypothetical ancestor.

activity (Barry et al. 1962; Howell, Bourlière, et al. 1963; Bakker 1967) is in the process of rapidly complementing our knowledge. Judged mainly from the discontinuities of vertebrates, the dynamism of the Pleistocene affected even the tropical African vegetational (altitudinal, latitudinal, and longitudinal) zones. Australian vertebrate zoogeographers emphatically ascribe the present distributional picture to Pleistocene alternations between pluvial and arid periods, with corresponding expansions and retreats of mesic and xerophil faunas, and to latitudinal oscillations of the temperature zones.

392 Dynamic Zoogeography

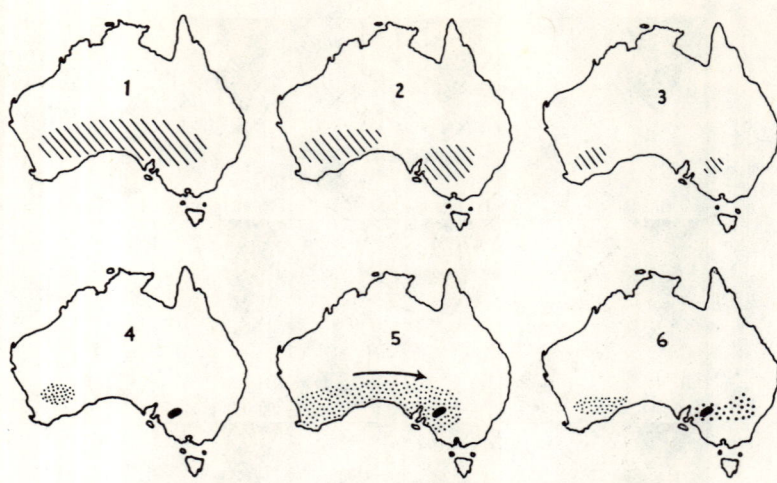

PACHYCEPHALA RUFOGULARIS–
INORNATA

Figure 6-32 Hypothetical distributional and evolutionary history of the Australian birds of the genus *Pachycephala* [Muscicapinae, Aves].

1–3: A widely ranging parental form became split and isolated in the western and eastern sector of the continent as a result of the north-south shift of the rainfall belts. These isolates became speciated; with subsequent habitat shift, the new western form *P. inornata* colonized eastward to the area of the eastern form *P. rufogularis*. More recently, the widespread species *P. inornata* has been disconnected again by habitat shift, and now has two disjunct subspecies. Cf. Figure 6-11. From Keast 1961. (*Courtesy of A. Keast and the Museum of Comparative Zoology, Harvard University.*)

Paleozoogeography

Historic zoogeography encompasses events during the whole geological time scale, and therefore its aims and methods are radically different from those of the dynamic zoogeography of present proximate causations. In the course of millions of years, it is not the species but the higher taxa that fluctuate. The geographic entities involved are continental or worldwide. If we would assert that the cornerstones of dynamic biogeography of the present are species area and climate, then we should perhaps state that historic biogeography is based on macroevolution and continental history, with an added emphasis—but only in the third place—on climate.

In this context, we mean by "macroevolution" the process of phylogenetic unfolding—in time as well as in space—of a higher taxon from a single, ancestral species by means of cladogenesis, i.e., by cleavage of the ancestral species into one or more daughter species, among which certain "parental" characteristics are retained, some are altered, and others are abandoned. Thus phylogeny and historic biogeography complement one another on a continental and worldwide scale (Hennig 1966). When gradually changing morphological characteristics are found to occur in a number of related taxa, the phenomenon is called "character progression." When character progression shows an order paralleling the orderly distribution of the same taxa, Hennig and his followers perceive certain inferences to be drawn respecting the immigration

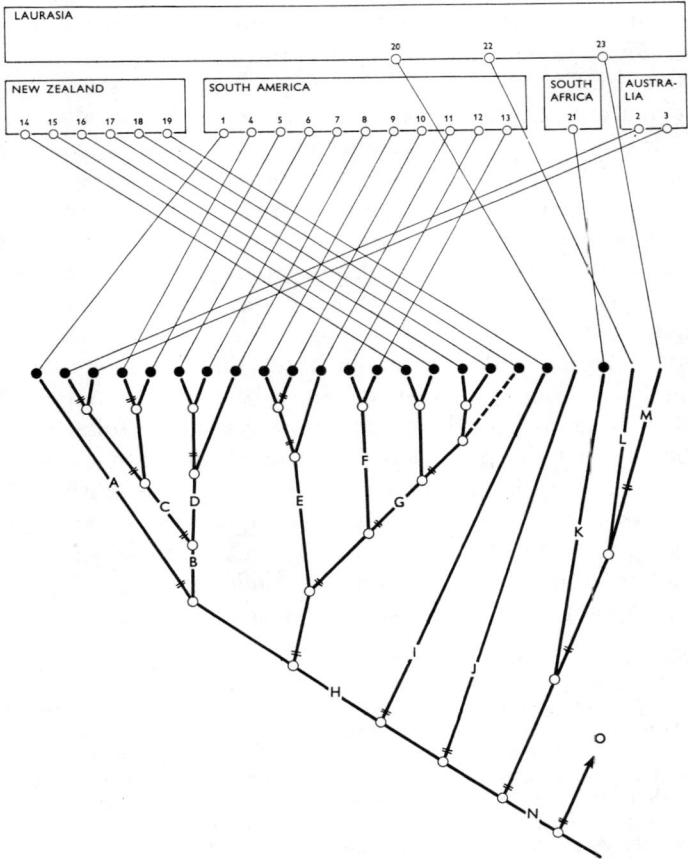

Figure 6-33 Geographic distribution and phylogenetic relationships as shown by the southern hemisphere tribes of the subfamily Diamesinae (Chironomidae, Diptera). From Brundin 1966. (*Courtesy of L. Brundin and the Royal Swedish Academy of Sciences*)

394 Dynamic Zoogeography

route of a spreading and evolving major taxon. A good example is elaborated in Brundin's (1966) monograph on the chironomid midge taxonomy and geography of the Southern Hemisphere (Figure 6–33).

Questions of continental history are answered by the geologist and not by the biogeographers. However, the zoogeographer needs to know how the faunas of the past were interconnected, how the animals and faunas arrived at those places where the fossil record and taxonomic indications place them. Knowledge of the past geography and physiognomy of land and sea is the key to the past location of faunal connections and barriers. Paleoclimatological knowledge is a clue to the availability of these localities for the contemporaneous biota. Historic zoogeographers of the nineteenth and early twentieth centuries were often geologist-paleontologists; they were especially trained in the recognition and study of marine invertebrates (because geological strata of marine deposits were identified by the accumulated shells of so-called leading fossils) and also in the palaeontology of vertebrates. In land formations, the leading animal fossils are herpetological and mammalogical specimens, with a few freshwater fish and almost no birds. The only extensively known Tertiary insect fauna is that of the Oligocene amber. As long as paleobiogeography and historic geology developed side by side, circular reasoning was common and often hampered progress. To many it seemed reasonable that while the geologist described wide introgressions of the sea upon the presently continental and even mountaneous areas (based on layer upon layer of fossilized marine shells), the paleozoogeographer postulated wide land connections across the present oceans (based on the negative evidence of wide disjunctions of primitive or relict taxa). Geologists, not having access to the ocean bottoms, believed and quoted the zoological speculators, and these again invoked the authority of the same geologists.*

The rather indiscriminate land-bridge building (Figure 6–34) ended with two important new scientific tendencies: emphasis on the roles of

---

*It is interesting to compare two arguments taken from the zoogeographical literature; the first author supports, the second rejects, the theory of land bridges across the ocean. Apart from this, their reasoning is strikingly similar. Zimmerman (1963, p. 477) states that the same genus of moth is present on the small Polynesian island of Rapa and also in Australia; Meyrick (1926, p. 271) believes that it reached Rapa by means of a land bridge. Meyrick writes: "A rise of 12,000 feet in the sea-bottom of the South Pacific *is required to show these results,* but I entertain no doubt that such an elevation must have existed since the Eocene period, because *it is absolutely the only explanation possible*" (italics added). Croizat (1958, p. 1624), on the contrary, draws attention to the following, from Darlington (1957, p. 323–324): "It is possible—it has been suggested before—that the ancestor of the South American hystricomorph rodents came across the Atlantic from Africa, perhaps on a raft. . . . I do not say that this happened, but just that *it may have happened, if the distribution of the hystricomorphs requires it.*" (Italics added.)

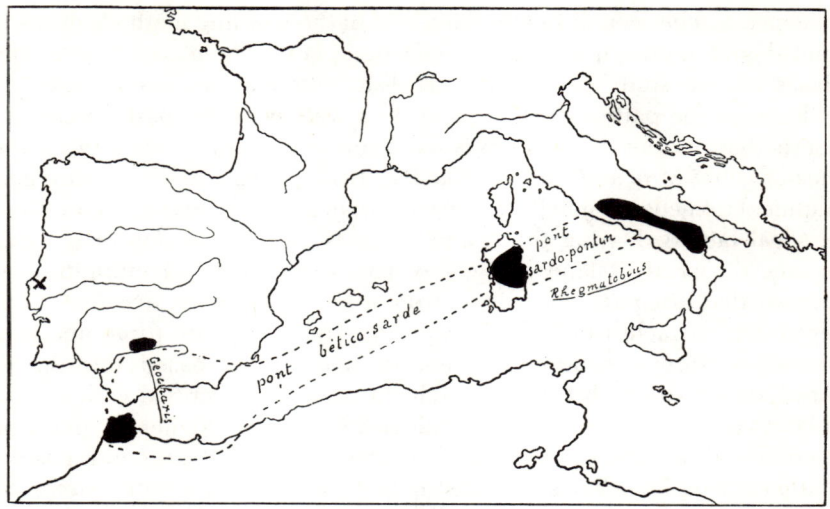

Figure 6-34 The postulation of land bridges on the basis of recent distribution. Two beetle genera of the tribe Anillini [Carabidae] are here connected with a postulated Oligocene land bridge across the eastern Mediterranean, and the disjunct area of *Rhegmatobius* is connected with a surmised Miocene bridge between Sardinia and the Apennine peninsula. From Jeannel 1942 *Courtesy of Presses Universitaires de France.* Jeannel notes that after his map was drawn a species of *Geocharis* was found in the vicinity of Lisbon, Portugal (marked X). Should we now extend the bridge to this area?

climate and of the existing continental connections and barriers as projected into the past, and the emergence of geological theories which gradually led to the separation of these from the zoogeographical evidence.

In this respect, we should direct our attention to the activities of two forceful zoogeographers active during the first third of the twentieth century: Anton Handlirsch and William D. Matthew (cf. p. 304). Not only did these scientists take basically the same stand, each in his own language, on the critical aspects of their subject matter but they arrived, though by different routes, at almost identical conclusions. Yet the impact of these two men became different, as we shall shortly see.

Handlirsch was at once an insect paleontologist and an insect geographer. He analyzed the occurrence of species and genera, and compared it with the fossil record and with the geological evidence. His sound zoogeographical principles (published first in 1913 and reprinted in 1939) included many new ideas which have since become so well known that they are seldom attributed to him. Handlirsch's

396 Dynamic Zoogeography

influence can be seen in the thinking and in the working methods of many central and northern European biogeographers. He placed major emphasis on the study of the present distribution of species in order to understand the patterns of the present as well as of the past faunas. He asserts that present distribution is the result of *evolution with equal influence of geohistoric and ecologic factors* (1913, p. 465), and that pure geographical influence is much less decisive than climatic, ecological, or physiological factors. During the heated debates on the question of whether oceanic dispersal or land bridges explain wide areal discontinuities, he stressed that *discontinuities,* apart from a few instances of accidental dispersal, *are* characteristic of *receding,* formerly widespread *forms* and bear witness to former continuity. Handlirsch evidently based this on his knowledge of insect distributions, which can now be verified with respect to land vertebrates. Not being a geologist, he did not comment upon continental and oceanic permanence, but pointed out that *presently and recently existing land bridges* across shallow seas, together with *chance extinctions and climatic factors, explain the origins of present distributions*

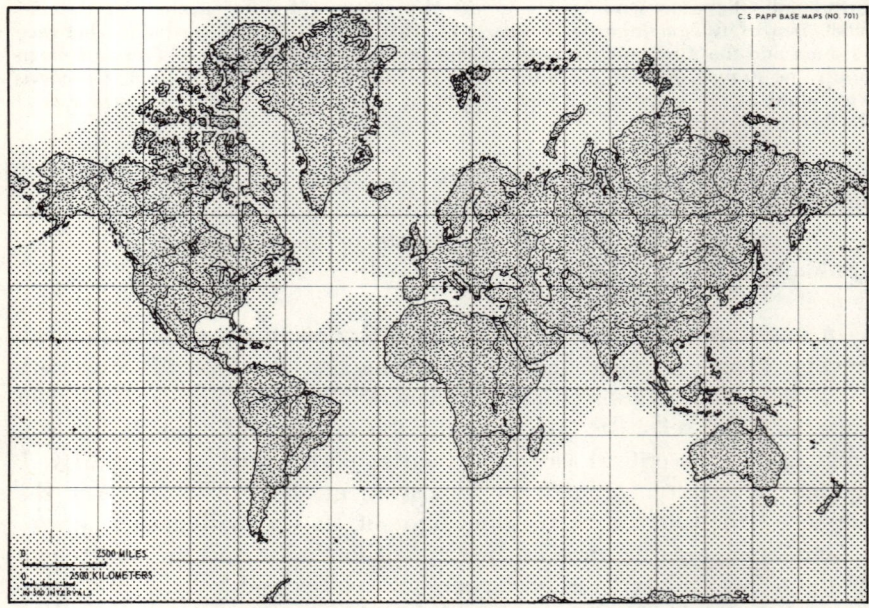

Figure 6-35 Map of the world, showing the combined area of land bridges that several authors postulated for the Cretaceous and for the Tertiary, up to 1913. As can be seen, hardly any ocean bottom remained after all the postulated land bridges had been superimposed on the map. After Handlirsch 1913.

without the need to speculate on intercontinental connections (Figure 6-35). Most important for our present discussion were his conclusions, summarized below, regarding climate and evolutionary centers (Handlirsch 1913, a summary of pp. 465–468).

(a) The present distribution area of many systematic groups is not their area of origin. Frequency of occurrence of member taxa is not an a priori key to the place of origin of a higher taxon. Those members of a group which live farthest from the center of frequency are rarely pioneer forms but usually survivors of a time when the group was more widespread.

(b) It is unlikely that there ever was a single, uniform evolutionary center for all land animal groups. On the contrary, it is likely that new groups may have evolved in all parts of the world.

(c) It is likely that the greater number of those groups which arose since the Cretaceous stem from the northern circumpolar land mass which stretched across all climatic zones and was exposed to all climatic fluctuations, rather than from the three southern continents, which do not extend as close to the pole and which have evidently not been in connection with Antarctica during the time period since the Cretaceous.

This last trend of thought also forms the central theme of Matthew's theses, which he published in final form in 1915 and which he summarized as follows.

1. Secular climatic change has been an important factor in the evolution of land vertebrates and the principal known cause of their present distribution.

2. The principal lines of migration in later geological epochs have been radial from Holarctic centers of dispersal.

3. The geographic changes required to explain the present distribution of land vertebrates are not extensive, and for the most part do not affect the permanence of the oceans as defined by the continental shelf.

4. The theories of alternations of moist and uniform with arid and zonal climates . . . are in exact accord with the course of evolution of land vertebrates. . . .

5. The numerous hypothetical land bridges in tropical and southern regions, connecting continents now separated by deep oceans . . . are improbable and unnecessary to explain geographic distribution. On the contrary, the known facts point distinctly to a general permanency of continental outlines during the later epochs of geologic time, provided that due allowance be made for the known or probable gaps in our knowledge. . . . (p. 3.)

Many students of geographic distribution proceed on what appear to me to be wholly false premises. They assume that the habitat (i.e., geographic area of domicile) of the most primitive living member of a race is the original habitat of the race, the most advanced forms inhabiting the limits of its migration. It seems to me that we should assume directly the reverse of this. (p. 10.)

398 Dynamic Zoogeography

Thus we see that Handlirsch and Matthew both advocated the influence of *climate* on *evolution*, and that important centers of origin and radiation of certain land animal groups—"land vertebrates" of Matthew; "a greater number of groups" of Handlirsch—were on the northern circumpolar land mass (Handlirsch) and, more specifically, in the Holarctic (Matthew). Both based their theories on the presence of primitive survivors of such groups on the peripheries of the continental mass (Figures 6-36, 6-37), pointing out in their documentation that these had become extinct in the intervening areas. Handlirsch, who had a clear and firsthand experience of the antiquity as well as of the often more southern distribution of many insect groups, was more noncommittal than the geologist-paleomammalogist Matthew, who conjectured heavily from fossil data.

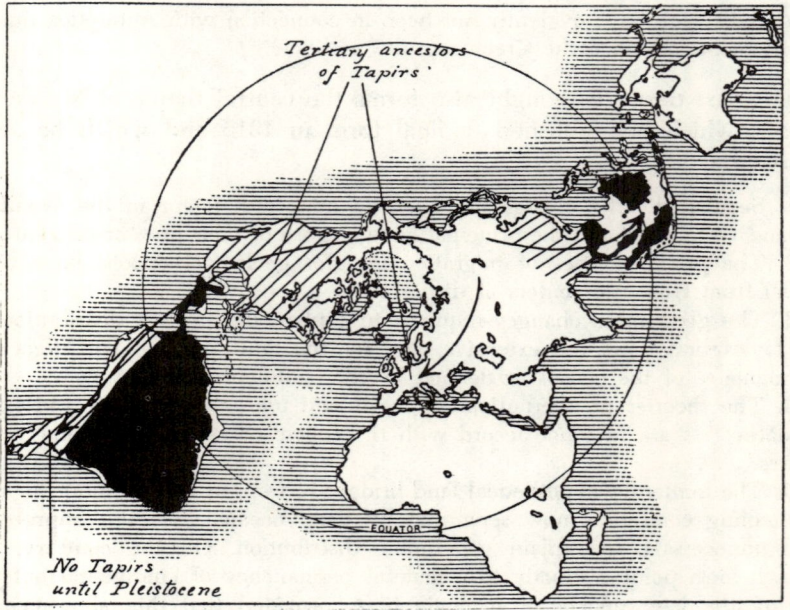

Figure 6-36 Distributional history of tapirs.
Radial "migration" of the tapirs (Tapiridae, Perissodactyla, Mamm.) from a surmised Holarctic center, acccording to Matthew (1915), who points out that Tertiary ancestors of the tapirs lived in Europe and North America. Hatching indicates Pleistocene distribution; present distribution in black. This figure is reproduced for its historic value; we did not check the present interpretation of the paleogeography of tapirs. After W. D. Matthew, *Climate and Evolution,* 1939. (*Courtesy of New York Academy of Sciences*)

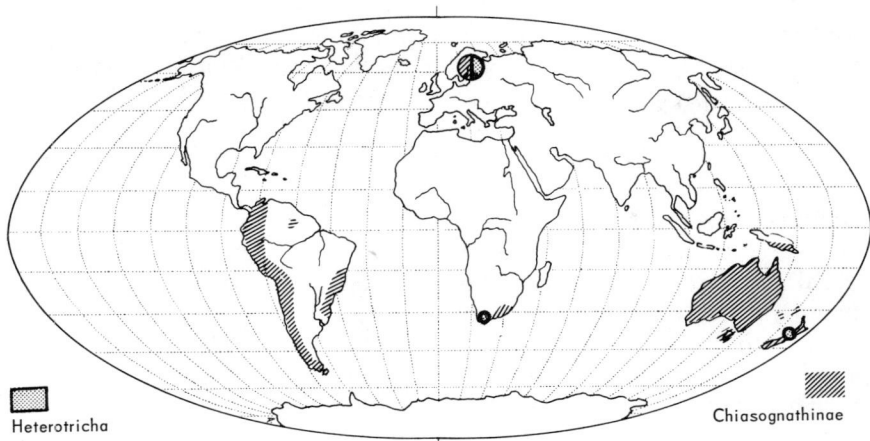

Figure 6-37 Recent and fossil distributions of widely disjunct, relict insect taxa. The family Chiasognathinae [Coleopt.], recent in South America, southern South Africa, and in the Australian region. The genus *Heterotricha* (Dipt.), is recent only in the Cape and locally in New Zealand. Both fossil in the Baltic amber of Oligocene age. (After Ander 1942.) Without the fossil evidence of northern, historic occurrence, such distributions are still used as arguments for the former existence of a large, joint southern land—Gondwanaland.

The "climate and evolution" theory—named after Matthew's 1915 title—became the favorite basis for the historic zoogeography of vertebrates, especially in North America, and of the angosperm phytogeography of the northern continents. Geologists believing in the *permanency* of continents also use this theory for support of their thesis. While Handlirsch's version was presented in an equally appealing and rational form, his European opponents soon found a new basis—elaborated upon by Wegener—for modification of the land bridge theories in the *continental drift* hypothesis. Geologic conformities of the southern continents, geographic resemblances of the South Atlantic coastlines, and other phenomena—not the least of which are the faunal and floral conformities—convinced many scientists that the southern continents had been unified at some point during the geologic past. Wegener (1915) and others (see du Toit 1937) suggested the possibility that all continents originated as one coherent land mass which later split up and drifted apart—and that they are continuing to drift.

The available paleobiological and biogeographical evidence of the former unity of the southern continents remains incomplete and often ambiguous. Historic biogeographers, having disagreed for many decades, are still divided into two camps. To an outsider, it would seem that since much vertebrate distribution and evolution occurred in the not too remote Tertiary, it might well be interpreted satisfactorily on the basis of the permanence of continents and of ocean basins during this period. Other major taxa, especially among plants and land invertebrates, had evidently dispersed earlier, but their phylogenetic and paleogeographic relations are not well known. Decisive evidence must lie buried under the mighty ice sheet of Antarctica, which, even without continental drift, is in the center of the Southern Hemisphere and certainly must have had faunal and floral connections—whether through land bridges or stepping stones is irrevelant—during periods when it was not glaciated, and when it harbored a fauna and flora of its own. Analysis of the floras, and especially of the land faunas of the southern continents and particularly of southern South America, has not yet advanced enough to add new evidence; but such will be forthcoming in the near future because pertinent research is being carried out (cf., for example Pantin 1960, Brundin 1966). Darlington's (1965) synthesis of "the southern end of the world" uses the old and widespread carabid beetle family as the thread of his argument, and he finds much evidence of southern evolutionary and biogeographical connections. Van Steenis (1962) presents a thorough summary of the botanical evidence, of which he is responsible for a large share, in favor of the land bridge theory.

The opponents of continental stability found a new weapon in Runcorn's (1962) studies of the changes in the magnetic orientation of rocks. These studies seem to indicate that the continents at one time did not have the same position as they now have relative to the magnetic poles, and that if the poles themselves did not shift, the continents may perhaps have drifted, or both the continents and the magnetic poles have moved. By the 1960's, accurate mapping of the ocean bottom has been completed; the charts reveal the Mid-Atlantic and other ridges and structures suggestive of a rift whence the continents broke apart. Many biogeographers have already accepted the possibility of continental drift as proved, since the evidence derived from the distribution of animals points to closer connections between the southern continents in pre-Tertiary times (e.g., Darlington 1965). Still others are waiting until the geologists and paleogeographers come to an agreement about the validity of paleomagnetism as evidence and until other alternatives can be reinterpreted.

CONCLUSIONS

Climatic and environmental changes that cause distributional changes are presented. The rate of vegetational change is slow, though single

Conclusions 401

plant species are able to expand or retreat as fast as the most conspicuously dynamic animal species.

The concept of faunal dynamism is introduced. Every local fauna at any one time has a *dynamic potential*, i.e., a certain proportion of its members are actually or potentially expansive or regressive. The magnitude of the dynamic potential of a fauna depends on the geographic and ecologic position of its members.

Examples of rapid expansion are found mainly among rather mobile forms (birds, mammals, butterflies, carabid beetles). The dynamism of most land invertebrate taxa is largely unknown, and synthetic as well as experimental work is needed.

Evolutionary rates and rates of distributional (and faunal) changes have an intimate relationship. Environmental changes induce distributional and adaptational changes i.e., geographic movement and speciation. This is substantiated by the example of the North American land vertebrate fauna. Between 25 and 43 percent of its species show morphological and areographic evidences of recently completed or presently progressing dynamism.

Four kinds of extinction are distinguished: pseudoextinction, singular, group (or synthetic), and phyletic. It is postulated that the ultimate cause of the extinction of a species (singular extinction) lies in the rate of adaptation and/or dispersal being slower than the rate of environmental changes adversely affecting the population. Synthetic extinctions occur when the responsible environmental changes reach major proportions, such as in a glaciation or by the recent overwhelming dominance of man. Competition is a factor too often stated to have caused extinction of faunas or of phyletic lines. It is recommended that since competitive replacement of phyletic groups in the course of geologic times is difficult to prove—and at any rate is not the same phenomenon as intra-specific or interspecific competition—the more objective term *phyletic replacement* or *faunal replacement* be used.

The mode of community dynamism is comparable to that of the member species of communities. Communities in time exhibit geographic movements; while doing this, they undergo dynamic, "evolutionary" changes in their structure and function by acquiring new elements and dropping old ones and by rearranging "dominance" relationships. Thus, in the course of their geographic history, they change location, become extinct, or transform (evolve) into new communities.

Man's role in recent faunal dynamism cannot be too strongly emphasized. The most important role of man is not as a carrier but in habitat alteration; the detrimental role of man is especially demonstrated in juvenile and imbalanced island biota.

Historic zoogeography blends into paleontology, which in turn, blends with historic geology. Geology's guiding principle is uniformitarianism; for zoogeography, it is stated in the reverse—*All of those dynamic processes which have been active in the geologic past are also acting at the present time.* Another prinicple, that of the *constancy of ecological valency,* is stressed with emphasis on its application to faunas or other group entities.

Size, isolation, and environmental dynamism cause the peculiarities of *smallness* and *faunal disharmony* characteristic of island faunas. Theories of dispersal to, and colonization and extinction on, islands are discussed. Similarities and differences between islands and ecological archipelagoes on continents are pointed out.

The immediate causation of the present distributional picture of the continents occurred at the time of Pleistocene glaciations. Pleistocene biogeographical history of any one geographic area consists of periodic alternation of antagonistic major ecosystems which simultaneously expand and shrink.

In the extensive field of paleozoogeography, the thesis of land bridge crossings and its antithesis about the role of climatic changes and evolution are discussed. These questions also involve the controversies about continental stability versus continental drift. Solution to these questions is, in the first place, in the hands of the physical scientist; therefore, zoogeography should follow rather than precede the findings of geology and geophysics.

References

Agassiz, L. 1850. *The Christian Examiner* 48, 181 (Fide Schilder 1954).
Ahlquist, H., and E. Fabricius. 1938. Die Vögel der äusseren Schärenhofes zwischen Tvärminne und Jussarö. *Ornis Fennica* 15, 21–32.
Aldrich, J. W. 1967. Taxonomy, distribution and present status. Chapter 2 in: O. H. Hewitt (ed.) The wild turkey and its management. Wildlife Soc. Washington. pp. 589.
Allen, J. A. 1871. On the mammals and winter birds of East Florida . . . *Bull. Museum Comp. Zool.* II, 3. pp. 161–450.
Allen, J. A. 1878. The geographic distribution of mammals. *Bull. U. S. Geol. Geogr. Survey* 4, 339–343.
Allen, J. A. 1892. The geographic distribution of North American mammals. *Bull. Am. Mus. Nat. Hist.* 4, 199–242.
Alvariño, A. 1964. Bathymetric distribution of Chaetognaths. *Pacific Sci.* 18, 64–82.
Ander, K. 1942. Die Insektenfauna des baltischen Bernsteins nebst damit verknüpften zoogeographischen Problemen. *Kung. Fysiogr. Sällsk. Handl. N.S.* 53, No. 4. pp. 83.
Andrewartha, H. G. 1961. Introduction to the study of animal populations. Univ. Chicago Press, Chicago. pp. xii+281.
Andrewartha, H. G., and L. C. Birch. 1954. The distribution and abundance of animals. Univ. of Chicago Press, Chicago. pp. xv+782.
Anonymous. 1893. *Nat. Sci.* 3, 289 (Fide Schilder 1954).
Antevs, E. 1931. Late-glacial chronology of North America. *Ann. Rep. Smiths. Inst.* 1931, 313–324.
Antevs, E. 1947. Biogeographical principles. *Oregon State Coll. Biol. Colloqu.* 8, 7–9.
Auer, V. 1951. Consideraciones cientificas sobre la conservación de los recursos naturales de la Patagonia. *Idia* 4, 1–36.
Auer, V. 1958. The Pleistocene of Fuego-Patagonia II. *Ann. Acad. Sci. Fennicae, Ser. A. III.* 50, pp. 239.

Auer, V. 1960. The Quaternary history of Fuego-Patagonia. *Proc. Roy. Soc.* **152**, 507–516.
Auer, V. 1963. Die geographische Gebiete der Moore Feuerlands. *Mitt. Fränk. Geogr. Ges.* **10**, 31–38.
Auer, V. 1965. The Pleistocene of Fuego-Patagonia. IV. *Ann. Acad. Sci. Fennicae Ser. A. III.* **80**, pp. 160.
Axelrod, D. 1950. Evolution of desert vegetation in western North America. *Carnegie Inst. Wash. Publ.* **590**, 215–306.
Axelrod, D. 1952. Variables affecting the probabilities of dispersal in geologic time. *Bull. Am. Mus. Nat. Hist.* **99**, 177–188.
Axelrod, D. I., and W. S. Ting. 1960. Late Pliocene floras east of the Sierra Nevada. *Univ. Calif. Publ. Geol. Sci.* **39**, 1–118.

Baker, H. B. 1958. Land snail dispersal. *Nautilus* **71**, 141–148.
Baker, H. G. 1953. Race-formation and reproductive method in flowering plants. *Symp. Soc. Exptl. Biol.* **7**, 114–145.
Baker, H. G. 1955. Self-compatibility and establishment after "long-distance" dispersal. *Evolution* **9**, 347–349.
Baker, H. G. 1959. The contribution of autecological and genecological studies to our knowledge of the past migrations of plants. *Am. Naturalist* **93**, 255–272.
Baker H. G., and G. L. Stebbins (eds.). 1965. The genetics of colonizing species. Academic Press, New York. pp. xv+588.
Baker, J. R. 1938. The evolution of breeding seasons. pp. 161–177 in: G. R. de Beer (ed.) *Evolution,* essays presented to E. S. Goodrich. Oxford Univ. Press, Oxford.
Bakker, E. M. van Zinderen. 1967. Upper Pleistocene and Holocene stratigraphy and ecology on the basis of vegetation changes in Sub-Saharan Africa. pp. 125–147 in: W. W. Bishop and J. D. Clark (eds.) Background of evolution in Africa. Univ. Chicago Press, Chicago.
Balogh, J. 1958. Lebensgemeinschaften der Landtiere. Vg. Ungar. Akad. Wiss., Budapest. pp. 560.
Banfield, A. W. F. 1954. Preliminary investigation of the Barren Ground caribou. I. Wildlife Manag. Bull. Can. Wildl. Service Ser. 1. No. 10A. pp. 79.
Barrera, A. 1962. La peninsula de Yucatán como provincia biotica. *Rev. Soc. Mex. Hist. Nat.* **23**, 71–105.
Barry, T. H. (ed.). 1962. Proceedings of a symposium on the causes and problems of animal distribution with special reference to Southern Africa. *Ann. Cape Prov. Museums* **2**, 1–317.
Bartenew, A. 1932. Über einige Grundfragen der Zoogeographie. *Zool. Zh.* **11**, 23–38.
Bartholomew, J. G., W. E. Clark, and P. H. Grimshaw. 1911. Atlas of zoogeography. Bartholomew's, London. pp. 67+36 plates.
Bary, B. McK. 1963. Distributions of Atlantic pelagic organisms in relation to surface water bodies pp. 51–67 in: M. J. Dunbar (ed.) Marine distributions. Univ. Toronto Press, Toronto.
Bates, M. 1954. The natural history of mosquitoes. Macmillan, New York. pp. 379.
Beaufort, L. F. de 1951. Zoogeography of the land and inland waters. Sidgwick and Jackson, London. pp. viii+208.
Beament, J. W. L. 1962. The surface properties of insects. *Proc. Linnean Soc. London* **173**, 115–119.
Beebe, W. 1951. Migration of insects (other than Lepidoptera) through Portachuello pass, Rancho Grande, North-Central Venezuela. *Zoologica* **36**, 255–266.
Beebe, W., and H. Fleming. 1951. The migration of day-flying moths through Portachuello Pass, Rancho Grande, North-Central Venezuela. *Zoologica* **36**, 4, 17–23, 243–254.
Beeson, C. F. C., and M. Bhatia. 1939. Indian forest records. *Entomology* **5**, 1–235 (Fide Linsley 1959).
Benson, C. W., and M. P. S. Irwin. 1965. Some west-east distributional gaps in birds

of the evergreen forest in South-Central Africa. *Proc. Centr. Afr. Sci. & Med. Congr.*, Pergamon Press, Oxford. pp. 309–320.

Benson, C. W., M. P. S. Irwin, and C. M. N. White. 1962. The significance of valleys as avian zoogeographical barriers. *Ann. Cape Prov. Museums* **2**, 155–189.

Berg, L. S. 1950. Natural regions of the U.S.S.R. Macmillan, New York. pp. xxxi+436.

Bergman, G. 1946. Der Steinwälzer, *Arenaria i. interpres* (L.) in seiner Beziehung zur Umwelt. *Acta Zool. Fenn.* **47**, 1–151.

Bernardi, G. 1961. Biogéographie et spéciation des lépidoptères rophalocères des îles méditerranéennes. *Colloq. Intern. Centre Nat. Rech. Sci.* **94**, 181–215.

Birch, L. C. 1957. The meanings of competition. *Am. Naturalist* **91**, 5–18.

Birch, L. C. 1960. The genetic factor in population ecology. *Am. Naturalist* **94**, 5–24.

Bird, R. D. 1937. Records of northward migration of southern insects during drought years. *Can. Entomol.* **69**, 119–120.

Blair, W. F. 1953. Factors affecting gene exchange between populations in the *Peromyscus maniculatus* group. *Texas J. Sci.* **5**, 17–33.

Blair, W. F. 1965. Amphibian Speciation. pp. 543–556 in H. E. Wright, Jr., and D. G. Frey (eds.) The Quaternary of the United States. Princeton Univ. Press, Princeton.

Blanford, W. T. 1890. Anniversary address to the Geological Society. *Proc. Geol. Soc. London* 1890, 43–110.

Bleakney, J. S. 1958. A zoogeographic study of the amphibians and reptiles of Eastern Canada. *Bull. Nat. Mus. Can.* **155**, pp. 119.

Blyth, E. 1871. A suggested new division of the earth into zoological regions. *Nature* **3**, 427.

Bobrinskij, N. A. 1927. Zoogeografie i ekologie. Moscow (Fide Schilder 1956).

Bobrinskij, N. A. 1946. Geografija zhivotnykh. Sovjetskaia Navka, Moscow. pp. 453.

Bodenheimer, F. S. 1930. Über die Grundlagen einer allgemeinen Epidemiologie der Insektkalamitäten. *Z. Angew. Entomol.* **16**, 433–450.

Bodenheimer, F. 1938. Problems of animal ecology. Oxford Univ. Press, London. pp. vi+183.

Boness, M. 1953. Die Fauna der Wiesen unter besondere Berücksichtigung der Mahd. *Z. Morphol. Ökol. Tiere* **42**, 225–277.

Boyd, E. 1962. A half century's changes in the bird-life around Springfield, Massachusetts. *Bird-Banding* **33**, 137–148.

Bradshaw, G. V. 1961. New Arizona locality for the dwarf shrew. *J. Mammal.* **42**, 96.

Brattstrom, B. H. 1963. Barcena Volcano, 1952: its effect on the fauna and flora of San Benedicto Island, Mexico. pp. 499–524 in: J. L. Gressitt (ed.) Pacific Basin Biogeography. Bishop Museum Press, Honolulu.

Braun-Blanquet, J. 1932. Plant sociology: the study of plant communities. McGraw-Hill, New York. pp. xviii+439.

Braun-Blanquet, J. 1955. Die Vegetation des Piz Languard, ein Masstab für Klimaänderungen. *Svensk Bot. Tidsskr.* **49**, 1–8.

Brink, F. H. van den. 1958. Alla Europas däggdjur. Elsevier, Bonnier, Stockholm. pp. 245.

Brinkmann, A., Jr. 1951. En kjempesnegl på plyndringstog rundt jorden. *Naturen* **75**, 385–392.

Bristowe, W. S. 1939. The comity of spiders. Vol. I. Ray Society, London. pp. X+228+40.

Broadbrooks, H. E. 1965. Ecology and distribution of the pikas of Washington and Alaska. *Am. Midland Naturalist* **73**, 299–335.

Brooks, C. E. P. 1949. Climate through the ages. 2d ed. McGraw-Hill, New York. pp. 395.

Brown, R. C., and R. A. Sheals. 1944. The present outlook on the gipsy moth problem. *J. Forestry* **42**, 393–407.

Brundin, L. 1965. On the real nature of Transatlantic relationships. *Evolution* **19**, 496–505.

Brundin, L. 1966. Transatlantic relationships and their significance, as evidenced by chironomid midges. *Kungl. Svenska Vet. Akad. Handl.* Ser. 4, Vol. 11. No. 1. pp. 472.

Brundin, L. 1967. Insects and the problem of austral disjunctive distribution. *Ann. Rev. Entomol.* **12**, 149–168.

Buchanan, G. D., and R. V. Talmage. 1954. The geographic distribution of the armadillo in the United States. *Texas J. Sci.* **6**, 142–150.

Burges, A. 1960. Time and size in ecology. *J. Ecol.* **48**, 273–285.

Burt, W. H. 1958. The history and affinities of the recent land mammals of western North America. pp. 131–154 in: C. L. Hubbs (ed.) Zoogeography. Amer. Assoc. Advanc. Science, Washington.

Burt, W. H., and R. P. Grossenheider. 1964. A field guide to the mammals. 2d ed. Houghton Mifflin, Boston. pp. xxv+284.

Cain, S. A. 1944. Foundations of plant geography. Harper, New York. pp. xiv+556.

Candolle, A. P. de. 1813. Théorie élémentaire de la botanique, ou exposition des principles de la classification naturelle et de l'art de décrire et d'étudier les végétaux. (2nd ed., 1819. Fide Tschulok 1910).

Candolle, A. P. de. 1855. Géographie botanique . . . 2 vols. Masson, Paris and Geneva. pp. xxxii+1366.

Carcasson, R. H. 1964. A preliminary survey of the zoogeography of African butterflies. *East Afr. Wildl. J.* **2**, 122–157.

Carlquist, S. 1965. Island life. Doubleday, New York. 451 pp.

Carlquist, S. 1966a. The biota of long-distance dispersal I. Principles of dispersal and evolution. *Quart. Rev. Biol.* **41**, 247–270.

Carlquist, S. 1966b. The biota of long-distance dispersal. II. Loss of dispersibility in Pacific Compositae. *Evolution* **20**, 30–48.

Carrick, R. 1963. Ecological significance of territory in the Australian Magpie, *Gymnorhina tibicen*. Proc. XIII Intern. Ornith. Congr. Vol. **II**, 740–753.

Carson, H. L. 1959. Genetic conditions which promote or retard the formation of species. *Cold Spring Harbor Symp. Quant. Biol.* **24**, 87–105.

Caughley, G. 1964. Does the New Zealand vertebrate fauna conform to zoogeographical principles? *Tuatara* **12**, 49–56.

[Chambers, R.] published anonymously, 1844. Vestiges of the natural history of creation. London, Chambers. pp. 353.

Chaney, R. W. 1940. Tertiary forests and continental history. *Bull. Geol. Soc. Am.* **51**, 469–488.

Chapin, J. P. 1932. The birds of the Belgian Congo. *Bull. Am. Mus. Nat. Hist.* **65**, 1–723.

Chapman, F. M. 1940. The post-glacial history of *Zonotrichia capensis*. *Bull. Am. Mus. Nat. Hist.* **76**, 381–438.

Chapman, R. N. 1931. Animal ecology. McGraw-Hill, New York. pp. 464.

Chitty, D. 1965. Qualitative changes within fluctuating populations, including genetic variability. pp. 384–386. in: *Proc. XII Intern. Congr. Entomol.*

Clements, F. E. 1905. Research methods in ecology. Univ. Publ. Co., Lincoln, Nebr. pp. 334.

Clements, F. E. 1916. Plant succession. *Carnegie Inst. Wash. Publ.* **242**, pp. 512.

Clements, F. E. & V. E. Shelford. 1939. Bio-ecology. Wiley, New York. pp. viii+825.

Codreanu, R. 1961. Sur le peuplement en Triclades et Asellides d'eau douce de quelques îles méditerranéennes. *Colloq. Intern. Centre Nat. Rech. Sci.* **94**, 163–179.

Colbert, E. H. 1955. Evolution of the vertebrates. Wiley, New York. pp. xiii+479.

Conant, R. 1958. A field guide to reptiles and amphibians of eastern North America. Houghton Mifflin, Boston. pp. xv+366.

Conant, R. 1960. The queen snake, *Natrix septemvittata*, in the interior highlands of Arkansas and Missouri, with comments upon similar disjunct distributions. *Proc. Acad. Nat. Sci. Phila.* **112**, 25–40.

Connell, J. H., and E. Orias. 1964. The ecological regulation of species diversity. *Am. Naturalist* **98**, 399–414.

Cook, L. M. 1961. The edge effect in population genetics. *Am. Naturalist* **95**, 295–307.
Cooke, M. T. 1928. The spread of the European starling in North America (to 1928). *U.S. Dept. Agr. Circ.* **40**, pp. 9.
Cowles, R. B. 1930. The life history of *Varanus niloticus* (Linn.) as observed in Natal, South Africa. *J. Entomol. Zool.* **22**, 1–31.
Cowles, R. B. 1939. Possible implications of reptilian thermal tolerance. *Science* **90**, 465–466.
Cowles, R. B. 1959. Zulu journal. Univ. Calif. Press, Berkeley and Los Angeles. pp. xiv+267.
Croizat, L. 1958. Panbiogeography. 3 vols. Caracas. pp. xxxi+1018+1731.
Croizat, L. 1966. Space, time, form: the biological synthesis. Caracas. pp. xix+881.
Crowell, K. L. 1963. On determinants of insular faunas. *Am. Naturalist* **97**, 194–196.
Curry-Lindahl, K. 1957. The distribution of some species of animals. Maps 45–46, pp.1–8 in: Atlas över Sverige. Generalstabens Litografiska Anstalts Förlag, Stockholm.
Curry-Lindahl, K. 1961. Landscape changes and the vertebrate fauna in Sweden during the last 150 years. *Bijdr. t.d. Dierkunde* **31**, 27–44.

Dahl, F. 1921–23. Grundlagen einer ökologischen Tiergeographie. 2 vols. G. Fischer, Jena, pp. VIII+113, VI+122.
Dale, F. H. 1939. Variability and environmental response of the kangaroo rat, *Dipodomys heermanni saxatilis*. *Am. Midland Naturalist* **22**, 703–731.
Dammerman, K. 1948. The fauna of Krakatau, 1883–1933. *Verhandel. Koninkl. Ned. Akad. Wetenschap. Afdel. Natuurk.* **44**, 1–594.
Danilevskii, A. S. 1965. Photoperiodism and seasonal development of insects. Oliver and Boyd, Edinburgh. pp. ix+283.
Dansereau, P. 1951. Description and recording of vegetation upon a structural basis. *Ecology* **32**, 172–229.
Dansereau, P. 1957. Biogeography, an ecological perspective. Ronald, New York. pp. xiii+394.
Darlington, P. J., Jr., 1938a. Was there an Archatlantis? *Am. Naturalist* **72**, 521–533.
Darlington, P. J., Jr., 1938b. The origin of the fauna of the Greater Antilles, with discussion of dispersal of animals over water and through the air. *Quart. Rev. Biol.* **13**, 274–300.
Darlington, P. J., Jr., 1943. Carabidae of the mountains and islands; data on the evolution of isolated faunas, and on atrophy of wings. *Ecol. Monographs* **13**, 37–61.
Darlington, P. J., Jr., 1948. Geographic distribution of cold-blooded vertebrates. *Quart. Rev. Biol.* **23**, 1–26, 105–123.
Darlington, P. J., Jr., 1957. Zoogeography: the geographic distribution of animals. Wiley, New York. pp. xiii+675.
Darlington, P. J., Jr., 1959. Area, climate and evolution. *Evolution* **13**, 488–510.
Darlington, P. J., Jr., 1961. Australian carabid beetles. V. Transition of wet forest fauna from New Guinea to Tasmania. *Psyche* **68**, 1–24.
Darlington, P. J., Jr., 1965. Biogeography of the southern end of the world. Harvard Univ. Press, Cambridge, Mass. pp. x+236.
Darwin, C. 1839. Journal of researches into the geology and natural history of the various countries visited by the H. M. S. Beagle, under the command of Captain FitzRoy from 1832 to 1836. Colburn, London. pp. XIV+615.
Darwin, C. 1859. On the tendency of species to form varieties; and on the perpetuation of varieties and species by natural means of selection (with A. R. Wallace). *Proc. Linnean Soc. (Zool.)* **3**, 45–62.
Darwin, C. 1859. On the origin of species. Murray, London. pp. ix+502.
Dasmann, R. F. 1964. Wildlife biology. Wiley, New York. pp. 231.
Daubenmire, R. F. 1938. Merriam's life zones of North America. *Quart. Rev. Biol.* **13**, 327–332.
Davis, D. H. S. 1962. Distribution patterns of southern Muridae, with notes on some of their fossil antecedents. *Ann. Cape Prov. Museums* **2**, 56–84.

Deevey, E. S. 1949. Biogeography of the Pleistocene I. Europe and North America. *Bull. Geol. Soc. Am.* **60**, 1315–1416.
Denman, N. S. 1965. Colonization of the islands of the Gulf of St. Lawrence by mammals. *Ecology* **46**, 340–341.
Dice, L. R. 1931. The relation of mammalian distribution to vegetation types. *Sci. Monthly* **33**, 312–317.
Dice, L. R. 1939. Variation in the deer-mouse *(Peromyscus maniculatus)* in the Columbia basin of southeastern Washington and adjacent Idaho and Oregon. Contrib. Lab. Vert. Genet. Univ. Mich. 12. pp. 22.
Dice, L. R. 1943. The biotic provinces of North America. Univ. Mich. Press, Ann Arbor. pp. 78.
Dice, L. R. 1949. Variation of *Peromyscus maniculatus* in parts of western Washington and adjacent Oregon. Contrib. Lab. Vert. Biol. Univ. Mich. 44. pp. 34.
Dice, L. R. 1952. Natural Communities. Univ. Mich. Press, Ann Arbor. pp. x+547.
Dice, L. R., and W. E. Howard. 1951. Distance of dispersal by prairie deermice from birthplace to breeding sites. Univ. Mich. Contr. Lab. Vert. Zool. 50. pp. 1–15.
Dickson, R. C. 1959. Aphid dispersal over southern California deserts. *Ann. Entomol. Soc. Am.* **52**, 368–372.
Dillon, L. S. 1956. Wisconsin climate and life zones in North America. *Science* **123**, 167–176.
Dobzhansky, T. 1935. Fecundity in *Drosophila pseudoobscura* at different temperatures. *J. Exp. Zool.*, **71**, 449–464.
Dobzhansky, T. 1950. Evolution in the tropics. *Am. Scientist* **38**, 209–221.
Dorf, E. 1959. Climatic changes of the past and present. Contrib. Mus. Paleont. Univ. Mich. **13**, 181–210.
Dorst, J. 1962. The migration of birds. Riverside Press, Cambridge, Mass. pp. xiv+476 (translation of Les migrations des oiseaux, 1956, Payot, Paris).
Drent, R., G. F. van Tets, F. Tompa, and K. Vermeer. 1964. The breeding birds of Mandarte Island, British Columbia. *Can. Field Nat.* **78**, 208–263.
Dunbar, M. J. 1960. The evolution of stability in marine environments—natural selection at the level of the ecosystem. *Am. Naturalist* **94**, 129–137.
Dunbar, M. J. (ed.) 1963. Marine distributions. Univ. Toronto Press, Toronto. pp. 110.

Ehrlich, P. R. 1958. Problems of Arctic-Alpine insect distribution as illustrated by the butterfly genus *Erebia* (Satyridae). *Proc. 10th Intern. Congr. Entomol.* **1**, 683–686.
Ehrlich, P. R. 1961. Intrinsic barriers to dispersal in Checkerpot butterfly. *Science* **134**, 108–109.
Ekman, S. 1915. Vorschläge und Erklärungen zur Reliktenfrage in der Hydrobiologie. *Arkiv för Zoologi.* **9**, 1–35.
Ekman, S. 1922. Djurvärldens utbredningshistoria på den skandinaviska halvön. Bonnier, Stockholm. pp. 614.
Ekman, S. 1935. Tiergeographie des Meeres. Akad. Verlagsges, Leipzig. pp. xii+542.
Ekman, S. 1940. Begründung einer statistischen Methode der regionale Tiergeographie. Nova Acta Reg. Soc. Sci. Uppsal. Ser. 4, Vol. 12, No. 2. pp. 117.
Ekman, S. 1953. Zoogeography of the sea. Sidgwick and Jackson, London. pp. xiv+418.
Ellenberg, H. 1956. Grundlagen der Vegetationsgliederung. I. Vol. IV, 1, pp. 136. in: Walter, H., Einführung i.d. Phytologie.
Elton, C. S. 1927. Animal ecology. Sidgwick and Jackson, London. pp. 207.
Elton, C. S. 1946. Competition and the structure of ecological communities. *J. Animal Ecol.* **15**, 54–68.
Elton, C. S. 1958. The ecology of invasions by animals and plants. Methuen, London. pp. 181.
Elton, C. S. 1966. The pattern of animal communities. Methuen, London. pp. 432.
Emerson, A. E. 1952. The biogeography of termites. *Bull. Am. Mus. Nat. Hist.* **99**, 217–225.

Emerson, A. E. 1955. Geographical origins and dispersions of termite genera. *Fieldiana Zool.* **37**, 465–521.
Engler, A. 1879–1882. Versuch einer Entwicklungsgeschichte der Pflanzenwelt, insbesondere der Florengebiete seit der Tertiärperiode. 2 Teile. Engelmann, Leipzig. Vol. 1. pp. XVIII+202, Vol. 2. pp. XIV+386.
Engler, A. 1899. Die Entwicklung der Pflanzengeographie in den letzten 100 Jahren. Humboldt-Centenarschrift d. Ges. Erdkunde, Berlin. pp. 247.
Erkamo, V. 1952. On plant-biological phenomena accompanying the present climatic changes. pp. 25–37 in: I. Hustich (ed.). The recent climatic fluctuations in Finland and its consequences. *Fennia* **75**, 1–128.
Erkamo, V. 1956. Untersuchungen über den pflanzenbiologischen und einige andere Folgeerscheinungen der neuzeitlichen Klimaschwankung in Finnland. *Ann. Botan. Soc. Zoo. Botan. Fennicae "Vanamo"* **28**, No. 3. pp. 290.
Errington, P. L. 1943. An analysis of mink predation upon muskrat in north-central United States. *Res. Bull. Iowa Agr. Exp. Sta.* **320**, 797–924.

Fabricius E., and K. J. Gustafson. 1958. Some new observations on the spawning behaviour of the pike, *Esox lucius* L. *Inst. Freshwater Res. Drottningholm* **39**, 23–54.
Falla, R. A. 1960. Oceanic birds as dispersal agents. *Proc. Royal Soc. Ser. B.* **152**, 655–659.
Farquharson, F. L. 1962. The distribution of Cyprinids in South Africa. *Ann. Cape Prov. Museums* **2**, 233–251.
Fejérváry, G. J. von. 1926. Über Erscheinungen und Prinzipien der Reversibilität in der Evolution und das Dollosche Gesetz. *Paläont. Zeitschr.* **7**, 173–184.
Fenner, F., and F. N. Ratcliffe. 1965. Myxomatosis. Cambridge Univ. Press, Cambridge. pp. xiv+371.
Findley, J. S., and S. Anderson. 1956. Zoogeography of the montane mammals of Colorado. *J. Mammal.* **37**, 80–82.
Firbas, F. 1949, 1952. Spät– und nacheiszeitliche Waldgeschichte Mitteleuropas. 2 vols. Fischer, Jena. pp. 480+256.
Fischer, R. H. 1887. Manuel de conchyliologie et de paléontologie conchyliologique. Savy, Paris. pp. xxiv+1369.
Fisher, A. G. 1960. Latitudinal variation in organic diversity. *Evolution* **14**, 64–81.
Fisher, J. 1955. The dispersal mechanisms of some birds. Proc. XI. Intern. Ornith. Congr. pp. 437–442.
Flint, R. F. 1957. Glacial and Pleistocene geology. Wiley, New York. pp. 553.
Ford, E. B. 1957. Butterflies. 3d ed. Collins, London. pp. XIV+368.
Ford, E. B. 1964. Ecological genetics. Methuen, London. pp. xv+335.
Formosov, A. N. 1946. Snow cover as an integral factor of the environment and its importance in the ecology of mammals and birds. Materials f. fauna and flora of the USSR.–*N. Ser. Zool.* **5** XX, 1–152. Translation (1964): Occas. Publ. No. 1. Boreal Inst. Univ. Alberta, Edmonton. pp. 184+xiv.
Formosov, A. N. 1966. Adaptive modifications of behavior in mammals of the Eurasian steppes. *J. Mammal.* **47**, 208–223.
Fosberg, F. R. 1963. Disturbance in island ecosystems. pp. 557–567 in: L. Gressitt (ed.) Pacific Basin Biogeography. Bishop Museum Press, Honolulu.
Francia, F. C. 1965. Studies on some aspects of behaviour in the ambrosia beetle (*Trypodendron lineatum* Oliver). Unpublished Ph.D. dissertation, Univer. British Columbia, Vancouver. pp. 89.
Fraenkel, G. 1932. Die Wanderungen der Insekten. *Ergeb. Biol.* **9**, 1–238.
Franz, H. 1936. Die thermophilen Elemente der mitteleuropäischen Fauna und ihre Beeinflussung durch die Klimaschwankungen der Quartärzeit. *Zoogeographica* **3**, 159–320.
Franz, J. 1949. Über die genetischen Grundlagen des Zusammenbruchs einer Massenvermehrung aus inneren Ursachen. *Z. Angew. Entomol.* **31**, 228–260.
Franz, J. M. 1964. Dispersion and natural-enemy action. *Ann. Appl. Biol.* **53**, 510–515.

Freitag, H. 1962. Einführung in die Biogeographie von Mitteleuropa. G. Fischer, Stuttgart. pp. xiv+214.
French, N. R., T. Y. Tagami and P. Hayden 1968. Dispersal in a population of desert rodents. *J. Mammal.* **49,** 272–280.
Frick, C. 1937. Horned ruminants of North America. *Bull. Am. Mus. Nat. Hist.* **69,** xxviii+669.
Friederichs, K. 1927. Grundsätzliches über die Lebenseinheiten höherer Ordnung und den ökologischen Einheitsfaktor. *Naturwiss.* **15,** 153–157, 182–186.
Frisch, K. von. 1951. Solved and unsolved problems of bee language. *Bull. Animal Behav.* **9,** 1–33.
Fry, F. E. J. 1947. Effects of the environment on animal activity. *Univ. Toronto Studies Biol. Ser.* **55,** pp. 62.
Furon, R. 1958. Causes de la répartition des êtres vivants, paléogéographie, biogéographie dynamique. Masson, Paris. pp. 164.

Gajl, J. 1924. Über zwei faunistische Typen aus der Umgebung von Warschau auf Grund von Untersuchungen an Phyllopoda und Copepoda (exkl. Harpacticidae). *Bull. Acad. Pol. Soc. Lett., Cl. Sci., Math. & Nat.* Ser B, **1924,** pp. 13–55 (Fide Thienemann 1950).
Gams, H. 1949. Variation des limites de la végétation alpine et variations des glaciers. *Terre Vie* **1949,** 178–193.
Gates, G. E. 1966. Requiem—for megadrile utopias. *Proc. Biol. Soc. Wash.* **79,** 239–254.
Gentilli, J. 1949. Foundations of Australian bird geography. *Emu* **49,** 85–130.
George, V. 1962. Animal geography. Heinemann, London. pp. vi+142.
Geptner, V. G. 1936. Onzhaija zoogeografia. Biomedgiz, Moscow, Leningrad, pp. 548.
Gerhard, B. 1883. Ueber die geographische Verbreitung der Makrolepidopteren auf der Erde. *Berliner entomol. Z.* **27,** 173–185.
Gill, E. 1955. The problem of extinction, with special reference to Australian marsupials, *Evolution* **9,** 87–92.
Glick, P. A. 1939. The distribution of insects, spiders and mites in the air. *U.S. Dept. Agr. Tech. Bull.* **673,** pp. 150.
Glover, R. S. 1961. Biogeographical boundaries: the shapes of distribution. pp. 201–228 in: M. Sears (ed.) Oceanography. Amer. Assoc. Advanc. Sci., Washington.
Goldman, E. A., and R. T. Moore. 1945. The biotic provinces of Mexico. *J. Mammal.* **26,** 347–360.
Good, R. 1931. A theory of plant geography. *New Phytologist* **30,** 149-171.
Good, R. 1936. On the distribution of the lizard orchid *Himantoglossum hircinum* (Koch). *New Phytologist* **35,** 142–170.
Good, R. 1964. The geography of flowering plants. 3d ed. Longmans, London. pp. xvi+518.
Goodhart, C. B. 1962. Variation in a colony of the snail *Cepaea nemoralis* (L.). *J. Animal Ecol.* **31,** 207–237.
Gorman, J. 1964. *Hydromantes brunus, H. platycephalus,* and *H. shastae.* p. 11 in: W. J. Riemer (ed.) Catalogue of American amphibians and reptiles. Am. Soc. Ichth. Herp., Kensington, Md.
Graham, K. 1960. Photic behaviour in the ecology of the ambrosia beetle *Trypodendron lineatum. Proc. XIth. Intern. Congr. Entomol. Vienna* **2,** 226.
Grassé, P. P. 1946. Sociétés animales et effet de groupe. *Experientia* **2,** 77–82.
Gressitt, L. 1956. Some distribution patterns of Pacific island faunae. *Syst. Zool.* **5,** 11–47.
Gressitt, L. 1961. Problems in the zoogeography of Pacific and Antarctic insects. Pac. Insects Monogr. 2. pp. 94.
Gressitt, J. L. 1963. Insects of Antarctica and subantarctic islands. pp. 435–450 in: J. L. Gressitt (ed.) Pacific Basin Biogeography, A Symposium, Bishop Museum Press, Honolulu.

Gressitt, L., and C. M. Yoshimoto. 1963. Dispersal of animals in the Pacific. pp. 283–292 in: J. L. Gressitt (ed.) Pacific Basin Biogeography. Bishop Museum Press, Honolulu.
Gressitt, L. et al. 1964. Insects of Campbell Island. Pac. Insects Monogr. 7. pp. 663.
Grinnell, J. 1914. The Colorado River as a hindrance to the dispersal of species. Univ. Calif. Publ. Zool. 12, 100–107.
Grinnell, J. 1917a. Field test of theories concerning distributional control. Am. Naturalist 51, 115–128.
Grinnell, J. 1917b. The niche-relationships of the California thrasher. Auk 34, 427–433.
Grinnell, J. 1926. Geography and evolution in the pocket gopher. Univ. Calif. Chronicle 28, 247–262. Reprinted (1943) pp. 167–182 in: Josef Grinnell's philosophy of nature. Univ. Calif. Press, Berkeley.
Grinnell, J., and A. H. Miller. 1944. The distribution of the birds of California. Cooper Ornith. Club, Berkeley. pp. 607.
Guilday, J. E. 1958. The prehistoric distribution of the opossum. J. Mammal. 39, 39–43.
Guilday, J. E., P. S. Martin, and A. D. McCrady. 1964. New Paris No. 4: a late Pleistocene cave deposit in Bedford County, Pennsylvania. Bull. Nat. Speleol. Soc. 26, 121–194.
Günther, A. 1858. On the zoogeographical distribution of the reptiles. Proc. Zool. Soc. London 21, 373–398.
Günther, A. 1870. Catalogue of the fishes in the British Museum. 8 vols. British Museum, London.

Haartman, L. von. 1949. Der Trauerfliegenschnäpper. I. Ortstreue und Rassenbildung. Acta Zool. Fenn. 56. pp. 104.
Haartman, L. von. 1956. Der Einfluss der Temperatur auf den Brutrhythmus experimentell nachgewiesen. Ornis Fennica 33, 100–107.
Haartman, L. von. 1960. The Ortstreue of the pied flycatcher. Proc. XII Intern. Ornith. Congr. pp. 266–273.
Haeckel, E. 1866. Generelle Morphologie der Organismen. 2 vols. Reimer, Berlin. pp. xxxii+576, clx+462.
Haffer, J. 1967. Speciation in Colombian forest birds west of the Andes. Am. Museum Novitates 2294, pp. 57.
Hagen, W. von. 1938. A contribution to the biology of Nasutitermes (sensu stricto). Proc. Zool. Soc. London 100, 39–49.
Hagmeier, E. M. 1966. A numerical analysis of the distributional patterns of North American mammals II. Re-evaluation of the provinces. Syst. Zool. 15, 279–299.
Hagmeier, E. M., and C. D. Stults. 1964. A numerical analysis of the distributional patterns of North American mammals. Syst. Zool. 13, 125–155.
Hairston, N. G., F. E. Smith, and L. B. Slobodkin. 1960. Community structure, population control and competition. Am. Naturalist 94, 421–425.
Halftter, G. 1964. La entomofauna americana, ideas acerca de su origen y distribución. Folia Entomol. Mex. 6, 1–108.
Hall, R. E., and K. R. Kelson. 1959. The mammals of North America. 2 vols. Ronald Press, New York. pp. 1083.
Hamilton, T. H., R. H. Barth, and I. Rubinoff. 1964. The environmental control of insular variation in bird species abundance. Proc. Nat. Acad. Sci. 52, 132–140.
Hamilton, T. H., and I. Rubinoff. 1963. Isolation, endemism and multiplication of species in the Darwin finches. Evolution 17, 388–403.
Hamilton, T. H., and I. Rubinoff. 1964. On models predicting abundance of species and endemics for the Darwin finches in the Galapagos archipelago. Evolution 18, 339–342.
Handlirsch, A. 1913. Beiträge zur exakten Biologie. Sitzber. Math.-Naturw. Kl. Akad. Wiss. Wien. 122, 361–481.

Hanna, G D. 1966. Introduced mollusks of western North America. Occas. Pap. Calif. Acad. Sci. 48. pp. 108.
Hardin, G. 1960. The competitive exclusion principle. *Science* 131, 1291–1297.
Hardy, A. C., and P. S. Milne. 1937. Insect drift over the North Sea. *Nature* 139, 510–511.
Hardy, J. W. 1963. Epigamic and reproductive behavior of the orange-fronted parakeet. *Condor* 65, 169–199.
Harris, V. T. 1952. An experimental study of habitat selection by prairie and forest deer mouse, *Peromyscus maniculatus*. Contrib. Lab. Vert. Biol. Mich. Univ. 56. pp. 53.
Hasler, A. D. 1966. Underwater guideposts. The homing of the salmon. Univ. Wisconsin Press, Madison. pp. 155.
Hedgepeth, J. W. (ed.) 1957. Treatise on marine ecology and paleoecology. Mem. Geol. Soc. Amer. 67. pp. 1296.
Hediger, H. (ed.) 1967. Die Strassen der Tiere. Vieweg. Braunschweig. pp. 313.
Heilprin, A. 1887. The geographical and geological distribution of animals. Kegan Paul, Trench, London. pp. xii+435.
Hennig, W. 1966. Phylogenetic systematics. Univ. Illinois Press, Urbana. pp. 263.
Hesse, R. 1924. Tiergeographie auf ökologischer Grundlage. G. Fischer, Jena. pp. xii+613.
Hesse, R., W. C. Allee, and K. P. Schmidt. 1937. Ecological animal geography. Wiley, New York. pp. xiv+597. 1951. 2d ed. pp. xiii+715.
Heusser, C. J. 1960. Late-Pleistocene environments of North Pacific North America. Amer. Geogr. Soc. Spec. Publ. 35, pp. xxiii+308.
Hibbard, C. W., D. E. Ray, D. E. Savage, D. W. Taylor and J. E. Guilday. 1965. Quaternary mammals of North America. pp. 509–525 in: H. E. Wright, Jr., and D. G. Frey (eds.) The Quaternary of the United States. Princeton Univ. Press, Princeton.
Hoffman, R. S., and R. D. Taber. 1960. Notes on *Sorex* in the northern Rocky Mountain alpine zone. *J. Mamm.* 41, 230–234.
Holdhaus, K. 1911. Über die Abhängigkeit der Fauna von Gestein. Verh. VIII. Zool. Kongr. Graz, pp. 726–744.
Holdhaus, K. 1929. Die geographische Verbreitung der Insekten. pp. 592–1058 in: C. Schröder (ed.) Handbuch d. Entom., Vol. II. G. Fischer, Jena.
Holdhaus, K. 1954. Die Spuren der Eiszeit in der Tierwelt Europas. Abhandl. Zool.-Bot. Ges. Wien 18. pp. 493+52 maps.
Holdridge, L. R. 1967. Life zone ecology. 2d ed. Trop. Res. Center, San José, Costa Rica. pp. 206.
Hoogstraal, H. *et al.*, 1964. Ticks (Ixodidae) on migrating birds in Egypt. *Bull. World Health Organ.* 30, 355–367.
Hopkins, D. M. 1959. Cenozoic history of the Bering land bridge. *Science* 129, 1519–1528.
Horváth, O. 1963. Contributions to nesting ecology of forest birds. Unpublished Master's Thesis, Univ. British Columbia. Vancouver. pp. viii+182.
Hovanitz, W. 1959. Distribution of butterflies in the New World. pp. 321–368 in: C. L. Hubbs (ed.) Zoogeography. A.A.A.S. Publ. 51.
Howard, W. E. 1960. Innate and environmental dispersal of individual vertebrates. *Am. Midland Naturalist* 63, 152–161.
Howden, H. F. 1963. Speculations on some beetles, barriers and climates during the Pleistocene and pre-Pleistocene periods in some nonglaciated portions of North America. *Syst. Zool.* 12, 178–201.
Howell, F. C., and F. Bourlière (eds.). 1963. African ecology and human evolution. Aldine, Chicago. pp. viii+666.
Hubbs, C. L. (ed.). 1958. Zoogeography. Amer. Ass. Advanc. Sci. Washington Publ. 51. pp. x+509.

Hubbs, C. L. 1960. Quaternary paleoclimatology on the Pacific coast of North America. *Rep. Calif. Coop. Oceanic Fish. Invest.* **9**, 105–112.
Hubendick, B. 1962. Aspects on the diversity of the fresh-water fauna. *Oikos* **13**, 249–261.
Hudson, R. 1965. The spread of the collared dove in Britain and Ireland. *Brit. Birds* **58**, 105–139.
Huheey, J. E. 1965. A mathematical method of analyzing biogeographical data. I. Herpetofauna of Illinois. *Am. Midland Naturalist* **73**, 490–500.
Hultén, E. 1937. Outline of the history of arctic and boreal biota during the Quaternary period. Thule, Stockholm. pp. 168+43 maps.
Hultén, E. 1950. Atlas of the distribution of vascular plants in northwest Europe. Generalstabens Litografiska Anstalts Fg., Stockholm. pp. 120+512 (1,847 maps).
Humboldt, A. von, and A. Bonpland. 1807. Ideen zu einer Geographie der Pflanzen nebst einem Naturgemälde der Tropenländer. Cotta, Tübingen; Neudruck Geest, and Portig, Leipzig (1960). pp. 180.
Hustich, I. 1952. Agricultural production in Finland and the recent climatic fluctuations. *Fennia* **75**, 99–105.
Hutchinson, G. E. 1951. The biogeochemistry of vertebrate excretion. Bull. Amer. Mus. Nat. Hist. 96. pp. 554.
Hutchinson, G. E. 1957. Concluding remarks. *Cold Spring Harbor Symp. Quant. Biol.* **22**, 415–427.
Hutchinson, G. E. 1959. Homage to Santa Rosalia or why are there so many different kinds of animals. *Am. Naturalist* **93**, 145–159.
Hutchinson, G. E. 1965. The ecological theater and the evolutionary play. Yale Univ. Press, New Haven. pp. 139.
Huxley, T. H. 1868. On the classification and distribution of the Alectoromorphae and Heteromorphae. *Proc. Zool. Soc. London* **1868**, pp. 294–319.

Illies, J. 1965. Phylogeny and zoogeography of the Plecoptera. *Ann. Rev. Entomol.* **10**, 117–140.
Illiger, C. W. 1815. Ueberblick der Säugetiere nach ihrer Verbreitung über die Weltteile. Abhandl. Kgl. Akad. Wiss. Berlin 1804/11. pp. 39–159.
Inger, R. F. 1954. Systematics and zoogeography of Philippine amphibia. *Fieldiana Zool.* **33**, 181–531.
Isakov, Ju. A., and A. N. Formosov (eds.) 1963. [Questions of ecological zoogeography]. Geogr. Inst. Acad. Sci. U.S.S.R., Moscow. pp. 183.

Jaccard, P. 1902. Lois de distribution florale dans la zone alpine. *Bull. Soc. Vaudoise Sci. Nat.* **38**, 69–130.
Jacobi, A. 1900. Lage und Form biogeographischer Gebiete. *Z. Ges. f. Erdkunde* **35**, 147–238.
Jacobi, A. 1939. Tiergeographie. 2d ed. Samlung Göschchen, Berlin. pp. 153.
Jägerskiöld, L. A. 1919. Om förändringar i Sveriges fågelvärld under de senaste 75 åren. *Sveriges Nat.* **1919**, 47–73.
Jeannel, R. 1942. La genèse des faunes terrestres. Presses Univ. France, Paris. pp. viii+513.
Jeannel, R. 1961. Le foissonement de certaines lignées dans les îles. *Colloq. Intern. Centre Nat. Rech. Sci. XCIV. Paris*, 291–294.
Jeuken, M. 1953. The concept 'individual' in biology. *Acta Biotheoret.* **10**, 57–86.
Johansen, H. 1955. Die Jenissei-Faunenscheide. *Zool. Jb. (Syst.)* **83**, 237–247.
Johnson, C. 1962. The Scarabeoid (Coleoptera) fauna of Lancashire and Cheshire and its apparent changes over the last 100 years. *Entomologist* **95**, 153–165.
Johnson, C. G. 1966. A functional system of adaptive dispersal by flight. *Ann. Rev. Entomol.* **11**, 233–260.
Jovetić, R. 1963. Vom Leben des Weissstorches, *Ciconia ciconia*, in Macedonien. *Larus* **15**, 28–99.

Kaisila, J. 1962. Immigration und Expansion der Lepidopteren in Finnland in den Jahren 1869–1960. Acta Entomol. Fenn. 18. pp. 452.

Kalabukhov, N. E. 1938. On ecological character of closely related species of rodents. 1. The peculiarities of the reaction of the wood-mice (*Apodemus sylvaticus* L. and *A. flavicollis* Melch.) and ground-squirrels (*Citellus pygmaeus* Pall. and *C. suslicus* Gueld.) to the intensity of illumination. *Zool. Zh.* 17, 521–532.

Kalela, E. K. 1941. Ueber die Holzarten und die durch die klimatischen Verhältnisse verursachten Holzartenwechsel in den Wäldern Ostpatagoniens. *Ann. Acad. Sci. Fennicae* Ser. A. IV. Biol. 2. pp. 151.

Kalela, O. 1940. Zur Frage der neuzeitlichen Anreicherung der Brutvogelfauna in Fennoskandien mit besonderer Berücksichtigung der Austrocknung in den früheren Wohngebieten der Arten. *Ornis Fennica* 17, 41–59.

Kalela, O. 1944. Zur Frage der Ausbreitungstendenz der Tiere. Ann. Zool. Soc. Vanamo 10. pp. 23.

Kalela, O. 1949. Über Fjeldlemming-invasionen und andere irreguläre Tierwanderungen. Ann. Zool. Soc. Vanamo 13. pp. 90.

Kalela, O. 1952. Changes in the geographic distribution of Finnish birds and mammals in relation to recent changes in climate. *Fennia* 75, 38–51.

Kangas, E. 1953. Zur Ausbreitung des *Carabus cancellatus* Ill. (Col. Carabidae) in jüngster Zeit in Finnland. *Suomen Hyönteistiet. Aikakauskirja* 19, 175–181.

Keast, A. 1961. Bird speciation on the Australian continent. *Bull. Museum Comp. Zool.* 123, 305–495.

Keast, A. 1962. Vertebrate speciation in Australia; Some comparisons between birds, marsupials and reptiles. pp. 380–407. in: G. W. Leeper (ed.) Evolution of living organisms. Melbourne Univ. Press, Melbourne.

Keferstein, W. M. 1866. in: Bronn, H. G. Klassen und Ordnungen der Thier-Reichs. Winter, Leipzig. 3, 1281 (Fide Schilder 1954).

Kelsall, J. P. 1957. Continued Barren-ground caribou studies. Canad. Wildlife Service, Wildlife Manag. Bull. Ser. 1. No. 12. pp. 148.

Kelsall, J. P. 1963. Barren-ground caribou and their management. *Canad. Audubon* 25, 144–149.

Kelson, K. R. 1951. Speciation in rodents of the Colorado River drainage. Univ. Utah Biol. Ser. 11/3. pp. 125.

Kendeigh, S. C. 1934. The rôle of environment in the life of birds. *Ecol. Monographs* 4, 299–417.

Kendeigh, S. C. 1961. Animal ecology. Prentice-Hall, Englewood Cliffs. pp. x+468.

Kendeigh, S. C. 1964. Regulation of nesting time and distribution in the house wren. *Wilson Bull.* 75, 418–427.

Kennedy, J. S. (ed.). 1961. Insect polymorphism. Symp. No. 1. Royal Entom. Soc. London. pp. vi+115.

Kennedy, J. S., and H. L. G. Stroyan. 1959. Biology of aphids. *Ann. Rev. Entomol.* 3, 139–160.

Kerner, A. 1863. Das Pflanzenleben der Donauländer. Translated into English by H. S. Conard. 1951. The background of plant ecology. Iowa State College Press, Ames. pp. x+238.

Keve, A. 1960. Der Buntspecht (*Dendrocopos syriacus* Hempr. & Ehrenb.) in Ungarn. *Vert. Hung.* 2, 243–260.

Keve, A., and M. D. F. von Udvardy. 1951. Increase and decrease of the breeding range of some birds in Hungary. Proc. XI. Intern. Ornith. Congr. pp. 468–476.

Kew, H. W. 1893. The dispersal of shells. Kegan Paul, Trench, Trubner. London. pp. xv+241.

Key, K. H. L. 1950. A critique of the phase theory of locusts. *Quart. Rev. Biol.* 25, 363–407.

Klomp, H. 1954. Habitat selection in the Lapwing, *Vanellus vanellus* (L.). *Ardea* 42, 1–139.

Klomp, H. 1964. Intraspecific competition and the regulation of insect numbers. *Ann. Rev. Entomol.* **9**, 17–40.
Klopfer, P. H. 1959. Environmental determinants of faunal diversity. *Am. Naturalist* **93**, 337–342.
Klopfer, P. H. 1962. Behavioral aspects of ecology. Prentice-Hall, Englewood Cliffs. pp. 166.
Koch, L. 1945. The East Greenland Ice. Copenhagen. *Medd. Grønland* **130**(3), 374.
Koskimies, J. 1956. Zur Charakteristik und Geschichte der nistökologischen Divergenz beim Mauersegler, *Apus apus* (L.), in Nordeuropa. *Ornis Fennica* **33**, 78–96.
Koskimies, J., and L. Lahti. 1964. Cold-hardiness of the newly hatched young in relation to ecology and distribution in the species of European ducks. *Auk* **81**, 281–307.
Krätzig, H. 1936. Der Frühsommerzug des Stars auf der Windenburger Ecke. *Vogelzug* **7**, 1–16.
Krogerus, R. 1939. Zur Ökologie nordischer Moortiere. Verhandl. VII. Intern. Congr. Entom. Berlin 1939 II, pp. 1213–1231.
Kühnelt, W. 1943. Über Beziehungen zwischen Tier- und Pflanzengesellschaften. *Biol. Gener.* **17**, 566–593.
Kühnelt, W. 1965. Grundriss der Ökologie. G. Fischer, Jena. pp. 402.
Kurtén, B. 1958. Life and death of the Pleistocene cave bear. A study in paleoecology. Acta Zool. Fenn. 95. pp. 59.
Kurtén, B. 1968. Pleistocene mammals of Europe. Aldine, Chicago, pp. vii+317,

Lack, D. 1933. Habitat selection in birds. *J. Anim. Ecol.* **2**, 239–262.
Lack, D. 1937. The psychological factor in bird distribution. *Brit. Birds* **31**, 130–136.
Lack, D. 1947. Darwin's finches. Cambridge Univ. Press, Cambridge. pp. 208.
Lack, D. 1954. The natural regulation of animal numbers. Clarendon Press, Oxford. pp. 343.
Landry, S. O., Jr. 1957. The interrelationship of the New and Old World hystricomorph rodents, *Univ. Calif. Publ. Zool.* **56**, 1–118.
Lattin, G. de. 1957. Die Ausbreitungszentren der holarktischen Landtierwelt. Verhandl. Deutsch. Zool. Ges. Hamburg 1956, pp. 380–410.
Lattin, G. de. 1958. Postglaziale Disjunktionen und Rassenbildung bei europäischen Lepidopteren. pp. 392–403 in: Verhandl. Deut. Zool. Ges. Frankfurt a.M. 1958. Akademische Verlagsgesellschaft Geest & Portig K.-G., Leipzig.
Lattin, G. de. 1960. Darwin als Klassiker der Tiergeographie. pp. 203–233 in: G. Heberer and F. Schwenitz (eds.) Hundert Jahre Evolutionsforschung. G. Fischer, Stuttgart. pp. 548.
Lattin, G. de. 1967. Grundriss der Zoogeographie. G. Fischer, Stuttgart. pp. 602.
Laux, W. 1962. Individuelle Unterschiede in Verhalten und Leistung des Ringelspinners, *Malacosoma neustria* (L.). *Z. Angew. Entomol.* **49**, 465–524.
Laux, W., and J. M. Franz. 1962. Über das Auftreten von Individualunterschiede beim Ringelspinner, *Malacosoma neustria* (L.). Angew. Entomol. **50**, 105–109.
Lawrence, D. B. 1958. Glaciers and vegetation in southeastern Alaska. *Am. Scientist* **46**, 89–122.
Leech, R. E. 1965. The spiders (Araneida) of the Hazen Camp area, Ellesmere Island, Northwest Territories, Canada (81°49′ N. 71°18′ W.). Unpublished Master's Thesis, Univ. Alberta, Edmonton. pp. xiv+126.
Lees, A. D. 1961. Clonial polymorphism in aphids. pp. 68–79 in: Kennedy 1961.
Lenz, F. 1931. Lebensraum und Lebensgemeinschaft. Math.-Naturwiss. Techn. Bücherei No. 27. Frankfurt a/M., Berlin (Fide Thienemann 1950).
Leopold, A. S. 1959. Wildlife of Mexico. Univ. California Press, Berkeley. pp. 568.
Leston, D. 1957. Spread potential and the colonization of the islands. *Syst. Zool.* **6**, 41–46.
Leunis, J. 1860. Synopsis der drei Naturreiche. 2d ed. Hahn, Hannover.

Lincoln, F. C. 1939. The migration of American birds. Doubleday, Doran, New York. pp. xii+189.
Lindroth, C. H. 1931. Die Insektenfauna Islands und ihre Probleme. *Zool. Bidr.* **13**, 105–589.
Lindroth, C. H. 1946. Inheritance of wing dimorphism in *Pterostichus anthracinus* Ill. *Hereditas* **32**, 37–40.
Lindroth, C. H. 1948. Vingdimorfismen inom familjen Carabidae. *Arch. Soc. Vanamo* **1**, 70–72.
Lindroth, C. H. 1949. Die fennoskandischen Carabidae. III. Göteborgs Vetenskaps- och Vitterhets-samhälles Handlingar 6. följd Ser. B. Vol. 4. pp. 911.
Lindroth, C. H. 1953. Some attempts toward experimental zoogeography. *Ecology* **34**, 657–666.
Lindroth, C. H. 1954. Experimentelle Beobachtungen an parthenogenetischen und bisexuellen *Otiorrhynchus dubius* Stroem (Col., Curculionidae). *Entom. Tidskr.* **75**, 111–116.
Lindroth, C. H. 1956a. Ecological zoogeography. Statens Forskningsråds Årsbok. Annals State Res. Council of Nat. Sci. Stockholm, 1955, pp. 199–203.
Lindroth, C. H. 1956b. Movements and changes of area at the climatic limit of terrestrial animal species. pp. 226–230 in: K. G. Wingstrand (ed.), Bertil Hanström zoological papers in honour of his sixty-fifth birthday. Zoological Institute, Lund.
Lindroth, C. H. 1957. The faunal connections between Europe and America. Almqvist and Wiksell/Gebers, Stockholm, and Wiley, New York. pp. 344.
Lindroth, C. H. 1960. The ground-beetles of the Azores (Coleoptera: Carabidae) with some reflexions on over-seas dispersal. *Bol. Museu Munic. Funchal* **13**(31), 5–48.
Lindroth, C. H. 1962. Foreword. pp. 3–5 in: D. Nichols (ed.) Taxonomy and geography. Syst. Assoc. Publ. No. 4. London.
Lindroth, C. H. 1963. The faunal history of Newfoundland. Opuscula Entomol. 23. pp. 112.
Linnaeus, C. 1737. Flora lapponica. Schouten, Amsterdam. pp. 372.
Linnaeus, C. 1745. Fauna svecica. Wishoff, Lugduni Batavorum. pp. 419.
Linsley, E. G. 1959. Ecology of Cerambycidae. *Ann. Rev. Entomol.* **4**, 99–138.
Linsley, E. G. 1963. Bering arc relationships of Cerambycidae and their host plants. pp. 159–178 in: J. L. Gressitt (ed.) Pacific Basin Biogeography. Bishop Museum Press, Honolulu.
Liversidge, R. 1962. Distribution of birds in relation to vegetation. *Ann. Cape Prov. Museums* **2**, 143–151.
Livingstone, D. A. *et al.* 1966. Biological aspects of weather modification. *Bull. Ecol. Soc. Amer.* **47**, 39–78.
Lloyd, H. G. 1962. The distribution of squirrels in England and Wales, 1959. *J. Animal Ecology* **31**, 157–166.
Löffler, H. 1963. Ein Kapitel Crustaceenkunde für den Ornithologen. *Vogelwarte* **22**, 17–20.
Löhrl, H. 1959. Zur Frage des Zeitpunktes einer Prägung auf die Heimatregion beim Halsbandschnapper *(Ficedula albicollis). J. Ornithol.* **100**, 132–140.
Löhrl, H. 1960–61. Vergleichende Studien über Brutbiologie und Verhalten der Kleiber *Sitta whiteheadi* (Sharpe) und *Sitta canadensis* L. *J. Ornithol.* **101**, 245–264; **102**, 111–132.
Longhurst, A. R. 1954. Reproduction in Notostraca (Crustacea). *Nature* **173**, 781–782.
Longhurst, A. R. 1955. Evolution in the Notostraca. *Evolution* **9**, 84–86.
Lönnberg, E. 1924. Ett bidrag till den svenska faunans utbredningshistoria. *Fauna o. Flora* **19**, 97–119.
Lönnberg, E. 1929. The development and distribution of the African fauna in connection with and depending on climatic changes. *Arkiv för Zool.* **21A**, 1–33.

Löve, A. and D. Löve. 1943. The significance of differences in the distribution of diploids and polyploids. *Hereditas* **29**, 145–163.
Lowe, C. H., Jr. 1955. An evolutionary study of island faunas in the Gulf of California, Mexico, with a method for comparative analysis. *Evolution* **9**, 339–344.
Lydekker, R. 1896. A geographical history of mammals. Cambridge Geogr. Series, Cambridge Univ. Press, pp. xii+400.

Macan, T. T. 1961. Factors that limit the range of freshwater animals. *Biol. Rev.* **36**, 151–198.
MacArthur, R. H. 1964. Environmental factors affecting bird species diversity. *Am. Naturalist* **98**, 387–397.
MacArthur, R. H., and J. W. MacArthur. 1961. On bird species diversity. *Ecology* **42**, 594–598.
MacArthur, R. H., and E. O. Wilson. 1963. An equilibrium theory of insular zoogeography. *Evolution* **17**, 373–387.
MacArthur, R. H., and E. O. Wilson. 1967. The theory of island biogeography. Princeton Univ. Press, Princeton. pp. xi+203.
McCabe, T. T., and B. D. Blanchard. 1950. Three species of *Peromyscus*. Rood, Santa Barbara. pp. v+136.
McCabe, T. T., and I. McT. Cowan 1945. *Peromyscus maniculatus macrorhinus* and the problem of insularity. *Trans. Roy. Can. Inst.* 1945, pp. 117–215.
Macfadyen, A. 1957. Animal ecology. Pitman, London. pp. xx+264.
MacGinitie, H. D. 1958. Climate since the late Cretaceous. A. A. A. S. Publ. 51, pp. 61–79.
MacKerras, I. M. 1962. Speciation in Australian Tabanidae. pp. 328–358 in: G. W. Leeper (ed.) Evolution of living organisms. Melbourne Univ. Press, Melbourne.
MacLaughlin, C. A. 1958. A new race of the pocket gopher *Geomys bursarius* from Missouri. Los Angeles County Mus. Contr. Sci. No. 19. pp. 4.
MacLulich, D. A. 1957. The place of chance in population processes. *J. Wildlife Management* **21**, 293–299.
Macpherson, A. H. 1965. The origin of diversity in mammals of the Canadian Arctic tundra. *Syst. Zool.* **14**, 153–173.
Maguire, B., Jr. 1963. The passive dispersal of small aquatic organisms and their colonization of isolated bodies of water. *Ecol. Monographs* **33**, 161–185.
Mail, G. A., and R. W. Salt. 1933. Temperature as possible limiting factor in the northern spread of the Colorado potato beetle. *J. Econ. Entomol.* **26**, 1068.
Mani, M. S. 1962. Introduction to high altitude entomology. Methuen, London. pp. xix+302.
Marcus, E. 1933. Tiergeographie. pp. 81–166 in: F. Klute (ed.) Handb. geogr. Wiss. —Allg. Geogr. II. Athenaion, Potsdam.
Marcström, V. 1956. Om kroppstemperaturen hos tjäderkycklingar *Tetrao u. urogallus* Lin., vid kläckningen och omedelbart därefter. (Body temperature of capercaillie chicks during and after hatching). *Viltrevy* **1**, 139–149.
Margalef, R. 1961. Modalités de l'évolution en rapport avec la simplification des biocenoses insulaires. *Colloq. Intern. Centre Natl. Rech. Sci.* **94**, 313–320.
Margalef, R. 1963. On certain unifying principles in ecology. *Am. Naturalist* **97**, 357–374.
Martin, P. S. 1958. Pleistocene ecology and biogeography of North America. pp. 375–420 in: C. Hubbs (ed.) Zoogeography. Amer. Assoc. Adv. Sci. Symp., 51.
Martin, P. S., and P. J. Mehringer. 1965. Pleistocene pollen analysis and biogeography of the Southwest. pp. 433–451. in: H. E. Wright and D. G. Frey (eds.) The Quaternary of the United States.
Matthew, W. D. 1915. Climate and evolution. *Ann. N. Y. Acad. Sci.* **24**, 171–318. Reprinted 1939, Spec. Publ. N. Y. Acad. Sci. 1. pp. 223.

Matvejev, S. D. 1954. Relict and relictity in biology. Recueil des Travaux de l'Inst. d'Écol. et de Biogéogr. Acad. Serbe des Sci. Vol. 5. No. 4. pp. 9.

Matvejev, S. 1961. Biogeography of Yugoslavia. Biol. Inst. Monogr. Belgrade No. 9. pp. 232.

Mayr, E. 1926. Die Ausbreitung des Girlitz. *J. Ornithol.* **74,** 571–671.

Mayr, E. 1942. Systematics and the origin of species. Columbia Univ. Press, New York. pp. xvii+334.

Mayr, E. 1943. The zoogeographic position of the Hawaiian Islands. *Condor* **45,** 45–48.

Mayr, E. 1944. Wallace's line in the light of recent zoogeographic studies. *Quart. Rev. Biol.* **19,** 1–14.

Mayr, E. 1946. History of the North American bird fauna. *Wilson Bull.* **58,** 3–41.

Mayr, E. 1954. Change of genetic environment and evolution. pp. 157–180 in: J. Huxley and A. C. Hardy (eds.) Evolution as a process. Allen & Unwin, London.

Mayr, E. 1959. Isolation as an evolutionary factor. *Proc. Am. Phil. Soc.* **103,** 221–230.

Mayr, E. 1963. Animal species and evolution. Harvard Univ. Press, Cambridge, Mass. pp. xvi+797.

Mayr, E. 1963a. The fauna of North America, its origin and unique composition. Proc. XVI. Intern. Congr. Zool. **4,** 3–11.

Mayr, E. 1964. Inferences concerning the Tertiary American bird fauna. *Proc. Nat. Acad. Sci.* **51,** 280–288.

Mayr, E. 1965. Avifauna: turnover on islands. *Science* **150,** 1587–1588.

Mayr, E., and W. H. Phelps, Jr. 1967. The origin of the bird fauna of the south Venezuelan highlands. *Bull. Am. Mus. Nat. Hist.* **136,** 269–328.

Meigs, P. 1966. Geography of coastal deserts. Columbia Univ. Press, New York. pp. 150.

Menard, H. W., and E. L. Hamilton. 1963. Paleography of the Tropical Pacific. pp. 193–218 in: J. L. Gressitt (ed.) Pacific Basin Biogeography. Bishop Museum Press, Honolulu.

Mengel, R. M. 1964. The probable history of species formation in some northern wood warblers (Parulidae). *Living Bird* **3,** 9–43.

Menzbier, M. 1882, 1892. Ornitologicheskaya geografiya europeiskoy Rossiye. Part 2. 1892. Univ. Moscow, Moscow, pp. 53–244.

Merikallio, E. 1931. Kanalintujemme munamäärät. *Ornis Fennica.* **8,** 1–10.

Merikallio, E. 1958. Finnish birds, their distribution and numbers. *Fauna Fenn.* **5,** 1–181.

Merriam, C. H. 1892. The geographic distribution of life in North America with special reference to Mammalia. *Proc. Biol. Soc. Wash.* **7,** 1–64.

Merriam, C. H. 1894. Laws of temperature control of geographic distribution of terrestrial mammals and plants. *Nat. Geogr. Mag.* **6,** 229–238.

Merriam, C. H. 1898. Life zones and crop zones in the United States. U.S. Dept. Agr. Bull. 10. pp. 79.

Mertens, R., and H. Wermuth. 1960. Die Amphibien und Reptilien Europas. Kramer, Frankfurt. pp. xi+264.

Messenger, P. S. 1959. Bioclimatic studies with insects. *Ann. Rev. Entomol.* **4,** 182–206.

Meusel, H. 1943. Vergleichende Arealkunde. 2 vols. Borntraeger, Berlin. pp. xii+466; xii+92+90 maps.

Meyrick, E. 1926. On Micro-Lepidoptera from the Galapagos Islands and Rapa. Trans. Entom. Soc. London, 1926, pp. 269–278.

Michaelsen, J. W. 1903. Die geographische Verbreitung der Oligochaeten. Friedländer, Berlin. pp. vi+186.

Miller, A. H. 1942. Habitat selection among higher vertebrates and its relation to intraspecific variation. *Am. Naturalist* **76,** 25–35.

Miller, A. H. 1951. An analysis of the distribution of the birds of California. *Univ. Calif. Publ. Zool.* **50**, 531–643.
Miller, A. H. 1963. Desert adaptations in birds. Proc. XIII. Intern. Ornith. Congr. **2**, pp. 666–674.
Millot, J. 1952. La faune malgache et le mythe gondwanien. *Mem. Inst. Sci. Madagascar A.* **7**, 1–36.
Milne, A. 1950. The ecology of the sheep tick, *Ixodes ricinus* L. Spatial distribution. *Parasitology* **40**, 35–45.
Milstead, W. W. 1961. Competitive relationships in lizard populations. pp. 460–497 in: Blair, W. F. (ed.) Vertebrate Speciation. Univ. Texas Press, Austin.
Mitchell, R. 1962. Storm induced dispersal in the damselfly *Ischnura verticalis* (Say). *Am. Midland Naturalist* **68**, 199–202.
Möbius, K. 1877. Die Auster und die Austernwirtschaft. Parey, Berlin. pp. 126. (English trans., The oyster and oyster culture. U.S. Fish. Comm. Rept. 1880, pp. 683–751.)
Möbius, K. 1891. Die Tiergebiete der Erde, ihre kartographische Abgrenzung und museologische Bezeichnung. *Arch. Naturgesch.* **57**, 277–291.
Moore, J. A. 1952. An analytical study of the geographical distribution of *Rana septentrionalis*. *Am. Naturalist* **86**, 5–22.
Moore, J. C. 1960. Squirrel geography of the Indian subregion. *Syst. Zool.* **9**, 1–17.
Moreau, R. E. 1952. Africa since the Mezozoic. *Proc. Zool. Soc. London* **121**, 869–913.
Moreau, R. E. 1954. The distribution of African evergreen-forest birds. *Proc. Linncan Soc. London* **165**, 35–46.
Moreau, R. E. 1955. Ecological changes in the Palaearctic region since the Pliocene. *Proc. zool. Soc. London* **125**, 253–295.
Moreau, R. E. 1963. Vicissitudes of the African biomes in the late Pleistocene. *Proc. zool. Soc. London* **141**, 395–421.
Moreau, R. E. 1966. The bird faunas of Africa and its islands. Academic Press, New York and London. pp. ix+424.
Morrison, R. B. 1965. Quaternary geology of the Great Basin. pp. 265–285 in: H. E. Wright and D. G. Frey (eds.) The Quaternary of the United States. Princeton Univ. Press, Princeton.
Munroe, E. 1953. The size of island faunas. *Proc. 7th Pac. Sci. Congr.* **4**, 52–53.
Munroe, E. 1957. Comparison of closely related faunas. *Science* **126**, 437–439.
Munroe, E. 1958. The geographic distribution of the Scopariinae (Lepidoptera: Pyralidae). *Proc. 10th Intern. Congr. Entomol.* **1**, 831–837.
Murchie, W. R. 1956. Survey of the Michigan earthworm fauna. *Papers Mich. Acad. Sci.* **41**, 53–72.
Myers, G. S. 1953. Ability of amphibians to cross sea barriers, with special reference to Pacific zoogeography. *Proc. 7th Pac. Sci. Congr.* **4**, 19–27.

Naumov, N. P. 1955. [Animal ecology.] Sovjetskaia Nauka, Moscow. pp. 533.
Nehring A. 1890. Über Tundren und Steppen der Jetzt und Vorzeit, mit besonderer Berücksichtigung ihrer Fauna. Dümmler, Berlin. pp. VIII+257.
Neill, W. T. 1958. The occurrence of amphibians and reptiles in saltwater areas, and the bibliography. *Bull. Marine Sci. Gulf Caribbean* **8**, 1–97.
Neill, W. T. 1964. Viviparity in snakes: some ecological and zoogeographical considerations. *Am. Naturalist* **98**, 35–55.
Newbigin, M. I. 1936. Plant and animal geography. Methuen, London. pp. viii+298.
Nicol, D. 1961. Biotic associations and extinction. *Syst. Zool.* **10**, 35–41; **12**, 38–39.
Niering, W. A. 1956. Bioecology of Kapingamarangi Atoll. Atoll Res. Bull. 49. pp. 32+33 figs.
Niethammer, G. 1951. Arealveränderungen und Bestandsschwankungen mitteleuropäischer Vögel. *Bonner zool. Beitr.* **2**, 17–54.

Niethammer, G. 1953a. Zur Vogelwelt Boliviens. *Bonner zool. Beitr.* **4**, 73–78.
Niethammer, G. 1953b. Zum Transport von Süsswassertieren durch Vögel. *Zool. Anz.* **151**, 41–42.
Niethammer, G. 1958a. Tierausbreitung. Orion-Bücher Vol. 115. Lux, Murnau, München. pp. 89.
Niethammer, G. 1958b. Tiergeographie (Bericht über die Jahren 1950–56). *Fortschr. Zool.* **11**, 35–141.
Niethammer, G. 1963. Die Einbürgerung von Säugetiere und Vögeln in Europa. Parey, Hamburg/Berlin. pp. 319.
Niethammer, G. 1966. Tiergeographie (Bericht über die Jahren 1957–64). *Fortschr. Zool.* **18**, 1–138.
Nikolsky, G. 1947. On biological peculiarities of faunistic complexes and on the value of their analysis for zoogeography. *Zool. Zh.* **26**, 231.
Norris, K. S. 1958. The evolution and systematics of the iguanid genus *Uma* and its relation to the evolution of other North American desert reptiles. *Bull. Amer. Mus. Nat. Hist.* **114**, 247–326.
Norris, R. A. 1958. Comparative biosystematics and life history of the nuthatches *Sitta pygmaea* and *Sitta pusilla*. *Univ. Calif. Publ. Zool.* **56**, 119–300.
Nuorteva, P., and H. Hoogstraal. 1963. The incidence of ticks (Ixodoidea, Ixodidae) on migratory birds arriving in Finland during the spring of 1963. *Ann. Med. Exp. Biol. Fenniae* **41**, 457–468.

Odum, E. P. 1959. Fundamentals of ecology. 2d. ed. Saunders, Philadelphia. pp. xvii+546.
Odum, E. P. 1964. Biological productivity at the heterotrophic levels. *Proc. 16th Intern. Congr. Zool.* **4**, 145–149.
Ognev, S. I. 1940. [The mammals of the U.S.S.R. and adjoint countries.] 5 vols. Acad. Sci. U.S.S.R., Moscow Leningrad.
Öhrn, B. 1961. Fågelregioner. Norstedts, Stockholm. pp. 149.
Økland, F. 1955. Generell Dyregeografi. Aschehoug, Oslo. pp. 166.
Olstad, O. 1943. Rådyrets utbredelse i Norge. *Naturen* **67**, 65–74.
Omodeo, P. 1951. Problemi zoogeografici ed ecologici relativi a lombrichi peregrini, con particolare riguardo al tipo di riproduzione ed alla struttura cariologica. *Boll. Zool.* **18**, 117–122.
Omodeo, P. 1963. Distribution of the terricolous Oligochaetes on the two shores of the Atlantic. pp. 127–151 in A. Löve and Löve (eds.) North Atlantic biota and their history. Macmillan, New York. pp. xii+430.
Orians, G. H. 1962. Natural selection and ecological theory. *Am. Naturalist* **96**, 257–263.
Osche, G. 1963. Ökologie des Parasitismus und der Symbiose. *Fortschr. Zool.* **15**, 125–164.
Otterlind, G. 1954. Flyttning och utbredning. *Vår Fågelvärld* **13**, 1–31, 83–113.
Oughton, J. 1948. A zoogeographical study of the land snails of Ontario. Univ. Toronto Stud. Biol. Ser. 57. pp. 58.

Paine, R. T. 1966. Food web complexity and species diversity. *Am. Naturalist* **100**, 65–75.
Palmén, E. 1944. Die anemohydrochore Ausbreitung der Insekten als zoogeographischer Faktor, mit Berücksichtigung der baltischen Einwanderungsrichtung als Ankunftsweg der fennoskandischen Käferfauna. Ann. Zool. Soc. Vanamo 10(1). pp. 259.
Palmén, E. 1949a. A migration of *Vanessa io* L., with remarks on its occurrence in Finland. *Ann. Entomol. Fenn.* **14** (Suppl.), 160–168.
Palmén, E. 1949b. The diplopoda of E. Fennoscandia. Ann. Soc. Zool. Vanamo 13. pp. 54.

Palmgren, P. 1927. Die Haubenmeise *(Parus cristatus)* auf Åland. Ein Beitrag zur Kenntniss der Verbreitungsökologie der Vögel. Acta Soc. Flor. Faun. Fenn. 56. pp. 12.
Palmgren, P. 1930. Quantitative Untersuchungen über die Vogelfauna in den Wäldern Südfinnlands. Acta Zool. Fenn. 7. pp. 218.
Palmgren, P. 1932. Zur Biologie von *Regulus r. regulus* (L.) und *Parus atricapillus borealis* Selys. Acta Zool. Fenn. 14. pp. 113.
Palmgren, P. 1936. Bemerkungen über die ökologische Bedeutung der biologischen Anatomie des Fusses bei einigen Kleinvogelarten. *Ornis Fennica* 13, 53–58.
Palmgren, P. 1938. Zur Kausalanalyse der ökologischen und geographischen Verbreitung der Vögel Nordeuropas. *Arch. Naturgesch.* 7, 235–269.
Palmgren, P. 1949. Welche Faktoren bedingen die geographische und topographische Verbreitung der Vögel? *Folia Biotheoret.* 4, 23–40.
Pantin, C. F. A. (ed.) 1960. A discussion on the biology of the southern cold temperate zone. *Proc. Roy. Soc. Ser. B.* 152, 429–682.
Paulian, R. 1961. L'environment insulaire de Madagascar et le peuplement des îles océaniques. *Colloq. Intern. Centre Nat. Rech. Sci.* 94, 261–271.
Pax, F. 1930. Die Tierwelt Vol. II. Pt. 2. Sec. 6. pp. 164–251 in: A. Supan and E. Obst, Grundzüge der physischen Erdkunde. De Gruyter, Berlin.
Pearson, T. 1955. Djurgeografi. Bonnier, Stockholm. pp. 171.
Peitzmeier, J. 1942. Die Bedeutung der ökologischen Beharrungstendenz für faunistische Untersuchungen. *J. Ornithol.* 90, 331–322.
Peitzmeier, J. 1951. Beobachtungen über Klimaveränderungen und Bestandeveränderung einiger Vogelarten in Nordwestdeutschland Proc. 10th Intern. Ornith. Congr. 1950, pp. 477–483.
Perring, F. H., and S. M. Walters (eds.) 1962. Atlas of the British Flora. Nelson, London. pp. xxiv+432.
Petersen, B. 1954. Some trends of speciation in the cold-adapted Holaractic fauna. *Zool. Bidrag* 30, 233–314.
Peterson, R. L. 1955. North American moose. Univ. Toronto Press, Toronto. pp. xi+288.
Péwé, T. L., D. M. Hopkins, and J. L. Giddings. 1965. The Quaternary geology and archeology of Alaska. pp. 355–374 in: H. E. Wright and D. G. Frey (eds.) The Quaternary of the United States. Princeton Univ. Press, Princeton.
Pimentel, D. 1961. Animal population regulation by genetic feedback mechanism. *Am. Naturalist* 95, 65–79.
Pimentel, D. 1963. Natural population regulations and interspecies evolution. Proc. XVI Inten. Congr. Zool. 3, 329–336.
Pianka, E. R. 1966. Latitudinal gradients in species diversity: a review of concepts. *Am. Naturalist* 100, 33–46.
Pitelka, F. A. 1951. Speciation and ecologic distribution in American jays of the genus *Aphelocoma. Univ. Calif. Publ. Zool.* 50, 195–464.
Polunin, N. 1960. Introduction to plant geography. McGraw-Hill, New York. pp. xix+640.
Pompper, H. 1841. Die Säugethiere, Vögel und Amphibien nach ihrer geographischen Verbreitung. Hintlich, Leipzig. pp. iv+37.
Poore, M. E. D. 1962. The method of successive approximation in descriptive ecology. *Adv. Ecol. Res.* 1, 35–68.
Prenant, M. 1933. Géographie des animeaux. Colin, Paris. pp. 199.
Preston, F. W. 1962. The canonical distribution of commonness and rarity. *Ecology* 43, 185–215, 410–432.
Proctor, V. W., and C. A. Malone. 1965. Passive dispersal of small aquatic organisms via the intestinal tracts of birds. *Ecology* 46, 728–729.
Proctor, V. W., C. R. Malone, and V. L. Devlaming. 1967. Dispersal of aquatic organisms: viability of disseminules recovered from the intestinal tract of captive killdeer. *Ecology* 48, 672–676.

Rand, A. L. 1948. Glaciation, an isolating factor in speciation. *Evolution* **2**, 314–321.
Rand, A. L., and L. J. Brass. 1940. Results of the Archbold expeditions No. 29. Summary of the 1936–1937 New Guinea expedition. *Bull. Am. Mus. Nat. Hist.* **77**, 341–380.
Reichenow, A. 1888. Die Begränzung zoogeographischer Regionen vom ornithologischen Standpunkte aus. *Zool. Jahrb. (Syst.)* **3**, 661–704.
Reid, J. L., Jr. 1965. Intermediate waters of the Pacific Ocean. J. Hopkins Oceanogr. Studies 2. Johns Hopkins Univ. Press, Baltimore. pp. 85.
Reinig, W. F. 1937. Die Holarctis. Fischer, Jena. pp. vi+124.
Reinig, W. F. 1938. Elimination und Selektion. Fischer, Jena. pp. 146.
Reinig, W. F. 1950. Chorologische Voraussetzungen für die Analyse von Formenkreisen. pp. 346–378 in: Syllegomena Biol. (Festschr. Kleinschmidt). Geest and Portig, Leipzig.
Remane, A. 1943. Die Bedeutung der Lebensformtypen für die Ökologie. *Biol. Gener.* **17**, 164–182.
Rensch, B. 1931. Tiergeographie (Literatur von 1908–1930). *Geograph. Jahrb.* **45**, 51–132.
Rensch, B., M. Knipper, and E. Rieh. 1937. Die tiergeographische Publikazionen der Jahre 1933 und 1934. *Zoogeographica* **3**, 321–388.
Rensch, B. 1947. Tiergeographie. Fortschr. Zool. (N.F.) 1–8.
Rensch, B. 1950. Die Verteilung der Tierwelt in Raum. pp. 125–172 in: L. von Bertalanffy (ed.) Handbuch der Biologie, Vol. 5. Akademieverlag, Berlin-Potsdam.
Rensch, B. 1959. Evolution above the species level. Methuen, London. pp. xvii+419.
Ressl, F. 1963. Können Vögel als passive Verbreiter von Pseudoscorpioniden betrachtet werden? *Vogelwelt* **84**, 114–119.
Ricker, W. E. 1964. Distribution of Canadian stoneflies. *Gewässer Abwässer* **34/35**, 50–71.
Romer, A. S. 1961. Paleozoological evidence of climate. (1) Vertebrates. pp. 183–206 in: A. E. M. Nairn (ed.) Descriptive paleoclimatology. Interscience Publishers, New York and London.
Ross, H. H. 1962. A synthesis of evolutionary theory. Prentice-Hall, Englewood Cliffs. pp. xiii+387.
Ross, H. H. 1965. Pleistocene events and insects. pp. 583–596 in: H. E. Wright and D. F. Frey (eds.) The Quarternary of the United States. Princeton Univ. Press, Princeton.
Ross, H. H. 1965. A textbook of entomology. 3d ed. Wiley, New York. pp. ix+539.
Rübel, E. 1930. Pflanzengesellschaften der Erde. Huber, Bern and Berlin. pp. viii+464.
Rübel, E. 1935. The replaceability of ecological factors and the law of the minimum. *Ecology* **16**, 336–341.
Rubtsov, I. A. 1937. [The theoretical basis of the distribution of harmful insects and of predictions as to their mass multiplication.] *Vestn. Zashchita rastenii* **14**, 3–13.
Runcorn, S. K. (ed.) 1962. Continental drift. Academic Press, New York. pp. 338.
Rütimeyer, L. 1867. Ueber die Herkunft unserer Tierwelt. Eine zoogeographische Skizze. Program der Gewerbeschule. Reprinted, 1898 pp. 137–224 in: L. Rütimeyer, Gesammelte kleine Schriften allgemeinen Inhalts aus dem Gebiete der Naturwissenschaft, Vol. I. Basel, von Georg. pp. VI+400.
Ryan, R. M. 1963. The biotic provinces of Central America. Acta Zool. Mex. 6. pp. 54.
Rylow, V. M. 1921. Zur Frage der Glazialrelikte in der Süsswasserfauna. *Bull. Inst. Hydrologique de Russie* **1921**, 47–113 (Fide Thienemann 1950).

Salomonsen, F. 1931. Diluviale Isolation und Artenbildung. Proc. VII. Intern. Ornith. Congr., pp. 413–438.
Salomonsen, F. 1948. The distribution of birds and the recent climatic change in the North Atlantic area. *Dansk Ornith. Foren. Tidsskr.* **42**, 85–99.

Salomonsen, F. 1951. Immigration and breeding of fieldfare in Greenland. Proc. X Intern. Ornith. Congr., pp. 515–526.
Salomonsen, F. 1955. The evolutionary significance of bird-migration. *Dan. Biol. Medd.* **22**, 1–62.
Salt, G. W. 1952. The relation of metabolism to climate and distribution in three finches of the genus *Carpodacus. Ecol. Monographs* **22**, 121–152.
Savile, D. B. O. 1959. Limited penetration of barriers as a factor in evolution. *Evolution* **13**, 333–343.
Schaller, F. 1960. Das Phoresie-Phänomen vergleichend-ethologisch gesehen. *Forsch. Fortschr.* **34**, 1–7.
Schick, R. X. 1955. The crab spiders of California (Araneida, Thomisidae). *Bull. Am. Mus. Nat. Hist.* **129**, 1–180.
Schilder, F. A. 1943. Zur Verwandtschaft der Litoralfaunen. *Arch. Molluskenk.* **75**, 68–82.
Schilder, F. A. 1947-48. Probleme und Methoden der Biostatistik. *Biol. Zentralbl.* **66**, 186–197.
Schilder, F. A. 1954. Die Klassifikation der Faunengebiete des Festlandes. *Wiss. Z. Univ. Halle, Math.-Naturw.* **3**, 1153–1169.
Schilder, F. A. 1955. Statistische Methoden in der Biogeographie. *Wiss. Z. Martin-Luther-Univ., Halle-Wittenberg* **4**, 711–716.
Schilder, F. A. 1956. Lehrbuch der allgemeinen Zoogeographie. Fischer, Jena. pp. viii+150.
Schmarda, L. K. 1853. Die geographische Verbreitung der Tiere. Gerold, Vienna. pp. vii+755.
Schmid, E. 1936. Die Reliktföhrenwälder der Alpen. Beitr. zur geobotanischen Landaufnahmen der Schweiz. 20. pp. 190.
Schmidt, K. P. 1945. Evolution, succession, and dispersal. *Am. Midland Naturalist* **33**, 788–790.
Schmidt, K. P. 1950. The concept of geographic range with illustration from amphibians and reptiles. *Texas J. Sci.* **2**, 326–334.
Schmidt, K. P. 1954. Faunal realms, regions, and provinces. *Quart. Rev. Biol.* **29**, 322–331.
Schmidthüsen, J. 1959. Allgemeine Vegetationsgeographie. De Gruyter, Berlin. pp. xviii+261.
Schneider, F. 1952. Untersuchungen über die optische Orientierung der Maikäfer (*Melolontha vulgaris* F. und *M. hippocastani* F.) sowie über die Entstehung von Schwärmbahnen und Befallskonzentrationen. *Mitt. Schweiz. Entomol. Ges.* **25**, 269–340.
Schwerdtfeger, F. 1963. Ökologie der Tiere. Autökologie. Paul Parey, Hamburg-Berlin. pp. 461.
Sclater, P. L. 1858. On the general geographical distribution of the members of the class Aves. *J. Linn. Soc. (Zool.)* **2**, 130–145.
Sclater, W. L. 1894–1895. Geography of mammals. *Geogr. Journ.* **3**, 95–105; **4**, 35–52; **5**, 471–483.
Seibert, H. C. 1951. Light intensity and the roosting flight of herons in New Jersey. *Auk* **68**, 63–74.
Seiler, J. 1961. Untersuchungen über die Entstehung der Parthenogenese bei *Solenobia triquetrella* F. R. (Lepidoptera, Psychidae). *Z. Vererbungslehre* **92**, 261–316.
Severtsov, N. A. 1877. O zoologicheskikh (primushchestvenno ornitologicheskikh) oblastakh vnetropicheskikh chestey nashevo materika. Izv. Russk. Geogr. Obshch. *13*.
Shelford, V. E. 1911. Physiological animal geography. *J. Morphol.* **22**, 551–618.
Shelford, V. E. 1963. The ecology of North America. Univ. Illinois Press, Urbana. pp. xxii+610.
Short L. L., Jr. 1965. Hybridization in the flickers (*Colaptes*) of North America. *Bull. Am. Mus. Nat. Hist.* **129**, 307–428.

Sibley, C. G. 1957. The evolutionary and taxonomic significance of sexual dimorphism and hybridisation in birds. *Condor* 59, 166–191.
Sibley, C. G., and L. L. Short, Jr. 1959. Hybridisation in the buntings *(Passerina)* of the Great Plains. *Auk* 76, 443–463.
Sibley, C. G., and L. L. Short, Jr. 1964. Hybridisation in the orioles of the Great Plains. *Condor* 66, 130–150.
Siivonen, L. 1943. Artenstatistische Daten über die Veränderungen in der Vogelfauna Finnlands während der letzten Jahrzehnte. *Ornis Fennica* 20, 1–16.
Siivonen, L. 1952. Über den Einfluss regionaler Bestandesverschiebungen auf die lokale Vogeldichte. *Ornis Fennica* 29, 37–44.
Siivonen, L. 1953. [Roe deer spreading to Finland.] *Suomen Riista* 8, 110–113.
Siivonen, L. 1962. Die Schneemenge als überwinterungsökologischer Faktor. *Proc. Finnish Acad. Sci.* 1962, pp. 111–125.
Simpson, G. G. 1940. Mammals and land bridges. *Wash. Acad. Sci.* 30, 137–163.
Simpson, G. G. 1943. Mammals and the nature of continents. *Am. J. Sci.* 24, 1–31.
Simpson, G. G. 1944. Tempo and mode in evolution. Columbia Univ. Press, New York. pp. xviii+237.
Simpson, G. G. 1947. Evolution, interchange, and resemblance of North American and Eurasian Cenozoic mammalian faunas. *Evolution* 1, 218–220.
Simpson, G. G. 1950. History of the fauna of Latin America. *Am. Scientist* 38, 361–389.
Simpson, G. G. 1952. Probabilities of dispersal in geologic time. *Bull. Am. Mus. Nat. Hist.* 99, 163–176.
Simpson, G. G. 1953a. Evolution and geography. Oregon State Board of Education, Eugene. pp. 64.
Simpson, G. G. 1953b. The major features of evolution. Columbia Univ. Press, New York. pp. xxx+434.
Simpson, G. G. 1960. Notes on the measurement of faunal resemblance. *Am. J. Sci.* 258A, 300–311.
Simpson, G. G. 1961. Historic zoogeography of Australian mammals. *Evolution* 15, 431–446.
Simpson, G. G. 1964. Species density of North American mammals. *Syst. Zool.* 13, 57–73.
Simpson, G. G. 1965. The geography of evolution. Collected essays. Chilton, Philadelphia and New York. pp. xiv+249.
Simpson, G. G., and W. S. Beck. 1965. Life. 2d ed. Harcourt, Brace & World, New York. pp. xviii+869.
Sirén, G. 1961. [The tree-lined pine as indicator of climatic fluctuations in northern Fennoscandia during historic time.] Commun. Inst. Forest. Fenn. 54/2. pp. 66.
Slobodkin, L. B. 1961. Growth and regulation of animal populations. Holt, Rinehart, and Winston, New York. pp. 1–184.
Smith, P. W. 1957. An analysis of post-Wisconsin biogeography of the Prairie Peninsula region based on distributional phenomena among terrestrial vertebrate populations. *Ecology* 38, 205–218.
Smith, P. W., and H. M. Smith. 1962. The systematic and biogeographic status of two Illinois snakes. Occas. Pap. C. C. Adams Center Ecol. Stud. 5. pp. 10.
Sokal, R. R., and P. H. A. Sneath. 1963. Principles of numerical taxonomy. Freeman, San Francisco. pp. xvi+359.
Solem, A. 1959. Zoogeography of the land and freshwater mollusca of the New Hebrides. *Fieldiana Zool.* 43, 239–343.
Stantschinsky, V. V. 1923. [Contributions to ecological geography of birds.] *Smolensk Univ. Naushnye Izvestija* 1, 41–55.
Stantschinsky, V. V. 1927. Some climatic limits in the extension of birds in Eastern Europe. *Ecology* 8, 232–237.
Stebbins, G. L. 1950. Variation and evolution in plants. Columbia Univ. Press, New York. pp. xx+643.

Stebbins, R. C. 1966. A field guide to western reptiles and amphibians. Houghton Mifflin, Boston. pp. xiv+279.
Steenis, C. G. G. J. van. 1962. The land bridge theory in botany. *Blumea* XI, 235–542.
Stegmann, B. 1938. Principes généraux des subdivisions ornithogéographiques de la région paléarctique. *Faune de l'URSS. Acad. Sci. URSS.* Vol. I, No. 2. Moscow–Leningrad, pp. 156.
Stein, G. W. 1958. Über den Selektionswert der Simplex-Zahnform bei der Feldmaus, *Microtus arvalis* (Pallas). *Zool. Jahrb. (Syst.)* 86, 27–34.
Stresemann, E. 1919a. Über die Formen der Gruppe *Aegithalos caudatus* und ihre Kreutzungen. *Beitr. Zoogeogr. Paläarkt. Region* 1, 3–24.
Stresemann, E. 1919b. Über die europäischen Gimpel. *Beitr. Zoogeogr. Paläarkt. Region* 1, 25–56.
Stresemann, E., and L. A. Portenko (eds.). 1960. Atlas der Verbreitung palaearktischer Vögel. I. Lieferung. Akademie-Vg, Berlin. 20 maps.
Stresemann, E., L. A. Portenko, and G. Mauersberger. 1967. Atlas der Verbreitung palaearktischer Vögel II. Lieferung. Akademie-Vg, Berlin. 15 maps.
Sukachev, V. N. 1960. The correlation between the concept "forest ecosystem" and "forest biogeocoenose" and their importance for the classification of forests. *Silva Fennica* 105, 94–97.
Suomalainen, E. 1947. Parthenogenese und Polyploidie bei Rüsselkäfern (Curculionidae). *Hereditas* 33, 425–456.
Suomalainen, E. 1950. Parthenogenesis in animals. *Advan. Genet.* 3, 193–253.
Suomalainen, E. 1953. Die Polyploidie bei den parthenogenetischen Rüsselkäfern. *Zool. Anz. Suppl.* 17, 280–289.
Suomalainen, E. 1961. On morphological differences and evolution of different polyploid parthenogenetic weevil populations. *Hereditas* 47, 309–341.
Suomalainen, E. 1962. Significance of parthenogenesis in the evolution of insects. *Ann. Rev. Entomol.* 7, 349–366.
Suomalainen, E. 1965. Die Polyploidie bei dem parthenogenetischen Blattkäfer *Adoxus obscurus* L. (Coleoptera, Chrysomelidae). *Zool. Jahrb. (Syst.)* 92, 183–192.
Svensson, L. 1957. De senaste tio årens nytillskott av svenska storfjärilar. *Opuscula Entomol.* 22, 143–160.
Swainson, W. 1835. A treatise on the geography and classification of animals. Longman, Green, London. pp. vii+367.
Swärdson, G. 1957. The "invasion" type of bird migration. *British Birds* 50, 314–343.
Sylvester-Bradley, P. C. 1963. Post-tertiary speciation in Europe. *Nature* 199, 126–130.
Syroechkovskij, E. E. 1959. [The ornithological investigations of the Soviet Antarctic expedition in the year 1957 and some tasks of study the birds of Antarctic.] Vtor. Vsesoyuz. ornit. konfer. *Tezisy Dokl.* 3, 18–20. Moscow Univ. Press. (Abstracts of the Second All-Union Ornithol. Confer. 18 to 25 August, 1959, Moscow.)

Tansley, A. 1935. The use and abuse of vegetational concepts and terms. *Ecology* 16, 284–307.
Taylor, B. W. 1954. An example of long distance dispersal. *Ecology* 35, 569–572.
Tenuovo, R. 1963. Zur brutzeitlichen Biologie der Nebelkrähe (*Corvus corone cornix* L.) im äusseren Schärenhof Südwestfinnlands. Ann. Zool. Soc. Vanamo 25. pp. 147.
Tets, G. F. van. 1965. A comparative study of some social communication patterns in the Pelecaniformes. Ornith. Monogr. 2. pp. 88.
Thienemann, A. 1918. Lebensgemeinschaft und Lebensraum. *Naturw. Wochenschr.*, N. F. 17, 282–290, 297–303.
Thienemann, A. 1920. Die Grundlagen der Biocoenotik und Monards faunistische Prinzipien. *Festschr. Zschokke, Basel* 4, 1–14.
Thienemann, A. 1942. Vom Wesen der Ökologie. *Biol. Gener.* 15, 312–331.
Thienemann, A. 1950. Verbreitungsgeschichte der Süswassertierwelt Europas. Die Binnengewässer 18. pp. XVI+809.

Thomson, G. M. 1922. The naturalization of plants and animals in New Zealand. Cambridge Univ. Press, Cambridge. pp. x+607.
Timofeeff-Ressovsky, N.W. 1940. Mutation and geographic variation. pp. 73–126 in: Huxley, J. (ed.) The New Systematics, Oxford Univ. Press, Oxford.
Tinbergen, N. 1951. The study of instinct. Clarendon Press, Oxford. pp. xii+228.
Tinbergen, N. 1957. The functions of territory. *Bird Study* **4**, 14–27.
Tischler, W. 1955. Synökologie der Landtiere. G. Fischer, Stuttgart. pp. XVI+414.
Tischler, W. 1963. Ökologie der Landtiere. pp. 49–114 in L. von Bertalanffy and F. Gessner (eds.) Handbuch der Biologie, Bd. III. Athenaion, Konstanz.
Tischler, W. 1963. Weitere Untersuchungen zur Ökologie der Schmalwanze *Ischnodemus sabuleti* Fall. (Hem., Lygaeidae). *Zool. Anz.* **171**, 339–349.
Tischler, W. 1965. Agrarökologie. G. Fischer, Jena. pp. 499.
Tompa, F. S. 1964. Factors determining the numbers of song sparrows, *Melospiza melodia* (Wilson) on Mandarte Island, B.C., Canada. Acta Zool. Fenn. 109. pp. 73.
Toit, A. L. du. 1937. Our wandering continents: an hypothesis of continental drift. Oliver and Boyd, Edinburgh. pp. xiii+366.
Trouessart, E. L. 1890. La géographie zoologique. Bailliere, Paris. pp. xi+338.
Tschulok, S. 1910. Das System der Biologie in Forschung und Lehre. G. Fischer, Jena. pp. x+409.
Turbott, E. G. 1963. Three Kings Islands, New Zealand: a study in modification and regeneration. pp. 485–498 in J. L. Gressitt (ed.) Pacific Basin Biogeography. Bishop Museum Press, Honolulu.
Turček, F. J. 1966. The zoological significance of ecological and geographical borderlands. *Acta Zool. Acad. Sci. Hung.* **12**, 193–201.
Turrill, W. B. 1953. Pioneer plant geography: the phytogeographical researches of Sir Joseph Dalton Hooker. Lotsya 4. The Hague (Fide Good 1964, pp. 464).
Turrill, W. B. 1959. Plant geography. pp. 171–229 in: W. B. Turrill (ed.) Vistas in botany. Pergamon Press, New York.
Twitty, V. C. 1966. Of scientists and salamanders. Freeman, San Francisco. pp. ix+178.

Udvardy, M. D. F. 1947. Methods of bird sociological survey. *Arch. Biol. Hung.* Ser. II **17**, 61–88.
Udvardy, M. D. F. 1951. The significance of interspecific competition in bird life. *Oikos* **3**, 98–123.
Udvardy, M. D. F. 1954. Distribution of Appendicularians in relation to the Strait of Belle Isle. *J. Fisher. Res. Board Canada* **11**, 431–435.
Udvardy, M. D. F. 1956. Observations on the habitat and territory of the chaffinch, *Fringilla c. coelebs* in Swedish Lapland. *Arkiv Zool. Stockholm* Ser. 2 **9**, 499–505.
Udvardy, M. D. F. 1957. An evaluation of quantitative studies in birds. *Cold Spring Harbor Symp. Quant. Biol.* **22**, 301–311.
Udvardy, M. D. F. 1959. Notes on the ecological concepts of habitat, biotope and niche. *Ecology* **40**, 725–728.
Udvardy, M. D. F. 1963a. Zoogeographical study of the Pacific Alcidae. pp. 85–111 in: J. L. Gressitt (ed.) Pacific Basin Biogeography. Bishop Museum Press, Honolulu.
Udvardy, M. D. F. 1963b. Bird faunas of North America. *Proc. 13th Internat. Ornith. Congress* pp. 1147–1167.
Udvardy, M. D. F. 1969a. The concept of faunal dynamism and the analysis of an example. *Bonner zool. Beitr.* **20**, 1–10.
Udvardy, M. D. F. 1969b. In preparation.
Urquhardt, F. A. 1960. The monarch butterfly. Univ. Toronto Press, Toronto. pp. xxxiv+361.
Usinger, R. L. 1963. Animal distribution patterns in the tropical Pacific. p. 255–262 in: J. L. Gressitt (ed.) Pacific Basin Biogeography. Bishop Museum Press, Honolulu.

Uvarov, B. P. 1928. Insect nutrition and metabolism. *Trans. Entomol. Soc. London* 76, 255–343.
Uvarov, B. P. 1938. Ecological and biogeographical relations of Eremian Acrididae. *Mém. Soc. Biogéogr.* 6, 231–273.

Valentine, J. W. 1966. Numerical analysis of marine molluscan ranges on the extratropical Northeastern Pacific Shelf. *Limnol. Oceanog.* 11, 198–211.
Vandel, A. 1928. La parthénogenèse géographique. *Bull. Biologique* (Paris) 62, 164–281.
Vandel, A. 1960. Isopodes terrestres. Part 1. Fauna de France No. 64, Libraire de la Faculté des Sci., Paris. pp. 416.
Vandel, A. 1964. Biospéléologie. Gauthier-Villars, Paris. pp. 619.
Van Dyke, E. C. 1919. The distribution of insects in western North America. *Ann. Entomol. Soc. Am.* 12, 1–12.
Van Dyke, E. C. 1942. The origin and distribution of the coleopterous insect fauna of North America. *Proc. 6th Pacific Sci. Congr.* 4, 255–268.
Vaurie, C. 1959. The birds of the Palearctic fauna. A systematic reference. Order Passeriformes. Witherby, London. pp. 762.
Verbeek, A. M. 1965. Breeding biology, behavior and ecology of the water pipit (*Anthus spinoletta*). Unpublished Master's Thesis, Montana State Univ., Missoula, pp. 109.
Verheyen, R. 1951. La migration de la pie-grièche écorcheur, *Lanius c. collurio*. Gerfaut 41, 111–139.
Verhoeff, K. W. 1938. Diplopoden der Germania Zoogeographica im Lichte der Eiszeiten. *Zoogeographica* 3, 494–547.
Vestal, A. G. 1914. Internal relations of terrestrial associations. *Am. Naturalist* 48, 413–445.
Vierhapper, F. 1919. Über echten und falschen Vikarismus. *Oesterr. Botan. Z.* 68, 1–22.
Voous, K. H. 1955. Het probleem van de zöogeographische indeling van de landfauna. Inaugur. Vrije Univ. Volters, Groningen. pp. 20.
Voous, K. H. 1959. The relationship of the European and Aethiopian avifauna. *Ostrich* Suppl. 3, 34–39.
Voous, K. H. 1960. Atlas of European birds. Nelson, London, pp. 284.
Voous, K. H. 1963. The concept of faunal elements or faunal types. Proc. XIII. Internat. Ornith. Congr., pp. 1104–1108.
Voous, K. H. 1964. Het genusbegrip in de zöologie in theorie en praktijk. *Vakblad voor Biologen* 1964, 139–149.
Vos, A. de. 1964a. Range changes of birds in the Great Lakes region. *Am. Midland Naturalist* 71, 489–504.
Vos, A. de. 1964b. Range changes of mammals in the Great Lakes region. *Am. Midland Naturalist* 71, 216–231.

Wagner, A. 1844–46. Die geographische Verbreitung der Säugetiere. *Abhandl. Bayer. Akad. Wiss., Math. Phys. Cl.* IV(1), 1–146, (2), 1–108, (3), 3–114.
Wagner, G. 1950. Einführung in die Erd- und Landschaftsgeschichte. 2d ed. Rau, Öhringen, pp. 664.
Wagner, M. 1868. Die Darwin'sche Theorie und das Migrationsgesetz der Organismen. Duncker & Humblot, Leipzig. pp. 62.
Wakeland, C. 1958. The high plains grasshopper. *U.S. Dept. Agr. Tech. Bull.* 1167, 1–168.
Wallace, A. R. 1859. On the tendency of varieties to depart indefinitely from the original type. *Proc. Linn. Soc.* 3, 53–62.
Wallace, A. R. 1860. On the zoological geography of the Malay Archipelago. *Proc. Linnaean Soc. Zool. London* 4, 173–184.

Wallace, A. R. 1876. The geographical distribution of animals. 2 vols. Harper, New York. pp. xxiii+503, xi+553. Reprinted 1962 by Hafner, New York, and London.
Wallace, A. R. 1880. Island life. Macmillan, London, pp. 526.
Wallgren, H. 1954. Energy metabolism of two species of the genus *Emberiza* as correlated with distribution and migration. Acta. Zool. Fenn. 84. pp. 110.
Walter, H. 1954. Grundlagen der Pflanzenverbreitung. Part II. Arealkunde. pp. 245. Vol. III in: H. Walter (ed.) Einführung in die Phytologie. Ulmer, Stuttgart.
Walter, H. 1964. Die Vegetation der Erde in öko-physiologischer Betrachtung. I. Die tropischen und subtropischen Zonen. G. Fischer, Stuttgart, pp. 592.
Warming, E. 1909. Oecology of plants. Clarendon Press, Oxford, pp. xi+422. (Translation and revision of E. Warming, 1895. Plantesamfund. Philipsen, Copenhagen, pp. 335.)
Warnecke, G. 1934. Grundsätzliches zur Methodik zoogeographischer Untersuchungen in der Entomologie. *Intern. Entomol. Z.* **28**, 437–441, 451–545, 461–465.
Warnecke, G. 1936. Über die Constanz der ökologischen Valenz einer Tierart als Voraussetzung für zoogeographische Untersuchungen. *Entomol. Rundschau* **53**, 203–206, 217–219, 230–232.
Warnecke, G. 1961. Rezente Arealvergrösserungen bei Makrolepidopteren in Mittel- und Nordeuropa. *Bonn. Zool. Beitr.* **12**, 113–141.
Warner, R. E. 1968. The role of introduced diseases in the extinction of the endemic Hawaiian avifauna. *Condor* **70**, 101–120.
Webb, W. L. 1950. Biogeographic regions of Texas and Oklahoma. *Ecology* **31**, 426–433.
Wegener, A. 1915. Die Entstehung der Kontinente und Ozeane. Sammlung Vieweg No. 23, pp. 144. Brunswick. (Translation 1924. The origin of continents and oceans. Methuen, London. pp. xx+212.)
Welch, D'Alte A. 1938. Distribution and variation of *Achatinella mustelina* Mighels in the Waianae Mountains, Oahu. Bishop Mus. Bull. 152. pp. 64.
Wellington, W. G. 1960. Qualitative changes in natural populations during changes in abundance. *Can. J. Zool.* **38**, 289–314.
Wellington, W. G. 1964. Qualitative changes in populations in unstable environments. *Can. Entomologist* **96**, 436–451.
Whittaker, R. H. 1956. Vegetation of the Great Smoky Mountains. *Ecol. Monographs* **26**, 1–80.
Wigglesworth, V. B. (ed.). H. E. Hinton, C. G. Johnson, and D. Leston, 1963. The origin of flight in insects. A symposium. *Proc. Roy. Entomol. Soc. London* **28**, 23–32.
Williams, C. B. 1930. The migration of butterflies. Oliver & Boyd, Edinburgh. pp. 473.
Williams, C. B. 1957. Insect migration. *Ann. Rev. Entomol.* **2**, 163–180.
Willis, J. C. 1922. Age and area. Cambridge Univ. Press, Cambridge. pp. 259.
Wilson, E. O. 1959. Adaptive shift and dispersal in a tropical ant fauna. *Evolution* **13**, 122–144.
Winterbottom, J. M. 1965. Faunal groups in the avifauna of Southern Africa. *Rev. Zool. Bot. Afr.* **71**:157–170.
Wodzicki, K. 1965. The status of some exotic vertebrates in the ecology of New Zealand. pp. 425–460 in: H. G. Baker and G. L. Stebbins (eds.) The genetics of colonizing species. Academic Press, New York.
Wolda, H. 1963. Natural populations of the polymorphic land snail *Cepaea nemoralis* (L.). *Arch. Neerl. Zool.* **15**, 381–471.
Wolfenbarger, D. O. 1946. Dispersion of small organisms: distance dispersion rates of bacteria, spores, seeds, pollen and insects; incidence rates of diseases and injuries. *Amer. Midland Naturalist* **35**, 1–152.
Wood, A. E. 1950. Porcupines, paleogeography, and parallelism. *Evolution* **4**, 87–98.
Wood, S. L. 1963. A revision of the bark beetle genus *Dendroctonus* Erichson (Coleoptera: Scolytidae). *Great Basin Nat.* **23**, 1–116.

Woodward, S. P. 1851-56. A manual of the Mollusca. 3 vols. J. Weale, London. pp. 486.
Wright, H. E. 1961. Late Pleistocene climate of Europe: a review. *Bull. Geol. Soc. Am.* **72**, 933–984.
Wright, H. E., and D. G. Frey (eds.). 1965. The Quaternary of the United States. Princeton Univ. Press, Princeton. pp. 922.
Wright, S. 1943. Isolation by distance. *Genetics* **28**, 114–138.
Wulff, E. V. 1943. An introduction to historic plant geography. (Translation of 1933 Russian edition.) Chatham, Waltham, Mass. pp. 223.
Wynne-Edwards, V. C. 1962. Animal dispersion in relation to social behaviour. Oliver & Boyd, Edinburgh. pp. xi+653.

Yeatter, R. E. 1950. Effect of different preincubation temperatures on the hatchability of pheasant eggs. *Science* **112**, 529–530.

Zeuner, F. E. 1945. The Pleistocene period. Quaritch, London. pp. 322.
Zeuner, F. E. 1958. Dating the past. Methuen, London. pp. xx+516.
Zimmerman, E. C. 1948. Insects of Hawaii. Vol. I. Univ. Hawaii Press, Honolulu. pp. xx+206.
Zimmerman, E. C. 1963. A summary discussion. p. 477–484 in: J. L. Gressitt (ed.) Pacific Basin Biogeography. Bishop Museum Press, Honolulu.
Zimmermann, K. 1935. Zur Rassenanalyse der mitteleuropäische Feldmäuse. *Arch. Naturgesch.* **4**, 258–273.
Zimmermann, K. 1958. Selektionswert der simplex-Zahnform bei der Feldmaus? *Zool. Jahrb. (Syst.)* **86**, 35–40.

Name Index

Agassiz, L., 244
Ahlquist, H., 118
Aldrich, J. W., 249
Allee, W. C., 8, 361, 370
Allen, J. A., 244–247, 261, 262, 265, 301
Alvariño, A., 114
Ander, K., 399
Anderson, S., 19
Andrewartha, H. C., 23, 71, 93, 140, 141, 177, 183
Anonymous, 265
Antevs, E., 41, 317, 332
Archbold, R., 42
Aristotle, 1
Auer, V., 99, 315, 332–334, 359
Axelrod, D., 319, 320, 327, 329, 358

Baker, H. B., 48
Baker, H. G., 72, 124, 139, 142, 143
Baker, J. R., 117, 295
Bakker, E. M. van Zinderen, 391
Balogh, J., 93
Banfield, A. W. F., 157
Barrera, A., 278
Barry, T. H., 391
Bartenew, A., 4
Barth, R. H., 372
Bartholomew, J. G., 9, 182, 192, 203
Bary, B. McK., 238
Bates, M., 67
Bathia, M., 111, 112
Beament, J. W. L., 40
Beaufort, L. F. de, 9
Beck, W. S., 304
Beebe, W., 18
Beeson, C. F. C., 111, 112
Benson, C. W., 12, 27
Berg, L. S., 332
Bergman, C., 244
Bergman, G., 118, 121
Bernardi, G., 169
Birch, L. C., 23, 71, 93, 140–142, 177, 183, 355
Bird, R. D., 335
Blair, W. F., 142, 211, 213
Blanchard, B. D., 121
Blanford, W. T., 263, 265
Bleakney, J. S., 16
Blyth, E., 265
Bobrinskij, N. A., 9
Bodenheimer, F. S., 93, 130

Boness, M., 141
Bourlière, F., 391
Boyd, E., 339
Bradshaw, G. V., 164
Brass, L. J., 42
Brattstrom, B. H., 316
Braun-Blanquet, J., 55, 252, 270, 329
Brink, F. H. van den, 87
Brinkmann, A. Jr., 51
Bristowe, W. S., 38
Broadbrooks, H. E., 203
Brooks, C. E. P., 317
Brown, R. C., 52
Brundin, L., 231, 267, 393, 394, 400
Bryan, E. H. Jr., 363
Buchanan, G. D., 29
Buchanan, J. M., 131
Burges, A., 334
Burt, W. H., 220, 274, 277, 278

Cain, S. A., 168, 193, 238, 282
Candolle, A. P. de, 2, 153, 192, 244
Carcasson, R. H., 326, 339
Carlquist, S., 61, 370, 373, 378, 379
Carrick, R., 122, 123
Carson, H. L., 116, 142
Caughley, G., 374
Chambers, R., 258
Chaney, R. W., 329, 358
Chapin, J. P., 12
Chapman, F. M., 20
Chapman, R. N., 149
Chitty, D., 146
Clements, F. E., 252–254, 301
Codreanu, R., 170
Colbert, E. H., 182, 234
Conant, R., 210, 213
Connell, J. H., 295
Cook, L. M., 142
Cooke, M. T., 25
Cowan, I. McT., 40
Cowles, R. B., 131, 352
Croizat, L., 9, 34, 345, 366, 380, 387, 394
Crowell, K. L., 377
Curry-Lindahl, K., 76, 163, 168, 228, 339

Dahl, F., 9, 10
Dale, F. H., 232
Dammerman, K., 370
Danilevskii, A. S., 114, 115, 149, 224
Dansereau, P., vii., 9

430

Darlington, P. J. Jr., 9, 27, 35, 36, 53, 60, 80, 130, 267, 294, 296, 305, 354, 367, 371, 372, 374, 387, 394
Darwin, C., 2, 60, 258, 261, 354, 370
Dasmann, R. F., 220
Daubenmire, R. F., 247
Davis, D. H. S., 173
Deevey, E. S., 23, 212, 298, 317, 344
Denman, N. S., 41
Dice, L. R., 14, 67, 93, 116, 255–258, 278, 279
Dickson, R. C., 37
Dillon, L. S., 349, 350
Dmitriev, 158
Dobzhansky, T., 113, 150, 270, 279, 295
Dorf, E., 317, 322
Dorst, J., 154, 156
Drent, R., 315
Dunbar, M. J., 150, 238, 298
Du Rietz, G. E., 270

Ehrlich, P. R., 58, 67
Ekman, S., 8, 9, 18, 58, 156, 159, 163, 166, 205, 237, 274, 275, 277, 278, 281
Ellenberg, H., 250
Elton, C. S., viii, 12, 92, 93, 139, 149, 343, 357, 360
Emerson, A. E., 129, 192, 198, 360
Engler, A., 153, 263, 266, 267, 291
Erkamo, V., 329, 334
Errington, P. L., 26, 116, 137

Fabricius, E., 118, 120
Falla, R. A., 48
Farquharson, F. L., 237
Fejérváry, G. J. von, 93
Fenner, F., 45
Findley, J. S., 19
Firbas, F., 358
Fischer, R. H., 265
Fisher, A. G., 295
Fisher, J., 84
Fisher, R. A., 270
Fleming, H., 18
Flint, R. F., 316, 318
Ford, E. B., 64, 120, 336
Formosov, A. N., 9, 105, 254
Fosberg, F. R., 362
Fraenkel, G., 82
Francia, F. C., 120, 129
Franklin, J., 23
Franz, H., 219
Franz, J. M., 69, 146, 185
Freitag, H., 271, 317, 318, 369, 390
French, N. R., 67
Frey, D. G., 23, 344
Frick, C., 207
Friederichs, K., 147
Frisch, K. von, 120
Fry, F. E. J., 112
Furon, R., 9, 209, 234

Gajl, J., 308
Gams, H., 329
Gates, G. E., 128, 360
Gause, G. F., 270

Gentilli, J., 321, 331
George, V., 9
Geptner, V. G., 9
Gerhard, B., 265
Gill, E., 352
Glick, P. A., 36
Gloger, C. L., 244
Glover, R. S., 238
Goldman, E. A., 256
Good, R., 95, 141, 228, 238, 263, 267, 269, 296, 298, 306, 315, 334, 336, 369
Goodhart, C. B., 59
Gorman, J., 203
Graham, K., 129
Grassé, P. P., 135
Gray, A., 267
Gressit, J. L., 16, 36, 70, 99, 368, 380
Grinnell, J., 13, 14, 101, 113, 116, 149
Grossenheider, R. P., 220
Guilday, J. E., 227, 287, 344, 359, Plate 3
Günther, A., 244, 259, 264, 265
Gustafson, K. J., 120

Haartman, L. von, 65–67, 104, 143, 343
Haeckel, E., 2, 3, 96, 97, 246
Haffer, J. 324
Hagen, W. von, 131
Hagmeier, E. M., 243, 272, 278–281
Hairston, N. G., 149
Halftter, G., 272, 298, 367, 387
Hall, R. E., 164, 193, 220, 227, 232, 278
Hamilton, E. L., 379
Hamilton, T. H., 372, 377
Handlirsch, A., 33, 191, 265, 271, 272, 304, 395–399
Hanna, G D., 52
Hardin, G., 355
Hardy, A. C., 35
Hardy, J. W., 131
Harris, V. T., 121
Hasler, A. D., 67
Hedgepeth, J. W., 238
Heilprin, A., 263–265
Hennig, W., 393
Hesse, R., 8, 9, 93–96, 104, 134, 141, 143, 144, 361, 370
Heusser, C. J., 359
Hoffman, R. S., 164
Holdhaus, K., 42, 104, 211, 216, 265
Holdridge, L. R., 251
Hoogstraal, H., 45
Hooker, J. D., 153, 267
Hopkins, D. M., 387
Horváth, O., 113
Hovanitz, W., 114
Howard, W. E., 67, 71
Howden, H. F., 13, 113
Howell, F. C., 391
Hubbs, C. L., 327, 390
Hubendick, B., 236, 300
Hudson, R., 204
Huheey, J. E., 242, 278
Hultén, E., 167, 209, 283–285, 301, 352
Humboldt, A. von, 113, 244
Hustich, I., 362
Hutchinson, G. E., 144, 149, 295, 352; 355

432   Name Index

Hutton, J., 366
Huxley, T. H., 259, 261, 262, 264, 265, 267, 301

Illies, J., 310, 385
Illiger, C. W., 258
Inger R. F., 16, 39, 374
Irwin, M. P. S., 12, 27
Isakov, Ju. A., 9

Jaccard, P., 273
Jacobi, A., 9, 265
Jägerskiöld, L. A., 343
Jeannel, R., 9, 203, 214, 342, 383, 395
Jeuken, M., 152
Johansen, H., 26
Johnson, C., 336
Johnson, C. G., 59, 82
Jovetic, R., 84
Kaisila, J., 69, 337, 338, 340, 377
Kalabukhov, N. E., 101
Kalela, E. K., 333
Kalela, O., 71, 104, 159, 332, 338, 343, 377
Kangas, E., 44
Keast, A., 197, 311, 347, 387, 392
Keferstein, W. M., 258, 265
Kelsall, J. P., 157
Kelson, K. R., 14, 164, 193, 220, 227, 232, 278
Kendeigh, S. C., 93, 104, 254, 279
Kennedy, J. S., 37, 61
Kerner, A. von, 147, 248
Keve, A., 84, 332, 336, 338, 341
Kew, H. W., 48
Key, K. H. L., 71
Kleinschmidt, O., 270
Klomp, H., 71, 117
Klopfer, P. H., 295
Kochi L., 324
Koskimies, J., 134, 364
Kratzig, H., 156
Krogerus, R., 98
Kühnelt, W., 93, 112, 254
Kurtén, B., 198, 217, 352

Lack, D., 93, 116, 117, 145, 149, 377
Lahti, L., 134
Landry, S. O. Jr., 182
Lattin, G. de, 9, 67, 229, 285–287, 301, 390
Laux, W., 69, 115
Lawrence, D. B., 329
Leech, R. E., 39
Lees, A. D., 61
Lenz, F., 97
Leopold, A. S., 207
Leston, D., 374
Leunis, J., 244
Liebig, J., 95, 134
Lincoln, F. C., 154
Lindroth, C. H., Frontispiece, 7, 18, 19, 41, 49, 50, 55–57, 61–64, 80, 104, 106, 107, 109–112, 125, 128, 162, 171, 173, 176, 203, 215, 337, 360, 368, 376

Linnaeus, C., 4, 244, 258, 342
Linsley, E. G., 111, 203
Liversidge, R., 256, 257
Livingstone, D. A., 335
Lloyd, H. G., 172
Löffler, H., 48
Löhrl, H., 67, 179
Longhurst, A. R., 128
Lönnberg, E., 342, 390
Lorenz, K. Z., 118
Lotka, A. J., 270
Löve, Á., 124
Löve, D., 124
Lowe, C. H., Jr., 372, 373
Lydekker, R., 263, 265

Macan, T. T., 120
MacArthur, J. W., 295
MacArthur, R. H., 9, 279, 295, 370, 374, 375, 377
McCabe, T. T., 40, 121
McCoy, C. J. Jr., 21, 61
McCrady, A. D., 287, Plate 3
Macfadyen, A., 93
Macginitie, H. D., 318
MacKerras, I. M., 17
MacLachlan, G. R., 256, 257
MacLauchlin, C. A., 13, 25
Maclulich, D. A., 312
Macpherson, A. H., 208
Maguire, B. Jr., 48
Mail, G. A., 102
Malone, C. R., 48
Mani, M. S., 37, 368, 370, 381
Marcström, V., 132
Marcus, E., viii, 9
Margalef, R., 148, 372
Martin, P. S., 287, 327, 345, 359, 367, Plate 3
Matthew, W. D., 304, 395, 397–399
Matvejev, S., 209, 258
Mauersberger, G., 165
Mayr, E., 20, 21, 27, 70, 124, 139, 141, 142, 200, 230, 272, 277, 315, 347, 353, 354, 364, 368, 372, 378, 387, 400
Mehringer, P. J., 327, 345, 359
Meigs, P., 23
Menard, H. W., 379
Mengel, R. M., 322, 389
Menzbier, M., 268
Merikallio, E., 75, 132
Merriam, C. H., 247, 248, 301
Mertens, R., 383
Messenger, P. S., 28
Meusel, H., viii, 143, 219, 238, 282, 334, 339
Meyrick, E., 394
Michaelsen, J. W., 265
Miller, A. H., 24, 113, 116, 288, 291, Plate 4
Millot, J., 265
Milne, A., 119
Milne, P. S., 35
Milstead, W. W., 194
Mitchell, R., 38

Möbius, K., 147, 263, 265
Moore, J. A., 391
Moore, R. T., 256
Moreau, R. E., 321, 380, 382, 390
Morrison, R. B., 321
Munroe, E., 235, 371, 372
Murchie, W. R., 361
Myers, G. S., 14

Nansen, F., 23
Naumov, N. P., 81, 83, 93, 105, 133, 158
Nehring, A., 367
Neill, W. T., 16, 130, 315
Newbigin, M. I., 9
Nicol, D., 354
Niering, W. A., 372
Niethammer, G., 9, 17, 48, 53, 136–139, 339
Nikolsky, G., 282
Norris, K. S., 77, 108
Norris, R. A., 178
Nuorteva, P., 45

Odum, E. P., 93, 149, 254
Ognev, S. I., 203
Öhrn, B., 325
Økland, F., 9, 37, 40, 48, 51
Olstad, O., 76
Omodeo, P., 121, 128, 360
Orians, G. H., 141
Orias, E., 295
Osche, G., 47
Otterlind, G., 84
Oughton, J., 59, 104

Paine, R. T., 295
Palmén, E., 40, 42–44, 82, 127
Palmgren, P., 118, 120, 121, 143, 145, 253, 308, 317
Pantin, C. F. A., 400
Paulian, R., 377
Pax, F., 9
Pearson, T., 9
Peary, R. E., 23
Peitzmeier, J., 143
Perring, P. H., 171
Petersen, B., 101
Peterson, R. L., 385
Péwé, T. L., 321
Phelps, W. H. Jr., 20, 368
Pianka, E. R., 295, 298
Pimentel, D., 148, 149
Pitelka, F. A., 174, 175
Polunin, N., 238, 282
Pompper, H., 244
Poore, M. E. D., 194
Portenko, L. A., 165
Prenant, M., 9
Prescott, 197
Preston, F. W., 279, 372
Proctor, V. W., 48

Rand, A. L., 23, 42, 349
Ratcliffe, F. N., 45
Ray, C., 208

Reichenow, A., 265
Reid, J. L. Jr., 114
Reinig, W. F., 58, 139, 143, 201, 285, 352
Remane, A., 254
Rensch, B., 9, 101, 270, 277, 351, 352, 354
Ressl, F., 47
Ricker, W. E., 385
Romer, A. S., 364
Ross, H. H., 130, 148, 160, 210, 211, 217, 353
Ross, J. C., 23
Rübel, E., 98, 150, 270
Rubinoff, I., 372, 377
Rubtsov, I. A., 183
Runcorn, S. K., 400
Rütimeyer, L., 258, 293
Ryan, R. M., 278
Ryberg, O., 101
Rylow, V. M., 299

Salomonsen, F., 38, 39, 87, 180, 338, 377
Salt, G. W., 24, 102, 134
Savile, D. B. O., 355
Schaller, F., 46
Schick, R. X., 224
Schilder, F. A., 9, 139, 173, 265, 275–278, 295, 298
Schmarda, L. K., 258
Schmid, E., 252
Schmidt, K. P., 8, 175, 180, 263, 266, 313, 314, 361, 370
Schmidthüsen, J., 252
Schneider, F., 119
Schwerdtfeger, F., 111, 130, 150
Sclater, P. L., 258, 259, 261, 264–267, 270, 301
Sclater, W. L., 265
Scott, R. F., 23
Seibert, H. C., 102
Seiler, J., 79, 80, 125, 126
Severtsov, N. A., 268
Sheals, R. A., 52
Shelford, V. E., 95, 96, 116, 150, 254, 301, 313
Short, L. L., 27, 349, 350
Sibley, C. G., 27, 347
Siivonen, L., 76, 84, 132, 342
Simpson, G. G., 9, 53, 55, 66, 85, 89, 90, 143, 273, 277, 278, 296, 304, 353, 354, 387
Sirén, G., 329
Slobodkin, L. B., 93, 149
Smith, F. E., 149
Smith, H. M., 349
Smith, P. W., 218, 311, 349
Sneath, P. H. A., 279
Sokal, R. R., 279
Solem, A., 16, 21, 39
Stantchinsky, V. V., 103, 116, 138
Stebbins, G. L., 72, 124, 139, 143
Stebbins, R. C., 216
Steenis, C. G. G. J. van, 400
Stegmann, B., 88, 285, 301, 311, Plate 2

434 Name Index

Stein, G. W., 199
Stephan, B., 165
Stresemann, E., 87, 165
Stroyan, H. L. G., 37
Stults, C. D., 243, 278, 279
Sukachev, W. N., 147, 252
Suomalainen, E., 124, 127, 128
Svensson, L., 335
Swainson, W., 258
Swärdson, G., 84
Sylvester-Bradley, P. C., 347
Syroechkovskij, E. E., 99

Taber, R. D., 164
Talmage, R. V., 29
Tansley, A., 147
Taylor, B. W., 47, 48
Tenuovo, R., 315
Tets, G. F. van, 315
Thienemann, A., 9, 95, 97, 112, 151, 237, 294, 298, 308, 310, 367
Thomson, G. M., 361
Timofeeff-Ressovsky, N. W., 113, 141, 184
Tinbergen, N., 118, 120, 146
Ting, W. S., 320
Tischler, W., 61, 93, 147, 362
Toit, A. L. du, 399
Tompa, F. S., 71
Trouvelot, L., 51
Truessart, E. L., 234
Tschulok, S., 3
Turbott, E. G., 362
Turček, F. J., 362
Turrill, W. B., 153, 267
Twitty, V. C., 67

Udvardy, M. D. F., 30, 84, 99, 121, 149, 157, 192, 222, 223, 288, 295, 300, 332, 339, 341, 342, 355, Plate 4
Urquhardt, F. A., 82
Usinger, R. L., 377, 380
Uvarov, B. P., 71, 282

Valentine, J. W., 279
Vandel, A., 41, 42, 124, 127, 315
Van Dyke, E. C., 268
Vaurie, C., 179
Verbeek, A. M., 37
Verheyen, R., 154
Verhoeff, K. W., 67
Vestal, A. G., 255, 257

Vierhapper, F., 194
Volterra, V., 270
Voous, K. H., 83, 84, 87, 88, 179, 191, 194, 196, 210, 214, 231, 283, 288, 347
Vos, A. de, 339, 362

Wagner, A., 244
Wagner, G., 303
Wagner, M., 192
Wallace, A. R., 213, 258, 259, 261, 263–266, 268, 270, 294, 301, 370, 371, Plate 1
Wallgren, H., 112
Walter, H., 114, 238, 267, 271, 278, 282
Walters, S. M., 171
Warming, E., 249
Warnecke, G., 112, 141, 335, 338, 340, 367
Warner, R. E., 361
Webb, W. L., 278
Wegener, A., 399
Welch, D'Alte A., 21
Wellington, W. G., 17, 69, 115, 120, 144, 146
Wermuth, H., 383
White, C. M. N., 27
Whittaker, R. H., 334, 335, 358, 359
Wigglesworth, V. B., 59
Williams, C. B., 82
Willis, J. C., 198
Wilson, E. O., 9, 129, 370, 375, 377
Winterbottom, J. M., 288
Wodzicky, K., 361
Wolda, H., 13, 47, 69
Wolfenbarger, D. O., 36
Wood, A. E., 182
Wood, J. G., 197
Wood, S. L., 233
Woodward, S. P., 258
Wright, H. E., 23, 298, 318, 344, 390
Wright, S., 30, 270
Wulff, E. V., 238
Wynne-Edwards, V. C., 146, 184

Yeatter, R. E., 104

Zeuner, F. E., 23, 317
Zherbinovski, 81
Zimmermann, E. C., 15, 199, 200
Zimmermann, K., 36, 39, 48, 90, 378, 380, 394

Scientific Name Index

Acarina, 36
Acer platanoides, 329
Achatina fulica, 51
Achatinella, 21
Acris gryllus, 194
Acronycta rumicis, 115
Aegithalos caudatus, 87
Aegolius funereus, 189
Agonum dorsale, 109–111
Alauda arvensis, 194
Alces alces, 168, 385
Alectoris graeca, 138
Allocapnia mohri, 210, 211, 217
— pygmaea, 210, 211
— recta, 210, 211
— rickeri, 210
Alloperla diversa, 385
Alopex lagopus, 40
Amara erratica, 216
Amblyotus nilssoni, 105
Ambystoma tigrinum, 189
Anas acuta, 189
— crecca, 134
— penelope, 134
— platyrhynchos, 134
Anser albifrons, 84
Anthracus conspulus, 43
Anthus trivialis, 116
Antilocapra americana, 205, 207
Antilocapridae, 205
Anura, 16, 259
Aphelocoma coerulescens, 174, 175
Aphidae, 35, 74, 128
Apoda, 16, 259
Apodemus flavicollis, 101
— sylvaticus, 101
Apodidae, 38
Apoidea, 10
Apollophanes francesca, 224
Aporia crataegi, 336
Apus apus, 364
Araneae, 35, 36, 38
Aratinga canicularis, 131
Ashmunella, 21
Aulocara elliotti, 335
Austroicetes cruciata, 177
Aythya ferina, 134
— fuligula, 134

Barbus paludinosus, 237

Basommatophora, 300
Bathysmaphorus reuteri, 98
Bembidion aeneum, 106–108, 116
— minimum, 106, 108
Bias musicus, 12
Bison bison, 215
Blethisa eschscholtzi, 203
Boletina borealis, 98
Brachynus crepitans, 109–111
Bucephala clangula, 134
Bufo canorus, 189
— hemiophrys, 213
— microscaphus, 216
Butorides virescens, 102

Calathus mollis, Frontispiece, 62
Camelidae, 351
Canis dingo, 351
Canthon, 46
Capra domestica, 362
Capreolus capreolus, 76, 103, 105
Caprimulgidae, 38
Carabidae, 18, 27, 50, 61, 74, 193, 215, 367–368
Carabus cancellatus, 44
— chamissonis, 41
Carpodacus, 134
— purpureus, 189
Castor fiber, 215
Catops basilaris, 203
— sparcepunctatus, 203
— subfuscus, 203
Cepaea nemoralis, 14, 59, 69
Cepphus carbo, 192
— columba, 192
Cerorhyncha monocerata, 223
Certhia brachydactyla, 87
— familiaris, 87
Cervus canadensis, 220
— elaphus, 158
Chaerocampa celerio, 130
Chaetura pelagica, 364
— vauxi, 364
Chelifer cancroides, 47
Chiasognathinae, 399
Chironomidae, 231
Chiroptera, 103
Chloroceryle americana, 131
Chlorura chlorura, 161
Chrysemys picta, 189

Scientific Name Index

Chrysomelidae, 127
Chrysophanus dispar, 229
Cicindelidae, 116
Ciconia ciconia, 228
 — *nigra*, 83
Citellus tridecemlineatus, 218
Clangula hyemalis, 180
Cnemidophorus sexlineatus, 189
Coccioidea, 34
Colaptes, 349
 — *auratus*, 350
 — *cafer, chrysoides*, 350
Coleoptera, 18, 56, 74, 162
Colias, 114
 — *palaeno*, 214
Colinus virginianus, 105
Columba palumbus, 166
Corvus caurinus, 315
 — *corone cornix*, 315
Coturnix coturnix, 60, 138
Crocuta crocuta, 198
 — *sivalensis*, 198
Cryptorhynchidae, 90
Cuculus canorus, 196
Culex pipiens quinquefasciatus, 361
Culicidae, 120
Curculionidae, 127
Cyanopica cyana, 211

Danaus plexippus, 82
Daphnia, 74
Dasypus novemcinctus, 29
Dendrocopos Dryobates, 336
Dendroctonus, 233
Dendroica chrysoparis, 389
 — *kirtlandii*, 189
 — *nigrescens*, 389
 — *occidentalis*, 389
 — *townsendi*, 389
 — *virens*, 389
Dendrolimus pini, 130
Dermaptera, 18
Diamesinae, 393
Dictya umbrarum, 98
Didelphis marsupialis, 29, 227
Dilacnus piceae, 37
Diplopoda, 74, 127
Dipodomys, 232
Diptera, 36, 120
Dissosteira longipennis, 160
Dolichonyx oryzivorus, 154, 217
Dryobates major, 87
 — *syriacus*, 87, 336

Emberiza cia, 37
 — *cirlus*, 87
 — *citrinella*, 87, 99, 112, 113
Empidonax oberholseri, 289
Endothia parasitica, 355
Ensatina eschscholtzi, 189
Ephemeroptera, 120
Eremophila alpestris, 194
Erethizon dorsatum, 41
Erethizontoidea, 182
Eukronia hamata, 114

Eumeces fasciatus, 175
Eupithecia sinuosaria, 340

Falco peregrinus, 191
Ficedula albicollis, 87
 — *hypoleuca*, 65, 87
Formicoidea, 10, 74
Funambulus, 391

Galerida cristata, 102, 103
Garrulus glandarius, 87, 88
Geocharis, 395
Geomyidae, 12, 101
Geomys bursarius, 14
Gopherus polyphenus, 189
Gronocarus, 15
Gulo luscus, 189
Gymnorhina tibicen, 122

Harpalus, 105, 106
Hayoceros falkenbachi, 207
Hedera helix, 145
Hemidactylium scutatum, 211
Hemiptera, 35, 74, 374
Heterodon nasicus, 349
Heteroptera, 120
Heterotricha, 399
Himantoglossum hircinum, 336
Hipparchia allionii, 169
Hirundo rustica, 189
Homoptera, 36
Homo sapiens, 353
Hoplocerambyx spinicornis, 111, 112
Humboldtiana, 21
Hyaena hyaena, 198
 — *namaquensis*, 198
Hyalomma marginatum, 45
Hybopsis dissimilis, 210
Hydrodamalis stelleri, 351
Hydromantes, 203
Hymenoptera, 35, 36, 128
Hyppolais icterina, 87
 — *polyglotta*, 87
Hystricoidea, 182
Hystricomorpha, 394

Icterus galbula, 26
Ilingoceros alexandrae, 207
Ischnodemus sabuleti, 61
Isopoda, 74
Isoptera, 10, 74
Ixodes ricinus, 119

Kalotermes, 192

Lacerta pityusensis, 382
 — *muralis*, 383
 — *sicula*, 383
Lagopus, 132
Lanius collurio, 155
Larca lata, 47
Larix, 233
Larus ridibundus, 118
Lemmus lemmus, 159
Lemuroidea, 234

Scientific Name Index 437

Lepidoptera, 35, 113, 285, 337
Leptinotarsa decemlineata, 102
Leptura obliterata, 203
Leucosticte, 12
Libellula quadrimaculata, 82
Limenitis camilla, 336
Limosa limosa, 117
Loxia leucoptera, 196
Lumbricidae, 359–361
Luscinia luscinia, 87, 113, 173, 341, 342
 — *megarhynchos*, 87, 173
Lutra sp., 12
Lymantria dispar, 51
Lyrurus tetrix, 132

Macrolepidoptera, 335
Macronectes giganteus, 48
Malacosoma castrensis, 64
 — *neustria*, 69
 — *pluviale*, 17, 69, 74, 120
Malacothryx typica, 173
Mallophaga, 46
Maniola jurtina, 120
Marmota bobak, 368, 369
 — *marmota*, 368, 369
Megoura viciae, 61
Melanagria galathea, 338
Melanitta fusca, 134
Meliphaga chrysops, 197
 — *fascigularis*, 197
 — *fusca*, 197
 — *lewini*, 197
 — *ornata*, 197
 — *virescens*, 197
Meloe, 46, 47
Melolontha melolontha, 119
Melospiza melodia, 63, 71, 189
Mephitis mephitis, 41
Mergus merganser, 134
 — *serrator*, 134
Merriamoceros coronatus, 207
Metatetranychus, 71
Micralymma marinum, 55
 — *stimpsoni*, 55
Micrarionta, 21
Microtus arvalis, 14, 199, 200
 — *pennsylvanicus*, 189
Misumenops verityi, 224
Mollusca, 258
Momotus mexicanus, 131
Muridae, 192
Mustela nigripes, 189
 — *putorius*, 343
 — *vison*, 137
Mycotrupes, 14
Mylothris yulei ertli, 339
Myotis daubentonii, 105
 — *myotis*, 105
 — *nattereri*, 105
Mysis oculata relicta, 311
Myxoma, 45, 46

Nasutitermes trinerviformis, 131
 — *nigriceps*, 131
Natrix septemvittata, 210

Nectarinia famosa, 12
Nemoura arctica, 385
Neotermes, 192
Neuroptera, 35
Notophagus, 333
Notornis, 60
Notostraca, 74, 128

Ochotona, 203
Odonata, 18, 120
Oenanthe oenanthe, 30
Oikopleura labradoriensis, 157
 — *vanthoffeni*, 157
Oligochaeta, 74, 128
Ondatra zibethica, 26, 52, 136–138, 166
Oreophryne annulata, 40
Orthoptera, 18
Osbornoceros osborni, 207
Otiorrhynchus dubius, 125
Ovibos fossilis, 209
 — *moschatus*, 40, 209, 217

Pachycephala inornata, 392
 — *rufogularis*, 392
Pachymerium ferrugineum, 40
Paederus fuscipes, 43
Pandion haliaetus, 189
Pantala flavescens, 82
Paragnetina media, 385
Parus cristatus, 165
 — *montanus*, 145
Perisoreus infaustus, 196
Peromyscus, 142
 — *maniculatus*, 15, 189
Phaeton lepturus, 48
Phasianus, 105
Pheidole sitarches, 129
Pheucticus ludovicianus, 26
Philodromus rodecki, 224
Phoenicurus ochruros, 47, 364
Phymatoniscus tuberculatus, 41
Pica pica, 289
Picea, 233
Picoides tridactylus, 215
Picus canus, 87
 — *viridis*, 87
Pieris brassicae, 224
Pinicola enucleator, 222
Pinus, 233
Podiceps novaehollandiae, 283
 — *ruficollis*, 283
Podicipediformes, 33
Polyxenus lagurus, 127
Populus italica, 130
 — *trichocarpa*, 130
Potosia cuprea, 144
Precis sofia infracta, 339
Procyon lotor, 138, 166, 362
Progne subis, 364
Protonotaria citrea, 65
Pseudochazara anthelea, 169
Pseudopaludicola falcipes, 194
Pseudoscorpionida, 46
Pseudotsuga, 233

Psittirostra cantans, 379
Pterostichus anthracinus, Frontispiece
    minor 63
Pulmonata, 258
Pyrrhia umbra, 69
Pyrrhula pyrrhula, 87

Rana cancrivora, 16
— *catesbeiana*, 189, 225
— *palustris*, 211
— *septemtrionalis*, 225
— *sylvatica*, 189, 211, 213
Rangifer tarandus, 40, 157
Regulus ignicapillus, 87
— *regulus*, 87, 145
Rhegmatobius, 395
Rhynolophus ferrum-equinum, 105
Robinia pseudoacacia, 210

Sachalinobia rugipennis, 203
Salpinctes obsoletus exsul, 316
Sayornis saya, 289
Sciurus, 85
— *carolinensis*, 171
Scoparia, 235
Senecio vernalis, 339
Serpentes, 16
Sialia currucoides, 289
Sicista betulina, 163
— *subtilis*, 173
Sitta canadensis, 179
— *europea*, 87
— *pusilla*, 178
— *pygmaea*, 178
Solenobia triquetrella, 74, 77, 79, 80, 125–127
Somateria mollissima, 134
Sonorella, 21
Sorex nanus, 164, 217
Spermophilus, 101
Spizella breweri, 289
Stellaria palustris, 167
Sterna paradisea, 118
Stilpnotia salicis, 130
Stipa stenophylla, 219
Stockoceros onusrosagris, 207
Streptopelia decaocto, 28, 70, 75, 203, 204, 369
Sturnus vulgaris, 25, 156
Stylommatophora, 16
Succinea, 48
Sylvia atricapilla, 113
— *borin*, 99

Sylvilagus aquaticus, 193
— *palustris*, 193

Tabanidae, 17
Tapiridae, 398
Taricha rivularis, 68
Tetraonidae, 132
Tetrao urogallus, 132, 133
Thalarctos maritimus, 40
Thamnophis radix, 218
— *sirtalis*, 189
Therioaphis maculata, 37
— *riehmi*, 37
Thersamonia dispar, 229
Thomomys monticola, 101
— *umbrinus*, 25
Trechus, 383
Trichoniscus pusillus, 124
Tricladida, 170
Triops cancriformis, 128
Troglodytes aedon, 104
Tropidoclonion lineatum, 349
Trypodendron lineatum, 129
Tsuga heterophylla, 194
Turdus migratorius, 112
— *pilaris*, 38, 39
— *viscivorus*, 142
Typha latifolia, 329

Uma, 76, 108
— *inornata*, 108
— *notata*, 108
— *rufopunctata*, 108
— *scoparia*, 108
Uria aalge, 222
Urodela, 259
Ursus arctos horribilis, 189

Vaccinium vitis-idea, 178
Vanellus vanellus, 117
Vanessa io, 82
Varanus niloticus, 131
Vermivora crissalis, 390
— *luciae*, 189, 390
— *ruficapilla*, 12, 390
— *virginiae*, 390
Vespertilio murinus, 105
Vespertilionidae, 192
Vitrina, 48

Zonotrichia capensis, 20
Zosperops, 380, 382
— *japonica*, 361
Zygoptera, 37

Subject Index

Accidental visitor, 377, 378
Acclimatization, 97
Acidity as limiting factor, 98
Aclimatic dispersal, 201
Action, radius, 75, 78, 199
— circle, 75, 78
Adaptability, 97
Adaptations, microevolutionary, 141
Adaptive radiation, 207, 380
Adventive, 49, 52, 56, 128, 217, 360, 362
Aethiopian Region, 259
African Temperate Realm, 246
Age, and area, 198
Aggregation, 135, 144
Alateness, 74
Allochtonous, 217
Altitudinal, climatic belts, 244
— zone, 201
Amelioration, of climate, 343
American Tropical Realm, 246
Amphiamerican, 218, 221
Amphiatlantic, 55, 218, 221
Amphiecious, 100
Amphipacific, 203, 218, 221, 223
Amphithermy, 100
Amphitopy, 100, 113, 186
Amphitropical, 218, 221
Anemochore, 34, 36, 39, 52, 339
Anemohydrochore, 42, 43, 378
Antarctic Realm, 261
Anthropochore, 51, 55, 339, 359
Apomixis, 124
Aptery, 74
Arboreal Fauna, 286, 287
Archipelago, 380
— ecological, 54, 380–383
Arctic Realm, 246, 261
Arctic-alpine, 216, 219, 221
Arctogaea, 259, 264
Arctotertiary elements, 267
Area, distributional (= range), 134, 152, 153, 159, 226
— and age, 198
— analysis, 344–348
— of aquatic animals, 236–238
— breeding, 154, 155
— center, 195
— of class, 234
— of community, 194, 235
— composite, 240
— constant, 226

— continuous, 178–181, 210
— contracting, 226
— discontinuous, 178–181, 202, 204, 215, 222, 227
— disjunct, 178–181, 185, 206, 210, 211, 222, 237
— disperse, 178–181, 222
— dynamism of, 225–230
— epicontinental, 237
— equiformal progressive, 283, 284
— expanding, 226
— and extinction, 230
— of family, 231, 234
— floristic, 252
— of genus, 231–233, 235
— geography of, 217–225
— of higher taxa, 230–235
— history of, 195–203
— irruption, 154
— limit, 164, 176–177, 242–243
— — altitudinal, 169
— — dynamism of, 229–230, 347–349
— linear, 221–223
— marginal, 223
— migration, 154
— minimum, 236
— oceanic, 237, 238
— of order, 234
— plant, 238–239
— potential, 204, 228
— relict, 185, 195, 205–215
— reproduction, 154
— retracting, 228
— shape of, 177–182
— size of, 186
— of species, 194
— sterile expatriation, 157, 159
— structure of, 182–186
— temporary, 154
— and thermal resistance, 134
— vegetational, 252
— winter, 154, 156, 158
— zonal, 222
Area, geographic, 159
Areography, 153, 239, 281–293
Aridity, 318, 326, 327
Astasy, astatic, 308, 309
Australasian Region, 259
Australian, Kingdom, 267
— Realm, 246
Autecology, 94

440 Subject Index

Autochthonous, 217, 234

Barrier, 12, 91
— avoidance, 27
— biotic, 26–28, 360
— climatic, 28, 32
— deserts as, 28–30
— distance as, 28–30
— durability of, 30–32
— ecological, 18, 19, 22–28
— ethological, 27
— faunal, 30
— geographic, 32
— mountain, 17–19
— ocean, 31
— physical, 17–22
— time, 28
— topographic, 17, 18, 20–22
— water, 13–17, 31
Biocenters, 238
Biochore, 45, 47–49
Bioecology, 253–255
Biogenesis, 11
Biome, 26, 249, 254, 255, 301
Biophag, 145
Biosphere, 1
Biotic, community, 254, 255, 261, 357, 358
— province, 255–258, 271, 288, 301
Biotope, 96, 194, 236, 250, 252, 255
Biotype, 201
Bipolar, 218, 221
Borderland, 362
Boreal fauna, 286, 287
Boreo-alpine, 214, 216
Boreomontane, 215, 221, 222
Boundary, of area. See Limit, of area
Brachyptery, 61, 63, 74
Bridge, land. See Land bridge

Calciphilic animals, 105
Carriers, 91
Catching angle, 376
Center of, density, 195
— differentiation, 234
— dispersal, 195, 234, 397
— diversity, 222–223
— efficiency, 196, 197
— endemism, 235
— frequency, 397
— maximum variation, 195
— origin, 197, 198, 234, 285
— preservation, 195
— radiation, 285
Centrifugal dispersal, 201
Chance, 53, 351
— in habitat selection, 119, 121
— in settling, 226
Change, of ecological valency, 369
— climatic, 32, 198, 317–327, 332, 333, 338
— faunal, 274, 279, 334–340
— floral, 327
— vegetational, 327, 329–334
Character progression, 393
Chorology, 1, 2, 3, 183

Circadian changes, 307
Circumboreal, 215, 216, 218
Circumpolar, 203, 218, 221, 246
Cladogenesis, 392
Class, 234
Climatic, changes, 32, 198, 317, 332, 333, 338, 343
— belts, altitudinal and latitudinal, 244
— and evolution, 396–399
— zone, 244, 250, 252, 301
Climax, 358
Clinal variation, 190
Colonization, intrinsic factors of, 118
Colonizing, viii, 7
Colony, 184
Community, 254, 255, 261
— area, 196–197, 235
— climax, 253, 254
— dynamism of, 357–359
— evolution, 356
— plant, 250, 252, 329
Community, (continued)
— relict, 213
— successional, 253, 254
Compensation, law of, 150
Competition (and competitive exclusion), 86, 118, 145, 146, 149, 295, 354–356, 360–362
Complex, faunal, 282
Constancy, of ecological valency, 367–369, 402
— of means of dispersal, 367–369
Continental, 246
— dynamism, 386–387
— permanence, 31, 397, 400
— shelf, 31, 237
— stability, 400
Corridor, 87, 88, 90, 91, 368
Cosmic system, 225
Cosmopolitanism, 188, 190, 192, 234, 246
Counterpart, ecological, 194
Creatio, Neogeana, 259
— Palaeogeana, 259
Creation, special, 11
Cultural habitats, 24
Cycle, annual, 307
— multiannual, 308
Cytogenetic systems, 124

Demecology, 5
Density, of population, 71
Depauperation, of fauna, 372, 373
Deserts, 22–24
Diaspore, 32
Diluvium, 318
Dimorphy, of wing, 63, 64, 128
Disciplines, auxiliary, 5
Discontinuity, 203, 350
— apparent, 180
— fallacious, 182
— geographic, 349
— temporary, 203
Disharmony, faunal, 372, 402
Disjunction, 203, 348
Dispersability, 379

Dispersal, aclimatic, 201
— active, 32–34
— adaptive, 60
— aerial, 34, 36
— anemochore, 34–40, 52, 339
— anemo-hydrochore, 42–44
— anthropochore, 49–52, 55, 339
— biochore, 46–49
— center, 285
— chance, 9, 53–58, 71
— ecological, 71
— ectozoic, 48
— endophytic, 48
— endozoic, 48
— of faunas, 85–89, 92
— highway of, 12
— hydrochore, 40–42, 55
— innate, 71
— intrinsic factors of, 58
— long range, 203, 377
— modes of, 75
— passive, 32, 34, 91
— phoretic, 46
— pioneering, 157
— rate of 314, 315
— by water, 44
— over water, 370, 373
— by wind, 36, 44
Dispersion, 10, 71, 80, 138, 183
Dispersional systems, 72–75
Distance, isolation by, 30
Distribution, altitudinal and depth, 114
— changes of, 311–315
— disjunct, 218
— ecological, 71, 243–244
— homogeneity of, 160
— isoclimatic, 201
— linear, 223
— local, 183
— mapping, 153–168
— Maps Scheme, 161
— range (= distribution area), 153
— regional, 71
— zonal, 201, 221, 224
Diversification, 335, 356
Diversity, 300
— ecological, 295
— faunal, 243, 293–300
— species, 294, 295
Domicile, 75
Dominance, ecological, 357
Drift, continental, 399, 400
— dispersal by, 35, 39
— genetic, 58, 139, 230, 377
— visitor, 158
Dynamic potential, 401
— of fauna, 341–343
Dynamics, of population, 138
Dynamism, millennial, 325, 389

Ecofauna, 282, 288, 311
Ecogeographic, approach, 243
— faunal groups, 244–247, 288–289, 291
— floral groups, 248–253

Ecological, counterparts, 194
— diversity, 295
— indicators, 242
— potential, 150
— replacement, 354
— valency, 94, 96, 112, 150, 186, 187, 201, 352, 354
— — Change of; 112–116
— — constancy of, 367–369, 402
— vicar, 193, 194
Ecosystem, 144–150, 226, 258
— evolution, 151
— extinction, 151
Ecotype, 201
Ectotherm, 169–171
Element, dynamic, 347
— faunal, 282, 345
— floral, 267, 281
— marginal, 341
Elimination, 139, 190
Emigration, 69, 183
Endemic, 61, 215, 217, 234, 242, 270, 362, 368, 374
Endemicity, 215–217, 270, 272
Endemism, 217
Endotherm, 169
Environment, marginal, 142
— requirements, 97–99
— resistance, 96, 166
Epicontinental sea, 237, 307, 371
Epidemic, 217
Equatorial submergence, 113
Equiformal progressive area, 283–285
Eremial Fauna, 286, 287
Euryecy, 98, 140
Eurythermy, 100
Eurytopy, 113, 150, 186, 188–191, 195, 201
Eustasy, eustatic, 308, 309, 312, 313, 386
Evolution, 11, 141, 296, 314, 315, 346, 396–399
— of communities, 356
— of ecosystems, 151
— insular, 377, 380
— processes, 239
— rates, 298, 315, 401
Exclusion principle. See Competition
Expansion, 69, 95, 198, 340
— isoclimatical, 143
— and population dynamics, 137
— postglacial, 126, 127
Expatriation area, 157
Extinction, 228, 229, 333, 343, 349–353, 355–357, 372, 378, 396
— and area, 230
— competitive, 354
— of ecosystems, 151
— group, 351, 353, 401, 402
— insular, 353
— phyletic, 353, 354, 357, 401, 402
— pseudo-, 351, 401, 402
— rate of, 375, 377
— singular, 350, 401, 402
— synthetic, 353, 401, 402
Extralimital occurrence, 158, 161, 165

Subject Index

Family, 231, 234
Fauna, vii–ix, 6, 246, 282
Faunal, change, 274
— complex, 282
— disharmony, 402
— diversity, 293–301
— divider, 17
— dynamism, 341, 401
— elements, 282, 345
— groups, ecogeographic, ecofaunal, 288, 291
— insular, 353
— on islands, 402
— mixing, 35
— province, 280, 281
— region, 261, 281
— relict, 214
— replacement, 401
— resemblance, 273, 276–277
— size, 293–294, 370
— types, 285, 288
— unbalanced, 54, 362
Faunation, viii–ix, 86
— relict, 212
Faunistic transect, 241
Faunistics, 6, 153, 239
Filter, bridge, 89, 373
— route, 89, 91, 374
Fire, 313
Flightlessness, 61, 63, 69
Flora, viii–ix, 246
— elements, 267, 291
— provinces, 301
— (or floristic) region, 252
— — plant geography, 153
Fluctuation, of area, 339
Food, of settling young, 130
Forest, cultural, 24
Formation, 197
Fossil, leading, 394
Freshwater bodies, as barriers, 14

Geburtsortstreue, 65
Genetic drift, 58, 139, 230, 377
Genus, 231
Geographical, zoology, 3
Geography, zoological, 3
Gigantism, 352
Glacial relict, 219
Glaciation, 298, 318, 321, 326
Gossamer, 38, 39
Grid systems, 171–173
Group, attributes, 151
— effect, 135

Habitat, alteration, 71
— deterioration, 71
— ecology, 94, 131
— improvement of, 71
— selection, 27, 37, 116–121, 151
— vacant, 94
Habitus, biological 4, 7
— morphological, 4
Hermaphroditism, 124, 128
Higher taxa, 181, 230

Hilfswissenschaften, 5, 6
Holarctic Region, 264
Home range, 75, 78, 184
Homeostasis, of ecosystems, 150
Humidity, as limiting factor, 98, 111
Hydrochore, 40–42, 55
Hypothetical occurrence, 224
Hypsithermal, 321

Ice age, little, 321
Ice sheets, 316
Immigrant, 56, 378
Immigration, 69, 183
— rate of, 375, 376
Index, of faunistic change, 279
— percentage occurrence, 275
Indian Region, 259
Indigenous, 217
Indo-African Tropical Realm, 246
Innate Releasing Mechanism, 118
Insularity, 382
Interdependence of limiting factors, 111
Interglacials, 387
Intrinscic factors of existence, 116, 117
Introduction, 49–52, 56, 139, 360, 361, 365
Irregular visitor, 158
Irruption, 81, 123, 160
Island, continental, 201, 305, 308, 371, 372
— fauna, 402
— oceanic, 20, 36, 54, 201, 305, 371, 373, 379, 380
— as stepping stones, 91
— zoogeography, 370–380
Isoclimatic distribution, 201
— dispersal and expansion, 143, 201
Isolate, peripheral, 347
Isolating factors, in speciation, 12
Isolation, by distance, 30

Kingdom, floral, 261

Lake, 236
— as barrier, 14, 16, 31
Land bridge, 31, 54, 203, 360, 394–396, 399, 400, 402
Landing, in aerial dispersal, 35
Latitudinal climatic belts, 244
Law, of compensation and moderation, 111, 150
— Liebig's, 95, 134
— of relativity of ecological valency, 150
— of substitution, 98, 150
— of tolerance, 95, 150
Lemuria, 265
Liebig's law, 95, 134
Life form, 250, 255
Life processes, timing of, 121
Life Zone, 247–248, 301
Light as limiting factor, 98, 101, 102
Limit (border), advancing, 230
— of area, 168, 176, 177, 242, 243
— dynamism of, 341, 343

— fluctuating, 229
— retreating, 230
— stable, 229
Limiting factors, 95–101, 103, 105, 109–110
— interdependence of, 111–112
Line of faunal resemblance, 277
Little Ice age, 321
Long-distance pioneer, 204

Macroevolution, 392
Macroptery, 61, 63, 74, 188
Malagassy Region, 265
Mapping, of area, 167
Marginal environment, 142
Masking factors, 111
Mass effect, 135, 144
Microclimate, 106
Microevolutionary adaptations, 141
Migrants, 123
Migration, 81–84, 123
— and dispersal, 60
Minimum area, 236
Mixing, of faunas, 88, 356
Mobility, 187
Moderation, law of, 150
Monocentric, 287
Monophagy, 57, 130
Monotypic, 346, 347
Mortality, 183
Mountain chains, as barriers, 17
Myxomatosis, 45

Natality, 183
Natural history, 70
Neogaea, 264
Neotropical Elements, 267
Niche, 149, 150, 194, 195, 226, 293, 295
— sharing, 297
Nistortstreue, 65
Nival species, 370
Nomadizing, 65, 66
North Temperate Realm, 245, 246
Northern Extratropical Kingdom, 267
Northern Zone, 244
Notogaea, 259, 264

Occurrence, extralimital, 158, 161
— extremes of, 160
— hypothetical, 224
Ocean, as barrier, 16, 31, 91
Oceanic island. See island, oceanic
Oceanity, 374
Old-Oceanic Element, 267
Optimum, synecological, 183, 185
Order, 234
Ortstreue, 65
Overspecialisation, 352
Overwater dispersal, 380

Palaearctic 259
Palaeotropical Element, 267
Palaeotropical Kingdom, 267
Paleontology, 365
Paleozoogeography, 392–400
Panboreal, 218, 222

Pantropical, 191, 218, 221
Parthenogenesis, 57, 61, 74, 124–128
— cyclic, 128
— geographic, 126–128
Passes, 18, 19, 63
Penetration, slow, 59, 75, 77–80, 368
Percentage Occurrence Index, 275
Periodicity, seasonal, 307
Permanence, of continents, 31, 397, 400
— of oceans, 31, 399, 400
Philopatry, 34, 65–68, 71, 134, 138, 178, 186
Phoresy, 45–48
Phyletic replacement, 401
Phylogenetic extinction, 354
Phytogeography, floristic and vegetational, 153
Pioneer, long distance, 75, 204
Pioneering, viii, 70, 76
— marginal, 58
Plant formation, 26, 97, 250, 255, 257
Plasticity, ecological, 186
Pleistocene relict, 219
Pluvial, 321, 387
Polycentric, 387
Polymorphism, 61, 73
— dispersional, 72, 74, 92
Polyphasy, 75
Polyploidy, 124, 128
Polytopic, 179
Polytypic, 179, 346, 347
Population, 70, 71, 135–143
— density, 71, 161
— disjunct, 349
— dynamics, 138, 144–149
— genetics, 139
— incipient, 95
— local, 135
— marginal, 58, 312
— pressure, 70
— relict, 349
— size, 56, 136, 140
— species, 135
Postglacial climatic optimum, 321, 349
Potential, area, 204
— dynamic, 341, 342, 401
— ecological, 150
— spreading, 29, 166, 181, 374
Prairie, cultural, 24
Probability, of crossing, 374
— of dispersal, 54, 91
Province, 246
— biogeographic, 271
— biotic. See biotic, province
— faunistic, 272, 278
Proximate factor, 113, 117
Pseudovicar, 183, 194

Radiation, adaptive, 207
Rafting, 40–42
Range. See Area, distributional
Range, adaptive, 94
Rate, of evolution, 298, 315, 401
— of reproduction, 70
Realm, 246, 258, 261
Refuge, 209, 230, 324

— ecological, 209
Refugium, 66, 209, 335, 352
Region, faunal, 246, 258–266
 — floral, 266–269
 — glacial, 209, 219
Relic, 208
Relict, 205, 208, 209, 211–215, 221, 224, 230, 242, 372, 374, 390, 399
 — "anti-culture", 215
 — area, 185, 195, 205, 206, 216
 — biogeographical, 205
 — disharmonious, 209, 211
 — distribution, 380
 — evolutionary, 205
 — glacial, 212, 213, 345
 — heat, 213, 219
 — hypsithermal, 213
 — phylogenetic, 205, 206
 — Pleistocene (= glacial), 212, 213, 345
 — pluvial, 216
 — preglacial, 345
 — reductional, 206, 208, 217
 — refugional, 209–211
 — secondary, 214, 220
 — successional, 213
 — Tertiary, 212, 213
 — xerothermal, 213, 218, 219
Relictness, 205–215, 234
Replacement, 356
 — faunal, 401
 — phyletic, 401
Reproduction, asexual, 124
 — mode of, 129
 — monoecious, 124
 — rate, 70
 — self-fertilizing, 124
 — sexual, 124
 — success as limiting factor, 122–128, 133
 — timing of, 129
 — vegetative, 124
Resemblance, faunal, 273
Resistance, environmental, 96, 166
 — thermal, 134
River, 236
 — as barrier, 13, 14, 16, 31

Sea, water as barrier, 15, 27
 — permanence of, 16
Secondary area, 227
Secular trend, 308, 321
Selection of habitat, 27, 37, 116–121, 151
Sexual reproduction, 124
Shelf, continental, 31
Snow, as limiting factor, 103, 105
Sociability, 69
Soil, as limiting factor, 104–109
South American Kingdom, 267
South American Temperate Realm, 246
Southern Elements, 267
Spacing, 10
Spatial heterogeneity, 298
Specialization, progressive, 313, 314
Speciation, in archipelagoes 380–383
Species, 230

 — diversity, 294
 — polytopic, 181
 — rare, 183
Spreading, 71
 — potential, 166, 181, 374
Statistical methods, 270–279
Stenoecy, 98, 140
Stenohygric, 98
Stenothermy, 100
Stenotopy, 113, 150, 186, 188, 191, 201
Steppe, cultural, 24
Stepping stones, of dispersal, 31, 91, 370, 400
Sterile expatriation area, 157, 159, 160
Stones, stepping, 31, 91, 370, 400
Subclimax, fire, 313
Substitute, ecogeographical, 194
Substitution, law of, 98, 150
Subterranean animals, 14
Succession, 255, 313, 314, 317, 328, 358, 362
 — fire, 313
Survivor, 205–207, 209, 353
Sweepstakes route, 89–91
Sympatry, 297
Synecological optimum, 183, 185
Synecology, 5, 194–195
Syngeography, 240

Take-off, in aerial dispersal, 34
Taxa, higher, 181
Temperature, as limiting factor, 98, 102–104, 111, 133, 134
 — zonal, 100, 102
Territory, 184
Thermal resistance, and area, 134
Time, in dispersal, 28–30, 58
 — in population dynamics, 143
 — scale, geological, 304
 — in settling, 94
 — in zoogeography, 302, 307–309
Timing of life processes, 121
Tolerance, 97–99
 — law of, 95, 150
 — theory of, 95
Topography, 12
Tropicopolitan, 191
Typology, 270

Ubiquitism, ubiquitist, ubiquitious, 190, 191
Ultimate factors, 117
Uniformitarianism, 366, 367, 402
Uplift, 316, 317
Urbanization, 359, 364

Vagility, 138, 186–191, 311, 312
Valency, ecological. See Ecological, valency
Value, basic zoogeographical, 274
Vegetation, viii–ix
 — belts, 296
Vegetational botany, 153
 — provinces, 301
Vegetative reproduction, 124

Vicar, 78, 192–194, 210, 211
— ecological, 28, 193–194
Vicariant, 192
Vicariate, 193
Vicarious, 86, 87, 347, 368
Vicarism, 192–194
Visitor, accidental, 75
— drift —, 158
— irregular, 158
Vivipary, 130

Wandering, 81, 82, 84, 85
Winglessness, 60–63

Wisconsin glaciation, 321
Würm glaciation, 321

Xerotherm, 321

Zonal distribution, 184, 201
Zone, altitudinal, 201
Zoogeographical value, 274
Zoogeography, anthropogene, 365
— descriptive, 6
— historical, 365
Zoological geography, 3
Zoology, geographical, 3

L-6
1750NR

Group selection 67, 146, 379
Liebig's Law 95.
Geogr. parth. 126.
Homology 146, 194
Wallace 261
300 — Hubendick on f.w. snails
354 for stability evolving from diversity.